D1693644

Membrane Operations

Edited by
Enrico Drioli and Lidietta Giorno

Related Titles

Seidel-Morgenstern, A. (ed.)

Membrane Reactors

Case Studies to Improve Selectivity and Yields

2009

ISBN: 978-3-527-32039-4

Koltuniewicz, A., Drioli, E.

Membranes in Clean Technologies

Theory and Practice

2008

ISBN: 978-3-527-32007-3

Peinemann, K.-V., Pereira Nunes, S. (eds.)

Membrane Technology

Volume 1: Membranes for Life Sciences

2007

ISBN: 978-3-527-31480-5

Peinemann, K.-V., Pereira Nunes, S. (eds.)

Membrane Technology

Volume 2: Membranes for Energy Conversion

2007

ISBN: 978-3-527-31481-2

Li, K.

Ceramic Membranes for Separation and Reaction

2007

ISBN: 978-0-470-01440-0

Pereira Nunes, S., Peinemann, K.-V. (eds.)

Membrane Technology

in the Chemical Industry

2006

ISBN: 978-3-527-31316-7

Sammells, A. F., Mundschau, M. V. (eds.)

Nonporous Inorganic Membranes

for Chemical Processing

2006

ISBN: 978-3-527-31342-6

Freeman, B., Yampolskii, Y., Pinnau, I. (eds.)

Materials Science of Membranes for Gas and Vapor Separation

2006

ISBN: 978-0-470-85345-0

Ohlrogge, K., Ebert, K. (eds.)

Membranen

Grundlagen, Verfahren und industrielle Anwendungen

2006

ISBN: 978-3-527-30979-5

Membrane Operations

Innovative Separations and Transformations

Edited by
Enrico Drioli and Lidietta Giorno

WILEY-VCH Verlag GmbH & Co. KGaA

The Editors

Prof. Enrico Drioli
University of Calabria
Institute on Membrane Technology
Via P. Bucci 17 /C
87030 Rende (CS)
Italy

Prof. Lidietta Giorno
University of Calabria
Institute on Membrane Technology
Via P. Bucci 17 /C
87030 Rende (CS)
Italy

Library of Congress Card No.: applied for

British Library Cataloguing-in-Publication Data
A catalogue record for this book is available from the British Library.

Bibliographic information published by the Deutsche Nationalbibliothek
The Deutsche Nationalbibliothek lists this publication in the Deutsche Nationalbibliografie; detailed bibliographic data are available on the Internet at http://dnb.d-nb.de.

Composition Thomson Digital, Noida, India
Printing Betz-Druck GmbH, Darmstadt
Bookbinding Litges & Dopf GmbH, Heppenheim
Cover Design Formgeber, Eppelheim

Printed in the Federal Republic of Germany
Printed on acid-free paper

ISBN: 978-3-527-32038-7

Contents

Membrane Operations. Innovative Separations and Transformations. Edited by Enrico Drioli and Lidietta Giorno

ISBN: 978-3-527-32038-7

List of Contributors

Richard W. Baker
Membrane Technology and Research, Inc.
1360 Willow Road
Menlo Park, CA 94025
USA

Giuseppe Barbieri
University of Calabria
Institute on Membrane Technology (ITM-CNR)
Via P. Bucci, 17/C
87030 Rende (CS)
Italy

Loredana De Bartolo
University of Calabria
Institute on Membrane Technology (ITM-CNR)
Via P. Bucci, 17/C
87030 Rende (CS)
Italy

Rune Bredesen
SINTEF Materials and Chemistry
P.O. Box 124
Blindern
0314 Oslo
Norway

Carla Campana
University of Calabria
Institute on Membrane Technology (ITM-CNR)
Via P. Bucci, 17/C
87030 Rende (CS)
Italy

and

University of Calabria
Department of Chemical Engineering and Materials
Via P. Bucci, cubo 45/A
87030 Rende (CS)
Italy

Angela Caruso
University of Calabria
Department of Chemical Engineering and Materials
Via P. Bucci, cubo 45/A
87030 Rende (CS)
Italy

Membrane Operations. Innovative Separations and Transformations. Edited by Enrico Drioli and Lidietta Giorno

ISBN: 978-3-527-32038-7

Tzyy H. Chong
Nanyang Technological University
Singapore Membrane Technology Centre
School of Civil and Environmental Engineering
Singapore
639798

João G. Crespo
Universidade Nova de Lisboa
Faculdade de Ciências e Tecnologia
Requimte-CQFB
Departamento de Química
2829-516 Caparica
Portugal

Enrico Drioli
University of Calabria
Institute on Membrane Technology (ITM-CNR)
Via P. Bucci, 17/C
87030 Rende (CS)
Italy

and

University of Calabria
Department of Chemical Engineering and Materials
Via P. Bucci, cubo 44/A
87030 Rende (CS)
Italy

Alessandra Criscuoli
University of Calabria
Institute on Membrane Technology (ITM-CNR)
Via P. Bucci, 17/C
87030 Rende (CS)
Italy

Anthony G. Fane
University of New South Wales
UNESCO Centre for Membrane Science & Technology
School of Chemical Sciences and Engineering
Sydney, NSW 2052
Australia

and

Nanyang Technological University
Singapore Membrane Technology Centre
School of Civil and Environmental Engineering
Singapore
639798

Alberto Figoli
University of Calabria
Institute on Membrane Technology (ITM-CNR)
Via P. Bucci, 17/C
87030 Rende (CS)
Italy

Enrica Fontananova
University of Calabria
Institute on Membrane Technology (ITM-CNR)
Via P. Bucci, 17/C
87030 Rende (CS)
Italy

and

University of Calabria
Department of Chemical Engineering and Materials
Via P. Bucci, cubo 44/A
87030 Rende (CS)
Italy

Soccorso Gaeta
GVS S.P.A.
Via Roma 50
40069 Zola Predosa (Bo)
Italy

Lidietta Giorno
University of Calabria
Institute on Membrane Technology
(ITM-CNR)
Via P. Bucci, 17/C
87030 Rende (CS)
Italy

David Hasson
Technion – Israel Institute of
Technology
Stephen and Nancy Grand Water
Research Institute
Wolfson Chemical Engineering
Department
Rabin Desalination Laboratory
Technion City
Haifa, 32000
Israel

Tzyy Haur
Nanyang Technological University
Singapore Membrane Technology
Centre
School of Civil and Environmental
Engineering
Singapore
639798

Dieter Hofmann
GKSS Research Center
Center for Biomaterial Development
of the Institute of Polymer Research
Kantstr. 55
14513 Teltow
Germany

Shabbir Husain
Georgia Institute of Technology
School of Chemical & Biomolecular
Engineering
Atlanta, GA 30332-0100
USA

Michel Y. Jaffrin
UMR CNRS 6600
Technological University of Compiegne
60200 Compiegne
France

William J. Koros
Georgia Institute of Technology
School of Chemical & Biomolecular
Engineering
Atlanta, GA 30332-0100
USA

Adam Kratochvil
PRISM Membranes
Air Products and Chemicals, Inc.
St. Louis, Mo 63146
USA

Izumi Kumakiri
SINTEF Materials Technology
P.O. Box 124
Blindern
0314 Oslo
Norway

Pierre Le-Clech
University of New South Wales
UNESCO Centre for Membrane Science
& Technology
School of Chemical Sciences and
Engineering
Sydney, NSW 2052
Australia

Cécile Legallais
UMR CNRS 6600
Technological University of Compiegne
60200 Compiegne
France

TorOve Leiknes
NTNU - Norwegian University of Science and Technology
Department of Hydraulic and Environmental Engineering
S.P. Andersensvei 5
7491 Trondheim
Norway

Giorgio De Luca
University of Calabria
Institute on Membrane Technology (ITM-CNR)
Via P. Bucci, 17/C
87030 Rende (CS)
Italy

Rosalinda Mazzei
University of Calabria
Institute on Membrane Technology (ITM-CNR)
Via P. Bucci, 17/C
87030 Rende (CS)
Italy

and

University of Calabria
Department of Ecology
Via P. Bucci 6/B
87036 Rende (CS)
Italy

Raffaele Molinari
University of Calabria
Department of Chemical Engineering and Materials
Via P. Bucci
87030 Rende (CS)
Italy

Sabrina Morelli
University of Calabria
Institute on Membrane Technology (ITM-CNR)
Via P. Bucci, 17/C
87030 Rende (CS)
Italy

T.M. Murphy
The University of Texas at Austin
Department of Chemical Engineering
Austin, TX 78712
USA

Endre Nagy
University of Pannonia
Research Institute of Chemical and Process Engineering
P.O. Box 158
8201, Veszprém
Hungary

Grant T. Offord
The University of Texas at Austin
Department of Chemical Engineering
Austin, TX 78712
USA

Leonardo Palmisano
University of Palermo
Department of Chemical Engineering Processes and Materials
'Schiavello-Grillone' Photocatalysis Group
viale delle Scienze
90128 Palermo
Italy

Don R. Paul
The University of Texas at Austin
Department of Chemical Engineering
Austin, TX 78712
USA

Thijs Peters
SINTEF Materials Technology
P.O. Box 124
Blindern
0314 Oslo
Norway

Emma Piacentini
University of Calabria
Institute on Membrane Technology
(ITM-CNR)
Via P. Bucci, 17/C
87030 Rende (CS)
Italy

Antonella Piscioneri
University of Calabria
Institute of Membrane Technology
National Research Council of Italy
ITM-CNR
Via P. Bucci, cubo 17/C
87030 Rende (CS)
Italy

and

University of Calabria
Department of Cell Biology
via P. Bucci
87030 Rende (CS)
Italy

Maria Rende
University of Calabria
Institute on Membrane Technology
(ITM-CNR)
Via P. Bucci, 17/C
87030 Rende (CS)
Italy

and

University of Calabria
Department of Chemical Engineering
and Materials
Via P. Bucci, cubo 45/A
87030 Rende (CS)
Italy

Simona Salerno
University of Calabria
Institute on Membrane Technology
(ITM-CNR)
Via P. Bucci, 17/C
87030 Rende (CS)
Italy

Štefan Schlosser
Slovak University of Technology
Institute of Chemical and
Environmental Engineering
Radlinského 9
812 37 Bratislava
Slovakia

Raphael Semiat
Technion – Israel Institute of
Technology
Wolfson Chemical Engineering
Department
Rabin Desalination Laboratory
Stephen and Nancy Grand Water
Research Institute
Technion City
Haifa, 32000
Israel

Francesco Scura
University of Calabria
Institute on Membrane Technology
(ITM-CNR)
Via P. Bucci, 17/C
87030 Rende (CS)
Italy

Shu Shu
Georgia Institute of Technology
School of Chemical & Biomolecular
Engineering
Atlanta, GA 30332-0100
USA

Heinrich Strathmann
University of Stuttgart
Institute of Chemical Technology
Böblingerstr. 72
70199 Stuttgart
Germany

Heru Susanto
Universität Duisburg-Essen
Lehrstuhl für Technische Chemie II
45117 Essen
Germany

Elena Tocci
University of Calabria
Institute on Membrane Technology
(ITM-CNR)
Via P. Bucci, 17/C
87030 Rende (CS)
Italy

Mathias Ulbricht
Universität Duisburg-Essen
Lehrstuhl für Technische Chemie II
45117 Essen
Germany

Bart Van der Bruggen
K.U. Leuven, Department of Chemical
Engineering
Section Applied Physical Chemistry and
Environmental Technology
W. de Croylaan 46
3001 Heverlee (Leuven)
Belgium

Introduction

Membrane processes are state of the art technologies in various industrial sectors, including gas separation, wastewater treatment, food processing and medical applications.

Modelling methodologies are contributing significantly to the knowledge-based development of membrane materials and engineering.

Micro-ultrafiltration and reverse osmosis are mature technologies for separations based on molecular exclusion and solution-diffusion mechanisms, respectively. Cleaning and maintenance procedures able to control fouling to an acceptable extent have made these processes commercially suitable.

Some of the largest plants for seawater desalination, wastewater treatment and gas separation are already based on membrane engineering. For example, the Ashkelon Desalination Plant for seawater reverse osmosis (SWRO), in Israel, has been fully operational since December 2005 and produces more than 100 million m^3 of desalinated water per year. One of the largest submerged membrane bioreactor unit in the world was recently built in Porto Marghera (Italy) to treat tertiary water. The growth in membrane installations for water treatment in the past decade has resulted in a decreased cost of desalination facilities, with the consequence that the cost of the reclaimed water for membrane plants has also been reduced.

Membranes are growing significantly also in gas separation, for example, the current market size of carbon-dioxide separation from natural gas is more than 70 million Euro/year.

Medical applications are among the most important in the membrane market, with hemodialysis, blood oxygenators, plasma separation and fractionation being the traditional areas of applications, while artificial and bioartificial organs and regenerative medicine represent emerging areas in the field.

Nanofiltration has achieved a good stage of development, gaining attention in various applications for separations based on both molecular exclusion and charge interaction as well as on the solution-diffusion mechanism. In particular, nanofiltration is considered among the most suitable technologies for solvent separation. More recent processes such as membrane reactors, membrane contactors, and membranes in life science are also developing very rapidly. The optimal design of

Membrane Operations. Innovative Separations and Transformations. Edited by Enrico Drioli and Lidietta Giorno

ISBN: 978-3-527-32038-7

chemical transformation processes with control of reagent supply and/or product removal through catalytic membranes and membrane reactors is one of the most attractive solutions in process intensification. The catalytic action of biocatalysts is extremely efficient, selective and highly stereospecific when compared to conventional chemical catalysts. Membrane bioreactors are particularly attractive in terms of ecocompatibility, because they do not require additives, are able to operate at moderate temperature and pressure, reduce the formation of by-products, while permitting the production of high valuable coproducts. This may allow challenges in developing new production lines moving towards zero discharge to be faced. The development of catalytic membrane reactors for high-temperature applications became realistic more recently, with the development of high-temperature-resistant membranes.

The major market for membrane bioreactors is represented by wastewater treatment with the use of submerged modules configuration. These are considered among the best available technologies by the European Directives on Environment. Membrane bioreactors are also applied in food, red and white biotechnology. In these cases, the external loop configuration is used.

Membrane contactors, including membrane crystallizers and membrane emulsifiers, are among the most recent membrane operations with growing interest in various industrial sectors. For example, membrane emulsification has grown from the 1990s, when it was first developed in Japan, to nowadays with applications in food, chemical, pharmaceutical and cosmetic fields. In Europe, the research at the academic level has achieved a thorough knowledge both from experimental and theoretical points of view. This is fuelling the industrial interest towards the membrane emulsification technology, especially for those productions that involve labile bioactive molecules.

In general, nowadays the attention towards membrane science and technology is increasing significantly. Drivers of this interest include the need for technologies to enable sustainable production, directives and regulations about the use of eco-friendly technologies, consumer demand for high-quality and safe products, public concern about environment, and stakeholder confidence in and acceptance of advanced technologies.

Current initiatives recognize that a sustainable solution to the increasing demand of goods and energy is in the rational integration and implementation of new technologies able to achieve concrete benefits for manufacturing and processing, substantially increasing process precision, reducing equipment size, saving energy, reducing costs, and minimizing environmental impact.

Membranes and membrane processes are best suited in this context as their basic aspects well satisfy the requirements of process intensification for a sustainable industrial production. In fact, they are precise and flexible processing techniques, able to maximize phase contact, integrate conversion and separation processes, with improved efficiency and with significantly lower energy requirements compared to conventional techniques.

This multiauthor book highlights the current state and advances in membranes and membrane operations referring to three major roles of the membrane: molecular separation, (bio)chemical transformation and phase contactors. Each topic includes fundamentals and applications of membranes and membrane operations.

The largest section is constituted by membranes in molecular separation, which is the most traditional application of membranes. Significant advances of membrane science and technologies are expected in transformation processes and membrane contactors for conventional and innovative applications.

Part One
Molecular Separation

This Part will be focused on the fundamentals and applications of membranes and membrane operations for separation at the molecular level. Both liquid (including organic solvents) and gaseous streams will be discussed.

The book opens with a chapter on molecular modeling to highlight the powerful instruments for designing appropriate membrane materials with predicted properties.

This is followed by a chapter on polymeric membranes that discusses the current achievements and challenges on membranes for molecular separation in liquid phase.

Subsequent individual chapters discuss membranes in organic solvent separation, gas separation and electrochemical separation. A whole chapter is focused on the fundamentals of fouling molecular separation by membranes are completed by a chapter focused on fouling. and another on energy and environmental issues.

The application part of this section illustrates the membrane-assisted molecular separation in (i) gases, with a separate chapter dedicated to the CO_2 capture using inorganic membrane; (ii) water desalination; (iii) downstream processing of biological products. A chapter on integrated membrane operations illustrates new strategies in water treatment and chemical production.

Membrane separation in the medical field has been included in a chapter focused on medical extracorporeal devices, which illustrates the use of membranes for separation of biological fluids and for preparation of bioartificial organs able to accomplish *ex vivo* biological transformation (Part headed 'Transformation').

The overall aim of the 'molecular separation' section is to illustrate the current capability of membranes and membrane operations in assisting and governing molecular separations and the future perspectives they offer for a more sustainable industrial growth through innovative process design. Their implementation will lead to concrete benefits in manufacturing and processing, substantially shrinking equipment size, boosting plant efficiency, saving energy, reducing capital costs, minimizing environmental impact, and using remote control and automation.

Membrane Operations. Innovative Separations and Transformations. Edited by Enrico Drioli and Lidietta Giorno

ISBN: 978-3-527-32038-7

Membrane operations have the potential to replace conventional energy-intensive separation techniques, such as distillation and evaporation, to accomplish the selective and efficient transport of specific components, to improve the performance of reactive processes and, ultimately, to provide reliable options for a sustainable industrial growth.

This is in line with the strategy of process intensification and it is expected to bring substantial improvements in chemical and many other manufacturing and processing industries.

Many membrane operations are based on similar materials and structures, while differing in the method by which they carry out the separation process. Step forward innovations can be promoted by appropriate integration of traditional membrane operations (reverse osmosis, micro-, ultra- and nanofiltration, electrodialysis, pervaporation, etc.) among them and with innovative membrane operations. In fact, while being already widely used in many different applications, they can be combined with new membrane systems such as catalytic membrane reactors and membrane contactors. Nowadays, redesign of industrial production cycles by combining various membrane operations suitable for separation, conversion and concentration units is an attractive opportunity because of the synergic effects that highly integrated membrane processes can promote.

1
Molecular Modeling, A Tool for the Knowledge-Based Design of Polymer-Based Membrane Materials

Dieter Hofmann and Elena Tocci

1.1
Introduction

Most important macroscopic transport properties (i.e., permeabilities, solubilities, constants of diffusion) of polymer-based membranes have their foundation in microscopic features (e.g., free-volume distribution, segmental dynamics, distribution of polar groups, etc.) which are not sufficiently accessible to experimental characterization. Here, the simulation of reasonably equilibrated and validated atomistic models provides great opportunities to gain a deeper insight into these microscopic features that in turn will help to develop more knowledge-based approaches in membrane development.

The mentioned transport properties for small and medium-sized molecules in polymers are decisive in many technologically important processes, for example, in biotechnology and biomedicine, in pharmacological and chemical industries but also in integrated environmental protection. The respective penetrants can be anything from rather small hydrogen or oxygen molecules to chemicals like benzene up to relatively large drug molecules.

Membrane processes for the separation of gaseous and liquid mixtures are important examples. In these cases there are already large numbers of applicable materials and processes. Further improvements (mostly concerning better selectivities at acceptably high permeabilities), often needing real jumps in performance, are, however, still needed in many cases. This applies, although in the opposite sense, also to barrier materials where permeations at least of certain types of molecules will be extremely small. Other areas concern biomaterials or material systems for the controlled release of drugs.

More specific examples for the need to develop new materials with tailored transport properties are:

- The separation of methane from higher hydrocarbons in natural gas for safer and more economical transport through pipelines, or for better exploitation;

Membrane Operations. Innovative Separations and Transformations. Edited by Enrico Drioli and Lidietta Giorno

ISBN: 978-3-527-32038-7

- The design of packaging materials for conservation of fresh fruits and vegetables, which means good specific permeation and selectivity properties in order to maintain a modified/controlled atmosphere;
- The control of migration of additives, monomers or oligomers, from packaging materials, for example, into food (important for the enforcement of a high level of food quality and safety) or other consumer products;
- The resistance of resins used in composites for aircraft construction to ageing caused by water absorption;
- Small but continuous fuel loss by permeation through polymeric parts of the fuel system;
- Separation of CO_2 from flue gases, and separation of NO_x from vehicle emissions;
- Efficient and inexpensive proton-conducting membranes for fuel cells;
- components in polymer electronics (such as for light-emitting diodes or display components) with extremely low permeabilities for oxygen and water;
- Optimum controlled drug release systems, for example, for medical applications, cosmetics or agriculture;
- Transport problems in artificial or bio-hybrid organs;
- Optimum biocompatibility of polymers in contact with cells and blood;
- Optimum chemical degradation behavior (often to a large extent a water-permeation problem) for surgical sutures, scaffold materials for tissue engineering, degradable screws in orthopaedic surgery and so on.

In the near future, the use of multifunctional polymer-based materials with separation/selective transport capabilities is also to be expected in the design of production systems with integrated environmental protection or in the combination of chemical reactions and separation by attaching a catalytic functionality to the respective material [1]. Thus, those multifunctional materials should contribute materially to the development of clean energy and/or energy saving and therefore sustainable production technologies. In connection with these perspectives, there is considerable interest in new/modified polymer-based materials with tailored transport/catalytic properties. Also, many sensor applications are based on controlled permeation.

Amorphous polymers or respective composites with inorganic components are an important class of materials to solve many of the above-mentioned problems. However, the design of these multifunctional materials, based on experimentation and correlative thinking alone is unreliable, time consuming, expensive and often not successful. Systematic multiscale computer-aided molecular design (CAMD) offers a very attractive alternative, insofar as these techniques allow for the very elaborate investigation of complex material behavior with regard to the links between structure, dynamics and relevant properties of the discussed multifunctional polymer-based materials on the length and time scales (from Angstroms to micrometers and from picoseconds to milliseconds, respectively) which are most important for the penetrant transport and other relevant processes (e.g., selective transport, separation, catalysis, biodegradation, sensor applications) of interest. In the present chapter, molecular modeling tools (i.e., quantum chemistry (QM), atomistic- and mesoscale modeling) will be in the focus of interest. Consequently, the microscopic properties to be related with macroscopically determined transport parameters are, for example, chain stiffness parameters, free volume and its distribution, mobility measures for chain segments, energy densities describing interactions of chain segments with penetrants, microscopic effects of swelling and so on.

Over the last 15 years particularly atomistic molecular modeling methods have found widespread application in the investigation of small-molecule permeation [2–15].

1.2 Basics of Molecular Modeling of Polymer-Based Membrane Materials

The permeation of small molecules in amorphous polymers is typically following the solution diffusion model, that is, the permeability P_i of a feed component i can be envisioned as the product of the respective solubility S_i and constant of diffusion D_i. Both parameters can be obtained experimentally and in principle also by atomistic simulations.

The molecular modeling of these polymers typically starts with the construction of normally rectangular packing models. There, the related chain segments of the respective polymer will be arranged in realistic, that is, statistically possible, way. To do this, first the involved atoms are considered to be spheres of the respective atomic radius R_i (as obtainable from QM) and atomic weight m_i. The bonded interactions between atoms resulting in bonds, bond angles and conformation angles are then described by mechanic springs or torsion rods with spring constants related to, for example, experimentally known bond strengths. So-called nonbond interactions between atoms that either belong to different molecules or that in one and the same molecule are further apart from each other than about three bonds are considered via, for example, Lennard-Jones (to describe van der Waals interactions) and Coulomb potentials (to describe electrostatic interactions). The sum of all interatomic interactions written as the potential energy of a packing model is then called a forcefield. Forcefields form the core of all atomistic molecular modeling programs. Equation 1.1 shows the principal structure of a typical forcefield for a system of N atoms with the Cartesian atomic position vectors $\vec{r}_i$.

$$
\begin{aligned}
V(\vec{r}_1, \vec{r}_2, \ldots, \vec{r}_N) = & \sum_{\text{Covalent bonds}} K_b (l - l_0)^2 + \sum_{\text{Bond angles}} K_\Theta (\Theta - \Theta_0)^2 \\
& + \sum_{\text{Dihedral angles}} K_\varphi [1 + \cos(n\varphi - \delta)] \\
& + \sum_{\text{nonbonded atom pairs } i,j} \left[\left(\frac{a_{ij}}{r_{ij}^{12}} \right) - \left(\frac{b_{ij}}{r_{ij}^{6}} \right) + \frac{q_i q_j}{\varepsilon_0 \varepsilon_r r_{ij}} \right]
\end{aligned}
\tag{1.1}
$$

with the following parameters:

l = actual length of a bond
l_0 = length of a bond in equilibrium
K_b = force constant for a bond length deformation
Θ = actual value for a bond angle
Θ_0 = value for a bond angle in equilibrium
K_Θ = force constant for a bond-angle deformation
φ = actual value for a conformation angle
n = periodicity parameter in a conformation potential
δ = constant to fix trans-state in a conformation potential
K_φ = force constant for a conformation potential
R_{ij} = distance between atoms i and j with $(j-i)>3$
a_{ij} = constant describing repulsive interactions in the *Lennard-Jones* Potential
b_{ij} = constant describing attractive interactions in the *Lennard-Jones* Potential
q_i = partial charge of the ith atom
ε_0 = vacuum permittivity
ε_r = dielectric constant.

The parameters l_0, K_b, Θ_0, K_Θ, K_φ, n, δ, a_{ij}, b_{ij}, q_i, q_j and ε_r belong to the fit parameters, which can be determined by fitting of Equation 1.1 to a sufficient set of data calculated by QM and/or determined experimentally (e.g., X-ray scattering, IR spectroscopy, heats of formation). From a numeric point of view the pair interaction terms (van der Waals and Coulomb) are most demanding. In this connection the typical size of polymer packing models is limited to typically 3000–10 000 atoms (leading to lateral sizes of bulk models of a few nm), although in other connections now also models with up to 100 000 atoms have been used.

Forcefields may be utilized in two directions:

Model systems can be, on the one hand, subjected to a static structure optimization. There, the fact is considered that the potential energy of a relaxed atomistic system (cf. Equation 1.1) should show a minimum value. Static optimization then means that by suited numeric procedures the geometry of the simulated system is changed as long as the potential energy reaches the next minimum value [16]. In the context of amorphous packing models, the main application for this kind of procedure is the reduction of unrealistic local tensions in a model as a prerequisite for later molecular dynamic (MD) simulations.

It is, on the other hand, possible to use the potential energy of a model system as described by Equation 1.1 to calculate the forces $\vec{F}_i$ acting on each atom of the model via the gradient operation:

$$\vec{F}_i = -\frac{\partial V(\vec{r}_1, \vec{r}_2, \ldots, \vec{r}_N)}{\partial \vec{r}_i} \tag{1.2}$$

Then, Newton's equations of motion can be solved for every atom of the investigated system:

$$\vec{F}_i = m_i \frac{d^2\vec{r}_i(t)}{dt^2} \tag{1.3}$$

The necessary starting positions $\vec{r}_i(0)$of the atoms are in the given case usually obtained from methods of chain-packing procedures (see below). The starting velocities $\vec{v}_i(0)$ of all atoms are assigned via a suited application of the well-known relation between the average kinetic E_{kin} energy of a polyatomic system and its temperature T:

$$E_{kin} = \sum_{i=1}^{N} \frac{1}{2} m_i \vec{v}_i^{\,2} = \frac{3N-6}{2} k_b T \tag{1.4}$$

k_B is the Boltzmann constant. $(3N-6)$ is the number of degrees of freedom of an N-atom model considering the fact that in the given case the center of mass of the whole model with its 6 translation and rotation degrees of freedom does not move during the MD simulation. Using Equations 1.2–1.4 it is then possible to follow, for example, the motions of the atoms of a polymer matrix and the diffusive movement of imbedded small penetrant molecules at a given temperature over a certain interval of time.

Equation 1.3 represents a system of usually several thousand coupled differential equations of second order. It can be solved only numerically in small time steps Δt via finite-difference methods [16]. There always the situation at $t + \Delta t$ is calculated from the situation at t. Considering the very fast oscillations of covalent bonds, Δt must not be longer than about 1 fs to avoid numerical breakdown connected with problems with energy conservation. This condition imposes a limit of the typical maximum simulation time that for the above-mentioned system sizes is of the order of several ns. The limited possible size of atomistic polymer packing models (cf. above) together with this simulation time limitation also set certain limits for the structures and processes that can be reasonably simulated. Furthermore, the limited model size demands the application of periodic boundary conditions to avoid extreme surface effects.

The already mentioned limited lateral dimensions of packing models of just several nm makes it impossible to simulate complete membranes or other polymer-based samples. Therefore, on the one hand, bulk models are considered that are typically cubic volume elements of a few nanometers side length that represent a part cut out of the interior of a polymer membrane (cf. Figure 1.1). On the other hand interface models are utilized, for example, for the interface between a liquid feed mixture and a membrane surface or between a membrane surface and an inorganic filler (cf. Figure 1.2).

1.3 Selected Applications

1.3.1 Hard- and Software

The InsightII/MaterialsStudio/Discover software of Accelrys [18, 19] was utilized for the amorphous packing model construction, equilibration and the atomistic

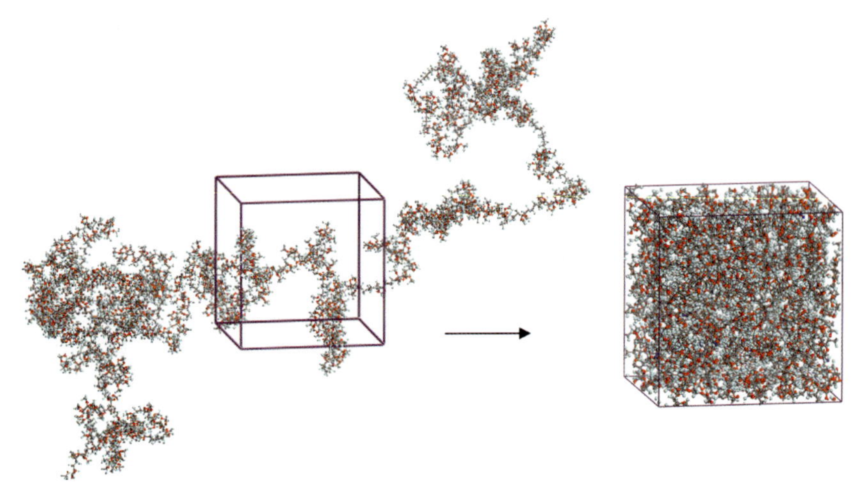

Figure 1.1 Atomic representation of a typical 3-dimensional packing model (thickness about 3 Å) starting with a single Hyflon AD60X polymer chain. Atom colors: gray = carbon, red = oxygen, light blue = fluorine [15].

simulations. In most of the following examples the COMPASS forcefield was applied [20, 21].

For data evaluation also self-programmed software (mostly in BTCL, Fortran, C) was applied. Data production runs were performed on a 74 processor Opteron Linux Cluster, a SGI Origin 2100 and on SGI Onyx workstation.

1.3.2
Simulation/Prediction of Transport Parameters and Model Validation

The quality of atomistic packing models is typically validated via comparisons between measured and simulated properties like wide-angle X-ray scattering (WAXS)

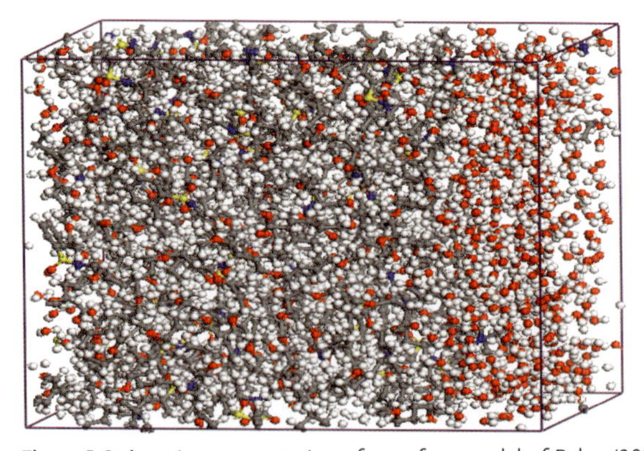

Figure 1.2 Atomic representation of a surface model of Pebax/30%KET with water [17].

curves, densities, transport parameters for small and medium sized penetrants. In the latter case both validating (if a polymer is already existing and experimentally characterized) and predictive (if a polymer has not been synthesized yet or if no transport parameters are available experimentally) applications are possible.

1.3.2.1 Prediction of Solubility Parameters

Here, hitherto in most cases the transition-state method of Gusev and Suter [22, 23] was utilized to first determine calculated solubility values S_{calc} values. There, a fine 3D-grid with a grid spacing of about 0.03 nm is layered over a completely refined detailed-atomistic amorphous polymer bulk packing model (cf. Figure 1.1). Then a small virtual test molecule of the intended kind (e.g., O_2) in a united atom representation is inserted in the polymer matrix at each lattice point of the grid. The resulting nonbonded interaction energy E_{ins} between the inserted molecule and the whole polymer matrix is calculated for each position of the respective inserted molecule. Only the van der Waals interactions are considered, that is, the method would not work for highly polar penetrants like water. Furthermore, since the polymer matrix can not locally relax to accommodate larger inserted penetrants it only works for small molecules (typically just up to O_2, N_2, etc.). From the insertion energy data via Equation 1.5 the chemical excess potential μ_{ex} for infinite dilution can be calculated and converted in the respective solubility using Equation 1.6.

$$\mu_{ex} = RT \times \ln < \exp(-E_{ins}/kT) > \tag{1.5}$$

$$S_{calc} = \frac{T_0}{p_0 T} \exp\left(-\frac{\mu_{ex}}{kT}\right) \tag{1.6}$$

with R being the universal gas constant and T_0 and p_0 being temperature and pressure under standard conditions ($T_0 = 273.15$ K; $p_0 = 1013 \times 10^5$ Pa).

Table 1.1 contains typical solubility prediction data for an ultrahigh free-volume polymer (PTMSP) and a polymer with more conventional transport properties (PTMSS).

As already mentioned the Gusev–Suter method normally only works for small penetrant molecules like oxygen or nitrogen. For a long time no really generally applicable alternative method was available to overcome the problem, but a few years ago Boulougouris, Economou Theodorou *et al.* [27, 28] suggested a new inverse Widom method based on the particle-deletion algorithm "DPD" to overcome this problem in principle. The related computer code was, however, only applicable to special, relatively simple model systems. Based on DPD also a generalized version of this algorithm was presented in the literature [29] permitting the calculation of solubility coefficients for molecules as large as, for example, benzene in polymers for which reasonable forcefield parameters exist. Table 1.2 contains solubility data for a number of penetrants of different size in PDMS obtained in this way.

1.3.2.2 Prediction of Diffusion Constants

The following description again follows the already quoted papers of Gusev and Suter. Using the E_{ins} values mentioned in the foregoing section, the whole packing model in

Table 1.1 Results of application of the Gusev–Suter method to the solubility of N_2 in PTMSP and PTMSS.

Polymer	Structure formula	Average simulated N_2 solubility coefficient S_{calc} [cm^3(STP)/(cm^3 atm)]	Average measured N_2 solubility coefficient S_{exp} [cm^3(STP)/(cm^3 atm)]
PTMSP	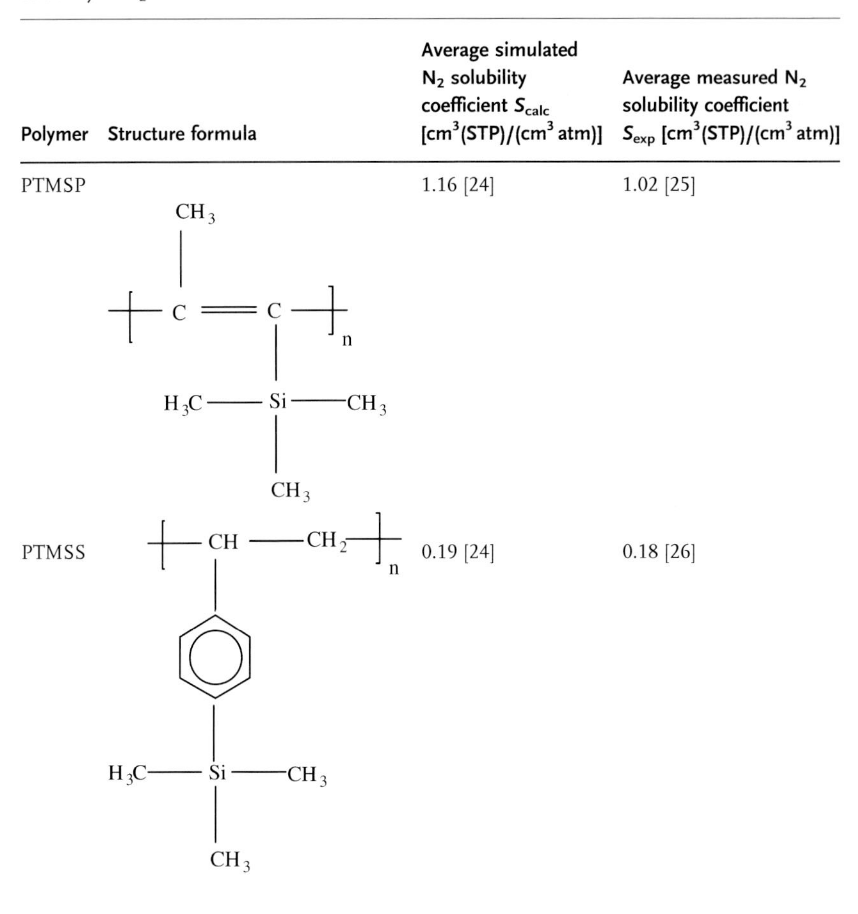	1.16 [24]	1.02 [25]
PTMSS		0.19 [24]	0.18 [26]

question is separated into regions of free volume (low interaction energy) and regions of densely packed polymer (high interaction energy; cf. Figure 1.3). The borders between the energetically attractive regions $E_{ins}(x, y, z)$ around the resulting local insertion energy minima are given as crest surfaces of locally maximum insertion energy. In the two-dimensional analogy of a cratered landscape a minimum energy region would be represented by a crater, while the crest surface of locally maximum insertion energy would be reduced to the crest line separating one crater from the adjacent ones. From this identification of energetically separated sites where a penetrant would typically sit (approximately the centers of holes) and jump probabilities between adjacent sites (which can be calculated by proper integration over the mentioned crest lines and "craters" of the insertion energy function $E_{ins}(x, y, z)$ an efficient Monte Carlo simulation method for the jump-like diffusion of small

Table 1.2 Results of application of a generalized DPD method to different penetrants in PDMS.

Solute	S_{calc} [cm^3(STP) cm^{-3} bar^{-1}]	S_{exp} [cm^3(STP) cm^{-3} bar^{-1}]
Oxygen	0.32[a]	0.224[b]
Nitrogen	0.13[a]	0.127[b]
Acetone	69[a]	33–66[c]
Benzene	495[a]	275–624[d]

[a][29, 30].
[b][31].
[c][32].
[d][33].

molecules in a polymer matrix can be developed (cf. Figure 1.4). With this algorithm the simulation range can almost extend in the ms range. That is, in most cases the normal diffusive regime can be reached and the respective constant of diffusion D_i can be obtained via the Einstein equation from the slope of the mean squared displacement $s_i(t)$:

$$\vec{s}_i(t) = \langle|\vec{r}_i(t) - \vec{r}_i(0)|^2\rangle \tag{1.7}$$

$$\rightarrow D_i(t) = \frac{\langle|\vec{r}_i(t) - \vec{r}_i(0)|^2\rangle}{6t} \tag{1.8}$$

Here, $\vec{r}_i(t)$is the position vector of penetrant i and <> is the average over all possible time origins $t = 0$ and all simulated trajectories of a penetrant of a given kind. Again, as with the solubilities the Gusev–Suter method can only handle small penetrants in this way, because the respective polymer matrix cannot conformationally adjust to larger penetrants. Table 1.3 contains a comparison between experimental and

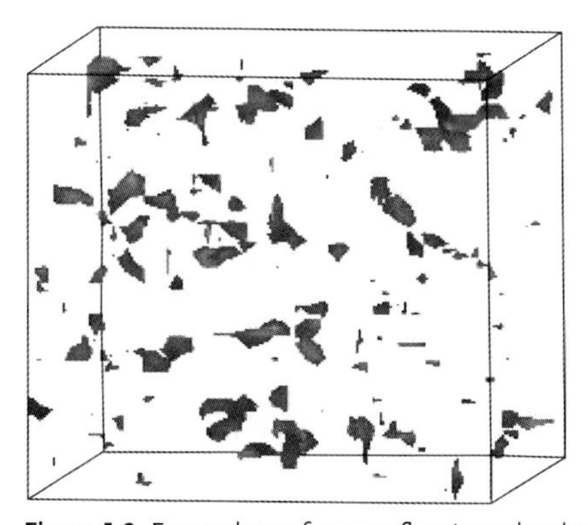

Figure 1.3 Free volume for a perfluorinated polymer in red indicating into the densely packed polymer.

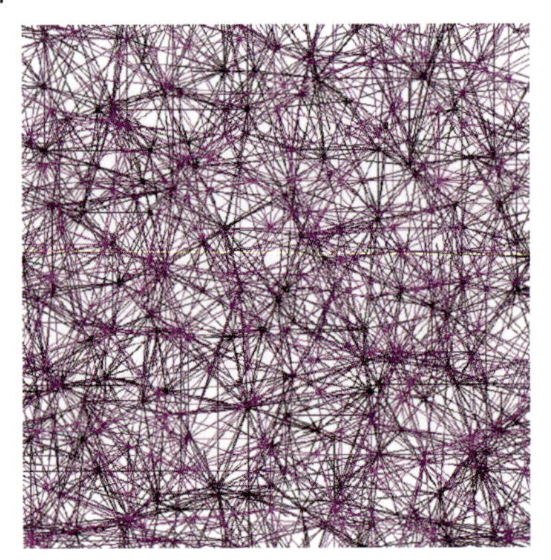

Figure 1.4 Jump-like diffusion of oxygen molecules in a perfluorinated polymer matrix.

calculated values, D_{exp} and D_{calc}, respectively for a number of gases in PTMSP. Here, for methane and carbon dioxide it has to be considered that these molecules are normally already too large to lead to reasonable results with the Gusev–Suter method.

In comparing simulated and experimentally measured transport parameters one has to be aware that experimental data in the literature depending, for example, on sample preparation conditions and the chosen measurement methodology can show a considerable scatter, often reaching a factor of two or even more. It is, for example well-known that polyimides often contain residual solvent filling a part of the free volume and thus leading to systematically lower S and D values from experiments than from simulations [34].

1.3.3
Permeability of Small Molecules and Free-Volume Distribution

The distribution of free volume in amorphous polymers is of paramount importance for the respective material's transport behavior towards small and medium-sized penetrants.

Table 1.3 Results of application of the Gusev–Suter method for the diffusion constants of different penetrants in PTMSP.

Solute	D_{calc} [10^{-5} cm^2/s]	D_{exp} [10^{-5} cm^2/s]
Nitrogen	7.7[a]	3.50[b]
Oxygen	7.5[a]	4.66[b]
Methane	8.2[a]	2.64[b]
Carbon dioxide	9.2[a]	8.02[b]

[a][24].
[b][25].

While in rubbery polymers differences in the segmental mobility can be more important than differences in the free-volume distribution for glassy polymers often certain basic correlations can be found between the permeability of small molecules and free-volume distribution. Other important factors are the molecular mobility of chain segments and the local chemical composition.

Experimentally, the free-volume distribution can be best characterized with positron annihilation lifetime spectroscopy (PALS). There, in organic glasses ortho-positronium (o-Ps) which has a lifetime of 142 ns in vacuo shows a strong tendency to localize in heterogeneous regions of low electron density (holes). In polymeric materials the vacuum lifetime is cut short via the "pick-off" mechanism, where o-Ps prematurely annihilates with one of the surrounding bound electrons. This lifetime can (under certain assumptions) be converted in an average hole radius [35, 36], while the intensity of the lifetime signal may permit conclusions about the overall contents of free volume. There are, however, a number of shortcomings with common PALS methodology. Often, the holes forming the free volume are assumed to be just spheres and the shape of calculated hole radius distribution peaks is set to Gaussian. Furthermore, positrons in their limited lifetime seem not to be capable of probing large holes of complex topology (cf. in particular PTMSP and other ultrahigh free-volume polymers) [24, 37]. Finally the size of the positronium molecule does only permit probing of the accessible free volume for molecules about the size of hydrogen.

Atomistic molecular modeling utilizing bulk models on the other hand can provide additional even more detailed information about free-volume distributions in amorphous polymers. In this way, glassy polymers, where individual differences in chain segment mobility do not have an as distinct influence on transport properties than in rubbery polymers, can be roughly grouped into three classes regarding their small molecule permeability, as will be outlined in the following for the example of oxygen.

1.3.3.1 Examples of Polymers with Low Permeability of Small Molecules (e.g., $PO_2 \leq 50$ Barrer)

Figure 1.5(a) shows as a typical example a computer-tomography-like atomic monolayer representation of a bulk model for diisopropyldimethyl PEEK WC (DIDM-PEEK). In this case the oxygen-accessible free volume is obviously organized in relatively small isolated holes and the respective size distribution (cf. Figure 1.5(b)) is monomodal and extending only to hole radii of about 5 Å.

1.3.3.2 Examples of Polymers with High Permeability of Small Molecules (e.g., 50 Barrer $\leq PO_2 \leq$ 200 Barrer)

Similarly to Figure 1.5(a), Figure 1.6(a) displays an atomic monolayer representation for a so-called high-performance polymer (here PPrSiDPA with a PO_2 of 230 Barrer [38]). Already in this view larger holes are visible than for the case of low-performance polymers (cf. Figure 1.5(a)) and the hole-size distribution (Figure 1.6(b)) reveals a much wider range of radii (here extending to 10 Å and being bi-modal). This situation is quite typical for polymers with high gas transport capacity. A more systematic study

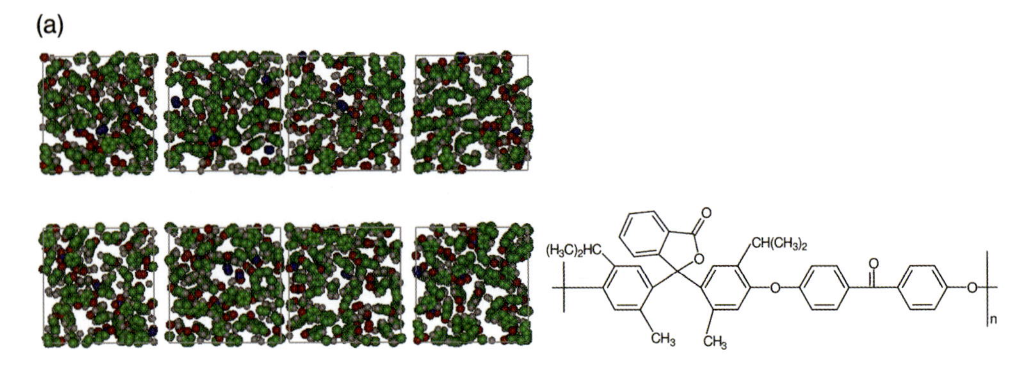

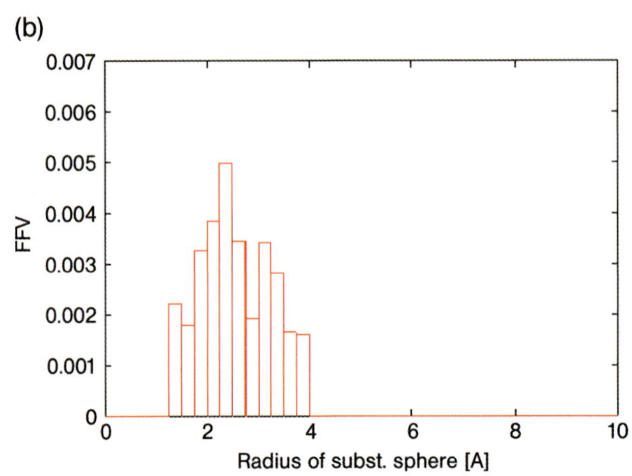

Figure 1.5 (a) Atomic monolayer representation (thickness about 3 Å) of a typical packing model and structure formula for DIDMPEEK. (b) Hole-size distribution for the packing model shown in Figure 1.5(a).

on polyimides [34] did, for example, reveal that the major difference between low-performance and high-performance polyimides with about the same overall contents of free volume lies in the distribution of the (e.g., oxygen) accessible free volume. Low-performance polyimides show just a monomodal distribution extending up to about 5 Å, while high-performance polyimides behave more or less similar to the example illustrated in Figure 1.6.

1.3.3.3 **Examples of Polymers with Ultrahigh Permeability of Small Molecules (e.g., $PO_2 \geq 1000$ Barrer)**

Figure 1.7 then shows respective data for an ultrahigh free-volume and performance polymer, Teflon AF2400 of DuPont ($PO_2 = 1140$ Barrer; [39]). One can recognize that

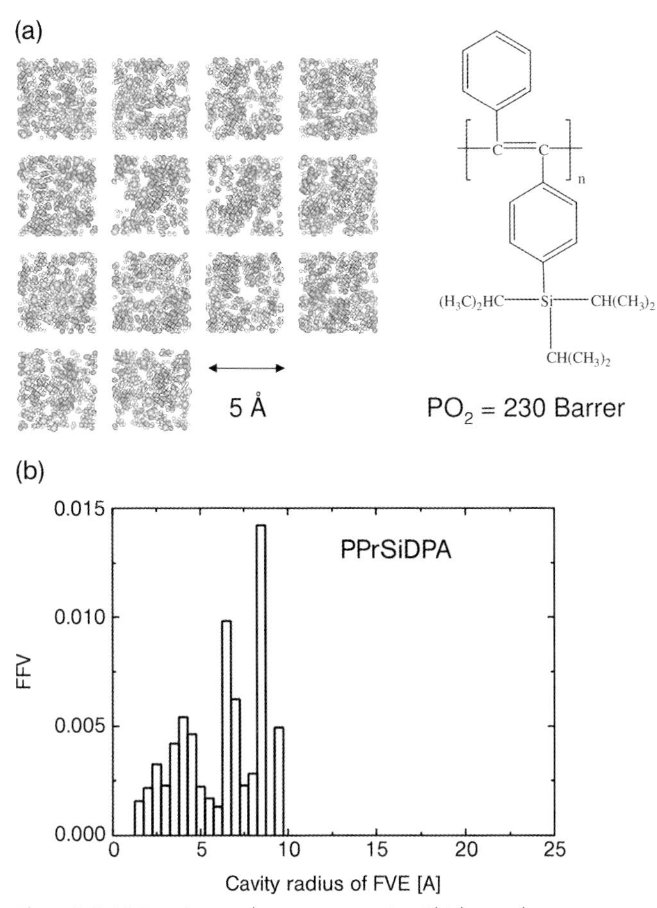

Figure 1.6 (a) Atomic monolayer representation (thickness about 3 Å) of a typical packing model and structure formula for PPrSiDPA. (b) Hole-size distribution for the packing model shown in Figure 1.6(a).

in this case there is "conventional" free volume organized in isolated holes in the radius range below 10 Å existing in parallel with a partly continuous phase of much larger holes that in this case are visible as a peak between 15 and 20 Å. The effect is even more pronounced for PTMSP, the polymer of this kind with the highest oxygen permeability so far measured (about 9000 Barrer; [38]). There, the continuity for the large-hole phase is more clearly visible already in atomic monolayer representations of respective packing models [37] and the ratio between the area under the "conventional" free-volume peak and the continuous hole phase peak in the hole-size distribution is even smaller than for Teflon AF2400.

The fact that for the mentioned ultrahigh free-volume polymers the continuous hole-phase peak appears at rather limited values is related with the limited size of the

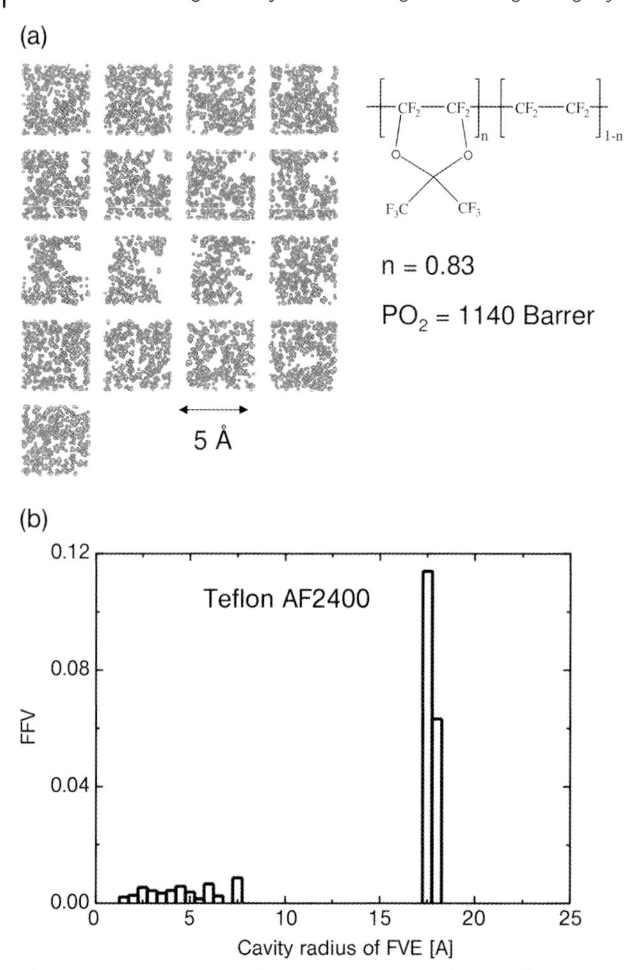

Figure 1.7 (a) Atomic monolayer representation of a typical packing model and structure formula for Teflon AF2400. (b) Hole-size distribution for the packing model shown in Figure 1.7(a).

investigated models (45–50 Å) while the thickness of real polymer membranes can extend into the micrometer range.

1.4 Summary

Atomistic molecular modeling techniques have proven to be a very useful tool for the investigation of the structure and dynamics of dense amorphous membrane polymers and of transport processes in these materials. By utilizing these methods, information can be obtained that is not accessible by experimental means.

Acknowledgments

The work was in part supported by the European projects "Growth" Program, "PERMOD – Molecular modeling for the competitive molecular design of polymer materials with controlled permeability properties." Contract #G5RD-CT-2000-200, the 6 Framework Programme project "MULTIMATDESIGN- Computer aided molecular design of multifunctional materials with controlled permeability properties", Contract no.: 013644, INTAS - RFBR 97–1525 grant.

References

1 Drioli, E., Natoli, M., Koter, I. and Trotta, F. (1995) *Biotechnology and Bioengineering*, **46**, 415.

2 Boyd, R.H. and Krishna Pant, P.V. (1991) *Macromolecules*, **24**, 6325.

3 Müller-Plathe, F. (1991) *Journal of Chemical Physics*, **94**, 3192.

4 Sok, R.M., Berendsen, H.J.B. and van Gunsteren, W.F. (1992) *Journal of Chemical Physics*, **96**, 4699.

5 Krishna Pant, P.V. and Boyd, R.H. (1993) *Macromolecules*, **26**, 679.

6 Gusev, A.A., Müller-Plathe, F., van Gunsteren, W.F. and Suter, U.W. (1994) *Advances in Polymer Science*, **16**, 207.

7 Tamai, Y., Tanaka, H. and Nakanishi, K. (1994) *Macromolecules*, **27**, 4498.

8 Chassapis, C.S., Petrou, J.K., Petropoulos, J.H. and Theodorou, D.N. (1996) *Macromolecules*, **29**, 3615.

9 Fried, J.R., Sadad-Akhavi, M. and Mark, J.E. (1998) *Journal of Membrane Science*, **149**, 115.

10 van der Vegt, N.F.A., Briels, W.J., Wessling, M. and Strathmann, H. (1999) *Journal of Chemical Physics*, **110**, 11061.

11 Hofmann, D., Fritz, L., Ulbrich, J., Schepers, C. and Böhning, M. (2000) *Macromolecular Theory and Simulations*, **9**, 293.

12 Tocci, E., Hofmann, D., Paul, D., Russo, N. and Drioli, E. (2001) *Polymer*, **42**, 521.

13 Neyertz, S. and Brown, D. (2001) *Journal of Chemical Physics*, **115**, 708; Neyertz, S. Brown, D. Douanne, A. Bas, C. and Albérola, N.D. (2002) *The Journal of Physical Chemistry. B*, **106**, 4617.

14 van der Vegt, N.F.A. (2002) *Journal of Membrane Science*, **205**, 125; van Gunsteren, W.F. and Berendsen, H.J.C. (1990) *Angewandte Chemie*, **29**, 992.

15 Macchione, M., Jansen, J.C., De Luca, G., Tocci, E., Longeri, M. and Drioli, E. (2007) *Polymer*, **48**, 2619–2635.

16 Haile, J.M. (1992) *Molecular Dynamics Simulation Elementary Methods*, Wiley Interscience, New York.

17 Tocci, E., Gugliuzza, A., De Lorenzo, L. and Drioli, E. in preparation.

18 (1999) *Polymer User Guide, Amorphous Cell Section, Version 400p+*, Molecular Simulations Inc., San Diego, CA.

19 (1999) *Discover User Guide, Polymer User Guide, Amorphous Cell Section, Version 4.0.0p+*, Molecular Simulations Inc., San Diego, CA.

20 Sun, H. and Rigby, D. (1997) *Spectrochim Acta*, **53A**, 1301.

21 Rigby, D. Sun, H. and Eichinger, B.E. (1997) *Polymer International*, **44**, 311.

22 Gusev, A.A. Arizzi, S. and Suter, U.W. (1993) *Journal of Chemical Physics*, **99**, 2221.

23 Gusev, A.A. and Suter, U.W. (1993) *Journal of Chemical Physics*, **99**, 2228.

24 Hofmann, D., Heuchel, M., Yampolskii, Yu., Khotimskii, V. and Shantarovich, V. (2002) *Macromolecules*, **35**, 2129–2140.

25 Ichiraku, Y., Stern, S.A. and Nakagawa, T. (1987) *Journal of Membrane Science*, **34**, 5–18; Masuda, T., Iguchi, Yu., Tang, B.-Z. and Higashimura, T. (1988) *Polymer*, **29**, 2041–2049; Bondar, V.I. (1995) Ph.D.

Thesis, Institute of Petrochemical Synthesis, Moscow.

26 Khotimskii, V.S., Filippova, V.G., Bryantseva, I.S., Bondar, V.I., Shantarovich, V.P. and Yampolskii, Y.P. (2000) *Journal of Applied Polymer Science*, **78**, 1612–1620.

27 Boulougouris, G.C., Economou, I.G. and Theodorou, D.N. (2001) *Journal of Chemical Physics*, **115**, 8231–8237.

28 Boulougouris, G.C., Voutsas, E.C., Economou, I.G., Theodorou, D.N. and Tassios, D.P. (2001) *The Journal of Physical Chemistry B*, **105**, 7792–7798.

29 Siegert, M.R., Heuchel, M. and Hofmann, D. (2007) *Journal of Computational Chemistry*, **28**, 877–889.

30 Simulated data from, Siegert, M.R. (2006) PhD Thesis, Berechnung von Löslichkeitskoeffizienten in Polymer-Materialien, FU-Berlin.

31 Kamiya, Y., Naito, Y., Hirose, T. and Mizoguchi, K. (1990) *Journal of Polymer Science, Part B, Polymer Physics*, **28** (8), 1297–1308.

32 Singh, A., Freeman, B.D. and Pinnau, I. (1998) *Journal of Polymer Science, Part B, Polymer Physics*, **36** (2), 289–301; Frahn, J. (GKSS), personal communication.

33 Summers, W.R., Tewari, Y.B. and Schreiber, H.P. (1972) *Macromolecules*, **5** (1), 12–16; Lichtenthaler, R.N., Liu, D.D. and Prausnitz, J.M. (1974) *Berichte der Bunsen-Gesellschaft — Physical Chemistry Chemical Physics*, **78** (5), 470–477; Grate, J.W., Kaganove, S.N. and Bhethanabotla, V.R. (1997) *Faraday Discussions*, **107**, 259–283; Frahn, J. (GKSS), personal communication.

34 Heuchel, M., Hofmann, D. and Pullumbi, P. (2004) *Macromolecules*, **37**, 201.

35 Gregory, R.B. and Yongkang, Zhu (1991) *Positron and Positron Chemistry* (ed. Y.C. Jean), World Scientific, Singapore, p. 136.

36 Provensher, S.W. (1982) *Computer Physics Communications*, **27**, 229.

37 Hofmann, D., Entrialgo-Castano, M., Lerbret, A., Heuchel, M. and Yampolskii, Yu. (2003) *Macromolecules*, **36**, 8528–8538.

38 Yampolskii, Yu.P., Korikov, A.P., Shantarovich, V.P., Nagai, K., Freeman, B.D., Masuda, T., Teraguchi, M. and Kwak, G. (2001) *Macromolecules*, **34**, 1788–1796.

39 Alentiev, A.Yu., Yampolskii, Yu.P., Shantarovich, V.P., Nemser, S.M. and Platé, N.A. (1997) *Journal of Membrane Science*, **126**, 123–132.

2
Polymeric Membranes for Molecular Separations

Heru Susanto and Mathias Ulbricht

2.1
Introduction

In this chapter we describe the state-of-the-art and the challenges in preparation and manufacturing of polymeric membranes for molecular separations in liquid phase. The processes include separation of aqueous solutions, that is, pressure-driven desalination using reverse osmosis and nanofiltration, fractionations of small and larger molecules using ultrafiltration and removal of organic substances by pervaporation (e.g., for shifting equilibria for (bio)chemical reactions). Separations in nonaqueous organic systems such as pervaporation and nanofiltration will also be covered. The preparation of charged membranes for electromembrane processes is another important application area for special polymers. Surface modification of membranes has become an important tool to reduce fouling or increase biocompatibility, but it can also be used to change membrane selectivity by combining separation mechanisms (e.g., based on size and charge).

2.2
Membrane Classification

Synthetic membranes for molecular liquid separation can be classified according to their selective barrier, their structure and morphology and the membrane material. The selective barrier – porous, nonporous, charged or with special chemical affinity – dictates the mechanism of permeation and separation. In combination with the applied driving force for transport through the membrane, different types of membrane processes can be distinguished (Table 2.1).

Selective barrier structure. Transport through porous membranes is possible by viscous flow or diffusion, and the selectivity is based on size exclusion (sieving mechanism). This means that permeability and selectivity are mainly influenced by membrane pore size and the (effective) size of the components of the feed: Molecules

Membrane Operations. Innovative Separations and Transformations. Edited by Enrico Drioli and Lidietta Giorno

ISBN: 978-3-527-32038-7

Table 2.1 Overview of main polymer membrane characteristics and membrane-based processes for molecular separations in liquid phase.

		Transmembrane gradient		
Selective barrier	**Typical structure**	**Concentration difference**	**Pressure difference**	**Electrical potential**
Nonporous	anisotropic, thin-film composite	Pervaporation	Reverse Osmosis Nanofiltration	
Microporous $d_p \leq 2\,nm$	anisotropic, thin-film composite	Dialysis	Nanofiltration	Electrodialysis
Non- or microporous, with fixed charge	isotropic	Dialysis		Electrodialysis
Mesoporous $d_p = 2 \ldots 50\,nm$	anisotropic, isotropic track-etched	Dialysis	Ultrafiltration	Electro-ultrafiltration
Carrier in liquid	immobilized in isotropic porous membrane	Carrier-mediated separation		
Affinity ligand in solid matrix	isotropic, anisotropic			

with larger size than the largest membrane pore will be completely rejected, and molecules with smaller size can pass through the barrier; the Ferry–Renkin model can be used to describe the effect of hindrance by the pore on rejection in ultrafiltration (UF) [1]. Transport through nonporous membranes is based on the solution-diffusion mechanism [1, 2]. Therefore, the interactions between the permeand and the membrane material dominate the mass transport and selectivity. Solubility and chemical affinity on the one hand, and the influence of polymer structure on mobility on the other hand serve as selection criteria. However, the barrier structure may also change by uptake of substances from the feed (e.g., by plastification), and in those cases real selectivities can be much lower than ideal ones obtained from experiments using only one component in the feed or at low feed activities. Separation using charged membranes, either nonporous (swollen gel) or porous (fixed charged groups on the pore wall), is based on charge exclusion (Donnan effect; ions or molecules having the same charge as the fixed ions in the membrane will be rejected, whereas species with opposite charge will be taken up by and transported through the membrane). Therefore, the kind of charge and the charge density are the most important characteristics of these membranes [1]. Finally, molecules or moieties with special affinity for substances in the feed are the basis for carrier-mediated transport through the membrane; very high selectivities can be achieved; the diffusive fluxes are higher for (immobilized) liquid membranes than for polymer-based fixed-carrier membranes [1].

Concentration polarization can dominate the transmembrane flux in UF, and this can be described by boundary-layer models. Because the fluxes through nonporous barriers are lower than in UF, polarization effects are less important in reverse osmosis (RO), nanofiltration (NF), pervaporation (PV), electrodialysis (ED) or carrier-mediated separation. Interactions between substances in the feed and the membrane surface (adsorption, fouling) may also significantly influence the separation performance; fouling is especially strong with aqueous feeds.

Cross-section structure. An anisotropic membrane (also called "asymmetric") has a thin porous or nonporous selective barrier, supported mechanically by a much thicker porous substructure. This type of morphology reduces the effective thickness of the selective barrier, and the permeate flux can be enhanced without changes in selectivity. Isotropic ("symmetric") membrane cross-sections can be found for self-supported nonporous membranes (mainly ion-exchange) and macroporous microfiltration (MF) membranes (also often used in membrane contactors [1]). The only example for an established isotropic porous membrane for molecular separations is the case of track-etched polymer films with pore diameters down to about 10 nm. All the above-mentioned membranes can in principle be made from one material. In contrast to such an integrally anisotropic membrane (homogeneous with respect to composition), a thin-film composite (TFC) membrane consists of different materials for the thin selective barrier layer and the support structure. In composite membranes in general, a combination of two (or more) materials with different characteristics is used with the aim to achieve synergetic properties. Other examples besides thin-film are pore-filled or pore surface-coated composite membranes or mixed-matrix membranes [3].

Membrane materials. Polymeric membranes are still dominating a very broad range of industrial applications. This is due to their following advantages: (i) many different types of polymeric materials are commercially available, (ii) a large variety of different selective barriers, that is, porous, nonporous, charged and affinity, can be prepared by versatile and robust methods, (iii) production of large membrane area with consistent quality is possible on the technical scale at reasonable cost based on reliable manufacturing processes, and (iv) various membrane shapes (flat sheet, hollow-fiber, capillary, tubular, capsule; Figure 2.1) and formats including membrane modules with high packing density can be produced. However, membrane polymers also have some limitations. A very well-defined regular pore structure is difficult to achieve, and the mechanical strength, the thermal stability and the chemical resistance (e.g., at extreme pH values or in organic solvents) are rather low for many organic polymers. In that regard, inorganic materials can offer some advantages, such as high mechanical strength, excellent thermal and chemical stabilities, and in some cases a very uniform pore shape and size (e.g., in zeolites). However, some inorganic materials are very brittle, and due to complicated preparation methods and manufacturing technology, the prices for many inorganic membranes (especially those for molecular separations) are still very high. An overview of inorganic membranes for separation and reaction processes can be found elsewhere [4, 5].

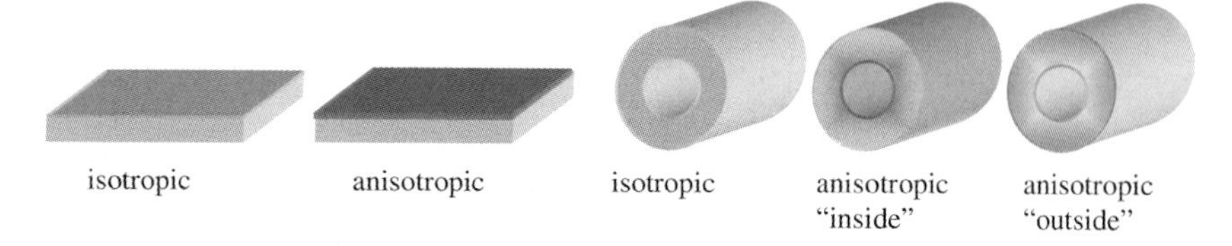

Figure 2.1 Polymeric membrane shapes and cross-sectional structures. Tubular membranes are similar to flat sheet membranes because they are cast on a macroporous tube as support. Capillary membranes are hollow fibers with larger diameter, that is, >0.5 mm.

2.3 Membrane Polymer Characteristics

2.3.1 Polymer Structure and Properties

Polymers for membrane preparation can be classified into natural and synthetic ones. Polysaccharides and rubbers are important examples of natural membrane materials, but only cellulose derivatives are still used in large scale for technical membranes. By far the majority of current membranes are made from synthetic polymers (which, however, originally had been developed for many other engineering applications). Macromolecular structure is crucial for membrane barrier and other properties; main factors include the chemical structure of the chain segments, molar mass (chain length), chain flexibility as well as intra- and intermolecular interactions.

Macromolecule chain flexibility is affected by the chemical structure of the main chain and the side groups. A macromolecule is flexible when unhindered rotation around single bonds in the main chain is possible. This flexibility can be reduced by several means, for example, by introducing double bonds or aromatic rings in the main chain, by forming ladder structures along the main chain or by incorporation of bulky side groups. Even larger effects with respect to the possible macroconformations can be imparted by changes of the chain architecture, that is, the transition from linear to branched or network structures. Polymer molar mass and its polydispersity have an influence on chemical and physical properties via the interactions between chain segments (of different or even the same molecule), through noncovalent binding or entanglement. For stability, high molar mass is desirable because the number of interaction sites increases with increasing chain length. However, the solubility will decrease with increasing molar mass.

The preceding structural characteristics dictate the state of polymer (rubbery vs. glassy vs. semicrystalline) which will strongly affect mechanical strength, thermal stability, chemical resistance and transport properties [6]. In most polymeric membranes, the polymer is in an amorphous state. However, some polymers, especially those with flexible chains of regular chemical structure (e.g., polyethylene/PE/, polypropylene/PP/or poly(vinylidene fluoride)/PVDF/), tend to form crystalline domains. This will lead to higher mechanical stability (high elastic modulus) as well as higher temperature and chemical resistance than for the same polymer in amorphous state, but the free volume (and hence permeability) will be much smaller. For semicrystalline polymers, the melting temperature (T_m) is important, because at this temperature a transition between crystalline and liquid state will occur. The glass transition temperature (T_g) is a much more important parameter to characterize amorphous polymers, because at this temperature a transition between solid (glass) and supercooled melt (rubber) takes place. In the glassy state molecules are frozen, therefore, chain mobility of a polymer is very limited. Heating this polymer over its T_g leads to a much more mobile and more flexible state, with lower elastic modulus and higher permeability. So-called "glassy polymers" have a T_g beyond room temperature, and "rubbery polymers" (or elastomers) have a T_g below room temperature. Polymer selection will be more important for membranes with nonporous selective barrier, because flux and selectivity depend on the solution-diffusion mechanism. For membranes with a porous selective barrier, the mechanical stability will be crucial to preserve the shape and size of the pores.

Block- or graft copolymers, which contain two or more different repeating units within the same polymer chains, are often used instead of homopolymers in order to obtain high-performance polymeric membranes; the overriding aims are synergies between properties of the different components. In addition, blending of polymers or copolymers is also performed. In these cases, compatibility and miscibility of both (co)polymers in one solvent are required in order to get a homogenous solution (cf. Section 2.4.2). The resulting solid membrane can be a homogenous polymer blend, as indicated by one T_g value between those for the two (co)polymers. A heterogeneous (phase separated) polymer blend will be characterized by two (or more) T_g values for the individual phases. Extensive existing knowledge from polymer blending can also be adapted to membrane preparation [7].

Chemical or physical cross-linking of the polymer is applied in order to control membrane swelling, especially for separations of organic mixtures. In addition, this can also enhance mechanical strength as well as chemical stability of a membrane. However, crosslinking decreases polymer solubility, therefore it is often done after membrane formation (cf. Sections 2.4–2.6).

The hydrophilicity–hydrophobicity balance of the membrane polymer is another important parameter that is mainly influenced by the functional groups of the polymer. Hydrophilic polymers have high affinity to water, and therefore they are suited as a material for nonporous membranes that should have a high permeability and selectivity for water (e.g., in RO or hydrophilic PV). In addition, hydrophilic membranes have been proven to be less prone to fouling in aqueous systems than hydrophobic materials.

2.3.2 Membrane Polymer Selection

2.3.2.1 Polymers for Porous Barriers

The selection of the polymer for a porous membrane is based on the requirements of the manufacturing process (mainly solubility for controlled phase separation;

cf. Section 2.4.2), and the behavior and performance under application conditions. The following material properties are important to be considered:

(i) *Film-forming properties* indicate the ability of a polymer to form a cohesive film, and the macromolecular structure, especially molar mass and attractive interactions between chain segments, is crucial in this regard (cf. Section 2.3.1). Poly(ether sulfones) (PES), polysulfones (PSf), polyamides (PA) or polyimides (PI) are examples for excellent film-forming materials [8].

(ii) *Mechanical properties* involve film strength, film flexibility and compaction stability (especially of a porous structure). The latter is most important for high-pressure processes (e.g., for the porous substructure of an integrally anisotropic RO membrane). Because hollow fiber membranes are self-supporting, the mechanical stability will be especially relevant. Many commercial flat-sheet membranes are prepared on a nonwoven support material (Figure 2.2).

(iii) *Thermal stability* requirements depend very much on the application. In order to ensure the integrity of a pore structure in the nanometer dimension, the T_g of the polymer should be higher than the process temperature.

(iv) *Chemical stability* requirements include the resistance of the polymer at extreme pH values and other chemical conditions. Cleaning agents such as strong acids or bases, or oxidation agents are usually used to clean a fouled membrane. The stability in special solvents is also important in selected cases, that is, when processes with nonaqueous mixtures are considered.

(v) *The hydrophilicity–hydrophobicity balance* correlates with the wettability of the material. This can be important in order to use all the pores in UF, or

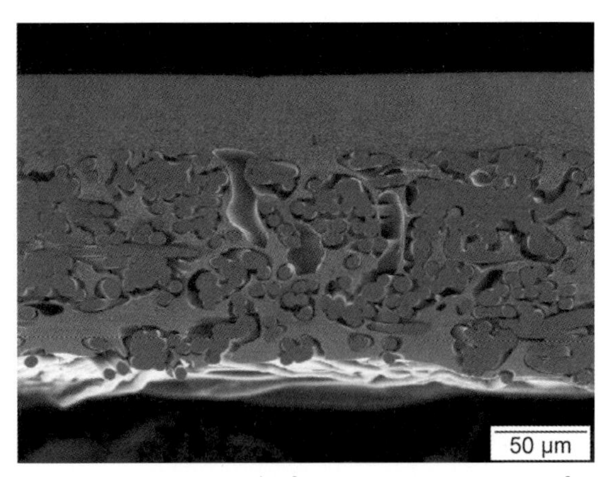

Figure 2.2 SEM micrograph of a microtome cross-section of a porous polymer membrane with an anisotropic structure on a nonwoven as mechanical support (reprinted from [9], with permission from Wiley-VCH, 2006).

when a porous membrane is applied as a contactor between a liquid and a gas phase, and the phase boundary is stabilized because the liquid will not wet the dry pores of the membrane. With aqueous liquid phases, the first case will require a more hydrophilic (e.g., polyacrylonitrile, PAN), and the second case a hydrophobic membrane polymer (e.g., PP). Surface wettability is also critical for fouling; cellulose is an excellent example of a hydrophilic polymer as material for low fouling UF membranes. Nevertheless, hydrophobic polymers, for instance PVDF or PES, show better chemical and thermal stability.

Considering all the above criteria, PSf, PES, PAN, PVDF and cellulose-based polymers (cellulose acetates/CA/and regenerated cellulose) are mostly used for commercial UF membranes (see also Section 2.6.1).

2.3.2.2 Polymers for Nonporous Barrier

The separation performance of membranes with nonporous barriers is – because of the transport via solution-diffusion (cf. Section 2.2) – predominantly influenced by the polymer material itself. Therefore, the material selection is directly related to the intrinsic (bulk) properties of the polymer, but – as for porous membranes – film-forming properties, mechanical and thermal stability form the basis of applicability (cf. Section 2.3.2.1). The following characteristics should be considered:

(i) *Glassy or rubbery state of the polymer.* Thermal analysis to know the T_g value is essential. The state of the nonporous polymer will determine the available free volume and the segmental mobility, and those have a decisive influence on the diffusion of molecules through the polymer. Size-based diffusion selectivity will only be possible with polymers in the rigid amorphous state.

(ii) *Free volume* will depend on the interchain distance in the bulk of the polymer. Somewhat independent of the state (cf. above), pronounced rigidity of the main chain and very bulky side groups can lead to larger free volume and, consequently, high permeabilities.

(iii) *The hydrophilicity–hydrophobicity balance* or other more special affinities can lead to (selective) dissolution (sorption) of molecules in the membrane. When the membrane is in contact with a liquid feed, swelling can become quite large, and this effect is often dominant for selectivity (see hydrophilic vs. organophilic PV; cf. Section 2.6.3).

(iv) *Chemical stability* requirements are similar to those for porous materials. Cleaning-related instability against active chlorine is a special problem for PA-based TFC membranes for RO. Due to the increasing number of nonaqueous applications (especially in PV and NF), polymer resistance to various organic solvents is gaining particular importance.

CA, PA, PI, poly(vinyl alcohol) (PVA) and polydimethylsiloxane (PDMS) are examples of selective polymers frequently used for nonporous barriers (see also Sections 2.6.2, 2.6.3).

2.3.2.3 Polymers for Charged Barrier

A charged (ion-exchange) membrane is prepared from a polyelectrolyte, that is, a polymer that contains ionic side groups. An anion exchange membrane contains fixed positively charged ions (e.g., $-NR_2H^+$, $-NR_3^+$), and this membrane will bind any anions from the feed stream. A cation exchange membrane contains fixed negatively charged ions (e.g., $-SO_3^-$, $-COO^-$), binding any cation from the feed. Exclusion of ions with the same charge depends strongly on the fixed-charge density in the membrane and the electrolyte concentration outside the membrane. The basic criteria of polymer selection – film-forming properties, mechanical and thermal stability as well as high chemical stability (extreme pH, oxidizing agents) – are similar in porous and nonporous membranes (cf. Sections 2.3.2.1 and 2.3.2.2). The following more specific important properties for ion-exchange membranes should be considered in addition:

(i) *High charge density* is the basis for high permselectivity. An ion-exchange membrane should be highly permeable to counterions for the fixed ions, but should be impermeable to co-ions (same charge as fixed ions).

(ii) *Low electrical resistance* is achieved when the permeability of an ion-exchange membrane for the counterions with an electrical potential gradient as the driving force is high.

(iii) *Controlled swelling* and low susceptibility to changes in external salt concentration are essential in order to keep charge density (and hence permselectivity) high, and are thus the basis for sufficient stability and constant separation performance. Due to the high affinity of polyelectrolytes to water, swelling is strong in ion-exchange membranes. To limit excessive swelling, chemical crosslinking is usually performed. An alternative are phase-separated polymers with ion-exchange clusters continuously distributed in a continuous hydrophobic phase.

Perfluorosulfonic acid polymers, for example, Nafion, or ionic and cross-linked polystyrene derivatives, are the best known examples of ion-exchange membrane materials (see also Section 2.6.4).

2.4 Membrane Preparation

2.4.1 Track-Etching of Polymer Films

Membranes with very regular pores of sizes down to around 10 nm can be prepared by track-etching [10], and, in principle, those membranes can be used for the fractionation of macromolecules in solution. A relatively thin (<35 μm) polymer film (typically from poly(ethylene terephthalate)/PET/or aromatic polycarbonate/PC/) is first bombarded with fission particles from a high-energy source. These particles pass through the film, breaking polymer chains and creating damaged "tracks." Thereafter, the film is immersed in an etching bath (strong acid or alkaline), so that the film is preferentially etched along the tracks, thereby forming pores. The pore density is determined by irradiation intensity and exposure time, whereas etching time determines the pore size. The advantage of this technique is that uniform and cylindrical pores with very narrow pore-size distribution can be achieved. In order to avoid the formation of double or multiple pores, produced when two nuclear tracks are too close together, the membrane porosity is usually kept relatively low, that is, typically less than 10%.

2.4.2 Phase Separation of Polymer Solutions

Polymer membranes by phase separation. The method is often called "phase inversion," but it should be described as a phase-separation process: a one-phase solution containing the membrane polymer is transformed by a precipitation/solidification process into two separate phases (a polymer-rich solid and a polymer-lean liquid phase). Before the solidification, usually a transition of the homogeneous liquid into two liquids (liquid–liquid demixing) occurs. The "proto-membrane" is formed from the solution of the membrane polymer by casting a film on a suited substrate or by spinning through a spinneret together with a bore fluid. Based on the way the polymer solution is solidified, the following techniques can be distinguished:

(i) *Nonsolvent-induced phase separation (NIPS)* – the polymer solution is immersed in a nonsolvent coagulation bath (typically water); demixing and precipitation occur due to the exchange of solvent (from polymer solution) and nonsolvent (from coagulation bath), that is, the solvent and nonsolvent must be miscible.

(ii) *Vapor-induced phase separation (VIPS)* – the polymer solution is exposed to an atmosphere containing a nonsolvent (typically water); absorption of nonsolvent causes demixing/precipitation.

(iii) *Evaporation-induced phase separation (EIPS)* – the polymer solution is made in a solvent or in a mixture of a volatile solvent and a less volatile nonsolvent, and solvent is allowed to evaporate, leading to precipitation or demixing/precipitation.

(iv) *Thermally induced phase separation (TIPS)* – a system of polymer and solvent is used that has an upper critical solution temperature; the solution is cast or spun at high temperature, and cooling leads to demixing/precipitation.

By far the majority of polymeric membranes, including UF membranes and porous supports for RO, NF or PV composite membranes, are produced via phase separation. The TIPS process is typically used to prepare membranes with a macroporous barrier, that is, for MF, or as support for liquid membranes and as gas–liquid contactors. In technical manufacturing, the NIPS process is most frequently applied, and membranes with anisotropic cross-section are obtained. Often,

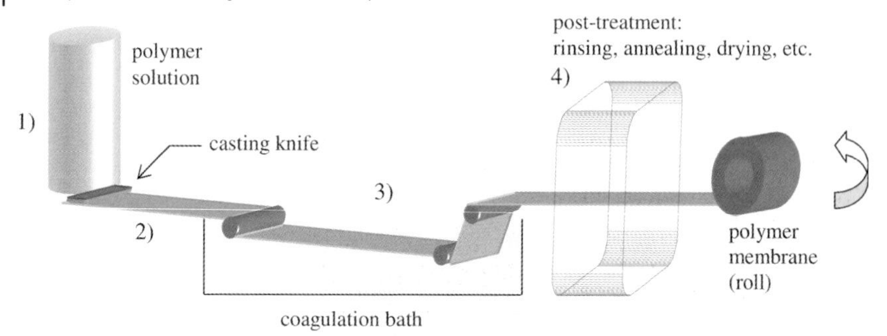

Figure 2.3 Schematic depiction of the continuous manufacturing process of polymeric membranes by the NIPS process.

the time before contact with the coagulation bath is used to "fine tune" membrane pore structure; some of these processes can thus be described as combinations of VIPS followed by NIPS.

Integrally anisotropic polymer membranes via NIPS process. The cross-sectional structure of an anisotropic membrane is crucial in order to combine the desired selectivity (by a barrier with pores in the lower nm range or by a nonporous polymer) with high fluxes: the top layer acts as a thin selective barrier and a porous sublayer provides high mechanical strength. Such integrally "asymmetric" membranes were first discovered by Loeb and Sourirajan [11]. This finding was the first breakthrough for commercial membrane technology, that is, such RO membranes from CA showed much higher fluxes than the previously produced ones from the same polymer. This method involves (Figure 2.3): (1) polymer dissolution in single or mixed solvent, (2) casting the polymer solution as a film ("proto-membrane") on suited substrate (or spinning as free liquid film, for hollow fiber), (3) precipitation by immersion in a nonsolvent coagulation bath, and (4) post-treatments such as rinsing, annealing and drying. The membranes resulting from this process have typically a very thin (<1 μm, often even less than 100 nm) top skin layer (selective barrier), which is either nonporous or porous (Figure 2.4).

The selection of the materials and the discussion of mechanisms for phase separation are based on ternary phase diagrams with the three main components polymer, solvent and nonsolvent; a pronounced miscibility gap (instable region) is an essential precondition. Besides thermodynamics aspects, the onset and rate of precipitation in the liquid film (both are different depending on the distance to the plane of first contact with the coagulation bath) are also important; the mass transfer (nonsolvent in-flow, and solvent out-flow) can have tremendous influence. Two mechanisms are distinguished: (i) instantaneous liquid–liquid demixing, which will result in a porous membrane, (ii) delayed onset of liquid–liquid demixing, which can result in a membrane with a nonporous barrier skin layer [2]. The rate of precipitation decreases from the top surface (in most cases, this plane of first contact with the coagulation bath will be the barrier in the final membrane) to the bottom surface of the cast film. As precipitation slows down, the resulting pore sizes increase because

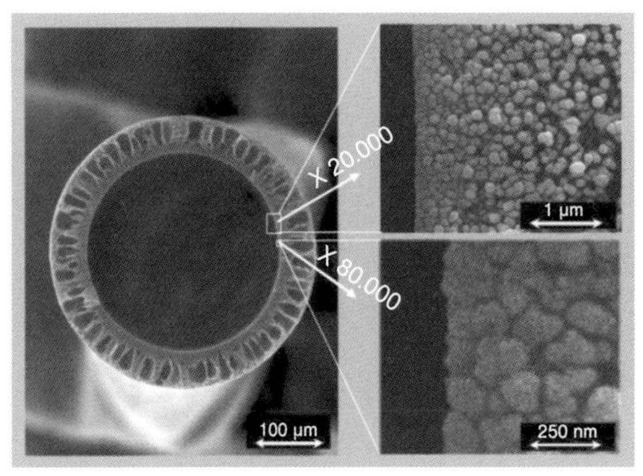

Figure 2.4 SEM micrograph of a cross-section of a hollow-fiber dialysis membrane (Polyflux, Gambro) with an anisotropic structure and macrovoids in the support layer (*left*), and details of the inner porous separation layer in two different magnifications (*right*; reprinted from [12], with permission from Wiley-VCH, 2003).

the two phases have more time to separate. In practice, most systems for membrane preparation contain more than three components (e.g., polymer blends as materials and solvent mixtures for casting solution and coagulation bath). Consequently, the mechanisms can be very complex and are still under intense scientific investigation and discussion [13, 14]. Important variables to control membrane characteristics will be outlined below.

Characteristics of the casting solution. Most important is the selection of a suitable solvent for the polymer, that is, the strength of mutual interactions is inversely proportional to the ease of precipitation by the nonsolvent (cf. below). Polymer concentration also plays a vital role to determine the membrane porosity. Increasing polymer concentration in the casting solution leads to a higher fraction of polymer and consequently decreases the average membrane porosity and pore size. In addition, increasing the polymer concentration could also suppress macrovoid formation and enhance the tendency to form sponge-like structure. However, this can also increase the thickness of the skin layer. Even though details depend on the properties of the membrane polymer, UF membranes can be obtained within a range of polymer concentrations of 12–20 wt%, whereas RO membranes are typically prepared from casting solutions with polymer concentrations $\geq$20 wt% (in order to increase salt rejection, a thermal annealing step is often added to the manufacturing scheme).

Solvent/nonsolvent system. The solvent must be miscible with the nonsolvent (here an aqueous system). An aprotic polar solvent like N-methyl pyrrolidone (NMP), dimethyl formamide (DMF), dimethyl acetamide (DMAc) or dimethylsulfoxide (DMSO) is preferable for rapid precipitation (instantaneous demixing) upon immersion in the nonsolvent water. As a consequence, a high porosity anisotropic membrane can be achieved. For slow precipitation, yielding low porosity or nonporous membrane, solvents having a relatively low Hildebrand solubility parameter [15], like tetrahydrofuran (THF) or acetone are preferable.

Additives. For certain purposes, additive or modifier is added in the casting solution. Indeed, this additive can determine the performance of the ultimate membrane and is often not disclosed for commercial membranes. Usually, additives include (i) cosolvent with relatively high solubility parameter (such a solvent can slow down the precipitation rate, and higher rejection is achieved), (ii) pore-forming agents such as poly(vinyl pyrrolidone) (PVP) or poly(ethylene glycol) (PEG) (these hydrophilic additives can enhance not only membrane pore size but also membrane hydrophilicity; at least partially, these polymers form stable blends with membrane polymers such as PSf or PES), (iii) nonsolvent (should be added only in such amounts that demixing of the casting solution does not occur; promotes formation of a more porous structure and could also reduce macrovoid formation), (iv) addition of cross-linking agent into casting solution (is less frequently used, but could also reduce macrovoid formation).

Characteristics of coagulation bath. The presence of a fraction of solvent in the coagulation bath can slow down the liquid–liquid demixing rate. Consequently, a less porous barrier structure should be obtained. However, the opposite effect can also occur, that is, addition of solvent can decrease polymer concentration (in the proto-membrane) leading to a more open porous structure. The amount of the solvent to be added strongly depends on the solvent–nonsolvent interactions. As the mutual affinity of solvent and nonsolvent increases, more solvent is required to achieve an effect on the membrane structure. For example, in preparation of CA membranes, the content of solvent needed in a coagulation bath for a DMSO/water system is higher than for a dioxan/water system. Instantaneous demixing resulting in a porous structure can be obtained by better miscibility between solvent and nonsolvent. In contrast, a less miscible solvent/nonsolvent combination results in a more nonporous structure. Furthermore, addition of solvent into a coagulation bath could also reduce the formation of macrovoids leading to the desired, more stable sponge-like structure of the supporting layer.

Exposure time of proto-membrane before precipitation. The effect of exposure to atmosphere before immersion is dependent on the solvent property (e.g., volatility, water absorption) and atmosphere property (e.g., temperature, humidity). This step (i.e., combination of EIPS or VIPS with NIPS; cf. above) has significant effects on the characteristics of the skin layer and the degree of anisotropy of the resulting membrane [14].

2.4.3
Composite Membrane Preparation

Composite membranes combine two or more different materials with different characteristics to obtain optimal membrane performance. Basically, the preparation involves: (i) preparation of porous support that is usually made by a phase-separation process (cf. Section 2.4.2), and (ii) deposition of a selective barrier layer on this porous support. A number of methods are currently used for manufacturing asymmetric composite membranes, which will be briefly discussed below [16, 17].

(i) *Laminating.* An ultrathin film is cast and then laminated to a (micro)porous support. This method has been used for preparing early RO membranes for water desalination [18].

(ii) *Dip-coating* of a polymer solution onto a support microporous support is followed by drying, or a reactive prepolymer is applied and IR radiation is used for curing. As a result, a thin layer of the coated polymer on the substrate is obtained. In some cases, crosslinking is done during curing to increase mechanical or chemical stability. Two problems are often observed, that is, penetration of the dilute coating solution into the pores of the support and formation of defective coatings. The first problem can be reduce by precoating the support with a protective layer from a hydrophilic polymer, such as polyacrylic acid or by filling the pores with a wetting liquid such as water or glycerin. The latter problem can be reduced by introducing an intermediate layer between the selective polymer film and the porous substrate.

(iii) *Plasma polymerization.* Gas-phase deposition of the barrier layer on a porous support is conducted from glow-discharge plasma via plasma polymerization. This method has been successfully used for RO membrane preparation [19].

(iv) *Interfacial polymerization.* This method has been developed by Cadotte *et al.* [20], and it is now the most important route to RO and NF membranes. The selective layer is formed *in situ* by polycondensation or polyaddition of reactive (bis- and trisfunctional) monomers or prepolymers on the surface of a porous support (Figure 2.5). Post-treatment such as heating is often applied in order to obtain a fully cross-linked structure of the selective barrier.

Other methods derived from surface modification, including heterogeneous graft copolymerization or *in situ* radical polymerization and deposition of polyelectrolyte

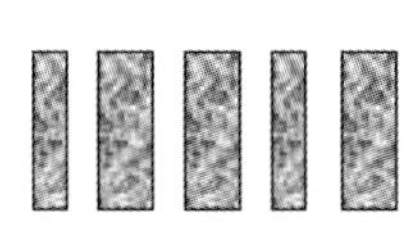

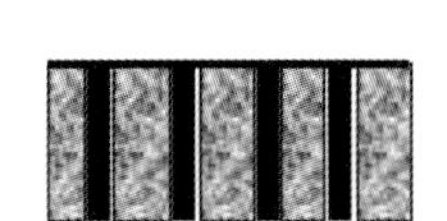

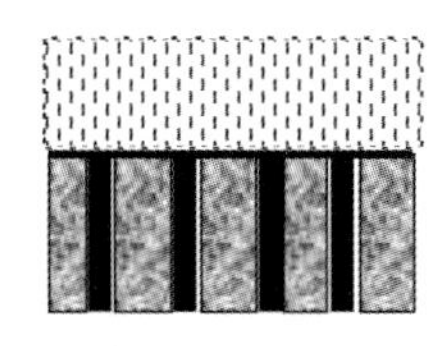

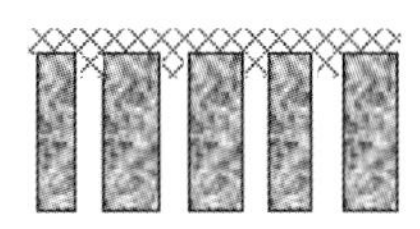

Figure 2.5 Schematic depiction of the preparation of TFC membranes by interfacial polymerization: The support membrane (e.g., from PES) is immersed in an aqueous monomer or prepolymer solution (e.g., bis- or trifunctional amine), and subsequently contacted with a second bath containing a water-immiscible solvent in which another reactive monomer or prepolymer has been dissolved (e.g., bis- or trifunctional carbonic acid chloride). The reaction takes place at the interface of the two immiscible solutions on the outer surface of the support membrane, and the thickness of the polymer layer (e.g., cross-linked polyamide) is limited by its barrier properties for further diffusion of reactants into the reaction zone.

layers, also gain more interest for manufacturing composite membranes with tailored selectivities (cf. Section 2.5).

2.4.4 Mixed-Matrix Membranes

Current polymeric materials are inadequate to fully meet *all* requirements for the various different types of membranes (cf. Section 2.2) or to exploit the new opportunities for application of membranes. Mixed-matrix membranes, comprising inorganic materials (e.g., metal oxide, zeolite, metal or carbon particles) embedded in an organic polymer matrix, have been developed to improve the performance by synergistic combinations of the properties of both components. Such improvement is either with respect to separation performance (higher selectivity or permeability) or with respect to membrane stability (mechanical, thermal or chemical).

One should note that the methods to prepare such mixed-matrix membranes and the resulting properties are strongly dependent on the interactions between the different materials, and a homogeneous, regular distribution and interface compatibility are the key issues. Techniques to prepare mixed-matrix materials have been reviewed recently [21]. Mixed-matrix membranes are typically prepared by: (1) separate preparation of a polymer solution and a suspension of inorganic material, (2) mixing of both resulting in a mixed-matrix solution, (3) casting (or spinning) this solution, and (4) inducing phase separation, typically in the framework of the NIPS process (cf. Section 2.4.2). A common alternative for the preparation of mixed-matrix membranes containing inorganic oxides (e.g., silica) is the *in situ* synthesis of nanoparticles within a polymer solution via the sol-gel method followed by phase separation. The above techniques are mainly applied for preparation of advanced RO, NF and PV membranes (cf. Section 2.6). Only in special cases is the separation performance of the barrier really determined by the added inorganic (nano)materials, for example, by zeolites or carbon nanotubes [22].

2.5 Membrane Modification

Because most of the established membrane polymers can not meet all the performance requirements for a membrane dedicated to a particular application, membrane modifications are gaining rapidly increasing importance. Membrane modification is aimed either to minimize undesired interactions, which reduce membrane performance (e.g., membrane fouling), or to introduce additional interactions (e.g., affinity, responsive or catalytic properties) for improving the selectivity or creating an entirely novel separation function [3]. Three general approaches can be distinguished:

(i) Chemical modification of the membrane polymer (for membrane formation),
(ii) Blending of the membrane polymer with (an)other polymer(s) (before membrane formation), and
(iii) Surface modification after membrane preparation.

Because the first two approaches can involve significant changes in composition of the casting or spinning solution, membrane structure formed during the phase separation (cf. Section 2.4.2), and, consequently, membrane properties can be quite different from the unmodified reference material. An important example of polymer modification before membrane formation is sulfonation or carboxylation, for example, of PSf or PES, to obtain a more hydrophilic ultrafiltration membrane from a very stable membrane polymer [23]. The most well-known example for blending with the membrane polymer is the use of the water-soluble PVP during manufacturing of flat-sheet or hollow-fiber membranes from PSf or PES [24]. Even though during the coagulation and washing steps, some of the added modified or other polymer can leach out from the membrane matrix, a fraction remains on the pore surface and thus enhances the membrane hydrophilicity. Recently, amphiphilic graft or block-copolymers – containing functional (surface active) macromolecule segments and other segments that are compatible with the bulk of the membrane polymer – have been introduced as "tailored" macromolecular additives to render the final membrane surface hydrophilic or hydrophobic [25, 26].

Surface modification of commercially established membranes. This approach is of greatest interest in academic research but also in development within membrane companies [3, 27]. An increasing number of methods and technologies investigated for polymer surfaces in general are now being adapted to surface functionalization of polymeric membranes [28]. Highly attractive are technologies that can be integrated as another step into the continuous membrane manufacturing process (cf. Figure 2.3). A key feature of a successful (i.e., "tailored") surface modification is a synergy between the useful properties of the base membrane and the novel functional layer. In order to achieve a stable effect, chemical modification is preferable over physical modification. Attachment of functional moieties onto a membrane surface by physical principles can be done via the following ways [3]:

(i) Adsorption/adhesion – the functional layer is only physically fixed on the base material, and the binding strength can be increased via multiple interactions between functional groups in the macromolecular layer and on the solid surface,
(ii) Interpenetration via mixing between the added functional polymer and the base polymer in an interphase, and
(iii) Mechanical interpenetration (macroscopic entanglement) of an added polymer layer and the pore structure of a membrane.

In order to achieve membrane surface modification by chemical reactions, the following approaches have been proposed [3]:

(i) Heterogeneous (polymer-analogous) reactions of the membrane polymer,
(ii) "Grafting to" (attachment of functional macromolecular moieties in one step), and
(iii) "Grafting from" (heterogeneous graft copolymerization of functional monomers).

Photografting technologies, that is, the control of chemical surface functionalization by highly selective excitation with UV light, can be used for "grafting to" and "grafting from" and has been intensively explored for controlled functionalization of polymeric membranes [29].

Fouling resistance in aqueous systems. With self-assembled monolayers, structure–property relationships for nonadsorptive and nonadhesive surfaces have been identified on a molecular level [30]. Characteristics of materials that resist the adsorption of protein should be: (a) hydrophilic/polar, (b) overall electrically neutral, (c) hydrogen-bond acceptor, and (d) not hydrogen-bond donor. In these regards, PEG, zwitterionic moieties and other materials that display "kosmotropes" on their surface were identified as nonfouling materials that resist protein adsorption. This principle has successfully implemented to develop fouling-resistant polymer membranes for UF and NF in aqueous systems, via membrane formation from blends with tailored graft copolymers [25] or via controlled photoinitiated "grafting from" [31].

2.6
Established and Novel Polymer Membranes for Molecular Separations

2.6.1
Ultrafiltration

Because the mechanisms are based on pore flow and size exclusion (cf. Section 2.2), the polymer material itself does not have direct influence on flux and selectivity in UF. The UF membranes usually have an integrally asymmetric structure, obtained via the NIPS technique, and the porous selective barrier (pore size and thickness ranges are 2–50 nm and 0.1–1 μm, respectively) is located at the top (skin) surface supported by a macroporous sublayer (cf. Section 2.4.2). However, the pore-size distribution in that porous barrier is typically rather broad (Figure 2.6), resulting in limited size selectivity.

It should be noted that the adaptation of the NIPS or TIPS method to the fabrication of hollow-fiber membranes is straightforward for most systems (cf. Figure 2.4). Characterization of UF membranes is typically done via sieving

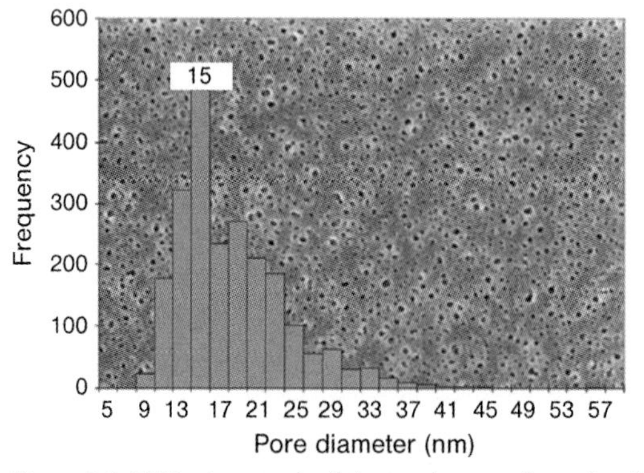

Figure 2.6 SEM micrograph of the top layer surface of an UF membrane from PAN and pore-size distribution from computerized image analysis (reprinted from [9], with permission from Wiley-VCH, 2006).

Table 2.2 Membrane polymers for UF and some characteristics.

Membrane polymer	Common solvent	T_g (°C)	pH range
Polysulfone	DMAc, DMF, DMSO, NMP	198	2–13
Poly(ether sulfone)	DMAc, DMF, DMSO, NMP	225	2–13
Poly(vinylidene fluoride)	DMAc, DMF, NMP, DMSO,	−40, (T_m ~ 175)	2–11
Polyacrylonitrile	DMAc, DMF, nitric acid	100	2–10
Cellulose acetate	Acetone, dioxan, DMAc, DMF, DMSO, THF	Around 135[a]	3–7
Regenerated cellulose	Stable in most organic solvents (typically prepared from cellulose acetate as precursor)	High crystalline content	4–9

[a]Depending on degree of acetylation.

experiments, that is, UF of a solution of macromolecular test solutes and subsequent analysis of the changed molar mass distribution (usually by size-exclusion chromatography) are performed. The specification of a commercial UF membrane is not the pore size, but mostly the "cutoff" value, that is, the molar mass for which more than 90% rejection have been observed. Isoporous track-etched membranes are a special case and typically made from PC or PET (UF-relevant pore sizes between 10 and 50 nm, and thicknesses between 8 and ~25 μm; cf. Section 2.4.1). The polymers most frequently used for commercial UF membranes, along with information about their stability, are summarized in Table 2.2.

Since fouling is the biggest problem for industrial application of UF, many developments in membrane materials address solutions to this problem. The following aspects should be considered in optimization of polymeric membranes for UF: (i) high permeability and rejection, (ii) hydrophilicity and fouling resistance, (iii) high maximum temperature and wide pH operating ranges, (iv) good mechanical properties, and (v) high chemical stability, especially towards cleaning agents. None of the established polymers satisfies all the above criteria. Chemical cross-linking during manufacturing via "regeneration" after NIPS of cellulose acetate can enhance cellulosic membrane performance with respect to chemical resistance and temperature stability, and a hydrophilic low-fouling UF membrane is obtained. Alternatively, surface modifications, mainly to increase hydrophilicity of membranes from hydrophobic materials, have been performed to enhance wettability and antifouling character (cf. Section 2.5).

Adopted from the state-of-the-art in RO, TFC membranes have become increasingly interesting for UF as well. One of the first examples of a commercial membrane of this type is composed of a thin barrier layer from regenerated cellulose on a porous polyolefine support [32]. Significant increase in selectivity in protein UF via electrostatic exclusion in addition to size exclusion has been achieved by introducing fixed charges into the barrier layer of a cellulose-based TFC membrane [33].

Similar to developments in NF and RO, solvent-resistant UF membranes could be the basis for a wide range of novel applications. Cross-linked integrally anisotropic membranes are explored with particular emphasis, and a very promising example are membranes made from poly(acrylonitrile-co-glycidyl methacrylate) that after NIPS had been cross-linked with ammonia or other tri- or difunctional amines [34].

Very uniform pore size leading to a more precise sieving would be interesting for improving UF performance. One promising approach towards this goal is pore formation via self-assembly of block-copolymers with "programmed" chemistry and architecture (and one crucial precondition is a low polydispersity of chain lengths). The feasibility of this approach has recently been demonstrated with the preparation of a composite membrane with 20-nm pores via formation of an ordered thin block-copolymer film on a macroporous support membrane and subsequent selective dissolution of the polymeric pore "template" [35]. Very recent work indicated that such tailored block-copolymers could also be processed via NIPS to integrally anisotropic polymer membranes with rather regular pore morphology [36].

2.6.2 Reverse Osmosis and Nanofiltration

RO and NF membranes for aqueous applications are quite similar in chemical composition and membrane preparation. However, mass transfer in NF is more complex than in RO because – in addition to solution-diffusion mechanism – size and charge exclusion are also usually involved. Ideally, polymeric membranes for RO and NF should be hydrophilic, resistant to chemicals (especially cleaning agents and chlorination) and microbial attack, and they should be structurally and mechanically stable over the long time of operation. Membranes with integrally asymmetric structure from the "first generation" material CA are currently still available (e.g., for NF in common applications like water treatment). However, TFC membranes dominate in the market (e.g., FT-30, SW30, ES10/ES15 for RO and DESAL 5 and NF 270 for NF). Most of commercial RO and NF composite membranes are polyamide-based although other composite membranes, for example, with sulfonated polysulfone as selective material, are also found. Interfacial polymerization is the standard method used for preparing the PA composite structure (cf. Figure 2.5), coating is occasionally applied for other selective polymers. For RO membranes, an ultrathin nonporous polymer layer (usually crosslinked PA) is formed on the top of a highly porous membrane with very small pores (e.g., from PES or PSf). Compared to PA-based materials, cellulosic membranes have a higher chlorine tolerance, but they are less solvent resistant and have only a narrow range of pH stability. Therefore, creating a selective material that is stable towards chlorine is still a motivation in the field of RO. Ion-exchange polymers, in particular copolymers of highly sulfonated polyethersulfone and polyethersulfone, originally developed as selective ion-conducting materials for fuel cells (cf. Sections 2.3.2.3 and 2.6.4), have recently been identified as promising candidates [37]. However, due to constraints in terms of low membrane prices and extensive process validation it would be more and more complicated to replace the established materials with such potential novel membranes.

Increasing the water flux considerably while keeping the selectivity high, is still a great challenge in further development of RO and NF. A very attractive TFC NF membrane with very high flux at promising rejection has been prepared via "layer-by-layer" (LBL) deposition of polyelectrolytes on porous UF membranes [38]. A mixed matrix RO membrane, composed of a water-selective zeolite in an ultrathin PA layer, with improved permeability and unchanged high salt rejection, has been developed recently [39]. A next generation of ultrahigh flux membranes could be based on composite membranes with an array of regular carbon nanotubes in the barrier layer [40].

In recent years, chemically stable membranes, which include oxidant and pH-stable and (organic) solvent-resistant materials, have been intensively developed in order to broaden application of NF or RO (for a review see [41]). PI and PAN derivatives or polyether-based materials are often used for preparing chemically stable NF membranes with integrally anisotropic structure. With composite membranes (with PSf, PAN, PI or PVDF as typical support materials), the permeability and selectivity for different solvents depend strongly on the barrier polymer. Selective layers from PA, polyureas, polyphenylene oxide or sulfonated PES are more suited for polar solvents, while silicone-based layers are preferred for nonpolar solvents (in that regard, SRNF membranes can be very similar to PV membranes; cf. Section 2.6.3). Mixed-matrix membranes are often proposed to increase the membrane stability. For instance, filling PDMS with porous zeolite yielded stable SRNF membranes with enhanced fluxes and selectivities allowing the use in nonpolar solvents and at high temperature [41]. A first large-scale success example for SRNF, with integrally anisotropic polyimide membranes, is the MAX-DEWAX process for crude-oil dewaxing [42]. However, to be practically useful, long-term stability and selectivity must be improved further. In addition, solvent–membrane interactions must be investigated in more detail in order to come up with satisfying and predictive models for transport and selectivity.

2.6.3 Pervaporation

As for RO and NF, most established PV membranes are composites with a nonporous polymeric barrier. In order to assure selectivity, the polymer should have preferential interactions with one of the components in the feed mixture. Integrity of the barrier is very important when separations of organic substances are concerned, and cross-linking is the preferred choice to limit swelling and improve stability. Three different types of selective barriers can be distinguished [43]: (i) hydrophilic, (ii) organophilic and (iii) organoselective ones. PAN is used as porous support for most PV membranes. This is due to its thermal stability and pronounced resistance to most organic solvents (cf. Table 2.2).

Hydrophilic polymers are used as selective barriers for dehydration of organic liquids via PV. The selective layer is typically from a glassy polymer; chemically cross-linked PVA is the established material for commercial membranes [43, 44]. Furthermore, poly(acrylic acid), other polyelectrolytes, PI or chitosan have also been explored.

In contrast, organophilic PV membranes are used for removal of (volatile) organic compounds from aqueous solutions. They are typically made of rubbery polymers (elastomers). Cross-linked silicone rubber (PDMS) is the state-of-the-art for the selective barrier [1, 43, 44]. Nevertheless, glassy polymers (e.g., substituted polyacetylene or poly(1-(trimethylsilyl)-1-propyne, PTMSP) were also observed to be preferentially permeable for organics from water. Polyether-polyamide block-copolymers, combining permeable hydrophilic and stabilizing hydrophobic domains within one material, are also successfully used as a selective barrier.

Organoselective membranes are used for separation of organic–organic liquid mixtures. Typical applications are separations of azeotropes or mixtures of substances that have close boiling points. An example is the commercial membrane PERVAP 2256, designed for the PV separation of methanol/MTBE or ethanol/ETBE. It had been reported that the selective barrier of this membrane is most likely PVA with incorporated polar moieties [45].

The challenge in further development of polymeric PV membranes is to create materials which can increase both selectivity and permeability and have high overall stability. To control swelling, many approaches have been proposed, for example, the use of rigid-backbone polymers (e.g., PI or highly aromatic polyurea/urethane copolymers), polymer blending or chemical cross-linking. Mixed-matrix membranes may be an alternative; silica-based nanoparticles have been added to the polymer matrix to reduce swelling and increase selectivity. Filling the membrane with an organophilic adsorbent (zeolite) was also used to increase selectivity [46].

An interesting pore-filled composite membrane, made by photograft copolymerization onto a solvent-stable PAN UF membrane, has been established [47]. High flux and selectivity for PV separation of organic–organic mixtures were achieved by a very thin selective barrier and prevention of swelling of the selective polymer in the pores of the barrier.

Novel polymers with "intrinsic microporosity" (PIMs) have recently been synthesized and characterized [48]. Their highly rigid, but contorted molecular structure leads to a very inefficient space filling. The polymers that are soluble in many common organic solvents form rather robust solids – including flat-sheet membranes – with very high specific surface areas (600–900 m^2/g). The first examples for their use as membrane materials, indicating a promising combination of high selectivities and fluxes in organoselective PV, have been reported recently.

2.6.4 Separations Using Ion-Exchange Membranes

Ion-exchange membranes are currently used not only for more or less "conventional" separation processes like membrane electrolysis (mainly the chlor-alkali process), electrodialysis, dialysis or electro-ultrafiltration (cf. Table 2.1), but also in various integrated processes such as fuel cells and catalytic reactors. The preparation of ion-exchange membranes and novel developments have been reviewed in recent publications [49–52].

Two types of membranes are distinguished, namely, homogeneous and heterogeneous ones. Heterogeneous ion-exchange membranes are prepared by dispersing anion- or cation-exchange particles into a polymer matrix and subsequent extrusion of the membrane film. Particle size significantly influences membrane swelling as well as mechanical strength. However, currently, homogenous anion- or cation-exchange membranes are preferred, composed of hydrocarbon (e.g., derivatives of styrene-divinylbenzene copolymers) or fluorocarbon (e.g., Nafion) polymers possessing ionic groups, and supported by backing materials. Such membranes can be prepared by the following routes [50]: (i) copolymerization of a monomer containing an ion-exchange group with a nonfunctionalized monomer, (ii) modification of a polymer film by introducing ionic groups (e.g., by "grafting from" of ion-exchange polymer or of nonfunctional polymer followed by chemical functionalization), and (iii) film casting and phase separation of solution of an ion-exchange polymer or its blend with another polymer.

Route (i) is most common for styrene-divinylbenzene copolymers. Route (ii) is typically applied for hydrocarbon- (PE, PP) or fluorocarbon-based membranes because it is difficult to find a suited solvent for the membrane polymers containing also highly polar ion-exchange groups. Examples of route (iii) are functionalized poly (ether ketone), polystyrene and PES; with these materials cross-linking is often performed to improve chemical stability. Overall, at moderate temperatures, ion-exchange functionalized perfluorocarbon polymers (e.g., Nafion) show still the best performance. These ion-exchange membranes have hydrophobic domains (providing a nonswelling matrix) as well as polar, charged domains (providing ion-selective water channels). In general, phase-separated polymers or polymer blends seem to be superior to one-phase materials with respect to a high conductivity at not too high water sorption (swelling), which leads to nonselective passage of solutes (e.g., methanol in fuel-cell systems) [53].

Hybrid organic–inorganic materials are promising to yield membranes with high chemical and mechanical stability and excellent conductivity, and the sol-gel process in conjunction with established membrane formation is the preferred preparation method.

Ion-exchange membranes with special structure and function have also been introduced. Amphoteric membranes consist of both positively (weak basic) and negatively (weak acidic) fixed-charge groups, chemically bound and randomly distributed to the polymer chains; the permselectivity of these membranes is pH responsive. Charge mosaic membranes possess both cation- and anion-exchange groups arranged in oriented parallel domains separated by neutral regions, each kind of ion-exchange group provides a continuous pathway from one side of the membrane to the other [49]. Bipolar membranes that contain cation-exchange groups at one side and anion-exchange groups at the other side are interesting for several applications, such as water splitting or chemical reactions [52].

2.7 Conclusion and Outlook

A wide variety of polymeric membranes with different barrier properties is already available, many of them in various formats and with various dedicated specifications. The ongoing development in the field is very dynamic and focused on further increasing barrier selectivities (if possible at maximum transmembrane fluxes) and/or improving membrane stability in order to broaden the applicability. This "tailoring" of membrane performance is done via various routes; controlled macromolecular synthesis (with a focus on functional polymeric architectures), development of advanced polymer blends or mixed-matrix materials, preparation of novel composite membranes and selective surface modification are the most important trends. Advanced functional polymer membranes such as stimuli-responsive [54] or molecularly imprinted polymer (MIP) membranes [55] are examples of the development of another dimension in that field. On that basis, it is expected that polymeric membranes will play a major role in process intensification in many different fields.

List of Abbreviations

CA	*cellulose acetate*
DMAc	*dimethyl acetamide*
DMF	*dimethyl formamide*
DMSO	*dimethyl sulfoxide*
d_p	*pore diameter*
ED	*electrodialysis*
EIPS	*evaporation-induced phase separation*
MF	*microfiltration*
NF	*nanofiltration*
NIPS	*nonsolvent-induced phase separation*
NMP	*N-methyl pyrrolidone*
PA	*polyamide*
PAN	*polyacrylonitrile*
PC	*polycarbonate*
PDMS	*polydimethylsiloxane*
PE	*polyethylene*
PEG	*poly(ethylene glycol)*
PES	*poly(ether sulfone)*
PET	*poly(ethylene terephthalate)*
PI	*polyimide*
PP	*polypropylene*
PSf	*polysulfone*
PV	*pervaporation*
PVA	*poly(vinyl alcohol)*

PVDF *poly(vinylidene fluoride)*
PVP *poly(vinyl pyrrolidone)*
RO *reverse osmosis*
SRNF *solvent-resistant nanofiltration*
TEP *triethyl phosphate*
TFC *thin-film composite*
T_g *glass transition temperature*
THF *tetrahydrofuran*
T_m *melting point temperature*
TIPS *thermally induced phase separation*
UF *ultrafiltration*
VIPS *vapor-induced phase separation*

References

1 Baker, R.W. (2004) *Membrane Technology and Applications,* 2nd edn, John Wiley & Sons, Ltd, Chichester, Chapters 2, 3, 5, 6 & 9.
2 Mulder, M. (1996) *Basic Principle of Membrane Technology,* 2nd edn, Kluwer Academic Publishers, Dordrecht, Chapters 2, 3 & 5.
3 Ulbricht, M. (2006) *Polymer,* **47**, 2217–2262.
4 Hsieh, P. (1996) *Inorganic Membranes for Separation and Reaction,* Elsevier, Amsterdam.
5 Marcano, J.G.S. and Tsotsis, T.T. (2002) *Catalytic Membranes and Membrane Reactors,* Wiley-VCH, Weinheim.
6 George, S.C. and Thomas, S. (2001) *Progress in Polymer Science,* **26**, 985–1017.
7 Utracki, L.A. (2002) *Polymer Blend Hand Book,* Kluwer Academic Publishers, Dordrecht.
8 Ho, W.S. and Sirkar, K.K. (1992) *Membrane Handbook,* Van Nostrand Reinhold, New York.
9 Abetz, V., Brinkmann, T., Dijkstra, M., Ebert, K., Fritsch, D., Ohlrogge, K., Paul, D., Peinemann, K.V., Nunes, S.P., Scharnagl, N. and Schossig, M. (2006) *Advanced Engineering Materials,* **8**, 328–358.
10 Fleischer, R.L., Alter, H.W., Furman, S.C., Price, P.B. and Walker, R.M. (1972) *Science,* **172**, 255–263.
11 Loeb, S. and Sourirajan, S. (1962) *Advances in Chemistry Series,* **38**, 117–132.
12 Krause, B., Storr, M., Ertl, T., Buck, R., Hildwein, H., Deppisch, R. and Göhl, H. (2003) *Chemie Ingenieur Technik,* **75**, 1725–1732.
13 Van de Witte, P., Dijkstra, P.J., van den Berg, J.W.A. and Feijen, J. (1996) *Journal of Membrane Science,* **117**, 1–31.
14 Khare, V.P., Greenberg, A.R. and Krantz, W.B. (2005) *Journal of Membrane Science,* **258**, 140–156.
15 Grulke, E.A. (1999) *Polymer Handbook,* 4th edn (eds J. Brandrup, E.H. Immergut, E.A. Grulke, A. Abe and D.R. Bloch), John Wiley & Sons, New York, p. VII/675, Chapter 7.
16 Cadotte, J.E. (1985) *Material Science of Synthetic Membranes* (ed. D.R. Lloyd), ACS Symposium Series 269, American Chemical Society, Washington DC, pp. 273–294.
17 Petersen, R.J. (1993) *Journal of Membrane Science,* **83**, 81–150.
18 Riley, R.L., Lonsdale, H.K., Lyons, C.R. and Merten, U. (1967) *Journal of Applied Polymer Science,* **11**, 2143–2158.

19 Yasuda, H. (1984) *Journal of Membrane Science*, **18**, 273–284.

20 Cadotte, J., King, R., Majerle, R. and Petersen, R. (1981) *Journal of Macromolecular Science-Chemistry*, **A15**, 727–755.

21 Kickelbick, G. (2003) *Progress in Polymer Science*, **28**, 83–114.

22 Sholl, D.S. and Johnson, J.K. (2006) *Science*, **312**, 1003–1004.

23 Möckel, D., Staude, E. and Guiver, M.D. (1999) *Journal of Membrane Science*, **158**, 63–75.

24 Boom, R.M., van den Boomgaard, Th. and Smolders, C.A. (1994) *Journal of Membrane Science*, **90**, 231–249.

25 Asatekin, A., Menniti, A., Kang, S., Elimelech, M., Morgenroth, E. and Mayes, A.M. (2006) *Journal of Membrane Science*, **285**, 81–89.

26 Rana, D., Matsuura, T. and Narbaitz, R.M. (2006) *Journal of Membrane Science*, **277**, 177–185.

27 Nunes, S.P. and Peinemann, K.V. (2006) *Membrane Technology in the Chemical Industry*, 2nd edn, (eds S.P. Nunes and K.V. Peinemann), Wiley-VCH, Weinheim, p. 1, Part 1.

28 Kato, K., Uchida, E., Kang, E.T., Uyama, Y. and Ikada, Y. (2003) *Progress in Polymer Science*, **28**, 209–259.

29 He, D.M., Susanto, H. and Ulbricht, M. (2009) *Progress in Polymer Science*, **34**, 62–98.

30 Kane, R.S., Deschatelets, P. and Whitesides, G.M. (2003) *Langmuir*, **19**, 2388–2391.

31 Susanto, H. and Ulbricht, M. (2007) *Langmuir*, **23**, 7818–7830.

32 Millipore Corp.; technical publications: http://www.millipore.com/techpublications/tech1/5f5nrn.

33 van Reis, R., Brake, J.M., Charkoudian, J., Burns, D.B. and Zydney, A.L. (1999) *Journal of Membrane Science*, **159**, 133–142.

34 Hicke, H.G., Lehmann, I., Malsch, G., Ulbricht, M. and Becker, M. (2002) *Journal of Membrane Science*, **198**, 187–196.

35 Yang, S.Y., Ryu, H., Kim, H.Y., Kim, J.K., Jang, S.K. and Russell, T.P. (2006) *Advanced Materials*, **18**, 709–712.

36 Peinemann, K.V., Abetz, V. and Simon, P.F.W. (2007) *Nature Materials*, **6**, 992–996.

37 Park, H.B., Freeman, B.D., Zhang, Z.B., Sankir, M., and McGrath, J.E. (2008) *Angew. Chem. Int. Ed.*, **47**, 6019–6024.

38 Malaisamy, R. and Bruening, M.L. (2007) *Langmuir*, **21**, 10587–10592.

39 Jeong, B.H., Hoek, E.M.V., Yan, Y., Subramani, A., Huang, X., Hurwitz, G., Ghosh, A.K. and Jawor, A. (2007) *Journal of Membrane Science*, **294**, 1–7.

40 Majumder, M., Chopra, N., Andrews, R. and Hinds, B.J. (2005) *Nature*, **438**, 44–44.

41 Vandezande, P., Gevers, L.E.M. and Vankelecom, I.F.J. (2008) *Chemical Society Reviews*, **37**, 365–405.

42 White, L.S. and Nitsch, A.R. (2000) *Journal of Membrane Science*, **179**, 267–274.

43 Jonquieres, A., Clement, R., Lochon, P., Neel, J., Dresch, M. and Chretien, B. (2002) *Journal of Membrane Science*, **206**, 87–117.

44 Feng, X. and Huang, R.Y.M. (1997) *Industrial & Engineering Chemistry Research*, **36**, 1048–1066.

45 Sharma, A., Thamphi, S.P., Suggala, S.V. and Bhattacharya, P.K. (2004) *Langmuir*, **20**, 4708–4714.

46 Vankelecom, I.F.J., Depre, D., De Beukelaer, S. and Uytterhoeven, J.B. (1995) *The Journal of Physical Chemistry*, **99**, 13193–13197.

47 Ulbricht, M. and Schwarz, H.H. (1997) *Journal of Membrane Science*, **136**, 25–33.

48 McKeown, N.B. and Budd, P.M. (2006) *Chemical Society Reviews*, **35**, 675–683.

49 Strathmann, H. (2004) *Ion Exchange Membrane Separation Processes*, Elsevier, Amsterdam.

50 Xu, T. (2005) *Journal of Membrane Science*, **263**, 1–29.

51 Nasef, M.M. and Hegazy, E.S.A. (2004) *Progress in Polymer Science*, **29**, 499–561.

52 Balster, J., Stamatialis, D.F. and Wessling, M. (2004) *Chemical Engineering Progress*, **43**, 1115–1127.

53 Robeson, L.M., Hwu, H.H. and McGrath, J.E. (2007) *Journal of Membrane Science*, **302**, 70–77.

54 Geismann, C., Yaroshchuk, A. and Ulbricht, M. (2007) *Langmuir*, **23**, 76–83.

55 Ulbricht, M. (2004) *Journal of Chromatography B*, **804**, 113–125.

3
Fundamentals of Membrane Solvent Separation and Pervaporation

Bart Van der Bruggen

3.1
Introduction: Separation Needs for Organic Solvents

Separation processes for product recovery and purification represent more than 40% of the energy needs in chemical production processes. These include removal of impurities from raw materials, separation of products and by-products after reaction, and separation of pollutants from water and process streams. Classical solutions to the former two are based on thermodynamic equilibria involving a phase transition; distillation and liquid–liquid extraction are typical examples [1]. All these processes use either energy (e.g., distillation) or mass (e.g., liquid–liquid extraction) as the separating agent, which is usually not sustainable in terms of energy consumption and/or waste generation. Membrane separations are a totally different class of processes. Although some may argue that a membrane can be considered a mass-separating agent, transport properties of compounds that have to be separated always determine the process efficiency, not the equilibrium between two contacting phases. This makes membrane processes a new and different class of separations. Thermodynamic calculations confirm that membrane separations have intrinsically a substantially lower energy consumption and a higher exergetic efficiency [2–4]. This requires the use of membranes as separating tools similar to the operating methods of classical processes. Nevertheless, membranes are still mainly used in areas where the profits that can be obtained (both the economical profits and the environmental profits) are relatively small, compared to the challenges in the process industry. Membranes are widely used in water treatment, for drinking-water production [5, 6], desalination [7], wastewater treatment [8, 9] and process water recycling [10–12]. The economical and environmental benefits are obvious for these applications. However, in these applications membranes are used only for purification, that is, to remove solutes from the (water) matrix. The desired purity determines which membrane is to be used. In the context of increasing demands for purity – drinking water free of any anthropogenic compound, wastewater purified to be reused – it is understandable that most attention is devoted to those pressure-driven filtration processes delivering

Membrane Operations. Innovative Separations and Transformations. Edited by Enrico Drioli and Lidietta Giorno

ISBN: 978-3-527-32038-7

a high-purity product (reverse osmosis and, to a certain extent, nanofiltration). Energy consumption is the main parameter to be optimized, low-pressure reverse-osmosis membranes or nanofiltration membranes with low cutoff being the most interesting compromise. In this context, nanofiltration is an exception since it is the only process where separation is achieved on purpose, not as a side effect: nanofiltration membranes should allow passage of monovalent salts and retain multivalent salts.

Separations between liquid matrix compounds is more challenging and less developed, in spite of the large benefits that can be (theoretically) attained. Environmental benefits of solvent separations are mainly in a drastic reduction of energy consumption, although considerable effects on wastewater generation can be obtained as well. The main reference process to achieve this is pervaporation [13]. This is not a new process, but it had for a long time difficulties in finding its position. Pervaporation is not the most suitable process for purification, but it is a powerful and underestimated tool for separation between solvents. Separation factors in pervaporation can be 100–200, whereas the separation factors between mono- and divalent salts in nanofiltration are usually 4–6 and never above 10. For organic solutes in solvents, both processes are surprisingly similar in terms of separation and transport, in spite of different operating principles. This will be made clear from a discussion of membrane materials and properties. The translation of operation principles to flux and separation prediction in both cases will also be discussed.

3.2 Pervaporation and Nanofiltration Principles

Pervaporation is a concentration-driven membrane process for liquid feeds. It is based on selective sorption of feed compounds into the membrane phase, as a result of differences in membrane–solvent compatibility, often referred to as solubility in the membrane matrix. The concentration difference (or, in fact, the difference in chemical potential) is obtained by applying a vacuum at the permeate side, so that transport through the membrane matrix occurs by diffusion in a transition from liquid to vapor conditions (Figure 3.1). Alternatively, a sweep gas can be used to obtain low vapor pressures at the permeate side with the same effect of a chemical potential gradient.

The precise point of transition is undetermined; most theories assume that this would occur inside the membrane during transport. This leads to a five-step

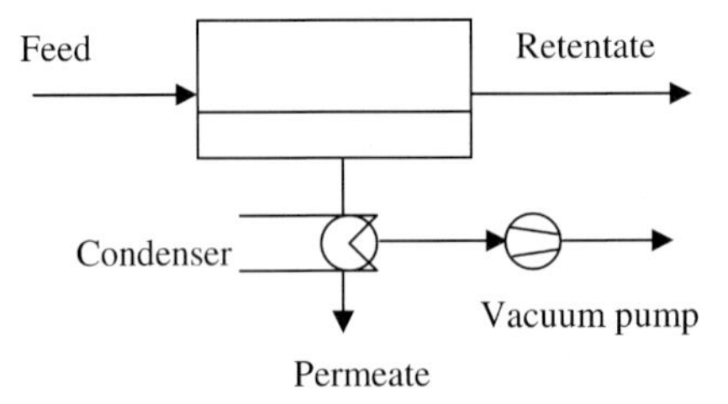

Figure 3.1 Schematic diagram of the pervaporation process.

description: sorption – liquid diffusion – vaporization – vapor diffusion – desorption. The physical relevance of this description may be questioned since solvent molecules permeating through the membrane are not a continuous phase but rather appear as individual molecules, or as small groups of molecules. Therefore, it is incorrect to consider the permeating substances as a liquid or a vapor; they are rather sorbed molecules. Separation is determined by factors affecting sorption (polarity, hydrophilicity/hydrophobicity differences, but also size) and molecular diffusion (sterical hindrance, interaction effects, dragging, competition). Therefore, typical separations in pervaporation are those where large differences in these parameters are present, mainly separations between water and an organic solvent, the latter preferably large and nonpolar. Hydrophilic pervaporation membranes are used for dehydration of organic solvents; in this application, the combination of a solvent and a polymer requires an optimal chemical stability. Hydrophobic membranes are used for the removal of (small) organic contaminants from water; these compounds can also be aggressive towards the membrane, in particular when present in relatively high concentrations. Because the driving force in pervaporation is a partial vapor pressure gradient, the process is only economical when the concentrations of contaminants to be removed are sufficiently high (no trace impurities) or when their driving force is high enough (possibly by operating at higher temperature), which again increases the need for membrane stability.

Solvent-resistant nanofiltration is based on a pressure gradient as the driving force. Pressure affects the chemical potential, so that the driving force again translates to a difference in chemical potential between feed and permeate. Solution-diffusion, similar to reverse osmosis and pervaporation, is often proposed as the determining transport mechanism [14]. This implies that nearly the same parameters are of importance as for pervaporation. Nevertheless, nanofiltration membranes may have a larger free space available for transport – some may denote this free space as pores, but this is probably incorrect, although there is no strict definition of a pore; 'nanovoids' is the scientifically correct description. When nanovoids become larger, they come close to the micropores observed in ultrafiltration; in this case, transport may be rather determined by viscous flow. Interactions with the membrane material are less intense here, and molar size remains as the only parameter determining separation.

Thus, (solvent resistant) nanofiltration is related to both pervaporation and ultrafiltration. Molar size emerges as the parameter to be used as the main discriminating factor in this case, although it is evident that other parameters will also play a role for membranes with a low cutoff. When organic solutes are considered, nanofiltration membranes are usually described by a single cutoff value, reflecting the large influence of molar size, but ignoring other interactions. A major problem that has been identified is the influence of the solvent itself on swelling, 'pore' solvation and solute solvation [15], leading to different cutoff values and fluxes depending on the solvent used [16, 17]. Applications are to be found in separation of relatively small organic solutes, in the range of 300–1000 g/mole, from any organic solvent [18–20]. Large-scale applications have been in operation since 1998 [21], the best known being the MAX-DEWAX process.

3.3
Membrane Materials and Properties for Solvent Separation

3.3.1
Solvent-Stable Polymeric Membrane Materials

Traditional polymers for separations in water have a limited chemical resistance and are not useful for solvent separations. Some may be applicable in nonaggressive solvents such as methanol and ethanol due to crosslinking, additives or additional interlayers, but not in any other solvent; modified polyamide membranes and poly (ethersulfone) membranes are typical examples.

Pervaporation membranes have always been intended for use in demanding conditions, including application in organic solvents. Therefore, a wide range of solvent stable membranes has been studied, usually with asymmetric structure, a dense top layer and several porous sublayers. Pervaporation membranes are prepared by dip-coating, plasma polymerization or interfacial polymerization [22]. Swelling can be a problem, because of the intense contact between liquid feed and the membrane top layer. The permeate side of the membrane is under vacuum and therefore not swollen. This results in an asymmetric structure that may cause stability problems. Nevertheless, swelling is needed to some extent in view of obtaining high fluxes, so that amorphous polymers with a sorption value of 5–25% by weight are to be preferred. Membrane thickness ranges from 100 nm to several μm, depending on how defect free the top layer can be made; often a somewhat thicker structure is preferred to avoid defects.

A list of typical commercial pervaporation membranes [23] is given in Table 3.1. Commercial hydrophilic membranes are very often made of polyvinyl alcohol (PVA), with differences in the degree of crosslinking. Commercial hydrophobic membranes often have a top layer in polydimethyl siloxane (PDMS). However, a wide variety of membrane materials for pervaporation can be found in the literature, including polymethylglutamate, polyacrylonitrile, polytetrafluoroethylene, polyvinylpyrrolidone, styrene-butadiene rubber, polyacrylic acid, and many others [24]. A comprehensive overview of membrane materials for pervaporation is given by Semenova *et al.* [25].

Solvent resistant nanofiltration membranes are a much more recent evolution. Historically, the membranes developed by Membrane Products Kyriat Weizmann (Israel) – now Koch – (MPF 44, MPF 50, MPF 60) were the first nanofiltration membranes intended for application in organic solvents, although other membranes (e.g., PES and PA membranes) also have a limited solvent stability. The Koch membranes are based on PDMS, similarly to pervaporation membranes, although the level of crosslinking is quite different.

Other membrane materials include mainly polyimide, polyacrylonitrile and polybenzimidazole. An overview of commercially available membranes is given in Table 3.2. These membranes are manufactured in procedures usually derived from practical experience; by using high-throughput screening, it was shown that optimization is possible [26]. Many other membrane materials are described in the scientific literature and in patents; an overview is given by Cuperus and Ebert [27].

Table 3.1 Pervaporation membranes used in commercial applications [23].

Brand name	Manufacturer	Material	Hydrophilic/ hydrophobic
PERVAP 2200	Sulzer Chemtech[a]	PVA cross-linked/PAN support	Hydrophilic
PERVAP 2201	Sulzer Chemtech[a]	PVA highly cross-linked/PAN	Hydrophilic
PERVAP 2202	Sulzer Chemtech[a]	PVA specially cross-linked/PAN	Hydrophilic
PERVAP 2205	Sulzer Chemtech[a]	PVA specially cross-linked/PAN	Hydrophilic
PERVAP 2210	Sulzer Chemtech[a]	PVA lightly cross-linked/PAN	Hydrophilic
PERVAP 2510	Sulzer Chemtech[a]	PVA specially cross-linked/PAN	Hydrophilic
CM-Celfa	CM-Celfa[b]	PVA cross-linked/PAN	Hydrophilic
GKSS Simplex	GKSS[c]	Complex polyelectrolytes/PAN	Hydrophilic
PERVAP 1060	Sulzer Chemtech[a]	PDMS cross-linked/PAN support	Hydrophobic
PERVAP 1070	Sulzer Chemtech[a]	PDMS cross-linked + silicalite/PAN	Hydrophobic
MTR 100	MTR[d]	PDMS cross-linked/porous support	Hydrophobic
MTR 200	MTR[d]	EPDM/PDMS cross-linked/ porous support	Hydrophobic
GKSS PEBA	GKSS[c]	PEBA/porous support	Hydrophobic
GKSS PDMS	GKSS[c]	PDMS cross-linked/porous support	Hydrophobic
GKSS PMOS	GKSS[c]	PMOS cross-linked/porous support	Hydrophobic

[a]Sulzer Chemtech Ltd, Winterthur, Switzerland.
[b]CM-Celfa, Seewen-Schwyz, Switzerland.
[c]GKSS, Geesthacht, Germany.
[d]Membrane Technology and Research Inc., Menlo Park, CA.

Among the various materials are crosslinked PAN, polyphosphazenes, polyphenylenesulfide, polyetheretherketone, and various polymer blends [28–31]. Particularly interesting is the use of zeolites as filler in organic polymers, which aims at improving the performance of (silicone-based) membranes for separations in nonpolar solvents, by adding more cross-links to the membrane material [32, 33].

3.3.2 Ceramic Membrane Materials

Ceramic membranes may overcome some of the disadvantages of polymeric membranes, particularly the chemical resistance. The higher cost of ceramic membranes may be compensated by the significantly higher fluxes, especially at high temperatures. A full comparison of polymeric and ceramic membranes is given in Table 3.3.

Some efforts have already been made to develop ceramic pervaporation membranes, especially silica and zeolite membranes, which are both hydrophilic membranes. Silica pervaporation membranes have been developed by ECN, The Netherlands. The membranes were tested in a pilot installation of 1 m^2 membrane surface at Akzo Nobel and other companies in the Netherlands [34, 35].

A-type zeolite pervaporation membranes have been developed by Mitsui Engineering & Shipbuilding Co Ltd, which have been implemented in an industrial

Table 3.2 Commercial solvent-resistant nanofiltration membranes with characteristics as specified by the manufacturers.

Membrane	Manufacturer	Material	MWCO (Da)	T_{max} (°C)	L (l/h m^2 bar)	R (%)
N30F	Nadir[a]	PES	400	95	1.0–1.8[h]	70–90[l]
NF-PES-010	Nadir[a]	PES	1000	95	5–10[h]	30–50[l]
MPF-44	Koch[b]	PDMS	250	40	1.3[h]	98[m]
MPF-50	Koch[b]	PDMS	700	40	1.0[i]	—
Desal-5-DK	Osmonics[c]	PA	150–300	90	5.4[h]	98[n]
Desal-5-DL	Osmonics[c]	PA	150–300	90	9.0[h]	96[n]
SS-030505	SolSep[d]	n.k.	—	90	1.0[j]	>90[p]
SS-169	SolSep[d]	n.k.	—	150	10[j]	95[p]
SS-01	SolSep[d]	n.k.	—	150	10[j]	97[q]
StarMem-120	MET[e]	PI	200	60	1.0[k]	—
StarMem-122	MET[e]	PI	220	60	1.0[k]	—
StarMem228	MET[e]	PI	280	60	0.26[k]	—

[a]Nadir Filtration GmbH, Wiesbaden, Germany;
[b]Koch Membrane Systems, Wilmington, MA, USA;
[c]GE Osmonics, Vista, CA, USA;
[d]SolSep BV, Apeldoorn, The Netherlands;
[g]Membrane Extraction Technology, London, UK;
[h]pure-water permeability;
[i]methanol permeability;
[j]ethanol permeability;
[k]toluene permeability;
[l]4% lactose (MW 342);
[m]5% sucrose (MW 342);
[n]$MgSO_4$;
[p]MW∼500 in ethanol;
[q]MW∼1000 in acetone;
n.k. not known;
— not specified.

Table 3.3 Comparison of polymeric and ceramic pervaporation membranes.

Polymeric membrane materials	Ceramic membrane materials
Low production cost	High production cost
Production upscaling easy	Production upscaling difficult
Variation in module form easy	Variation in module form difficult
Stability at long term unknown	Stability at long term expected good
Limited versatility in organics	Good versatility in organics
Vulnerable for unknown components in mixtures	Resistance to unknown component expected good
Thermal regeneration impossible	Thermal regeneration possible
High-temperature applications impossible	High-temperature applications possible

dehydration plant in Japan [36]. Pervaporation membranes with an active layer of zeolite and amorphous silica on porous supports of alumina or stainless steel have become commercially available [37]; research on other zeolite-type materials (e.g., silicalite, ZSM-5, T-type zeolite) is going on, but so far only at the laboratory scale.

Ceramic membranes might also be a significant improvement in solvent-resistant nanofiltration, although the cost of ceramic membranes is relatively high. To date, only a few ceramic membrane types are commercially available, in spite of the good performance of these membranes. Hydrophilic ceramic nanofiltration membranes in asymmetric multilayer configurations have been successfully developed since the late 1990s [38–40]. These consist of an open porous support, mesoporous interlayers, and defectless microporous top layers. Support layers may combine an extruded body containing coarse pores in the μm range, and a slip-casted layer on top with pores of the order of 50–100 nm. The most often used material is α-alumina, although titania is tending to become more popular due to its higher chemical stability [41]. The intermediate layers (usually more than one to prevent defects) are used to gradually decrease the pore size and the surface roughness of the membrane. The interlayers are prepared by the colloidal sol-gel procedure [42], often using alumina in the γ-alumina phase, or titania and even silica or zirconia. γ-alumina interlayers are relatively thick (about 2 μm) while titania or zirconia interlayers are thinner (about 0.5 μm). Therefore, γ-alumina interlayers are more suitable to cover irregularities and defects. The obvious disadvantage is that γ-alumina layers lead to lower solvent fluxes than titania layers, because of their higher fluid resistance. Furthermore, γ-alumina is chemically unstable in acid solutions ($pH < 3$) and in alkaline solutions ($pH > 11$). The pore size of the interlayers depends on the hydrolyzing/peptisizing process and the calcination temperature and should be about 3–5 nm to obtain nanofiltration membranes. Calcination temperatures should be below the phase-transition temperature of the used metal oxide; for titania this involves the anatase–rutile transition in the temperature region 500–700 °C [40].

The top layer contains the smallest pores and defines the membrane's nanoselectivity. Top layers can be made of alumina (boehmite), titania, zirconia, silica, or mixtures of these. Pore diameters are in the order of 1 nm and are obtained by applying the polymeric sol-gel method [42]. The calination temperature applied in this procedure should be low enough in order to avoid sintering effects, and consequent pore growth. The calcination temperature does not only determine the final pore size of the top layer, but also the phase structure of the top-layer material. For example, when the calcination temperature for a titania top layer is 200 °C, titania is amorphous with very small pores (1–2 nm), from 300 °C onwards titania is in an anatase phase having pores in the range of 2–4 nm. Amorphous titania, however, is less resistant to corrosion, and therefore has a smaller applicable pH range [40].

Metal oxides, used for manufacturing of ceramic nanofiltration membranes, are intrinsically hydrophilic. This limits the use of these membranes to polar solvents; filtration of nonpolar solvents (n-hexane, toluene, cyclohexane) usually yields zero fluxes. Attempts have been made to modify the pore structure by adding hydrophobic groups, for example, in a silane coupling reaction [38, 43]. This approach is similar to modifications of ultrafiltration and microfiltration membranes using chloroalkylsilanes [44–48], phosphonic acids [49, 50] or fluoroalkylsilanes [51], where the effect is twofold, that is, a combination of pore-size control and tuning of hydrophobicity. However, the precise interaction between the chlorosilanes and the membrane surface is very complex and there is in many cases no evidence that it involves a chemical reaction rather than an adsorption reaction [52, 53]. Due to the lack of a real chemical reaction between organochlorosilanes and the membrane surface, the stability of the modification with these reactants may be limited, and to date no commercial hydrophobic ceramic nanofiltration membranes have been developed.

3.3.3 Solvent Stability

A difficult problem that prevented the use of nanofiltration in organic solvents for a long time was the limited solvent stability of polymeric nanofiltration membranes, and the lack of ceramic nanofiltration membranes. For polymeric membranes, different problems occurred: zero flux due to membrane collapse [54], 'infinite' nonselective flux due to membrane swelling [54], membrane deterioration [55], poor separation quality [56], etc. In an early study of four membranes thought to be solvent stable (N30F, NF-PES-10, MPF 44 and MPF 50), it was observed that three of these showed visible defects after ten days exposure to one or more organic solvents, and the characteristics of all four membranes changed notably after exposure to the solvents [15]. This implies that these membranes should be denoted as semi-solvent-stable instead of solvent stable.

Less information is available about the stability of ceramic membranes. It is generally thought that ceramic membranes have excellent solvent stability. Acid conditions may be more problematic; it was shown [57] that an alumina nanofiltration membrane was very sensitive to corrosion effects in dynamic experiments, whereas the performance of a similar titania membrane was stable in the pH range from 1.5 to 13.

3.3.4 Structural Properties for Membranes in NF and PV

Due to recent advances in membrane development, nanofiltration membranes are nowadays increasingly used for applications in organic solvents [27, 58]. This narrows the gap between pervaporation and nanofiltration. It is even possible that the requirements for membrane structures completely overlap for the two processes: whereas membrane stability becomes more important for nanofiltration membranes, the performance of pervaporation membranes could be improved by using an optimized (thinner) structure for the top layers. It might even be possible to use the same membranes in both applications. At this moment it is not possible to define which membrane structure is necessary for nanofiltration or for pervaporation, and which membrane is expected to have a good performance in nanofiltration, in pervaporation or in both. Whereas pervaporation membranes are dense, nanofiltration membranes can be either dense or porous. For this reason, some nanofiltration and pervaporation membranes might be interchangeable. This was suggested for polymeric membranes [59]. Khayet and Matsuura [60] explored a similar relation between pervaporation and membrane distillation using polyvinylidene fluoride membranes.

Similar trends are developing for ceramic membranes applied in pervaporation and nanofiltration, although much slower because ceramic pervaporation and nanofiltration membranes are still sparsely available; more experimental observations and experience with applications are needed in this field. Promising results were obtained by Sekulic *et al.* [61] for titania membranes that can be used in pervaporation as well as nanofiltration.

3.4 Flux and Separation Prediction

3.4.1 Flux Models in NF

For relatively porous nanofiltration membranes, simple pore flow models based on convective flow will be adapted to incorporate the influence of the parameters mentioned above. The Hagen–Poiseuille model and the Jonsson and Boesen model, which are commonly used for aqueous systems permeating through porous media, such as microfiltration and ultrafiltration membranes, take no interaction parameters into account, and the viscosity as the only solvent parameter. It is expected that these equations will be insufficient to describe the performance of solvent resistant nanofiltration membranes. Machado *et al.* [62] developed a resistance-in-series model based on convective transport of the solvent for the permeation of pure solvents and solvent mixtures:

$$J = \frac{\Delta P}{\phi'[(\gamma_c - \gamma_l) + f_1\eta] + f_2\eta}$$

where f_1 and f_2 are solvent independent parameters characterizing the nanofiltration and ultrafiltration sublayers, ϕ' a solvent parameter, γ_c the critical surface tension of the membrane material and γ_l the surface tension of the solvent. This model is also based on the dependence of the flux on two parameters, namely the solvent viscosity and the difference in surface tension between the solid membrane material and the liquid solvent. However, this model does not cover the whole area of membranes and solvents, as shown by Yang *et al.* [63]. The model is developed for hydrophobic membranes, but seems inadequate for the description of fluxes through hydrophilic membranes. Moreover, for each solvent–membrane combination an empirical parameter ϕ' has to be determined as a measure for the interaction between a solvent and the membrane material. It will be attempted to replace this parameter by a combination of nonempirical parameters.

Polymeric membranes with a less porous structure, pervaporation membranes as well as nanofiltration membranes, can be described by a solution-diffusion mechanism, possibly corrected for the influence of convective transport [64]. A description of solvent transport in this case is necessarily based on the solution-diffusion (SD) model [65]. With respect to flux modeling of organic solvents, a diffusion based model was presented by Bhanushali *et al.* [66]:

$$J \propto \left(\frac{V_m}{\eta}\right)\left(\frac{1}{\phi^n \gamma_m}\right)$$

This model combines different approaches of existing models by introducing at the same time the solvent viscosity, the molar volume V_m (as a measure for the molecular size), the surface tension of the solid membrane material and a sorption value ϕ (as a measure for membrane–solvent interactions). Other SD-based transport models were presented by White [14], providing a predictive model for feed solutions with a high concentration of aromatics, by Scarpello *et al.* [67] and by Gibbins *et al.* [68]. A slightly modified equation was proposed by Geens *et al.* [69]:

$$J \propto \frac{V_m}{\eta \cdot \Delta\gamma}$$

where $\Delta\gamma$ is the difference in surface tension (mN/m), η is the dynamic viscosity (Pa s), and V_m is the solvent molar volume (m^3/mol).

Transport models for the description of *solute transport* in aqueous solution are the Spiegler–Kedem model and the solution-diffusion model [65]. The former model incorporates both viscous and diffusive flow, whereas the latter can only be used for transport through dense membranes by solute diffusion. White [14] presented an SD-based model for the permeation of several reference solutes, dissolved in toluene, through dense membranes. Bhanushali *et al.* [66] succeeded in describing experimental data with the Spiegler–Kedem model. Gevers *et al.* [70] and Vankelecom *et al.* [71] used the reformulated solution-diffusion model of Paul and the Kedem–Katchalsky model to explain solute fluxes; it was shown that solutes with a high molar volume were most influenced by diffusive transport, whereas solutes with a low molar volume are dominantly transported by convection. Matsuura and Sourirajan [72] developed a model for convective transport of dissolved components, incorporating a solvent-dependent pore diameter. Gibbins *et al.* [70] calculated the pore radii of MPF-50 and Desal-5-DK based on filtration experiments carried out in methanol, using several models for convective flow through porous membranes. The different attempts for the modeling of nonaqueous solute transport provide, however, models that are limited to specific experimental data.

A new approach is the application of chemometrics (and neural networks) in modeling [73]. This should allow identification of the parameters of influence in solvent-resistant nanofiltration, which may help in further development of equations. Development of a more systematic model for description and prediction of solute transport in nonaqueous nanofiltration, which is applicable on a wide range of membranes, solvents and solutes, is the next step to be taken. The Maxwell–Stefan approach [74] is one of the most direct methods to attain this.

3.4.2 Rejection in NF

Similar to the approach for solvents, both diffusive and convective transport of solutes can be modeled separately. For dense membranes, a solution-diffusion model can be used [14], where the flux J_i of a solute is calculated as:

$$J_i = D_i K_i \frac{\left(c_{f,i} - c_{p,i} \exp\frac{(-V_i(P_f - P_p))}{RT}\right)}{\Delta x}$$

with D_i the diffusivity of the solvent in the polymer matrix, K_i the partition coefficient between component i and polymer, $c_{f,i}$ and $c_{p,i}$ the concentration of component i in the feed or permeate (mol/l), V_c the molar volume of component c at the boiling point (m^3/mol), P_f the feed side pressure (bar), P_p the permeate-side pressure (bar), R the universal gas constant (J/mol K), T the absolute temperature (K), and Δx the membrane thickness (m). This equation yields a good description of solute transport, but it is not possible to predict separations because diffusivities and partition coefficients have to be related to measurable membrane/solvent parameters.

The transport equations of Spiegler and Kedem combine both diffusion and convection:

$$J_s = L(\Delta P - \sigma \Delta \pi)$$

$$J_c = P_s \Delta x \frac{dc}{dx} + (1 - \sigma) J_s c$$

The rejection of a given molecule can then be calculated as:

$$R = \frac{\sigma(1-F)}{1-\sigma F} \quad \text{with} \quad F = \exp\left(-\frac{1-\sigma}{P_s} J_s\right)$$

The permeability P_s is a measure of the transport of a molecule by diffusion. The reflection coefficient σ of a given component is the maximal possible rejection for that component (at infinite solvent flux). Various models have been proposed for the reflection coefficient [75–77]. In the lognormal model [78], a lognormal distribution is assumed for the pore size. No steric hindrance in the pores or hydrodynamic lag is taken into account, but it is assumed that a molecule permeates through every pore that is larger than the diameter of the molecule. Moreover, the diffusion contribution to the transport through the membrane is considered to be negligible. Therefore, the reflection curve can be expressed as:

$$\sigma = \int_0^{r_c} \frac{1}{S_p\sqrt{2\pi}} \frac{1}{r} \exp\left(-\frac{(\ln(r) - \ln(\bar{r}))^2}{2S_p^2}\right) dr$$

with $r_c = d_c/2$. This equation comprises two variables, S_p and $\bar{r}$, where S_p is the standard deviation of the distribution. This standard deviation is a measure for the

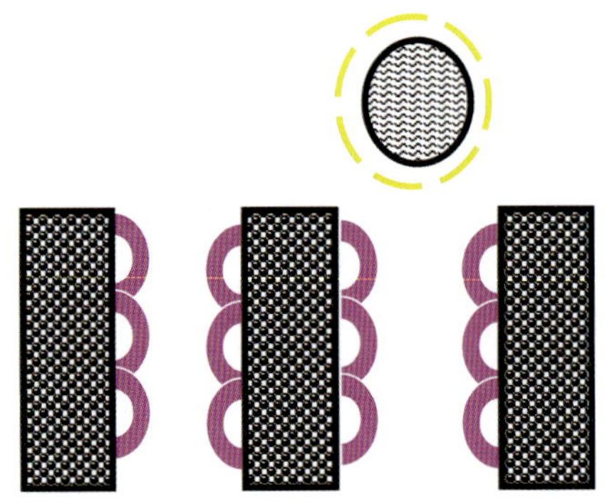

Figure 3.2 Pore solvation solute solvation may influence the rejection of solutes in organic solvents [80, 81].

distribution of the pore sizes. $\bar{r}$ is a mean pore size, namely the size of a molecule that is retained for 50%.

However, it is clear that other parameters than only solute size determine transport and rejection. Tarleton *et al.* [17, 79] showed that polarity has a major influence on permeation. Geens *et al.* [80, 81] showed that interaction effects between solvents, solutes and the membrane material determine the rejection of a given solute (Figure 3.2). Differences in solvation may result in lower rejections; this explains the differences in molecular weight cutoff that were observed.

These effects were observed for both polymeric and ceramic NF-membranes, showing that differences in rejection are not due to swelling. Nevertheless, swelling effects have been demonstrated by Tarleton *et al.* [82, 83] and are known to affect transport in polymeric membranes.

3.4.3
Models for PV: from Solution-Diffusion to Maxwell–Stefan

The transport mechanism in polymeric pervaporation is generally understood as a combined sorption–diffusion–desorption process. Simple sorption–diffusion models [84] can serve as a starting point for modeling the membrane separation. These models give the flux of a component through the membrane as a function of concentration or partial pressure differences over the membrane. This approach, however, does not incorporate coupling and interaction effects that are possible between the different components in a mixture. According to Lipnizki *et al.* [85], coupling effects can be expected during all three stages in the pervaporation process (sorption, diffusion, desorption). Current research focuses on thermodynamic models to describe sorption of different components into the membrane, and on the Maxwell–Stefan formulation for describing diffusion processes [86–88].

Pervaporation with ceramic membranes is less well understood in terms of transport mechanisms. Consequently, modeling of ceramic pervaporation is still less mature, although the performance of the process was reported to be good [89]. Nomura *et al.* [90] studied the transport mechanism of ethanol/water through silicalite membranes in pervaporation and vapor permeation and carried out single component and binary mixture adsorption experiments. It was shown that ethanol permeance was hardly influenced by the presence of water, whereas the water flux decreased substantially in the presence of ethanol. An adsorption–diffusion model was considered for the transport through the membrane. The high-selective permeation of ethanol was explained by the ethanol-selective adsorption to the silicalite membrane.

Krishna and Paschek [91] employed the Maxwell–Stefan description for mass transport of alkanes through silicalite membranes, but did not consider more complex (e.g., unsaturated or branched) hydrocarbons. Kapteijn *et al.* [92] and Bakker *et al.* [93] applied the Maxwell–Stefan model for hydrocarbon permeation through silicalite membranes. Flanders *et al.* [94] studied separation of C6 isomers by pervaporation through ZSM-5 membranes and found that separation was due to shape selectivity.

3.4.4
Hybrid Simulations

In industrial applications, pervaporation has to compete with conventional separation processes, such as distillation, liquid–liquid extraction, adsorption, and stripping. Pervaporation has attracted the interest of the chemical industry for separations that are difficult to achieve by distillation, for example, separations giving azeotropic mixtures and separations of components with a small difference in volatility.

Pervaporation as a standalone technique is still to be developed industrially, but as part of a hybrid process, combined with for example, distillation (Figure 3.3), it is very promising for difficult separations and may yield considerable energy savings.

Several authors have already developed methodologies for the simulation of hybrid distillation–pervaporation processes. Short-cut methods were developed by Moganti *et al.* [95] and Stephan *et al.* [96]. Due to simplifications such as the use of constant relative volatility, one-phase sidestreams, perfect mixing on feed and permeate sides of the membrane, and simple membrane transport models, the results obtained should only be considered qualitative in nature. Verhoef *et al.* [97] used a quantitative approach for simulation, based on simplified calculations in Aspen Plus/Excel VBA. Hömmerich and Rautenbach [98] describe the design and optimization of combined pervaporation–distillation processes, incorporating a user-written routine for pervaporation into the Aspen Plus simulation software. This is an improvement over most approaches with respect to accuracy, although the membrane model itself is still quite

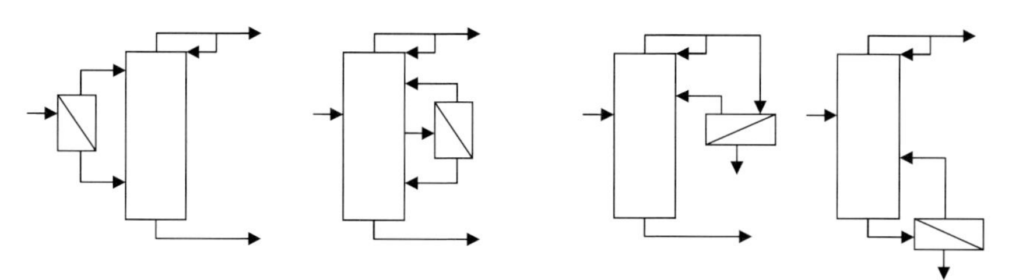

Figure 3.3 Possible configurations for distillation–pervaporation hybrid processes.

limited. Furthermore, most authors analyze and optimize the performance of only a particular, predetermined hybrid configuration. From the literature references cited and deriving from one of the conclusions of Lipnizki *et al.* [85], it appears that more adequate and accurate process design tools to optimize hybrid techniques involving pervaporation are strongly needed.

3.5 Conclusions

Solvent-resistant nanofiltration and pervaporation are undoubtedly the membrane processes needed for a totally new approach in the chemical process industry, the pharmaceutical industry and similar industrial activities. This is generally referred to as 'process intensification' and should allow energy savings, safer production, improved cost efficiency, and allow new separations to be carried out.

Problems to be solved are related to membrane stability (of polymeric membranes, but also the development of hydrophobic ceramic nanofiltration membranes and pervaporation membranes resistant to extreme conditions), to a lack of fundamental knowledge on transport mechanisms and models, and to the need for simulation tools to be able to predict the performance of solvent-resistant nanofiltration and pervaporation in a process environment. This will require an investment in basic and applied research, but will generate a breakthrough in important societal issues such as energy consumption, global warming and the development of a sustainable chemical industry.

Acknowledgment

The Research Council of the K.U. Leuven is gratefully acknowledged (OT/2006/37).

References

1 Seader, J.D. and Henley, E.J. (2006) *Separation Process Principles*, John Wiley & Sons, Hoboken, NJ.
2 Bernardo, P., Barbieri, G. and Drioli, E. (2006) *Chemical Engineering Research & Design*, **84** (A5), 405–411.
3 Macedonio, F., Curcio, E. and Drioli, E. (2007) *Desalination*, **203** (1–3), 260–276.
4 Mehdizadeh, H. (2006) *Desalination*, **191** (1–3), 200–209.
5 Reisch, M.S. (2007) *Chemical & Engineering News*, **85** (17), 22–24.
6 Rachwall, T. and Judd, S. (2006) *Water Environ Journal*, **20** (3), 110–113.
7 Van der Bruggen, B. and Vandecasteele, C. (2002) *Desalination*, **143** (3), 223–234.
8 Miller, G.W. (2006) *Desalination*, **187** (1–3), 65–75.
9 Wintgens, T., Melin, T., Schäfer, A., Khan, S., Muston, M., Bixio, D. and Thoeye, C. (2005) *Desalination*, **178** (1–3), 1–11.
10 Into, M., Jonsson, A.S. and Lengden, G. (2004) *Journal of Membrane Science*, **242** (1–2), 21–25.

11 Karakulski, K., Gryta, M. and Sasim, M. (2006) *Chemical Papers-Chemicke Zvesti*, **60** (6), 416–421.

12 Kremser, U., Drescher, G., Otto, S. and Recknagel, V. (2006) *Desalination*, **189** (1–3), 53–58.

13 Pribic, P., Roza, M. and Zuber, L. (2006) *Separation Science and Technology*, **41** (11), 2581–2602.

14 White, L.S. (2002) *Journal of Membrane Science*, **205**, 191–202.

15 Van der Bruggen, B., Geens, J. and Vandecasteele, C. (2002) *Separation Science and Technology*, **37** (4), 783–797.

16 Toh, Y.H.S., Loh, X.X., Li, K., Bismarck, A. and Livingston, A.G. (2007) *Journal of Membrane Science*, **291** (1–2), 120–125.

17 Tarleton, E.S., Robinson, C.P., Millington, C.R. and Nijmeijer, A. (2005) *Journal of Membrane Science*, **252** (1–2), 123–131.

18 Geens, J., De Witte, B. and Van der Bruggen, B. (2007) *Separation Science and Technology*, **42** (11), 2435–2449.

19 Aerts, S., Buekenhoudt, A., Weyten, H., Gevers, L.E.M., Vankelecom, I.F.J. and Jacobs, P.A. (2006) *Journal of Membrane Science*, **280** (1–2), 245–252.

20 Witte, P.T., Chowdhury, S.R., ten Elshof, J.E., Sloboda-Rozner, D., Neumann, R. and Alsters, P.L. (2005) *Chemical Communications*, **9**, 1206–1208.

21 White, L.S. (2006) *Journal of Membrane Science*, **286** (1–2), 26–35.

22 Mulder, M. (1996) *Basic Principles of Membrane Technology*, 2nd edn, Kluwer Academic Publishers.

23 Joncquières, A., Clément, R., Lochon, P., Néel, J., Dresch, M. and Chrétien, B. (2002) *Journal of Membrane Science*, **206**, 87–117.

24 Pereira, C.C., Ribeiro, C.C., Nobrega, R. and Borges, C.P. (2006) *Journal of Membrane Science*, **274** (1–2), 1–23.

25 Semenova, S.I., Ohya, H. and Soontarapa, K. (1997) *Desalination*, **110** (1–3), 251–286.

26 Vandezande, P., Gevers, L.E.M., Paul, J.S., Vankelecom, I.F.J. and Jacobs, P.A. (2005) *Journal of Membrane Science*, **250** (1–2), 305–310.

27 Cuperus, F.P. and Ebert, K. (2005) Non-aqueous applications of NF, Chapter 21 in *Nanofiltration, Principles and Applications* (eds A.I. Schäfer, A.G. Fane and T.D. Waite), Elsevier, Oxford, New York and Tokyo.

28 Linder, C., Perry, M. and Ketraro, R. (1988) US Patent 4, 761, 233.

29 Golemme, G. and Drioli, E. (1996) *Journal of Inorganic and Organometallic Polymers*, **6** (4), 341–365.

30 Lundgard, R.A. (1996) US Patent 5, 507, 984.

31 Shimoda, T. and Hachiya, H. (1999) US Patent 5, 997, 741.

32 Van Gemert, R.W., Bos, A.A.C.M., Maass, L. and Bakker, W.J.W. (2001) Proceedings of the conference Engineering with Membranes, June 2001, Granada Spain.

33 Gevers, L.E.M., Aldea, S., Vankelecom, I.F.J. and Jacobs, P.A. (2006) *Journal of Membrane Science*, **281** (1–2), 741–746.

34 Gevers, L.E.M., Vankelecom, I.F.J. and Jacobs, P.A. (2006) *Journal of Membrane Science*, **278** (1–2), 199–204.

35 Bakker, W.J.W., Bos, A.C.M., Rutten, W.L.P., Keurentjes, J.T.F. and Wessling, M. (1998) Proceedings of the 5th International Conference on Inorganic Membranes, Nagoya June 22–26, pp. 448–451.

36 Morigami, Y., Kondo, M., Abe, J., Kita, H. and Okamoto, K. (2000) Book of Abstracts of the 6th International Conference on Inorganic Membranes, Montpellier, June 26–30, pp. 55.

37 Verkerk, A.W., Van Male, P., Vorstman, M.A.G. and Keurentjes, J.T.F. (2001) *Journal of Membrane Science*, **1–2** (193), 227–238.

38 Tsuru, T., Miyawaki, M., Kondo, H., Yoshioka, T. and Asaeda, M. (2003) *Separation and Purification Technology*, **32** (1–3), 105–109.

39 Voigt, I., Fischer, G., Puhlfurss, P., Schleifenheimer, M. and Stahn, M. (2003) *Separation and Purification Technology*, **32** (1–3), 87–91.

40 Van Gestel, T., Vandecasteele, C., Buekenhoudt, A., Dotremont, C.,

Luyten, J., Leysen, R., Van der Bruggen, B. and Maes, G. (2002) *Journal of Membrane Science*, **207**, 73–89.

41 Voigt, I., Fischer, G., Puhlfürss, P. and Seifert, D. (1998) Proceedings of the 5th International Conference on Inorganic Membranes, Nagoya, Japan June 22–26, pp. 42–45.

42 Burggraaf, A.J. and Keizer, K. (1991) *Inorganic Membranes: Synthesis, Characterization and Applications* (ed. R. Bhave), Van Nostrand Reinhold, New York.

43 Van Gestel, T., Van der Bruggen, B., Buekenhoudt, A., Dotremont, C., Luyten, J., Vandecasteele, C. and Maes, G. (2003) *Journal of Membrane Science*, **224** (1–2), 3–10.

44 Leger, C., De, H., Lira, L. and Paterson, R. (1996) *Journal of Membrane Science*, **120** (1–2), 187–195.

45 Javaid, A., Hughey, M.P., Varutbangkul, V. and Ford, D.M. (2001) *Journal of Membrane Science*, **187** (1–2), 141–150.

46 Miller, J.R. and Koros, W.J. (1990) *Separation Science and Technology*, **25**, 1257–1280.

47 Simon, C., Bredesen, R. and Denonville, C. (1998) Proceedings of the 5th International Conference on Inorganic Membranes (ICIM), Nagoya, Japan June 22–26, pp. 416–419.

48 Picard, C., Larbot, A., Guida-Pietrasanta, F., Boutevin, B. and Ratsimihety, A. (2001) *Separation and Purification Technology*, **25**, 65–69.

49 Randon, J. and Paterson, R. (1997) *Journal of Membrane Science*, **134** (1–2), 219–223.

50 Caro, J., Noack, M. and Kölsch, P. (1998) *Microporous and Mesoporous Materials*, **22**, 321–332.

51 Picard, C., Larbot, A., Tronel-Peyroz, E. and Berjoan, R. (2004) *Solid State Sciences*, **6** (6), 605–612.

52 Impens, N.R.E.N., Van der Voort, P. and Vansant, E.F. (1999) *Microporous and Mesoporous Materials*, **28** (2), 217–232.

53 Hair, M.L. and Tripp, C.P. (1995) *Colloids and Surfaces A: Physicochemical and Engineering, Aspects*, **105** (1), 95–103.

54 Raman, L.P., Cheryan, M. and Rajagopalan, N. (1996) *Fett-Lipid*, **98** (1), 10–14.

55 Bridge, M.J., Broadhead, K.W., Hlady, V. and Tresco, P.A. (2002) *Journal of Membrane Science*, **195** (1–2), 51–64.

56 Subramanian, R., Nakajima, M. and Kawakatsu, T. (1998) *Journal of Food Engineering*, **38** (1), 41–56.

57 Van Gestel, T., Vandecasteele, C., Buekenhoudt, A., Dotremont, C., Luyten, J., Van der Bruggen, B. and Maes, G. (2003) *Journal of Membrane Science*, **214** (1), 21–29.

58 Ebert, K. and Cuperus, F.P. (1999) *Membrane Technology*, **107**, 5–8.

59 Van der Bruggen, B., Jansen, J.C., Figoli, A., Geens, J., Van Baelen, D., Drioli, E. and Vandecasteele, C. (2004) *The Journal of Physical Chemistry. B*, **108**, 13273–13279.

60 Khayet, M. and Matsuura, T. (2004) *AIChE Journal*, **50** (8), 1697–1712.

61 Sekulic, J., ten Elshof, J.E. and Blank, D.H.A. (2004) *Advanced Materials*, **16** (17), 1546.

62 Machado, D.R., Hasson, D. and Semiat, R. (2000) *Journal of Membrane Science*, **166**, 63–69.

63 Yang, X.J., Livingston, A.G. and Freitas dos Santos, L. (2001) *Journal of Membrane Science*, **190**, 45–55.

64 Silva, P., Han, S.J. and Livingston, A.G. (2005) *Journal of Membrane Science*, **262** (1–2), 49–59.

65 Paul, D.R. (2004) *Journal of Membrane Science*, **241**, 371–386.

66 Bhanushali, D., Kloos, S., Kurth, C. and Bhattacharyya, D. (2001) *Journal of Membrane Science*, **189**, 1–21.

67 Scarpello, J.T., Nair, D., dos Santos, L.M.F., White, L.S. and Livingston, A.G. (2002) *Journal of Membrane Science*, **203** (1–2), 71–85.

68 Gibbins, E., D'Antonio, M., Nair, D., White, L.S., Freitas dos Santos, L.M., Vankelecom, I.F.J. and Livingston, A.G. (2002) *Desalination*, **147**, 307–313.

69 Geens, J., Van der Bruggen, B. and Vandecasteele, C. (2006) *Separation and*

Purification Technology, **48** (3), 255–263, Dordrecht, The Netherlands.

70 Gevers, L.E.M., Meyen, G., De Smet, K., De Velde, P.V., Du Prez, F., Vankelecom, I.F.J. and Jacobs, P.A. (2006) *Journal of Membrane Science*, **274** (1–2), 173–182.

71 Vankelecom, I.F.J., De Smet, K., Gevers, L.E.M., Livingston, A., Nair, D., Aerts, S., Kuypers, S. and Jacobs, P.A. (2004) *Journal of Membrane Science*, **231** (1–2), 99–108.

72 Matsuura, T. and Sourirajan, S. (1981) *Industrial & Engineering Chemistry Process Design and, Development*, **20**, 272–282.

73 Santos, J.L.C., Hidalgo, A.M., Oliveira, R., Velizarov, S. and Crespo, J.G. (2007) *Journal of Membrane Science*, **300** (1–2), 191–204.

74 Dijkstra, M.F.J., Bach, S. and Ebert, K. (2006) *Journal of Membrane Science*, **286** (1–2), 60–68.

75 Zeman, L. and Wales, M. (1981) *Separation Science and Technology*, **16** (3), 275–290.

76 Li, W., Li, J., Chen, T., Zhao, Z. and Chen, C. (2005) *Journal of Membrane Science*, **258**, 8–15.

77 Nakao, S.I. and Kimura, S. (1982) *Journal of Chemical Engineering of Japan*, **15** (3), 200–205.

78 Van der Bruggen, B., Schaep, J., Vandecasteele, C. and Wilms, D. (2000) *Separation Science and Technology*, **35** (2), 169–182.

79 Tarleton, E.S., Robinson, J.P., Millington, C.R., Nijmeijer, A. and Taylor, M.L. (2006) *Journal of Membrane Science*, **278** (1–2), 318–327.

80 Geens, J., Hillen, A., Bettens, B., Van der Bruggen, B. and Vandecasteele, C. (2005) *Journal of Chemical Technology and Biotechnology (Oxford, UK, 1986)*, **80** (12), 1371–1377.

81 Geens, J., Boussu, K., Vandecasteele, C. and Van der Bruggen, B. (2006) *Journal of Membrane Science*, **281** (1–2), 139–148.

82 Tarleton, E.S., Robinson, J.P., Smith, S.J. and Na, J.J.W. (2005) *Journal of Membrane Science*, **261** (1–2), 129–135.

83 Tarleton, E.S., Robinson, J.P. and Salman, M. (2006) *Journal of Membrane Science*, **280** (1–2), 442–451.

84 Lee, C.H. (1975) *Journal of Applied Polymer Science*, **19** (1), 83–95.

85 Lipnizki, F., Hausmanns, S., Ten, P.K., Field, R.W. and Laufenberg, G. (1999) *Chemical Engineering Journal*, **73** (2), 113–129.

86 Lipnizki, F. and Tragardh, G. (2001) *Separation and Purification Methods*, **30** (1), 49–125.

87 Wang, H.Y., Ugomori, T., Tanaka, K., Kita, H., Okamoto, K. and Suma, Y. (2000) *Journal of Polymer Science Part B-Polymer Physics*, **38** (22), 2954–2964.

88 Jonquieres, A., Perrin, L., Arnold, S., Clement, R. and Lochon, P. (2000) *Journal of Membrane Science*, **174** (2), 255–275.

89 Van Veen, H.M., van Delft, Y.C., Engelen, C.W.R. and Pex, P.P.A.C. (2001) *Separation and Purification Technology*, **223** (1–3), 361–366.

90 Nomura, M., Yamaguchi, T. and Nakao, S. (1998) *Journal of Membrane Science*, **144** (1–2), 161–171.

91 Krishna, R. and Paschek, D. (2000) *Separation and Purification Technology*, **21** (1–2), 111–136.

92 Kapteijn, F., Bakker, W.J.W., Zheng, G.H., Poppe, J. and Moulijn, J.A. (1995) *Chemical Engineering Journal*, **57** (2), 145–153.

93 Bakker, W.J.W., Kapteijn, F., Poppe, J. and Moulijn, J.A. (1996) *Journal of Membrane Science*, **117** (1–2), 57–78.

94 Flanders, C.L., Tuan, V.A., Noble, R.D. and Falconer, J.L. (2000) *Journal of Membrane Science*, **176**, 43–53.

95 Moganti, S., Noble, R.D. and Koval, C.A. (1994) *Journal of Membrane Science*, **93**, 31–44.

96 Stephan, W., Noble, R.D. and Koval, C.A. (1995) *Journal of Membrane Science*, **99**, 259–272.

97 Verhoef, B.A., Huybrechs, B., van Veen, H., Pex, P., Degrève, J. and Van der Bruggen, B. (2008) Simulation of a hybrid pervaporation-distillation process, *Computers & Chemical Engineering*, **32** (6), 1135–1146.

98 Hömmerich, U. and Rautenbach, R. (1998) *Journal of Membrane Science*, **146**, 53–64.

4
Fundamentals of Membrane Gas Separation

Tom M. Murphy, Grant T. Offord, and Don R. Paul

4.1
Introduction

Research and technology innovations in the 1960s and 1970s led to the significant commercial practice of gas separations by membranes that exist today. These advances involved developing membrane structures that could produce high fluxes and modules for packaging large amounts of membrane area per unit volume. The discovery of asymmetric membrane structures for reverse osmosis was a key step in this evolution [1, 2]; such structures were eventually created in hollow fibers using solution spinning technology. Typical asymmetric membranes exhibit defects upon drying that limit their value for gas separations; however, this problem was eventually solved by the discovery that the defects could be effectively sealed by coating the membrane with a highly permeable polymer, such as silicone rubber [3, 4]. Composite membranes consisting of a thin separating layer coated onto a porous substrate or an intermediate layer have also been developed, which has expanded the types of materials that can be converted into high-flux membranes. For commercial use, high-flux flat-sheet membranes are packaged into spiral wound modules, while hollow-fiber membranes are assembled into modules resembling shell and tube heat exchangers. Today, such membrane modules are sold commercially by a number of companies for separating nitrogen from air, recovery of hydrogen from process streams, natural-gas processing, dehydration of gas streams, recovery of vapors from gases, and so on. A number of recent books summarize these developments and their industrial uses [5–8].

Most of the more recent research has focused on developing membrane materials with a better balance of selectivity and productivity (permeability) as that seems the most likely route for expanding the use of this technology. There appear to be natural upper bounds [9, 10] on this tradeoff that limit the extent of improvement that can be realized by manipulating the molecular structure of the polymer used for the selective layer of high-flux membranes, at least in many cases. This has led to interest in nonpolymeric and so-called mixed-matrix materials for membrane formation [8]; however, at this time, polymers remain the materials of choice for gas-separation

Membrane Operations. Innovative Separations and Transformations. Edited by Enrico Drioli and Lidietta Giorno

ISBN: 978-3-527-32038-7

membranes. The purposes of this chapter are to review briefly the fundamentals of gas permeation in polymeric materials and to explore in some detail two very different material issues of current interest. One of these relates to the physical aging of glassy polymer membranes that results from their nonequilibrium character. The other relates to the search for membrane materials that have exceptional selectivity for CO_2 relative to other gases, and it turns out that some of the most promising polymers are in the rubbery state.

4.2 Polymer Structure and Permeation Behavior

Most polymers that have been of interest as membrane materials for gas or vapor separations are amorphous and have a single phase structure. Such polymers are converted into membranes that have a very thin dense layer or skin since pores or defects severely compromise selectivity. Permeation through this dense layer, which ideally is defect free, occurs by a solution–diffusion mechanism, which can lead to useful levels of selectivity. Each component in the gas or vapor feed dissolves in the membrane polymer at its upstream surface, much like gases dissolve in liquids, then diffuse through the polymer layer along a concentration gradient to the opposite surface where they 'evaporate' into the downstream gas phase. In ideal cases, the sorption and diffusion process of one gas component does not alter that of another component, that is, the species permeate independently.

For rubbery polymers, that is, above the glass-transition temperature, T_g, the sorption of simple gases follows the relationship known as Henry's law

$$C = S p \tag{4.1}$$

where C is the equilibrium concentration of the gas dissolved in the polymer when its partial pressure in the gas phase is p and S is the solubility coefficient. At steady state, the diffusion process is described by a simple version of Fick's law

$$\text{Flux} = \frac{D\Delta C}{l} = \frac{DS\Delta p}{l} = \frac{P\Delta p}{l} \tag{4.2}$$

where l is the dense layer thickness, ΔC is the concentration difference of gas in the upstream and downstream faces of the dense layer and D is the diffusion coefficient. Since by Equation 4.1, $\Delta C = S\Delta p$, we can see that the permeability coefficient P is given by

$$P = D S \tag{4.3}$$

Polymers above their T_g are in a state of equilibrium much like simple liquids. However, upon cooling below T_g, polymers are not able to achieve an equilibrium state since the polymer chain segments lack sufficient mobility to reach this state in realizable time scales. Thus, glassy polymers exist in a nonequilibrium state that is a function of the prior history of the sample. It is useful to think of simple volumetric thermal expansion where at equilibrium the specific volume at a given temperature and pressure is $V_{eq}(T, p)$; the specific volume of a rubbery polymer is given by V_{eq}. The observed specific volume of a glassy polymer, V_g, will always be larger than V_{eq}; the excess volume of the glass ($V_g - V_{eq}$) affects many of the characteristics of the material and depends on the prior history of the sample. Because of the difference in segmental mobilities, glassy polymers are more than 10^3 times stiffer than rubbery polymers.

The sorption of simple gases in glassy polymers follows a more complex relation and is well described by the so-called dual sorption model [11–17]

$$C = k_D p + \frac{C'_H bp}{1 + bp} \tag{4.4}$$

where k_D, C'_H and b are parameters of the model. The second term on the right in Equation 4.4 represents an additional mode of sorption that can be linked quantitatively to the excess volume of the glassy state ($V_g - V_{eq}$) while the first term may be thought to represent an extension of the Henry's law mode seen above T_g [13, 14]. Thus, the extent of sorption of gases in glassy polymers is actually significantly greater than in rubbery polymers, which is counterintuitive considering that glasses are orders of magnitude more stiff than rubbers. In addition, the amount of sorption in the glass depends on the history of the sample; both effects being attributable to the nonequilibrium character [15–17].

The permeation of simple gases in glassy polymers is more complex than in rubbery polymers. An extension of the dual sorption model of permeation leads to a relation, when the downstream pressure is small, of the following form

$$P = k_D D_D + \frac{C'_H b D_H}{1 + bp_2} \tag{4.5}$$

where D_D and D_H are diffusion coefficients for gas molecules sorbed by each of the modes of sorption and p_2 is the upstream gas pressure [11–17]. This model predicts that the permeability coefficient decreases slightly as the upstream pressure is increased and generally describes experimental data quite well. Later, we will return to the issue of how history affects permeation behavior of glassy polymers of the type used to make gas-separation membranes like those shown in Table 4.1.

For gases or vapors that are quite soluble in polymers some of the simple relations described above break down. For example, the sorption isotherm for vapors in rubbery polymers may show upward curvature from the simple linear prediction of Henry's law, Equation 4.1, and this effect is actually expected from thermodynamic theories like the well-known Flory–Huggins equation [8]. In addition, the presence of the penetrant at high enough concentrations will affect the mobility of the polymer segments, which will be reflected as an increase in the penetrant diffusion coefficient; this is referred to as plasticization [8]. Similar effects can also be seen in glassy polymers; but as might be expected, they are even more complex to describe. In addition to plasticization, sorbing sizable quantities of penetrant into a glassy polymer alters the state of the glass such that after removal of the penetrant the glass does not return to its original state; this has been referred to as conditioning [17].

For now it is useful to take a more global view and not concern ourselves with the nonlinear effects described by Equations 4.4 and 4.5 or those caused by plasticization.

Table 4.1 Three glassy polymers used to form gas-separation membranes.

Polysulfone (PSF)	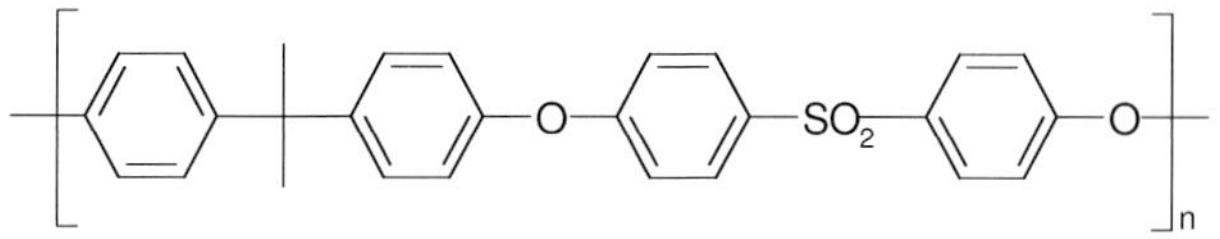 [structure: –O–, SO_2, n]
Matrimid®	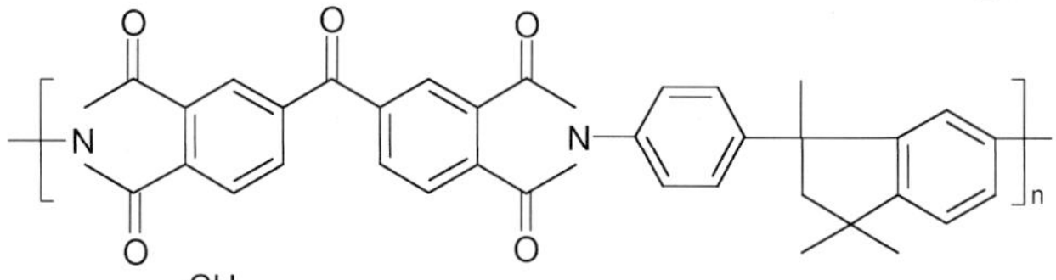 [structure: N, O, n]
Poly(2,6-dimethyl-1,4-phenylene oxide) (PPO)	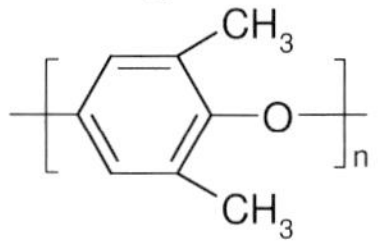 [structure: CH_3, O, CH_3, n]

Thus, we can use Equation 4.3 as representative of the solution–diffusion mechanism where S and D may not be constants but depend on the external conditions in the gas phases. With this in mind for a pair of gases A and B, we can construct the following useful relationship

$$\frac{P_A}{P_B} = \left(\frac{D_A}{D_B}\right)\left(\frac{S_A}{S_B}\right) \tag{4.6}$$

The ratio P_A/P_B is often referred to as the permselectivity of the membrane, and in simple cases allows one to determine the extent of separation that a given membrane can achieve in a given situation. It is a property of the membrane material and does not depend on the thickness of the separating layer. This permselectivity is the product of the 'diffusion selectivity' and the 'solubility selectivity.' To understand these terms, it is useful to know that in the simplest of cases, the diffusion coefficient of a penetrant in a given polymer decreases as the size of the penetrant molecule increases and that the solubility coefficient increases as the 'condensability' of the penetrant increases. This is illustrated schematically in Figure 4.1. Various molecular diameters, the van der Waals volume and the critical volume have been used to characterize penetrant size [7, 8]. The propensity of the penetrant to condense, that is, its condensability, may be characterized by its boiling point, critical temperature, or the Lennard-Jones potential well depth, ε/k. It is now well established that the dependence of D on penetrant size is much stronger for glassy polymers than for rubbery polymers, as suggested in Figure 4.1; that is, glasses may be said to be more size selective. Thus, glassy polymers like those shown in Table 4.1 have become the materials of choice for membranes to separate certain gas pairs. However, as we will explore more fully later, there are cases where rubbery polymers are more selective.

In many cases, but not all, the condensability of penetrants increases as size increases. This is the case for the gas pair i and j suggested in Figure 4.1. In the case shown there, the diffusion selectivity favors i over j but the solubility selectivity favors j over i.

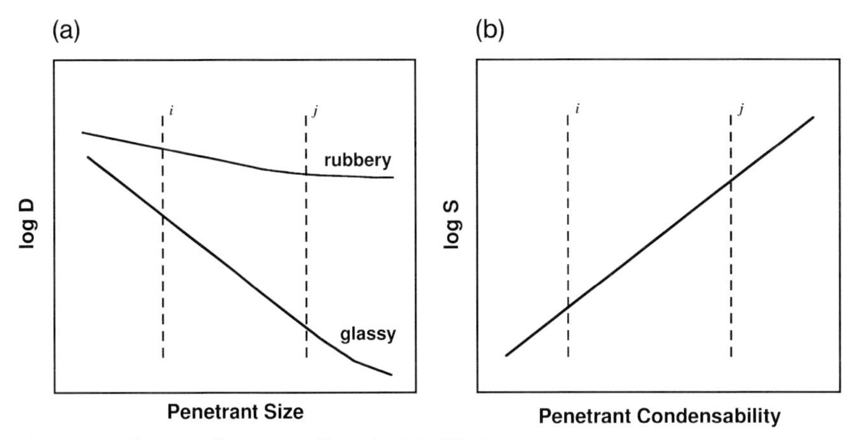

Figure 4.1 Schematic illustration of how the (a) diffusion coefficient of penetrants depend on their size in rubbery and glassy polymers and (b) solubility coefficients for penetrants depend on their condensability.

Table 4.2 illustrates the various selectivity factors for some typical rubbery polymers, that is, silicone rubber, poly(dimethyl siloxane), and natural rubber, polyisoprene, and a glassy polymer, polysulfone. Here, we consider the important O_2/N_2 pair and several pairs involving CO_2 that will be our focus later. In all the cases, the solubility selectivity is greater than unity and there is not a large difference between rubbery and glassy polymers. For most of these pairs, the diffusion selectivity is greater than unity, but there are some exceptions for CO_2/O_2 and CO_2/N_2 that reflect

Table 4.2 Permselectivity characteristics of selected rubbery and glassy polymers.

A/B	P_A/P_B	S_A/S_B	D_A/D_B	T °C	Reference
Poly(dimethyl siloxane)					
O_2/N_2	2.0	1.6	1.3	35,25	[18, 19]
CO_2/O_2	4.9	3.4	1.4	35,20	[18, 19]
CO_2/N_2	7.4	8.1	0.91	35	[18, 20]
CO_2/CH_4	3.1	2.9	1.1	35	[18]
Natural rubber					
O_2/N_2	2.9	2.0	1.4	25	[19]
CO_2/O_2	5.6	8.0	1.4	25	[19]
CO_2/N_2	16	16	1.0	25	[19]
CO_2/CH_4	4.5	3.6	1.2	25	[19]
Polysulfone					
O_2/N_2	5.6	1.6	3.5	35	[6]
CO_2/O_2	4.0	8.8	0.45	35	[6]
CO_2/N_2	22	14	1.6	35	[6]
CO_2/CH_4	22	3.7	5.9	35	[6]

subtle shape issues that become important when the sizes are similar and will not be pursued here. The important point is to see the much greater diffusion selectivity for O_2/N_2 and CO_2/CH_4 in polysulfone than in the two rubbery polymers; this translates into greater permselectivity of the glassy material than of the rubbery ones.

When the gas or vapor feed stream contains a component that is highly soluble in the polymer membrane and causes plasticization, then the selectivity as defined by Equation 4.6 will depend on the partial pressure or the amount of the plasticizing component sorbed into the membrane. Furthermore, pure-gas permeation measurements are generally not a good indicator of the separation performance, and mixed-gas permeation measurements will be needed [21–23]. Often, the mixed-gas selectivity is less than predicted from pure-gas measurements [8]; however, the opposite has been observed [24]. Competitive sorption effects can also compromise the prediction of mixed-gas behavior from pure-gas measurements [25]. For gas pairs where each component is less condensable than CO_2, like O_2/N_2, it is generally safe to conclude that the selectivity characteristics can be accurately judged from pure-gas permeabilities at all reasonable pressures. When the gas pair involves a component more condensable than CO_2, plasticization is likely to be a factor and pure-gas data may not adequately reflect mixed-gas selectivity. When CO_2 is a component, the situation depends on the partial pressures and the nature of the polymer.

Generally, polymers that crystallize are not considered good candidates for membrane materials; however, there are some exceptions [26, 27]. The presence of crystallinity reduces permeability [28, 29] and good membranes should be capable of high fluxes. The usual physical picture is to think of a semicrystalline polymer in terms of a simple two-phase model; one phase being amorphous and the other being crystalline. In the typical case, the crystals do not sorb or transmit penetrant molecules; the following relationship has been proposed [28, 29] to describe the extent to which crystallinity reduces permeability from that if the polymer were amorphous

$$P_c = \frac{P_a(1-\phi)}{\tau\beta} \tag{4.7}$$

where P_c is the penetrant permeability in the semicrystalline polymer, P_a is the permeability of the completely amorphous polymer, ϕ is the volume fraction of crystals, τ is the tortuosity factor to account for a more elongated path a penetrant molecule must take through the amorphous phase since it cannot go through the crystals, and β is the chain-immobilization factor that reflects the reduction in mobility that occurs in the amorphous phase caused by the presence of crystallites. Ideally, neither ϕ nor τ will depend on what the penetrant is; however, β clearly can [29, 30]. Thus, it is possible that in addition to reducing permeability, crystallinity can alter selectivity, that is,

$$(P_C)_A/(P_C)_B = [(P_a)_A/(P_a)_B]\left(\frac{\beta_B}{\beta_A}\right) \tag{4.8}$$

Polymer blends and block-copolymers have been considered as membrane materials as mentioned later. If the components are miscible and a single-phase material results, then no special considerations are needed for applying the concepts outlined above. However, these systems usually consist of separate phases of the components. In this case, the materials need to be treated as a composite and, then, their morphology becomes an important issue [31, 32]. Usually, the continuous phase dominates the permeation process, so if we want to build into a membrane material the permeation characteristics of one component, then this component must have some degree of phase continuity in the material [33].

4.3 Membranes from Glassy Polymers: Physical Aging

Glassy polymers are usually the preferred materials for practical gas-separation membranes because of their inherently better permeability/selectivity balance than is typically the case for polymers above their glass-transition temperature [5, 9]. In addition, the structural rigidity provided by the glassy state is essential for membranes that must be self-supporting (e.g., asymmetric hollow fibers) [5]. Glasses are not in a state of equilibrium; therefore, their properties are dependent on the details of their fabrication and time–temperature history [34–36]. Thus, it is not surprising to observe some variance in the reported properties, such as density, refractive index, gas permeability, and so on, of glassy polymers. At least for macroscopic specimens, the variability seems to be within a range small enough that meaningful property tabulations can be made for glassy polymers, as recorded in many handbooks [37]. However, recent research has shown that this variability may be considerably more pronounced for thin films because of their significantly more rapid evolution toward the equilibrium state, a process known as physical aging, most often observed in terms of volume relaxation or densification [38]. This densification, or physical aging, affects properties that are sensitive to free volume, such as permeability, and the associated changes can be quite significant, on time scales of weeks to years [38–52].

Practical membranes must be very thin to achieve the high fluxes needed for economical productivity; typically, 'skins' or separating layers with thicknesses of the order of 0.1 μm (or 100 nm) or less with minimal defects are essential for a viable technology. However, such thin layers of glassy polymers can be greatly affected by the physical aging issues mentioned previously. This brings into question the widely practiced approach of using relatively thick films for screening or selecting polymers as membrane materials. Indeed, the permeation properties of thick films are often used to calculate the effective thickness of the skin layer of asymmetric or composite membrane structures from observed fluxes.

Figure 4.2 is an attempt to classify glassy polymer films into different regimes of behavior according to thickness. To the far right of the thickness scale is the familiar case where properties, including those related to the departure from an equilibrium state, are expected to be independent of specimen size. This is clearly the expectation on the millimeter or centimeter scale and probably extends down to several micrometers; we might call this the 'bulk' regime. On the other extreme are ultrathin films where the thickness is of the same order of magnitude as the dimensions of the

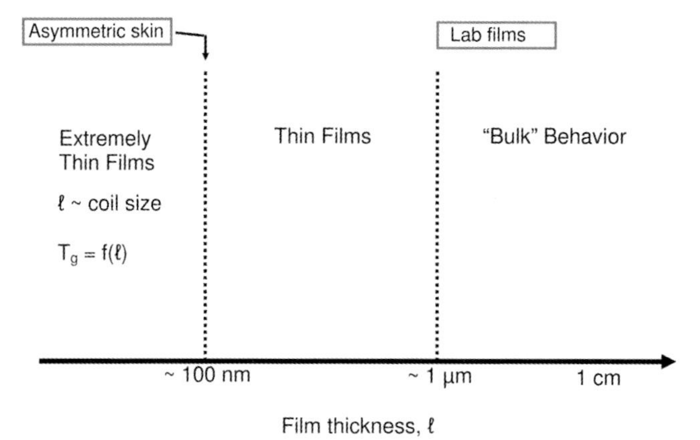

Figure 4.2 Approximate thickness regimes for the behavior of glassy polymer films [52]. Reproduced with permission of the American Chemical Society.

polymer chain coils (<100 nm, typically). In this region, the conformations of the polymer chains are perturbed by the boundaries imposed by the surfaces of the films, which results in the so-called 'confinement effects.' The recent literature on polymer physics contains many experimental and theoretical studies of such 'ultrathin films' [53–56]. The glass-transition temperature (T_g) and other characteristics of ultrathin films have been reported to be dependent on thickness; certain differences in behavior have been attributed to whether the film is freestanding or supported on some substrate, which may influence the polymer. It is beyond the scope of this chapter to review or explain this regime. It is sufficient for current purposes to note that characteristics such as T_g seem to reach a plateau or bulk value prior to thicknesses of the order of 100 nm, which is on the order of the thickness of the separating 'skin' layer of some commercial asymmetric membranes.

There is a region of 'thin films' with thicknesses between the two previously described extreme limits, ranging from ~100 nm to several micrometers, where volume relaxation processes – and, hence, the change in gas-permeability properties with time – are much more rapid than that expected based on observations of 'bulk' specimens as shown below.

The results shown next are for thin films prepared over a range of thicknesses from the polymers shown in Table 4.1 by spin casting that were then heated above T_g briefly to erase prior history and then cooled to 35 °C where aging was observed for more than a year [45]. Ellipsometry was used to measure the film thickness and refractive index accurately [38, 45]. Figure 4.3 shows how the permeability of O_2 in PPO decreases with aging time at 35 °C for films that range in thickness from 400 nm to 1 mil. The dashed line shows the single 'bulk' value for O_2 permeability reported in the literature for PPO measured on thick films. At short aging times, the permeability values reported lie well above the so-called bulk value. This is believed to reflect the state of relatively high free volume captured by the protocol of film preparation used in this work. At longer aging times, the current permeability values decrease well

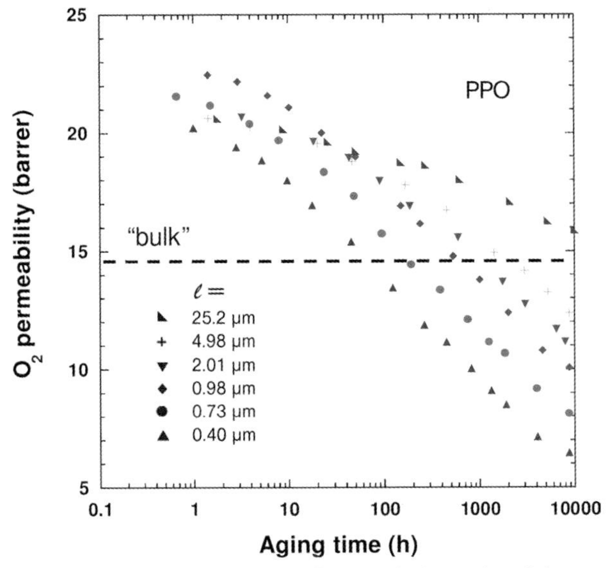

Figure 4.3 Oxygen permeability coefficients of poly (2,6-dimethyl-1,4-phenylene oxide) (PPO) films of various thicknesses, as a function of aging time at 35 °C [52]. Reproduced with permission of the American Chemical Society.

below the bulk values except for some thick films. Similar results have been observed for the other polymers shown in Table 4.1 as well as several novel polyimides [49, 50].

Thus, the central conclusion is that the bulk values may differ from the time-dependent values for thin films by very significant amounts. This results from the well-known fact that the state of a glassy polymer is dependent on its prior history, and the lesser-known fact that the rate of physical aging of thin films can be quite significant at temperatures well below the T_g value; in the present case, PSF is 150 °C, PPO is 175 °C, and Matrimid is 275 °C below their respective T_g values. The aging rate is clearly dependent on film thickness as shown in Figure 4.4. The thinnest films examined here are of the order of 0.4 μm (or 400 nm) in thickness. One would expect even faster rates at 100 nm; however, current techniques have not yet permitted probing such thin samples because of issues of manipulating such fragile structures. The present results are not due to any alteration of the T_g value by thickness because such effects seem to be significant only well below 100 nm [45, 53, 54]. Thus, it seems that different issues are at play here than in the regime we labeled 'ultrathin films' in Figure 4.2.

The rate of change in permeability during aging is dependent on the size of the gas molecule; the rate of change for all the polymers studied follows the order $O_2 < N_2$ CH_4 [48]. This is understandable because the underlying issue is the loss of free volume during aging and this has a larger effect on the permeation of larger molecules. Thus, the selectivity should increase on aging that is demonstrated to be the case in Figure 4.5 for PSF; the rate of increase is greater the thinner the film. Clearly, the use of 'bulk' values of gas permeability provides only a first-order

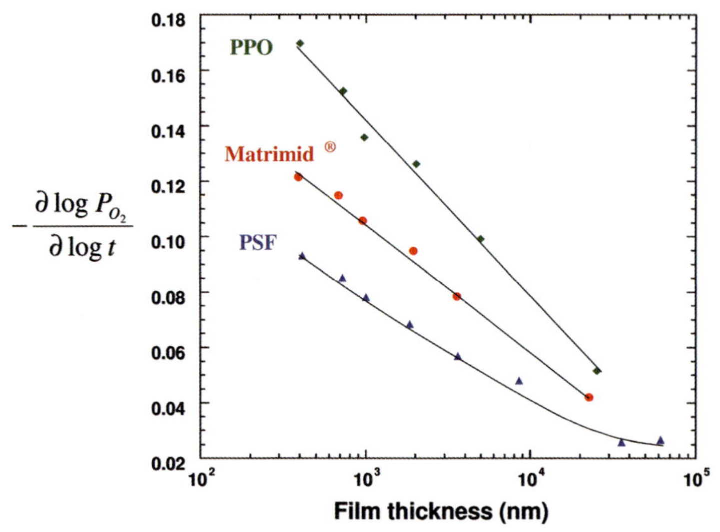

Figure 4.4 Effect of thickness on aging rate of glassy polymer films determined by change in oxygen permeability at 35 °C [46]. Reproduced with permission of Elsevier.

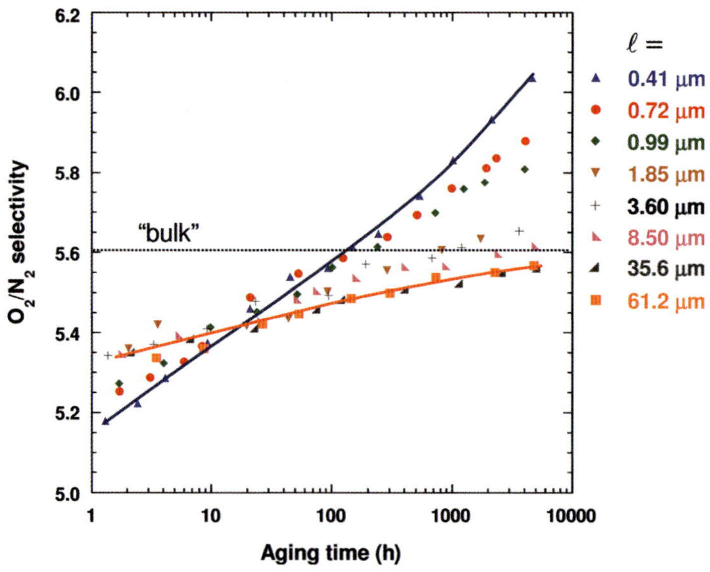

Figure 4.5 O_2/N_2 permselectivity of polysulfone films of various thicknesses versus aging time at 35 °C [46]. Reproduced with permission of Elsevier.

approximation to the productivity/selectivity balance observed in thin films because of the aging phenomenon.

In addition to thickness, ellipsometry techniques also give the refractive index of thin films [38, 45], which provides another useful way of tracking aging since refractive index can be related to density via relations like the Lorentz–Lorenz equation [38]. Figure 4.6 illustrates the change in refractive index (normalized by the initial value for thin films (~400 nm) of the three glassy polymers from Table 4.1. The increase in refractive index confirms that the aging process involves densification of the glass polymer. It is clear from these results that the aging rate of these three polymers is PPO > Matrimid>PSF, which is consistent with the results in Figure 4.4 where the aging response was tracked in terms of oxygen permeation. A more formal way to make the comparison between aging responses by permeation and optical properties is to define an aging rate as follows

$$r = -\frac{1}{V}\left(\frac{\partial V}{\partial \ln t}\right) = \left(\frac{\partial \ln \rho}{\partial \ln t}\right) \tag{4.9}$$

where V = specific volume and ρ = density of the polymer. The refractive-index data can be used to compute this aging rate using the Lorentz-Lorenz equation [38, 57]. Figure 4.7 compares the oxygen, nitrogen, and methane permeability reduction rates vs. the corresponding volumetric relaxation rate for thin films of the 6FDA-based polyimides aged at 35 °C with other thin glassy polymer films, viz., polysulfone (PSF), the commercial polyimide Matrimid, and poly (2,6-dimethyl-1,4-phenylene oxide) (PPO). Clearly, there is a strong correlation between the two measures of aging rate

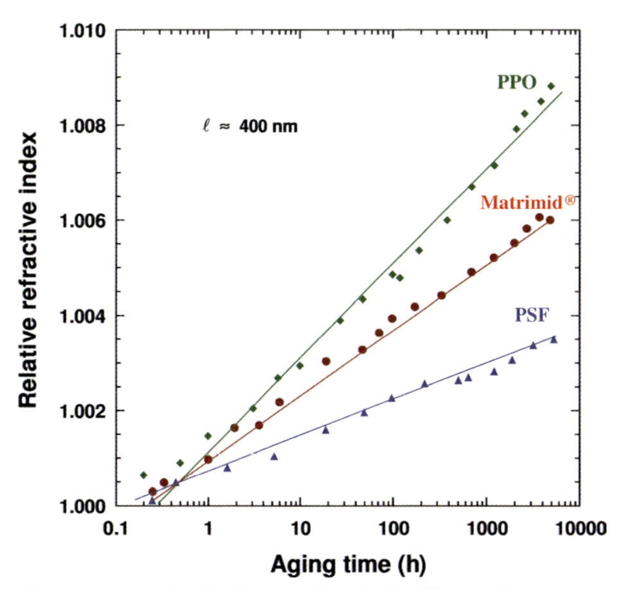

Figure 4.6 Normalized refractive indices for thin films (~400 nm) of three glassy polymers as a function of aging time [38]. Reproduced with permission of the American Chemical Society.

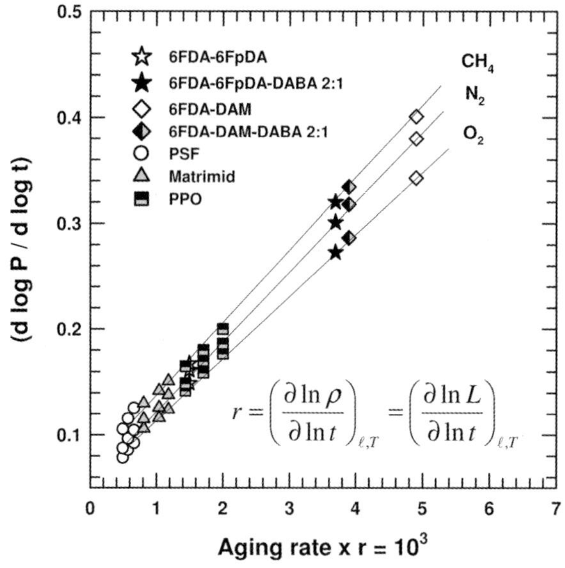

Figure 4.7 Correlation between O_2, N_2 and CH_4 permeability reduction rates and their volumetric relaxation rates for thin films of various glassy polymers [49]. Reproduced with permission of Elsevier.

that is consistent among all these polymers. Note that the multiple points for the latter three polymers from the literature correspond to films of different thickness. Interestingly, the 6FDA-based polyimide thin films show much larger aging rates than the other materials, which is consistent with the higher free volume of these polyimides.

The experimental results described above show that the gas-permeability properties of thin glassy polymer films (submicrometer in thickness) are more time- or history-dependent than much thicker films (the bulk state; for example, 50 μm or thicker) seem to be. This is manifested in terms of physical aging over a period of 1 year and more. The observed permeability values for the current thin films are all initially greater than the reported bulk values but approach or become less than these values after a few days or weeks, depending on the thickness. After a year, the thin films may be as much as four times less permeable than the reported bulk values. Selectivity increases with aging time, as might be expected from a densification process.

These observations have several practical consequences for membrane processes where the selective layers are as thin as or even thinner than the low end of the range studied here. First, it is clear that use of thick film data to design or select membrane materials only gives a rough approximation of the performance that might be realized in practice. Second, because the absolute permeability of a thin film may be severalfold different than the bulk permeability, use of the latter type of data to estimate skin thickness from flux observations on asymmetric or composite membranes structures is also a very approximate method. Finally, these data indicate that one could expect the productivity of commercial membrane modules to decline measurably over their lifetime of several years because of physical aging effects. Although there seems to be no published data showing such effects for gas-separation membranes based on polymer glasses, this is certainly the anecdotal experience of membrane suppliers and users. There is at least one documented report of such declines in reverse-osmosis membranes [58]. Often, these declines for asymmetric membranes have been attributed to 'compaction' of the porous substructure over time due to stress, effectively making the skin thicker; however, it is quite likely that at least some portion of this can be attributed to the more fundamental issue of physical aging of the glassy polymer skin. It should be said that one cannot precisely compare data such as that shown here to membrane module performance because the thin layers probably have experienced very different histories. In the present case, the films were heated above the glass-transition temperature (T_g) and then cooled to ambient conditions to give a well-defined starting state for the aging process. The skins of practical membranes are formed by a more complex process, and, generally, the online flux monitoring begins after considerable aging has already occurred.

4.4 Membranes from Rubbery Polymers: Enhanced CO_2 Selectivity

There are many examples of gas streams containing CO_2 as an impurity that must be removed from lighter gases like CH_4, N_2, and H_2. Examples include natural gas, where CH_4 is the desired product, refinery and reforming streams, where H_2 is the valued product, flue gases, where CO_2 needs to be removed from N_2 and sequestered, and others [59]. In these cases, membranes are needed that are much more permeable to CO_2 relative to these light gases than can be found in conventional polymers (see the data in Table 4.2). Membranes are also useful in modified atmosphere packaging of fruits and vegetables for extending shelf life, and there is a need in certain applications for membranes that are more permeable to CO_2, relative to O_2 and N_2, than current membrane materials [60]. This application will be explained more fully later.

One strategy for designing membranes that are more selective to CO_2 is to take advantage of its potential quadrapole interaction with the polymer to increase its solubility selectivity relative to the light gases, which cannot interact in this manner [59]. However, the advantages gained by building in this form of CO_2 selectivity would not be fully realized if it were counterbalanced with a large size selectivity favoring the light gas; that is, one can expect glassy polymers to favor the light gas because of their larger diffusive selectivity as illustrated in Figure 4.1. For rubbery polymers, the size selectivity is minimal, as shown in Table 4.2, which gives the solubility selectivity a chance to dominate. As it turns out, poly(ethylene oxide), PEO, segments have an excellent affinity for CO_2 relative to O_2, N_2, CH_4, and so on [20, 24, 59, 61–79]; however, poly(ethylene oxide) itself is highly crystalline, which reduces its permeability to all gases, it has a low melting point, $\sim$65 °C, and it is water soluble. Thus, PEO is not directly useful for most of the applications mentioned, but there has been

Table 4.3 Infinite dilution permselectivity characteristics of semicrystalline poly(ethylene oxide) at 35 °C [76].

A/B	P_A/P_B	S_A/S_B	D_A/D_B
O_2/N_2	2.7	1.14	2.4
CO_2/O_2	18	9	1.9
CO_2/N_2	48	10.3	4.7
CO_2/CH_4	20	4.7	4.3
CO_2/H_2	6.8	—	—

considerable research over the years directed at taking advantage of its desirable attributes, while minimizing its undesirable features [20, 24, 59, 61–79].

The various approaches to the problem outlined above have included blending PEO with polymers with which it may be miscible [61, 66], making block-copolymers of PEO with oligomers or polymers where the other segments were polyimides [64, 65, 67, 68], polyurethanes [71], polyamides [62, 70, 72–74] and polyesters [75], and making highly crosslinked structures containing ethylene oxide units [24, 76, 78, 79]. In general, these approaches can suppress PEO crystallization, prevent solubility in water, increase strength, and some lead to structures that can be converted to high-flux composite hollow-fibers [68] and flat-sheet [62] membranes. The block-copolymers were designed to have high PEO contents to give a PEO continuous phase. Selected results from these studies are presented here to show how effective this approach can be.

Table 4.3 shows the permselectivity characteristics of pure, semicrystalline PEO films [76]. The selectivity characteristics for O_2/N_2 are rather similar to those for silicone rubber and natural rubber shown in Table 4.2. However, the values of permselectivity for CO_2 relative to the various light gases shown are all much higher than Table 4.2 shows for the rubbery polymers listed there and even for polysulfone except for CO_2/CH_4. Comparison of the data in Tables 4.2 and 4.3 makes it clear that this high permselectivity of PEO stems from its high solubility selectivity for CO_2 versus other gases; this is augmented by modest values of diffusivity selectivity. Data in Table 4.4 for the CO_2/N_2 pair illustrate that this effect can be translated into various block-copolymer structures when the PEO content is high enough to ensure it is the continuous phase. In fact, nearly all these materials have higher permselectivity and solubility selectivity for CO_2/N_2 than does pure PEO (see Table 4.3); however, the diffusion selectivity for these copolymers is much closer to, or even less than, unity than seen for pure PEO. Furthermore, the copolymers all have much higher absolute permeability coefficients than does PEO.

Figure 4.8 shows graphically how the permselectivity for CO_2/N_2 is much higher for any other gas relative to N_2 in a poly(butylene terephthalate) block-copolymer containing 56 wt.% PEO segments [75]. The gas-solubility data in a polyamide block-copolymer containing 57 wt.% PEO given in Figure 4.9 clearly demonstrate the unusually high CO_2 solubility in these materials relative to other gases [70]. In fact, the CO_2 data fall higher than expected based on the trend line set by the other gases by a factor of about 6. Apparently this effect is unique to ethylene oxide segments since

Table 4.4 Carbon dioxide (A)/nitrogen (B) permselectivity characterization for polyimide, polyurethane and polyamide block-copolymers containing polyether segments.

Polymer	[PE][a] wt%	P_A/P_B	S_A/S_B	D_A/D_B	T °C	Reference
BP-ODA/DABA/PEO	57	69	77	0.92	25	[65]
BP-ODA/DABA/PPO	57	28	33	0.84	25	[65]
BP-ODA/DABA/PTHF	58	29	—	—	25	[65]
MDI-BPA/PEO	60	47	41	1.15	35	[71]
N6/PEO	70	71	26	2.8	25	[73]
N6/PEO	57	56	88	0.63	35	[72]
N12/PEO	55	51	63	0.81	35	[72]

[a]Weight per cent polyether in copolymer.
PEO = poly(ethylene oxide).
PPO = poly(propylene oxide).
PTHF = poly(tetramethylene oxide).
BP-ODA/DABA = polyimide (see Ref. 65 for structure).
MDI-BPA = polyurethane (see Ref. 71 for structure).
N6 = polyamide 6.
N12 = polyamide 12.

the data in Table 4.4 suggest that propylene oxide or tetramethylene oxide segments do not lead to such high CO_2/N_2 permselectivity characteristics.

Next, we show how membranes with very high CO_2 permselectivity relative to O_2 and N_2 would have value in preserving the shelf life of fruits and vegetables. It is well known that controlling the CO_2 and O_2 atmosphere around produce combined with refrigeration can extend the post-harvest viable life of produce; see Table 4.5 for recommended atmospheres for selected items of produce. The inset in Figure 4.10 illustrates how membranes are used to create a modified atmosphere inside a package of respiring produce. The produce consumes O_2 and gives off CO_2 in a

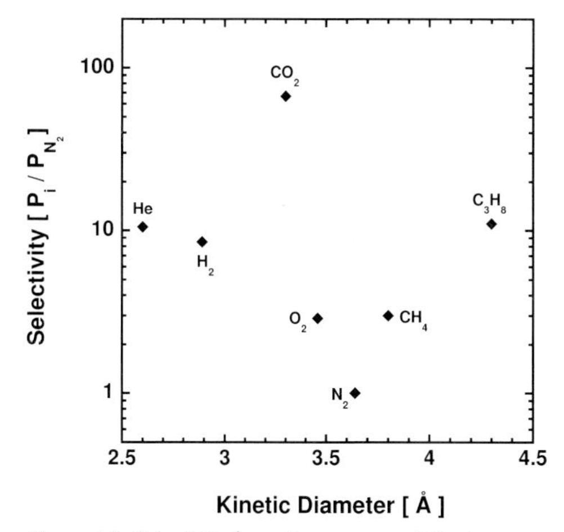

Figure 4.8 Selectivity for various gases relative to oxygen at 20 °C for a PBT-PEO block-copolymer [75].

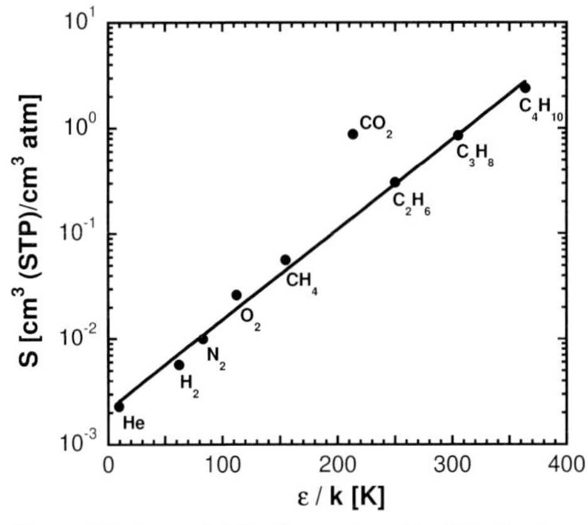

Figure 4.9 Gas solubility in a polyamide-PEO block-copolymer containing 57 wt.% PEO units [70].

certain molar ratio or respiratory quotient (RQ) that normally is about 1.2; a high-flux membrane patch affixed to the package regulates the inflow of O_2 and outflow of CO_2. A theory for this process shows that at steady state the content of O_2 and CO_2 will lie on a line like those illustrated in Figure 4.10 that depend only on the permselectivity characteristics of the membrane and RQ [60]. Figure 4.10 shows lines for membranes that are nonselective and ones that have the properties of silicone rubber and poly (ethylene oxide). Exactly where on these lines the atmosphere in a given package will lie depends on the ratio of respiration to permeation rates and is a design parameter controlled by factors like membrane area and permeance plus the type and amount of produce in the package. Clearly, a silicone-rubber membrane combined with some nonselective perforations will meet the optimum requirement for many items of produce in Table 4.5 [60]. However, some items, like apples and pears, require a lower content of CO_2 in the steady-state atmosphere than can be generated by such membranes. In these cases, a more CO_2-selective membrane is needed, and those based on PEO segments in a suitable form to meet other requirements appear promising.

Table 4.5 Recommended atmospheres for prolonging viable life of selected fruits and vegetables.[a]

Produce	T °C	% O_2	% CO_2
Broccoli	0–5	1–2	5–10
Cabbage	0–5	2–3	3–6
Celery	0–5	1–4	3–5
Oranges	5–10	5–10	0–5
Strawberries	0–5	5–10	15–20
Carrots (sliced)	0–5	2–5	15–20
Lettuce (Iceberg, chopped)	0–5	0.5–3	10–15
Apples (Braeburn)	0.7	1.8	1.0
Pears (Bartlett)	−1–0	1–2	0–1.5

[a]From http://postharvest.ucdavis.edu.

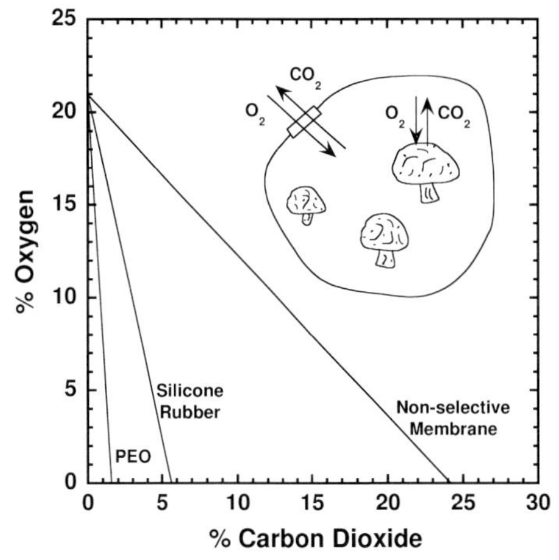

Figure 4.10 Steady-state relationships between O_2 and CO_2 content for a modified atmosphere package outfitted with different membranes. Lines calculated by theory [60].

4.5 Summary

The chemical structure and physical state of the polymer has a considerable effect on how a membrane formed from it performs for gas separations. The glassy state is preferred for high selectivity when there are considerable differences in penetrant sizes; however, thin, glassy polymer layers or skins may undergo substantial decline in permeability over time owing to physical aging that is much more rapid than observed in bulk. Rubbery polymers allow the opportunity to base selectivity on penetrant solubility, as illustrated for CO_2 relative to other gases. Polymers containing poly (ethylene oxide) segments have attractive CO_2 selectivities because of their interactions with the CO_2 quadrapole. However, to achieve practical membranes from such materials, it is necessary to suppress PEO crystallization (to achieve high fluxes) and to retard swelling by water using crosslinking, blending, or block-copolymer structures.

Acknowledgments

This work was supported by the National Science Foundation under Grant No. DMR0423914.

References

1 Merten, U. (ed.) (1966) *Desalination by Reverse Osmosis*, MIT Press, Cambridge.

2 Riley, R.L., Gardner, J.O. and Merten, U. (1964) *Science*, **143**, 801–803.

3 Henis, J.M.S. and Tripodi, M.K. (1980) *Separation Science and Technology*, **15**, 1059.

4 Henis, J.M.S. and Tripodi, M.K. (1980) U.S. Patent 4,230,236, Monsanto.

5 Kesting, R.E. and Fritzsche, A.K. (1993) *Polymeric Gas Separation Membranes*, John Wiley & Sons, New York.
6 Paul, D.R. and Yampol'skii, Y.P. (1994) *Polymeric Gas Separation Membranes*, CRC Press, Boca Raton.
7 Baker, R.W. (2004) *Membrane Technology and Applications*, 2nd edn, John Wiley, Chichester.
8 Yampol'skii, Y., Pinnau, I. and Freeman, B.D. (eds) (2006) *Materials Science of Membranes for Gas and Vapor Separation*, John Wiley, Chichester.
9 Robeson, L.M. (1991) *Journal of Membrane Science*, **62**, 165–185.
10 Freeman, B.D. (1999) *Macromolecules*, **32**, 375–380.
11 Paul, D.R. and Koros, W.J. (1976) *Journal of Polymer Science: Polymer Physics Edition*, **14**, 687.
12 Koros, W.J., Chan, A.H. and Paul, D.R. (1977) *Journal of Membrane Science*, **2**, 165.
13 Koros, W.J. and Paul, D.R. (1978) *Journal of Polymer Science: Polymer Physics Edition*, **16**, 1947 and 2171.
14 Koros, W.J. and Paul, D.R. (1981) *Journal of Polymer Science: Polymer Physics Edition*, **19**, 1655.
15 Chan, A.H. and Paul, D.R. (1979) *Journal of Applied Polymer Science*, **24**, 1539.
16 Chan, A.H. and Paul, D.R. (1980) *Polymer Engineering and Science*, **20**, 87.
17 Wonders, A.G. and Paul, D.R. (1979) *Journal of Membrane Science*, **5**, 63.
18 Stern, S.A., Shah, V.M. and Hardy, B.J. (1987) *Journal of Polymer Science, Part B: Polymer Physics Edition*, **25**, 1263–1298.
19 van Amerongen, G.J. (1964) *Rubber Chemistry and Technology*, **37**, 1065–1152.
20 Hirayama, Y., Kase, Y., Tanihara, N., Sumiyama, Y., Kusuki, Y. and Haraya, K. (1999) *Journal of Membrane Science*, **160**, 87–99.
21 Barbari, T.A., Koros, W.J. and Paul, D.R. (1989) *Journal of Membrane Science*, **42**, 69.
22 Puleo, A.C., Paul, D.R. and Kelley, S.S. (1989) *Journal of Membrane Science*, **47**, 301.
23 Raymond, P.C., Koros, W.J. and Paul, D.R. (1993) *Journal of Membrane Science*, **77**, 49.
24 Lin, H., Van Wagner, E., Freeman, B.D., Toy, L.G. and Gupta, R.P. (2006) *Science*, **311**, 639–642.
25 Koros, W.J., Chern, R.T., Stannett, V. and Hopfenberg, H.B. (1981) *Journal of Polymer Science: Polymer Physics Edition*, **19**, 1513–1530.
26 Puleo, A.C., Paul, D.R. and Wong, P.K. (1989) *Polymer*, **30**, 1357.
27 Mohr, J.M. and Paul, D.R. (1991) *Polymer*, **32**, 1236.
28 Michaels, A.S. and Bixler, H.J. (1961) *Journal of Polymer Science*, **50**, 393–412.
29 Michaels, A.S. and Bixler, H.J. (1961) *Journal of Polymer Science*, **50**, 413–439.
30 Mogri, Z. and Paul, D.R. (2001) *Polymer*, **42**, 7765–7780 and 7781–7789.
31 Bucknall, C.B. and Paul, D.R. (2000) *Polymer Blends: Formulation and Performance*, vols. **1** and **2**, John Wiley and Sons, New York.
32 Hadjichristidis, N., Pispas, S. and Floudas, G.A. (2003) *Block Copolymers: Synthetic Strategies, Physical Properties, and Applications*, John Wiley and Sons, Hoboken.
33 Newman, S. and Paul, D.R. (1978) *Polymer Blends*, vols. 1 and 2, Academic Press, New York.
34 Struik, L.C.E. (1978) *Physical Aging in Amorphous Polymers and Other Materials*, Elsevier Scientific Publishing Co, New York.
35 Hutchinson, J.M. (1997) *Progress in Polymer Science*, **125**, 23.
36 Curro, J.G., Lagasse, R.R. and Simha, R. (1982) *Macromolecules*, **15**, 1621.
37 van Krevelen, D.W. (1990) Thermophysical properties of polymers, in *Properties of Polymers: Their Correlation with Chemical Structure, Their Numerical Estimation and Prediction from Additive Group Contributions*, Elsevier, Amsterdam and New York.
38 Huang, Y. and Paul, D.R. (2006) *Macromolecules*, **39**, 1554.

39 Rezac, M.E., Pfromm, P.H., Costello, L.M. and Koros, W.J. (1994) *Industrial & Engineering Chemistry Research*, **32**, 1921.

40 Pfromm, P.H. and Koros, W.J. (1995) *Polymer*, **36**, 2379–2387.

41 Dorkenoo, K.D. and Pfromm, P.H. (1999) *Journal of Polymer Science Part B-Polymer Physics*, **37**, 2239–2251.

42 Chung, T.S. and Teoh, S.K. (1999) *Journal of Membrane Science*, **152**, 175.

43 Dorkenoo, K.D. and Pfromm, P.H. (2000) *Macromolecules*, **33**, 3747.

44 McCaig, M.S. and Paul, D.R. (2000) *Polymer*, **41**, 629–637 and 639–648.

45 Huang, Y. and Paul, D.R. (2004) *Journal of Membrane Science*, **244**, 167–178.

46 Huang, Y. and Paul, D.R. (2004) *Polymer*, **45**, 8377–8393.

47 Huang, Y. and Paul, D.R. (2005) *Macromolecules*, **38**, 10148–10154.

48 Huang, Y., Wang, X. and Paul, D.R. (2006) *Journal of Membrane Science*, **277**, 219–229.

49 Kim, J.H., Koros, W.J. and Paul, D.R. (2006) *Polymer*, **47**, 3094–3103 and 3104–3111.

50 Kim, J.H., Koros, W.J. and Paul, D.R. (2006) *Journal of Membrane Science*, **282**, 21–31 and 32–43.

51 Huang, Y. and Paul, D.R. (2007) *Journal of Polymer Science Part B-Polymer Physics*, **45**, 1390–1398.

52 Huang, Y. and Paul, D.R. (2007) *Industrial & Engineering Chemistry Research*, **46**, 2342–2347.

53 Kim, J.H., Jang, J. and Zin, W.C. (2000) *Langmuir*, **16**, 4064.

54 Pham, J.Q. and Green, P.F. (2003) *Macromolecules*, **36**, 1665–1669.

55 Forrest, J.A., Dalnoki-Veress, K., Stevens, J.R. and Dutcher, J.R. (1996) *Physical Review Letters*, **77**, 2002.

56 Park, C.H., Kim, J.H., Ree, M., Sohn, B.H., Jung, J.C. and Zin, W.C. (2004) *Polymer*, **45**, 4507.

57 Rowe, B.W., Freeman, B.D. and Paul, D.R. (2007) *Macromolecules*, **40**, 2806.

58 Baayens, L. and Rosen, S.L. (1972) *Journal of Applied Polymer Science*, **16**, 663.

59 Lin, H. and Freeman, B.D. (2005) *Journal of Molecular Structure*, **739**, 57–74.

60 Paul, D.R. and Clarke, R. (2002) *Journal of Membrane Science*, **208**, 269–283.

61 Kawakami, M., Iwanaga, H., Hara, Y., Iwamoto, M. and Kagawa, S. (1982) *Journal of Applied Polymer Science*, **27**, 2387–2393.

62 Blume, I. and Pinnau, I. (1990) U.S. Patent 4,963,165, Membrane Technology Research, Inc.

63 Qipeng, G., Hechang, X. and Dezhu, M. (1990) *Journal of Applied Polymer Science*, **39**, 2321–2330.

64 Okamoto, K., Umeo, N., Okamyo, S., Tanaka, K. and Kita, H. (1993) *Chemistry Letters*, **22**, 225–228.

65 Okamoto, K., Fujii, M., Okamyo, S., Suzuki, H., Tanaka, K. and Kita, H. (1995) *Macromolecules*, **28**, 6950–6956.

66 Li, J., Nagri, K., Nakagawa, T. and Wang, S. (1995) *Journal of Applied Polymer Science*, **58**, 1455–1463.

67 Okamoto, K., Yasugi, N., Kawabata, T., Tanaka, K. and Kita, H. (1996) *Chemistry Letters*, **25**, 613–614.

68 Suzuki, H., Tanaka, K., Kita, H., Okamoto, K., Hoshino, H., Yoshinaga, T. and Kusuki, Y. (1998) *Journal of Membrane Science*, **146**, 31–37.

69 Tsutsui, K., Yoshimizu, H., Tsujita, Y. and Kinoshita, T. (1999) *Journal of Applied Polymer Science*, **73**, 2733–2738.

70 Bondar, V.I., Freeman, B.D. and Pinnau, I. (1999) *Journal of Polymer Science Part B-Polymer Physics*, **37**, 2463–2475.

71 Yoshino, M., Ito, K., Kita, H. and Okamoto, K. (2000) *Journal of Polymer Science Part B-Polymer Physics*, **38**, 1707–1715.

72 Bondar, V.I., Freeman, B.D. and Pinnau, I. (2000) *Journal of Polymer Science Part B-Polymer Physics*, **38**, 2051–2062.

73 Kim, J.H., Ha, S.Y. and Lee, Y.M. (2001) *Journal of Membrane Science*, **190**, 179–193.

74 Barbi, V., Funari, S.S., Gehrka, R., Scharnagl, N. and Stribeck, N. (2003) *Macromolecules*, **36**, 749–758.

75 Metz, S.J., Mulder, M.H.V. and Wessling, M. (2004) *Macromolecules*, **37**, 4590–4597.

76 Lin, H. and Freeman, B.D. (2004) *Journal of Membrane Science*, **239**, 105–117.

77 Lin, H., Van Wagner, E., Swinnea, J.S., Freeman, B.D., Pas, S.J., Hill, A.J., Kalakknunnath, S. and Kalika, D.S. (2006) *Journal of Membrane Science*, **276**, 145–161.

78 Lin, H. and Freeman, B.D. (2006) *Macromolecules*, **39**, 3568–3580.

79 Lin, H., Van Wagner, E., Raharjo, R., Freeman, B.D. and Roman, I. (2006) *Advanced Materials*, **18**, 39–44.

5
Fundamentals in Electromembrane Separation Processes

Heinrich Strathmann

5.1
Introduction

Electromembrane processes such as electrolysis and electrodialysis have experienced a steady growth since they made their first appearance in industrial-scale applications about 50 years ago [1–3]. Currently desalination of brackish water and chlorine–alkaline electrolysis are still the dominant applications of these processes. But a number of new applications in the chemical and biochemical industry, in the production of high-quality industrial process water and in the treatment of industrial effluents, have been identified more recently [4]. The development of processes such as continuous electrodeionization and the use of bipolar membranes have further extended the range of application of electromembrane processes far beyond their traditional use in water desalination and chlorine-alkaline production.

The term 'electromembrane process' is used to describe an entire family of processes that can be quite different in their basic concept and their application. However, they are all based on the same principle, which is the coupling of mass transport with an electrical current through an ion permselective membrane. Electromembrane processes can conveniently be divided into three types: (1) Electromembrane separation processes that are used to remove ionic components such as salts or acids and bases from electrolyte solutions due to an externally applied electrical potential gradient. (2) Electromembrane synthesis processes that are used to produce certain compounds such as NaOH, and Cl_2 from NaCL due to an externally applied electrical potential and an electrochemical electrode reaction. (3) Eletectromembrane energy conversion processes that are to convert chemical into electrical energy, as in the H_2/O_2 fuel cell.

In this chapter only electromenbrane separation processes such as electrodialysis, electrodialysis with bipolar membranes and continuous electrodeionization will be discussed.

Membrane Operations. Innovative Separations and Transformations. Edited by Enrico Drioli and Lidietta Giorno

ISBN: 978-3-527-32038-7

5.2
The Structures and Functions of Ion-Exchange Membranes

The key components in electrodialysis and related processes are the ion-exchange membranes. There are three different types of ion-exchange membranes: (1) cation-exchange membranes that contain negatively charged groups fixed to the polymer matrix, (2) anion-exchange membranes that contain positively charged groups fixed to the polymer matrix, and bipolar membranes that are composed of an anion- and a cation-exchange layer laminated together.

In a cation-exchange membrane, the fixed negative charges are in electrical equilibrium with mobile cations in the interstices of the polymer as indicated in Figure 5.1, which shows schematically the structure of a cation-exchange membrane with negative charges fixed to the polymer matrix, and mobile cations and anions.

The mobile cations are referred to as counterions and the mobile anions that carry the same electrical charge as the polymer membrane that are more or less completely excluded from the membrane are referred to as co ions. Due to the exclusion of the co ions, a cation-exchange membrane is more or less impermeable to anions. Anion-exchange membranes carry positive fixed charges and exclude cations. Thus, they are more or less impermeable to cations. To what extent the co ions are excluded from an ion-exchange membrane depends on membranes as well as on solution properties. Bipolar membranes enhance the dissociation of water molecules into H^+ and OH^- ions and are used in combination with monopolar membranes for the production of acids and bases from the corresponding salts [5].

The most desired properties of ion-exchange membranes are: high permselectivity, low electrical resistance, good mechanical and form stability, and high chemical and thermal stability. In addition to these properties bipolar membranes should have high catalytic water dissociation rates.

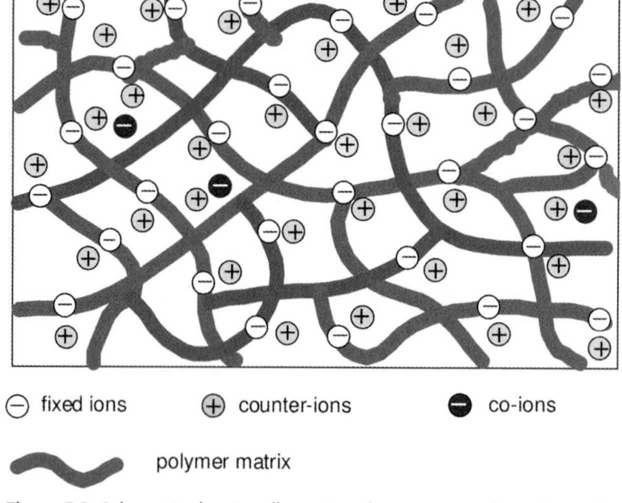

Figure 5.1 Schematic drawing illustrating the structure of a cation-exchange membrane.

5.2.1
Ion-Exchange Membrane Materials and Structures

Many of today's available membranes meet most of these requirements. In particular, the Nafion-type cation-exchange membrane has quite satisfactory properties for applications in the chlorine–alkaline electrolyses as well as in electrodialysis [6]. Anion-exchange membranes often show lower stability in strong alkaline solutions than cation-exchange membranes.

The properties of ion-exchange membranes are determined by two parameters, that is, the basic material they are made from and the type and concentration of the fixed ionic moiety. The basic material determines to a large extent the mechanical, chemical, and thermal stability of the membrane. Ion-exchange membranes are made today from hydrocarbon or partially halogenated hydrocarbon and perfluorocarbon polymers [7, 8].

The type and the concentration of the fixed ionic charges determine the permselectivity and the electrical resistance of the membrane, but they also have a significant effect on the mechanical properties of the membrane. The degree of swelling, especially, is effected by the type of the fixed charges and their concentration.

The following moieties are used as fixed charges in cation-exchange membranes:

$$-SO_3^- \quad -COO^- \quad -PO_3^{2-} \quad -PHO_2^- \quad -AsO_3^{2-} \quad -SeO_3^-.$$

In anion-exchange membranes fixed charges may be:

$$-\overset{+}{N}H_2R \quad -\overset{+}{N}HR_2 \quad -\overset{+}{N}R_3 \quad -\overset{+}{P}R_3 \quad -\overset{+}{S}R_2.$$

The sulfonic acid group is completely dissociated over nearly the entire pH range, while the carboxylic acid group is virtually undissociated in the pH range <3. The quaternary ammonium group again is completely dissociated over the entire pH range, while the secondary ammonium group is only weakly dissociated. Accordingly, ion-exchange membranes are referred to as being weakly or strongly acidic or basic in character.

Ion-exchange membranes can also be divided, according to their structure and preparation procedure, into homogeneous and heterogeneous membranes [4].

In homogeneous ion-exchange membranes the fixed-charged groups are evenly distributed over the entire membrane polymer matrix. Homogeneous membranes can be produced, for example, by polymerization or polycondensation of functional monomers such as phenolsulfonic acid, or by functionalizing a polymer such as polysulfone dissolved in an appropriate solvent by sulfonation.

Heterogeneous ion-exchange membranes have distinct macroscopic domains of ion-exchange resins in the matrix of an uncharged polymer. They can be produced by melting and pressing a dry ion-exchange resin with a polymer powder such as polyvinylchloride, or by dispersion of the ion-exchange resin in a polymer solution.

5.2.2
Preparation of Ion-Exchange Membranes

Ion-exchange membranes are ion-exchange resins in sheet form. There are, however, significant differences between ion-exchange resins and membranes as far as the mechanical properties and especially the swelling behavior are concerned. Ion-exchange resins are mechanically weak or tend to be brittle. Changes in the electrolyte concentration of an electrolyte in equilibrium with the ion-exchange resin may cause major changes in the water uptake and hence in swelling. These changes can not be tolerated in ion-exchange membranes that have to fit an apparatus under very different electrolyte concentrations and temperatures. The most common solution to this problem is the preparation of a membrane with a backing of a stable reinforcing material that gives the necessary strength and dimensional stability. Preparation procedures for making ion-exchange membranes are described in great detail in the literature [8–10] and are quite different for heterogeneous and homogeneous membranes.

5.2.2.1 Preparation Procedure of Heterogeneous Ion-Exchange Membranes

Ion-exchange membranes with a heterogeneous structure consist of fine ion-exchange particles embedded in an inert binder polymer such as polyethylene, phenolic resins, or polyvinylchloride. Heterogeneous ion-exchange membranes are characterized by the discontinuous phase of the ion-exchange material. The efficient transport of ions through a heterogeneous membrane requires either a contact between the ion-exchange particles or an ion-conducting solution between the particles. Heterogeneous ion-exchange membranes can easily be prepared by mixing an ion-exchange powder with a dry binder polymer and extrusion of sheets under the appropriate conditions of pressure and temperature or by dispersion of ion-exchange particles in a solution containing a dissolved film-forming binder polymer, casting the mixture into a film and then evaporating the solvent.

Heterogeneous ion-exchange membranes with useful low electrical resistances contain more than 65% by weight of the ion-exchange particles. Membranes that contain significantly less than 65 wt% ion-exchange particles have high electric resistance and membranes with significantly more resin particles have poor mechanical strength. Furthermore, heterogeneous membranes develop water-filled interstices in the polymer matrix during the swelling process that affects both the mechanical properties as well as the permselectivity.

The ion-exchange capacities of heterogeneous membranes are in the range of 1–2 equivalent per kilogram dry membrane and thus significantly lower than that of homogeneous membranes, which is between 2 and 3 equivalent per kilogram dry resin. In general, heterogeneous ion-exchange membranes have higher electrical resistances and lower permselectivity than homogeneous membranes.

5.2.2.2 Preparation of Homogeneous Ion-Exchange Membranes

Homogeneous ion-exchange membranes can be prepared by polymerization of monomers that contain a moiety that either is or can be made anionic or cationic, or by polymerization of a monomer that contains an anionic or a cationic moiety, or by introduction of anionic or cationic moieties into a polymer dissolved in a solvent by a chemical reaction, or grafting functional groups into a preformed polymer film [10].

A method of preparing both cation- and anion-exchange membranes, which is used for the preparation of commercial cation-exchange membranes, is the polymerization of styrene and divinylbenzene and its subsequent sulfonation according to the following reaction scheme:

$$H_2C{=}CH{-}C_6H_5 + H_2C{=}CH{-}C_6H_4{-}CH{=}CH_2 \xrightarrow[\text{sulfonation}]{\text{polymerization}} \text{cross-linked poly(styrene-divinylbenzene) with } {-}SO_3H \text{ groups}$$

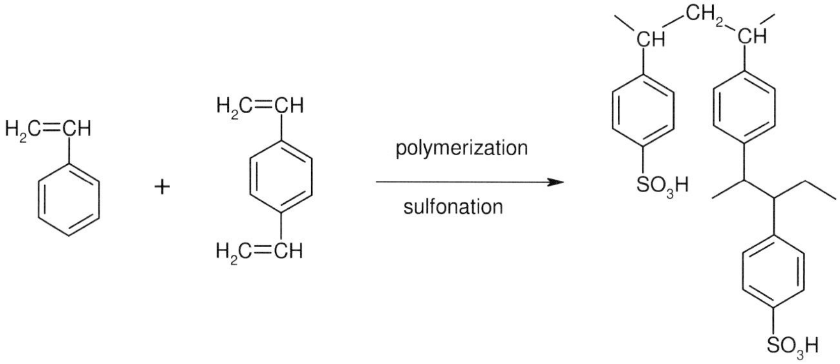

In a first step styrene is partially polymerized and cross-linked with divinylbenzene and then in a second step sulfonated with concentrated sulfuric acid. The obtained membranes show high ion-exchange capacity and low electrical resistance. To increase the mechanical strength the membrane is cast on a support screen.

A homogeneous anion-exchange membrane can be obtained by introducing a quaternary amine group into polystyrene by a chloromethylation procedure followed by an amination with a tertiary amine according to the following reaction scheme:

$$\text{polystyrene} \xrightarrow{CH_3CH_2OCH_2Cl} {-}C_6H_4{-}CH_2Cl \xrightarrow{(CH_3)_3N} {-}C_6H_4{-}CH_2{-}N^+(CH_3)_3\ Cl^-$$

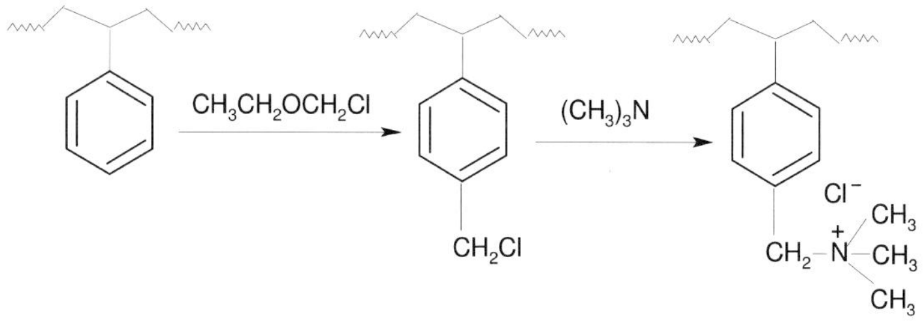

The membrane structures and their preparation described above are just two examples. There are many variations of the basic preparation procedure resulting in slightly different products. Instead of styrene, often substituted styrenes such as methylstyrene or phenyl-acetate are used instead of divinylbenzene monomers such as divinylacetylene or butadiene are used.

One of the technically and commercially most important cation-exchange membranes developed in recent years is based on perfluorocarbon polymers. Membranes of this type have extreme chemical and thermal stability and they are the key component in the chlorine–alkaline electrolysis as well as in most of today's fuel cells. They are prepared by copolymerization of tetrafluoroethylene with perfluorovinylether having a carboxylic or sulfonic acid group at the end of a side chain. There are several variations of a general basic structure commercially available today [11]. The various preparation techniques are described in detail in the patent literature.

Today's commercially available perfluorocarbon membranes have the following basic structure:

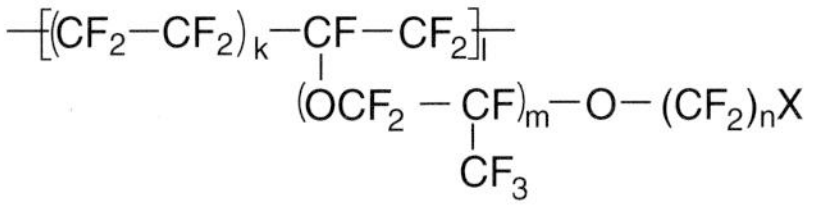

$$-\!\!\left[(CF_2-CF_2)_k-\underset{\displaystyle (OCF_2-\underset{\displaystyle CF_3}{\underset{|}{C}}F)_m-O-(CF_2)_nX}{\underset{|}{C}}F-CF_2\right]_l\!\!-$$

$k = 5\text{–}8$, $l = 600\text{–}1200$, $m = 1\text{–}2$, $n = 1\text{–}4$, $X = SO_3^-, COO^-$

The synthesis of the perfluorocarbon membranes is rather complex and requires a multistep process.

In addition to the various perfluorinated cation-exchange membranes an anion-exchange membrane has also been developed. The anion-exchange membrane has similar chemical and thermal properties to the perfluorinated cation-exchange membrane.

More recently cation-exchange membranes with good mechanical and chemical stability and well-controlled ion-exchange capacity are prepared by dissolving and casting a functionalized polymer such as sulfonated polysulfone, or sulfonated polyetheretherketone in an appropriate solvent, followed by casting the mixture into a film and then evaporation of the solvent [12].

To obtain membranes with different ion-exchange capacity the sulfonated polyetheretherketone or polysulfone can be mixed with unsulfonated polymer in a solvent such as *N*-methylpyrrolidone. By changing the ratio of the sulfonated to unsulfonated polymer the fixed-charge density can easily be adjusted to a desired value.

The sulfonated polysulfone as well as polyetheretherketone can be cast as a film on a screen. After the evaporation of the solvent a reinforced membrane with excellent chemical and mechanical stabilities and good electrochemical properties is obtained.

The anion-exchange membrane based on polysulfone can be prepared by halomethylation of the backbone polymer and subsequent reaction with a tertiary amine

5.2.2.3 **Special Property Membranes**

In addition to the monopolar membrane described above a large number of special property membranes are used in various applications such as low-fouling anion-exchange membranes used in certain wastewater treatment applications or composite membranes with a thin layer of weakly dissociated carboxylic acid groups on the surface used in the chlorine–alkaline production, and bipolar membranes composed of a laminate of an anion- and a cation-exchange layer used in the production of protons and hydroxide ions to convert a salt in the corresponding acids and bases. The preparation techniques are described in detail in numerous publications [13–15].

5.3 Transport of Ions in Membranes and Solutions

The transport rate of a component in a membrane and a solution is determined by its concentration, its mobility in a given environment and by the driving force or forces acting on the component. The concentration and mobility of a component are determined by its interaction with other components in its surrounding. The driving forces for the transport are gradients in the electrochemical potential. In electrolyte solutions the electrostatic forces must always be balanced, that is, the electroneutrality prevails on a macroscopic scale. For applying an electrical potential in an electrolyte solution two electron conductors must be in contact with an electrolyte. At the electrode/electrolyte interface the electron conductance is converted to an ionic conductance by an electrochemical reaction.

5.3.1 Electric Current and Ohm's Law in Electrolyte Solutions

In electromembrane processes the anions move towards the anode where they are oxidized by releasing electrons to the electrode in an electrochemical reaction. Likewise, the positively charged cations move towards the cathode where they are reduced by receiving electrons from the electrode in an electrochemical reaction. Thus, the transport of ions in an electrolyte solution and ion-exchange membrane between electrodes results in a transport of electrical charges, that is, an electrical current which can be described by the same mathematical relation as the transport of electrons in a metallic conductor, that is, by Ohm's law that is given by:

$$U = RI \tag{5.1}$$

Here, U is the electrical potential between two electron sources, for example, between two electrodes, I is the electrical current between the electron sources, and R the electrical resistance.

The resistance R is a function of the specific resistance of the material, the distance between the electron sources, and the cross-sectional area of the material through which the electric current passes. It is given by:

$$R = \rho \frac{l}{q} \tag{5.2}$$

Here, R is the overall resistance, ρ is the specific resistance, l is the length, and q the cross-sectional area of the conducting material.

The reversal of the resistance and of the specific resistance, respectively, is the conductivity and the specific conductivity, thus is:

$$S = \frac{1}{R} \quad \text{and} \quad \kappa = \frac{1}{\rho} \tag{5.3}$$

Here, S is the conductivity and κ the specific conductivity.

The conductivity of electrons in metal conductors, however, is generally 3–5 orders of magnitude higher than that of ions in electrolyte solutions. Furthermore, the conductivity of metals is decreasing with increasing temperature, while the conductivity of electrolyte solutions is increasing with temperature. The most important difference between electron and ion conductivity, however, is the fact that ion conductivity is always coupled with a transport of mass while, due to the very small mass of an electron, virtually no mass is transported in an electron conductor.

The conductivity of electrolyte solutions depends on the concentration and the charge number of the ions in the solution. It is expressed as the molar or equivalent conductivity or molar conductivity, which is given by:

$$\Lambda_{\text{mol}} = \frac{\kappa}{C_s} \quad \text{and} \quad \Lambda_{\text{eq}} = \frac{\kappa}{C_s \dfrac{(z_a v_a + z_c v_c)}{2}} \tag{5.4}$$

Here, Λ_{mol} and Λ_{eq} are the molar and the equivalent conductivity, C is the molar concentration of the electrolyte in the solution, z_a and z_c are the charge numbers of the anion and cation, respectively, and v_a and v_c are the stoichoimetric coefficients of the anion and cation, respectively.

The stoichoimetric coefficient gives the number of anions and cations in a mole electrolyte and the charge number gives the number of charges related to an ion. For example, for NaCl v_c and v_a are identical and 1 and also z_a and z_c are 1. However, for $MgCl_2$ v_c is 1 and v_a is 2, and z_c is 2 and z_a is 1.

The number of electrical charges carried by all the ions of an electrolyte under the driving force of an electrical potential gradient through a certain area A in the direction of transport is given by:

$$J_e = \sum_i z_i u_i v_i C F \frac{\Delta\phi}{\Delta z} = \sum_i z_i F J_i = \sum_i z_i v_i C \lambda_{\text{eq}} \frac{\Delta\phi}{\Delta z} \tag{5.5}$$

Here, J_e is the flux of electrical charges and J_i that of the individual ions, z, u, and v are the charge number, the ion mobility, and the stoichiometric coefficient, respectively, C is the concentration of the electrolyte, $\Delta\varphi$ and Δz are the potential difference and the distance between two points in the z direction, F is the Faraday constant, which is $F = 96\,485$ [C eq^{-1}], and λ_{eq} is the equivalent conductivity.

Thus, the flux of electrical charges represents an electrical current, which is according to Ohm's law given by:

$$I = \frac{U}{R} = \sum_i z_i F J_i A = \sum_i z_i v_i C \lambda_{\text{eq}} \frac{\Delta\varphi}{l} A = \kappa \frac{\Delta\varphi}{l} = iA \tag{5.6}$$

Here, I is the current, U is the applied voltage, R is the resistance, A is the area through which the current passes, $\Delta\varphi$ is the voltage difference between two points, and l is the distance between the two points, k is the conductivity, A is the cross-sectional area of the conducting media, and i is the current density

5.3.2
Mass Transport in Membranes and Solutions

To describe the mass transport in an electrolyte solution or in an ion-exchange membrane, three independent fluxes must be considered, that is, the fluxes of the cations the flux of anions, and the flux of the solvent [16]. The transport of ions is the result of an electrochemical potential gradient and the transport of the solvent through the membrane is a result of osmotic and electro-osmotic effects.

5.3.2.1 The Driving Force and Fluxes in Electromembrane Processes

The driving force for the flux of a component in electromembrane processes is a gradient in their electrochemical potential which is given at constant temperature by:

$$\frac{d\tilde{\mu}_i}{dz} = \frac{d\mu_i}{dz} + \frac{d\phi}{dz} = \bar{V}_i\frac{dp}{dz} + RT\frac{d\ln a_i}{dz} + z_i F\frac{d\phi}{dz} \tag{5.7}$$

Here $d\tilde{\mu}_i$, $d\eta_i$, $d\mu_i$, $d\ln a_i$, $d\varphi$ and dp are the gradients of the electrochemical potential, the chemical potential, the activity, the electrical potential and of the hydrostatic pressure, F is the Faraday constant and R the gas constant, and T the temperature.

The mass transport in electromembrane processes at constant pressure and temperature can be described as a function of the driving force by a phenomenological equation [17], that is,:

$$J_i = \sum_i L_{ik}\frac{d\tilde{\mu}_k}{dz} = \sum_i L_{ik}\left(RT\frac{d\ln a_i}{dz} + z_i F\frac{d\phi}{dz}\right) \tag{5.8}$$

Here, L_{ik} is a phenomenological coefficient relating the driving force to the corresponding flux, the subscripts i and k refer to various components in the system.

Assuming an ideal solution in which the activity of a component is identical to its concentration and no kinetic coupling occurs between individual fluxes, Equation 5.8 becomes identical with the Nernst–Planck flux equation [18], which is given by:

$$J_i = -D_i\left(\frac{dC_i}{dz} + \frac{z_i F C_i}{RT}\frac{d\varphi}{dz}\right) \tag{5.9}$$

Here, D_i is the diffusion coefficient of the component i which is related to the phenomenological coefficient by: $D_i = (L_{ii}/RT)$.

The first term $D_i(dC_i/dz)$ represents the diffusion, the second term $D_i(z_i C_i F/RT)$ $(d\varphi/dz)$ the migration of a component. Thus, the Nernst–Planck equation is an approximation of the more general phenomenological equation.

5.3.2.2 Electrical Current and Fluxes of Ions

The electric current in an electrolyte solution is transported by ions only. as described in Equation 5.6:

$$i = \frac{I}{A} = F\sum_i z_i J_i \tag{5.10}$$

Here, i is the current density, I the current, A the membrane surface, F the Faraday constant, J the flux, and z the valence, the subscript i refers to cations and anions.

Introducing Equation 5.8 and 5.9 and rearranging leads to:

$$i = F\sum_i z_i J_i = F^2 \sum_i z_i^2 \frac{C_i D_i}{RT}\left(\frac{RT}{z_i C_i F}\frac{\mathrm{d}C_i}{\mathrm{d}z} + \frac{\mathrm{d}\varphi}{\mathrm{d}z}\right) \tag{5.11}$$

Here, i is the current density, C is the concentration, F is the Faraday constant, φ is the electrical potential, z is the valence, D is the ion diffusivity, R is the gas constant, T is the absolute temperature, and the subscript i refers to anions and cations.

The term $(RT/z_iC_iF)(\mathrm{d}C_i/\mathrm{d}z)$ has the dimensions of an electrical potential gradient and represents the concentration potential that is established between two electrolyte solutions of different concentrations.

5.3.2.3 The Transport Number and the Membrane Permselectivity

In an electrolyte solution the current is carried by both ions. However, cations and anions usually carry different portions of the overall current. In ion-exchange membranes the current is carried preferentially by the counterions.

The fraction of the current that is carried by a certain ion is expressed by the ion transport number, which is given by:

$$T_i = \frac{z_i J_i}{\sum_i z_i J_i} \tag{5.12}$$

Here, T_i is the transport number of the component i, J_i is its flux, and z_i its valence.

The transport number T_i indicates the fraction of the total current that is carried by the ion i, the sum of the transport number of all ions in a solution is 1.

The membrane permselectivity is an important parameter for determining the performance of a membrane in a certain ion-exchange membrane process. It describes the degree to which a membrane passes an ion of one charge and retains an ion of the opposite charge. The permselectivity of cation- and anion-exchange membranes can be defined by the following relations [4]:

$$\Psi^{\mathrm{cm}} = \frac{T_{\mathrm{c}}^{\mathrm{cm}} - T_{\mathrm{c}}}{T_{\mathrm{a}}} \quad \text{and} \quad \Psi^{\mathrm{am}} = \frac{T_{\mathrm{a}}^{\mathrm{am}} - T_{\mathrm{a}}}{T_{\mathrm{c}}} \tag{5.13}$$

Here, Ψ is the permselectivity of a membrane, T is the transport number, the superscripts cm and am refer to cation- and anion-exchange membranes, and the subscripts c and a refer to cation and anion, respectively.

An ideal permselective cation-exchange membrane would transmit positively charged ions only, that is, for a transport number of a counterion in a cation-exchange membrane is $T_{\mathrm{c}}^{\mathrm{cm}} = 1$ and the permselectivity of the membrane is $\Psi^{\mathrm{cm}} = 1$. The permselectivity approaches zero when the transport number within the membrane is identical to that in the electrolyte solution, that is, for $T_{\mathrm{c}}^{\mathrm{cm}} = T_{\mathrm{c}}$ is $\Psi^{\mathrm{cm}} = 0$. For the anion-exchange membrane the corresponding relation holds.

The transport number of a certain ion in the membrane is proportional to its concentration in the membrane that again is a function of its concentration in the solutions in equilibrium with the membrane phase, due to the Donnan exclusion. Thus, the selectivity of ion-exchange membranes results from the exclusion of co ions from the membrane phase.

The concentration of a co ion in an ion-exchange membrane can be calculated from the Donnan equilibrium. For a monovalent salt and a dilute salt solution and assuming the activity coefficients of the salt in the membrane and the solution to be 1, the co ion concentration in the membrane is given to a first approximation by:

$$^{m}C_{co} = \frac{^{s}C_{s}^{2}}{C_{fix}} \tag{5.14}$$

Here C is the concentration, the subscripts co, s and fix refer to co ion, salt and fixed ion of the membrane, the superscripts s and m refer to membrane and solution.

Equation 5.14 indicates that the co ion concentration in the membrane and with that the permselectivity of the membrane is decreasing with salt concentration in the solution and will vanish when the salt concentration in the solution is identical to the fixed ion concentration of the membrane.

5.3.2.4 **Membrane Counterion Permselectivity**

The transport number of counterions in an ion-exchange membrane is always quite high compared to that of the co-ionco-ions. But the transport number of different counterions can be quite different, too. The transport rates of ions in a solution or through a membrane are determined by their concentration and mobility in the membrane. The concentration of the counterions is always close to the concentration of the fixed charges of the membrane. The mobility of the ions in the membrane depends mainly on the radius of the hydrated ions and the membrane structure. The mobility of different ions in an aqueous solution does not differ very much from each other. An exception is the H^+ and OH^- ions. Their mobility is about a factor 5 to 8 higher than that of other ions. This exceptionally high mobility of the H^+ ion can be explained by the transport mechanism of protons and hydroxide ions. Because of the molecular interaction of water dipoles with electrical charges, protons form hydronium ions. Common salt ions move with their hydrate shell through the solution. The proton, however, is transported mostly via a so-called tunnel mechanism from one hydronium ion to the next water molecule. This explains not only the extraordinarily high mobility of protons but it is also one of the reasons for the high permeability of anion-exchange membranes for protons, while these membranes generally have a very low permeability for salt cations. The same mechanism also holds true for the transport of hydroxide ions and thus the permeability of hydroxide ions in an aqueous solution and also in a cation-exchange membrane is much higher than that of other salt anions. Because protons and hydroxide ions are transported only to a small extent as individual ions surrounded by a hydration shell they contribute very little to the electro-osmotic transport of water, and their water-transport number is always quite low.

The permselectivity of an ion-exchange membrane for different counterions is determined by the concentration and the mobility of the different ions in the membrane as indicated earlier. The concentration of the different counterions in the membrane is determined mainly by electrostatic effects referred to as 'electroselectivity' [9]. The mobility depends on the size of the hydrated ion.

A typical counterion-exchange sequence of a cation-exchange membrane containing SO_3^- group as fixed charge is:

$$Ba^{2+} > Pb^{2+} > Sr^{2+} > Ca^{2+} > Mg^{2+} > Ag^{+} > K^{+} > NH_4^{+} > Na^{+} > Li^{+}$$

A similar counterion-exchange sequence is obtained for anions in an anion-exchange membrane containing quaternary ammonium groups as fixed charges:

$$I^{-} > NO_3^{-} > Br^{-} > Cl^{-} > SO_4^{2-} > F^{-}$$

The permselectivity is the product of ion-exchange selectivity and mobility selectivity. The mobility of different ions is determined mainly by steric effects, that is, the size of the ions and the cross-linking density of the membrane [4].

5.3.2.5 Water Transport in Electrodialysis

Water transport in electrodialysis from the diluate to the concentrate process stream can affect the process efficiency significantly. If a convective flux as a result of pressure differences between flow streams can be excluded there are still two sources for the transport of water from the diluate to the concentrate solution. The first one is the result of osmotic-pressure differences between the two solutions, and the second is due to electro-osmosis that results from the coupling of water to the ions being transported through the membrane due to the driving force of an electrical potential.

Each of the two fluxes may be dominant depending on the permselectivity of the ion-exchange membrane, the concentration gradient, and the current density. In a highly permselective membrane and with moderate differences in the salt concentration in the two solutions separated by the membrane the electro-osmotic flux is dominating and generally much higher than the osmotic solvent flux. In electrodialysis the water flux due to electro-osmosis can be expressed by a solvent transport number which gives the number of water molecules transported by one ion:

$$J_w = {}^{m}T_w \sum_i J_i \tag{5.15}$$

Here, ${}^{m}T_w$ is the water transport number, J_w is the water flux, and J_i is the flux of ions through a given membrane.

The water transport number thus is:

$${}^{m}T_w = \frac{J_w}{\sum_i J_i} \tag{5.16}$$

The water-transport number refers to the number of water molecules transferred by one ion through a given membrane. It depends on the membrane and on the electrolyte, that is, on the size of the ions, their valence, and their concentration in the solution. In aqueous salt solutions and commercial ion-exchange membranes the water transport number is of the order of 4–8, that is, one mole of ions transports about 4–8 moles of water through a typical commercial ion-exchange membrane.

5.4
The Principle of Electromembrane Processes

In this chapter only electromembrane separation processes such as electrodialysis, electrodialysis with bipolar membranes, and continuous electrodeionization will be discussed.

5.4.1
Electrodialysis

Electrodialysis is the most important electromembrane process and one of the first membrane processes used for desalination of brackish water to produce high-quality potable water at acceptable costs on a large commercial scale. Today, the process has found a multitude of applications in preconcentration of seawater for the production of table salt or in recovering valuable constituents from industrial effluents [19]. The principle of electrodialysis is illustrated in Figure 5.2 which shows a schematic diagram of an electrodialysis stack consisting of a series of anion- and cation-exchange membranes arranged in an alternating pattern to form individual cells between an anode and a cathode. If an ionic solution such as an aqueous salt solution is pumped through these cells and an electrical potential is established between the anode and cathode, the positively charged cations migrate towards the cathode and the negatively charged anions towards the anode. The cations permeate the cation-exchange membrane but are retained by the anion-exchange membrane. Likewise, the negatively charged anions permeate the anion-exchange membrane and are retained by the cation-exchange membrane. The overall result is an increase in the

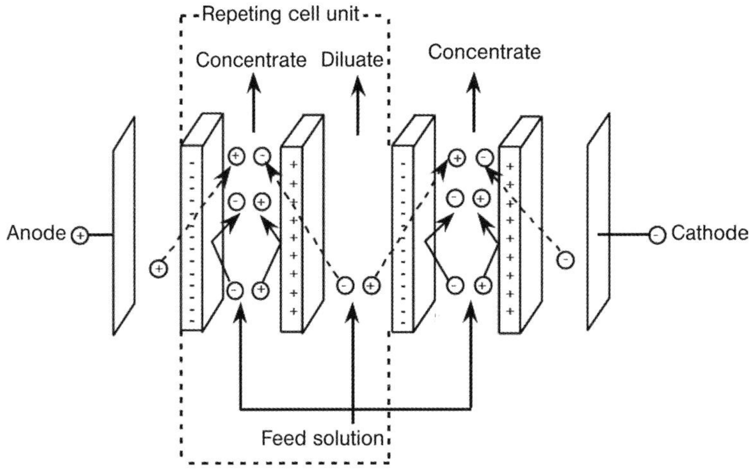

Figure 5.2 Schematic diagram illustrating the principle of desalination by electrodialysis in a stack with cation- and anion-exchange membranes in alternating series between two electrodes.

ion concentration in alternate compartments, while the other compartments simultaneously become depleted. The depleted solution is generally referred to as the diluate and the concentrated solution as the brine or the concentrate. The driving force for the ion transport in the electrodialysis process is the applied electrical potential between the anode and cathode. The total space occupied by the diluate and the concentrated solution and the anion- and cation-exchange membranes separating the solutions make up a cell pair that represents a repeating unit between the electrodes.

5.4.1.1 Electrodialysis System and Process Design

The efficiency of electrodialysis is determined to a large extent by the properties of the membranes. But it is also affected by the process and system design that determine the limiting current density, the current utilization, the concentration polarization and the overall efficiency and costs [20, 21].

The electrodialysis stack A key element in electrodialysis is the so-called stack, which is a device to hold an array of membranes between the electrodes that the streams being processed are kept separated. A typical electrodialysis stack used in water desalination contains 100–300 cell pairs stacked between the electrodes. The electrode containing cells at both ends of a stack are often rinsed with a separate solution which does not contain Cl^- ions to avoid chlorine formation.

The membranes in an electrodialysis cell are separated by spacer gaskets as indicated in Figure 5.3, which shows schematically the design of a so-called sheet flow electrodialysis stack. The spacer gasket consists of a screen that supports the membranes and controls the flow distribution in the cell and a gasket that seals the cell to the outside and also contains the manifolds to distribute the process fluids in

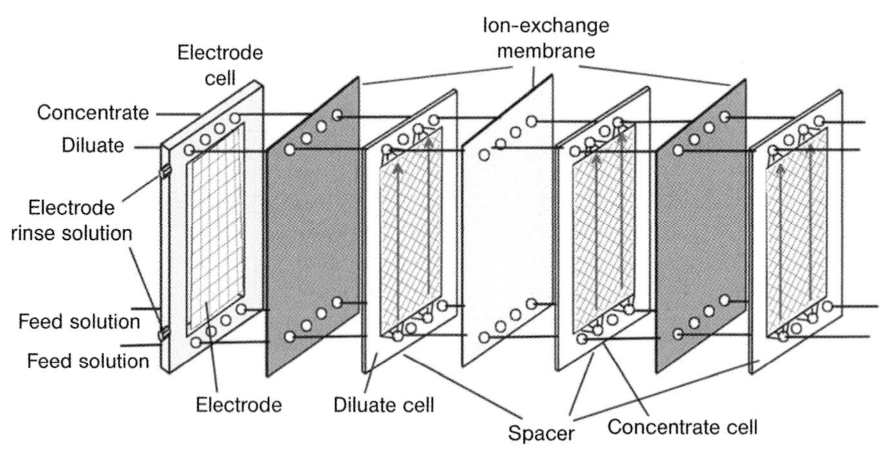

Figure 5.3 Exploded view of a sheet-flow-type electrodialysis stack arrangement indicating the individual cells and the spacer gaskets containing the manifold for the distribution of the different flow streams.

the different compartments. To minimize the resistance of the solution in the cell the distance between two membranes is kept as small as possible and is in the range of 0.5–2 mm in industrial electrodialysis stacks. A proper electrodialysis stack design provides a maximum effective membrane area per unit stack volume and ensures uniform flow distribution and mixing of the solutions to minimize concentration polarization at the membrane surfaces, but also minimizes the pressure loss of the solution flow in the stack.

concentration polarization and limiting current density The limiting current density is the maximum current that may pass through a given cell pair area without detrimental effects. If the limiting current density is exceeded, the electric resistance in the diluate will increase and water dissociation may occur at the membrane surface that can lead to pH changes in the solutions and effect the current utilization.

The limiting current density is determined by concentration-polarization effects at the membrane surface in the diluate containing compartment that in turn is determined by the diluate concentration, the compartment design, and the feed-flow velocity. Concentration polarization in electrodialysis is also the result of differences in the transport number of ions in the solution and in the membrane. The transport number of a counterion in an ion-exchange membrane is generally close to 1 and that of the co ion close to 0, while in the solution the transport numbers of anion and cations are not very different.

At the surface of a cation-exchange membrane facing the diluate solution the concentration of ions in the solution is reduced because of the lower transport number of the cations in the solution than in the membrane. Because of the electroneutrality requirements the number of anions is reduced in the boundary layer by migration in the opposite direction. The net result is a reduction of the electrolyte concentration in the solution at the surface of the membrane and a concentration gradient is established in the solution between the membrane surface and the well-mixed bulk. This concentration gradient results in a diffusive electrolyte transport. A steady-state situation is obtained when the additional ions that are needed to balance those removed from the interface due to the faster transport rate in the membrane are supplied by the diffusive transport. The other side of the cation-exchange membrane is facing the concentrate solution and here the opposite effect occurs and the electrolyte concentration at the membrane surface is increased accordingly. The concentration polarization is limited to the laminar boundary layer at the membrane surface, which is very thin due to turbulent mixing of the bulk solution. The effect of concentration polarization is illustrated in Figure 5.4 which shows the salt concentration profiles and the fluxes of cations and anions in the concentrate and diluate solution at the surface of a cation-exchange membrane.

The symbols J and C in Figure 5.5 denote the fluxes and the concentration of ions, the superscripts mig and diff refer to migration and diffusion, the superscripts d and c refer to diluate and concentrate solution, and the superscripts b and m refer to bulk phase and membrane surface, respectively, the subscripts a and c refer to anion and cation.

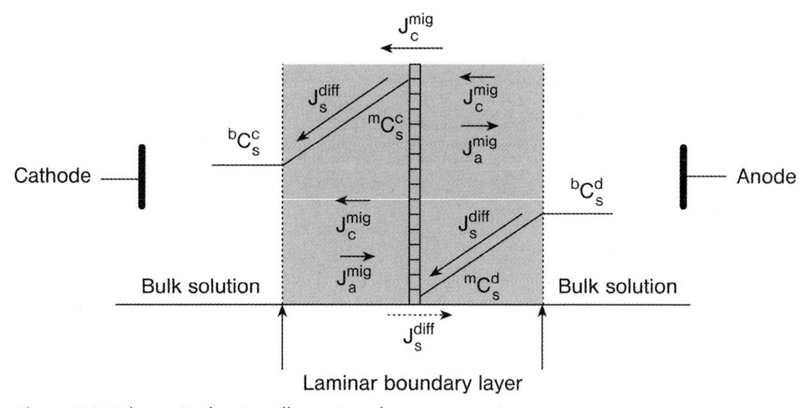

Figure 5.4 Schematic drawing illustrating the concentration profiles of a salt in the laminar boundary layer on both sides of a cation-exchange membrane and the flux of ions in the solutions and the membrane.

The concentration polarization occurring in electrodialysis, that is, the concentration profiles at the membrane surface can be calculated by a mass balance taking into account all fluxes in the boundary layer and the hydrodynamic conditions in the flow channel between the membranes. To a first approximation the salt concentration at the membrane surface can be calculated and related to the current density by applying the so-called Nernst film model, which assumes that the bulk solution between the laminar boundary layers has a uniform concentration, whereas the concentration in the boundary layers changes over the thickness of the boundary layer. However, the concentration at the membrane surface and the boundary layer thickness are constant along the flow channel from the cell entrance to the exit. In a practical electrodialysis stack there will be entrance and exit effects and concentration

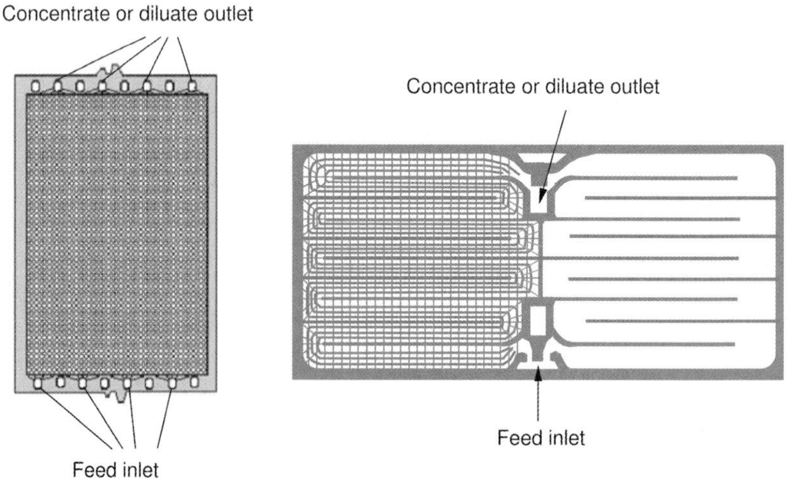

Figure 5.5 Schematic drawing illustrating the sheet-flow and a tortuous-path spacer concept.

differences between the solutions in the entrance and exit region of the cell, and the idealized model hardly exists. Nevertheless, the Nernst model provides a very simple approach to the mathematical treatment of the concentration polarization, which results in an expression for the current density as a function of the bulk solution concentration, the transport number of the ions, the diffusion coefficient of the electrolyte and the thickness of the laminar boundary layer [20].

$$i = \frac{z_i F D_i}{\left(T_i^{\mathrm{m}} - T_{i_i}^{\mathrm{s}}\right)} \frac{\Delta C_i^{\mathrm{d}}}{\Delta z} \tag{5.17}$$

Here, T is the transport number of the counterion, ΔC is the concentration difference between the solution in the diluate at the membrane surface and in the bulk, D is the diffusion coefficient, T is the transport number, F is the Faraday constant, z is the charge number, and Δz is the boundary layer thickness, the subscript i refers to cations or anions; the superscripts d, m and s refer to diluate, membrane and solution, respectively.

When the flow conditions are kept constant the boundary layer will be constant and the current density will reach a maximum value independent of the applied electrical potential gradient if the counterion concentration and thus the salt concentration at the membrane surface become 0. The maximum current density is referred to as the limiting current density. Thus is $i = i_{\mathrm{lim}}$ for ${}^{m}C_{\mathrm{s}}^{\mathrm{d}} \rightarrow 0$ and

$$i_{\mathrm{lim}} = \frac{z_i F D_{\mathrm{s}}}{\left(T_i^{\mathrm{m}} - T_i\right)} \left(\frac{{}^{b}C_{\mathrm{s}}^{\mathrm{d}}}{\Delta z}\right) \tag{5.18}$$

Here, i_{lim} is the limiting current density, ${}^{b}C_{\mathrm{s}}^{\mathrm{d}}$ is the salt concentration of the diluate in the bulk solution, Δz is the thickness of the laminar boundary layer, T^{m} and T^{s} are the transport numbers in the membrane and the solution, D_{s} is the salt diffusion coefficient in the solution, F is the Faraday constant, z is the charge number, and the subscript i refers to cation and anion.

Exceeding the limiting current density in practical applications of electrodialysis can affect the efficiency of the process severely by increasing the electrical resistance of the solution and causing water dissociation, which leads to changes of the pH values of the solution causing precipitation of metal hydroxide on the membrane surface.

Since the thickness of the laminar boundary in an electrodialysis stack is difficult to determine in an independent measurement, the limiting current density in practical application is generally not calculated by Equation 5.18 but by an experimentally determined relation which describes the limiting current density as a function to the feed-flow velocity in the electrodialysis stack [4]. The limiting current density is expressed by:

$$i_{\mathrm{lim}} = a\, u^{\mathrm{b}} F C_{\mathrm{s}}^{\mathrm{d}} \tag{5.19}$$

Here, $C_{\mathrm{s}}^{\mathrm{d}}$ is the concentration of the solution in the diluate cell, u is the linear flow velocity of the solution through the cells parallel to the membrane surface, F is the Faraday constant, and a and b are characteristic constants for a given stack design and must be determined experimentally. This is done in practice by measuring the limiting current density in a given stack configuration and constant feed solution salt concentrations as a function of the feed-flow velocity.

Current utilization In practical application electrodialysis is affected by incomplete current utilization. The reasons for the incomplete current utilization are poor membrane permselectivity, parallel current through the stack manifold, and water transport by convection and due to osmosis and electro-osmosis. In a well-designed stack with no pressure difference between diluate and the concentrate convective water transport is negligibly low and also the current through the manifold can be neglected. Under these conditions the overall current utilization is given by:

$$\xi = n\left(\psi^{cm} T_a^s + \psi^{am} T_c^s\right)\left(1 - \left[T_w^{cm} + T_w^{am}\right] \bar{V}_w \left(C_s^c - C_s^d\right)\right) \qquad (5.20)$$

Here, ξ is the current utilization, ψ is the membrane permselectivity, T is the transport number, n is the number of cell pairs in the stack, $\bar{V}_w$ is the partial molar volume of water, and C is the concentration, a, c, s and w refer to anion, cation, solution and water, respectively, and the superscripts cm, am, c, and d refer to cation-exchange membrane, anion-exchange membrane, concentrate and diluate.

Electrodialysis equipment and process design The performance of electrodialysis in practical applications is not only a function of membrane properties but is also determined by the equipment and overall process design. As far as the stack design is concerned there are two major concepts used on a large scale. One is the sheet-flow concept, which is illustrated in Figure 5.3 and the other is the so-called tortuous path concept, which is illustrated in Figure 5.5.

The main difference between the sheet-flow and the tortuous-path flow spacer is that in the sheet-flow spacer the compartments are vertically arranged and the process path is relative short. The flow velocity of the feed is between 2 and 4 cm/s and the pressure loss correspondingly low, that is, between 0.2 and 0.4 bars. In the tortuous-path flow stack, the membrane spacers are horizontally arranged and have a long serpentine cut-out that defines a long narrow channel for the fluid path. The feed-flow velocity in the stack is relatively high, that is, between 6 and 12 cm/s, which provides a better control of concentration polarization and higher limiting current densities, but the pressure loss in the feed-flow channels is quite high, that is, between1 and 2 bars. However, higher velocities help to reduce the deposition of suspended solids such as polyelectrolytes, humic acids, surfactants, and biological materials on the membrane surface.

In the practical application of electrodialysis there are two main process operation modes. The first one is referred to as the unidirectional electrodialysis and the second as electrodialysis reversal [22]. In a unidirectional operated electrodialysis system the electric field is permanently applied in one direction and the diluate and concentrate cells are also permanently fixed over the period of operation. Unidirectional operated electrodialysis plants are rather sensitive to membrane fouling and scaling and often require a substantial feed-solution pretreatment and stack-cleaning procedures in the form of periodical rinsing of the stack with acid or detergent solutions. The unidirectional operating concept is mainly used today for applications in the

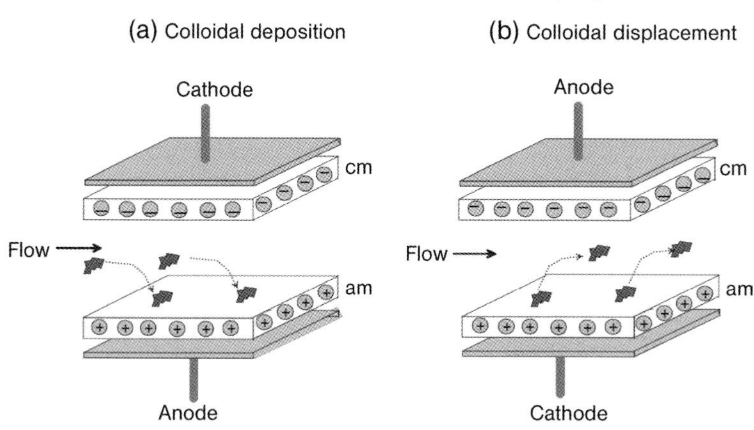

Figure 5.6 Schematic drawing illustrating the removal of deposited negatively charged colloidal components from the surface of an anion-exchange membrane by reversing the electric field.

food and drug industry where often solutions contain valuable components that must be recovered in the concentrate or in the diluate. In desalination of brackish or surface waters generally electrodialysis reversal is applied, which always results in some loss of the product water. In the electrodialysis reversal operating mode the polarity of the electric field applied to the electrodialysis stack is reversed in certain time intervals. Simultaneously the flow streams are reversed, that is, the diluate cell becomes the concentrate cell and vice versa with the result that matter being precipitated at the membrane surface will be redisolved and removed with the flow stream passing through the cell [22].

The principle of the electrodialysis reversal operating mode is illustrated in Figure 5.6 that shows an electrodialysis cell formed by a cation- and anion-exchange membrane between two electrodes. If an electric field is applied to a feed solution containing negatively charged particles or large organic anions these components will migrate to the anion-exchange membrane and be deposited on its surface to form a so-called 'fouling layer' that can increase the resistance of the membrane dramatically. If the polarity is reversed the negatively charged components will now migrate away from the anion-exchange membrane back into the feed stream and the membrane properties are restored. This procedure has been very effective not only for the removal of precipitated colloidal materials but also for removing precipitated salts and is used today in almost all electrodialysis water-desalination systems.

However, reversing the polarity of a stack has to be accompanied with a reversal of the flow streams. This always leads to some loss of product and requires a more sophisticated flow control. The flow scheme of an electrodialysis plant operated with reversed polarity is shown in Figure 5.7. In the reverse-polarity operating mode, the hydraulic flow streams are reversed simultaneously, that is, the diluate cell will become the brine cell and vice versa. In this operating mode, the polarity of the current is changed at specific time intervals ranging from a few minutes to several hours.

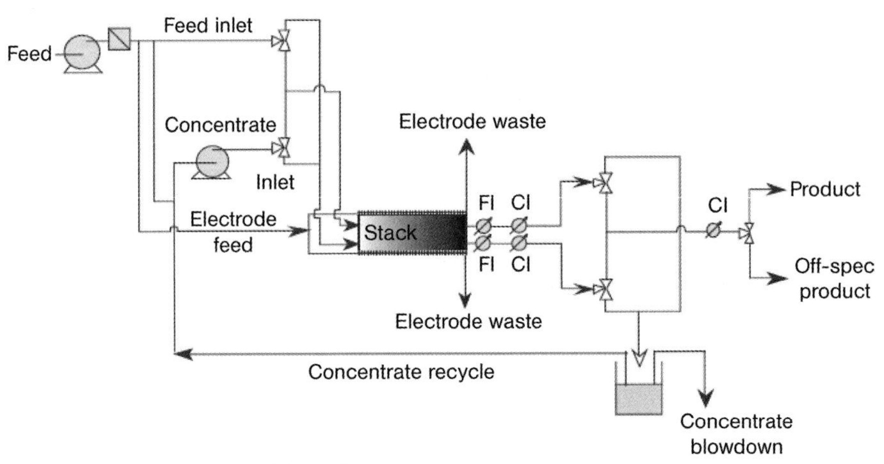

Figure 5.7 Flow scheme of electrodialysis reversal in a continuous operating mode with the feed solution also used as electrode rinse.

The advantage of the reverse-polarity operating mode is that precipitation in the brine cells will be re-dissolved when the brine cell becomes the diluate cell in the reverse operating mode. During the reversal of the polarity and the flow streams, there is a brief period when the concentration of the desalted product exceeds the product quality specification. The product water outlet has a concentration sensor that controls an additional three-way valve. This valve diverts highly concentrated product to waste and then, when the concentration returns to the specified quality, directs the flow to the product outlet. Thus, in electrodialysis reversal there is always a certain amount of the product lost to the waste stream. This is generally no problem in desalination of brackish water. It might, however, be not acceptable in certain applications in the food and drug industry when feed solutions with high value products are processed.

The degree of desalination that can be achieved in passing the feed solution through a stack is a function of the solution concentration, the applied current density, and the residence time of the solution in the stack. If the flow rates of diluate and concentrate through the stack are relatively high the degree of desalination or concentration that can be achieved in a single path is quite low and often not sufficient to meet the required product qualities.

If this is the case the electrodialysis can be operated as a process with feed and bleed in which the diluate or the concentrate or both are partially recycled as shown in Figure 5.8. In the feed and bleed mode both the brine and the product concentration can be determined independently and very high recovery rates can be obtained.

5.4.1.2 **Electrodialysis Process Costs**

The total costs in electrodialysis are the sum of fixed charges associated with the amortization of the plant capital costs and the plant operating costs. Both the capital costs as well as the plant operating costs per unit product are proportional to the number of ions removed from a feed solution, that is, the concentration difference

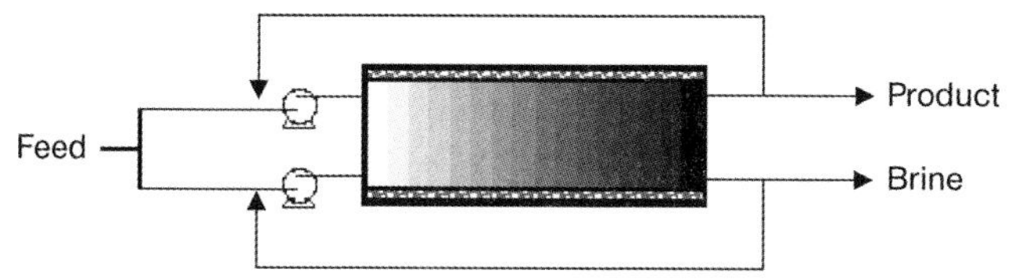

Figure 5.8 Flow scheme of an electrodialysis stack operated in a feed and bleed mode, that is, with partial recycling of the diluate and concentrate solutions.

between the feed and the product solution. But they are also strongly affected by the plant capacity and location and the composition of the feed water and the overall process design [23].

Capital-related costs The capital costs are determined mainly by the required membrane area for a certain plant capacity and feed and required product concentration. Other items such as pumps and process control equipment are considered as a fraction of the required membrane area. This fraction depends on the plant capacity. The same is true for the required land that also depends on the location of the plant.

The required membrane area for a given capacity plant can be calculated from the current density in a stack that again depends on feed and product solution concentration. It can be calculated for a solution containing a single monovalent salt such as NaCl from the total current passing through the stack which is given by:

$$I = \frac{Q^{d}_{cell} F(C^{f} - C^{d})}{\xi_{cell}} = Ai \tag{5.21}$$

Thus:

$$A = \frac{Q^{d}_{cell}\, F(C^{f} - C^{d})}{i\xi_{cell}} \tag{5.22}$$

Here, I and i are the electric current and the current density passing through a cell pair, A is the cell area, Q is the volume flow, C is the concentration expressed in equivalent per volume, F is the Faraday constant, and ξ the current utilization. The subscripts cell refers to the diluate cell, and the superscripts d and f refer to diluate and feed solution, respectively.

The voltage drop across a cell pair is constant over the entire length of a cell pair from the feed entrance to the product exit while the resistance of the cell pair is changing from the feed inlet to the product exit due to a decrease of the resistance of the diluate concentration. Therefore, the current density is also decreasing along the length of a cell pair.

The current density is related to the resistance and the voltage by:

$$i = \frac{U}{\bar{R}A} \tag{5.23}$$

Here, U is the voltage drop across a cell pair A and $\bar{R}$ is the average resistance in a cell pair.

The voltage drop in an electrodialysis cell pair is the result of the resistances of the membranes and the solutions and the concentration potential between the concentrate and diluate, which generally can be neglected. Most electrodialysis stacks used in practical applications consist of geometrically identical cells that are operated in cocurrent flow [23]. If, furthermore, it is assumed that to a first approximation the conductivity is independent of the concentration in the range of interest and the salt activity can be replaced by the concentration the voltage drop at any point for a single mono-valent salt solution across a cell pair length is given by.

$$U = i\left[\frac{\Delta}{\Lambda}\left(\frac{1}{C^{\mathrm{d}}}+\frac{1}{C^{\mathrm{c}}}\right)+r^{\mathrm{am}}+r^{\mathrm{cm}}\right] \quad (5.24)$$

Here, Δ is the cell thickness, Λ is the equivalent conductivity of the salt solution, r is the area resistance, the superscripts d, c, am and cm refer to the diluate, the concentrate, and the anion- and the cation-exchange membranes, respectively.

The electrical resistance at any point along the cell length is given by:

$$R = \frac{U}{I} = \frac{1}{A}\left[\frac{\Delta}{\Lambda}\left(\frac{1}{C^{\mathrm{d}}}+\frac{1}{C^{\mathrm{c}}}\right)+r^{\mathrm{am}}+r^{\mathrm{cm}}\right] \quad (5.25)$$

The average resistance over the entire length of the flow channel is determined by the integral average of the solution concentrations. Thus is:

$$\bar{R} = \frac{1}{A}\left[\frac{\Delta \ln \dfrac{C^{\mathrm{fd}}}{C^{\mathrm{fc}}}\dfrac{C^{\mathrm{c}}}{C^{\mathrm{d}}}}{\Lambda(C^{\mathrm{fd}}-C^{\mathrm{d}})}+r^{\mathrm{am}}+r^{\mathrm{cm}}\right] \quad (5.26)$$

Here, $\bar{R}$ is the average resistance and A the area of a cell pair, C^{fd} and C^{d} are the salt concentrations of the diluate at the inlet and outlet of the cell, C^{fc} and C^{c} are the salt concentrations of the concentrate cell at the inlet and outlet, r^{am} and r^{cm} are the area resistances of the anion- and cation-exchange membranes. The membrane area required for a certain plant capacity as a function of the feed and product concentration of a single mono-valent salt is obtained by combination of Equations 5.22–5.26 and rearranging:

$$A_{\mathrm{tot}} = N_{\mathrm{cell}} \frac{\left[\ln \dfrac{C^{\mathrm{fd}}}{C^{\mathrm{fc}}}\dfrac{C^{\mathrm{c}}}{C^{\mathrm{d}}} + \dfrac{\Lambda(r^{\mathrm{am}}+r^{\mathrm{cm}})(C^{\mathrm{fd}}-C^{\mathrm{d}})}{\Delta}\right]}{\left[\dfrac{C^{\mathrm{d}}}{C^{\mathrm{c}}}+1+\dfrac{\Lambda C^{\mathrm{d}}}{\Delta}(r^{\mathrm{am}}+r^{\mathrm{cm}})\right]} \frac{Q^{\mathrm{d}}_{\mathrm{cell}} F C^{\mathrm{d}}}{i_{\mathrm{lim}} \xi_{\mathrm{cell}}} \quad (5.27)$$

Here A_{tot} is the total membrane area in a stack and N_{cell} is the number of cell pairs in a stack and i_{lim} is the limiting current density that determines the maximum voltage that can be applied. All other symbols are identical as the ones in the Equations 5.21–5.27.

The total investment-related costs depend on the price of the membranes and their useful life under operating conditions, which is in practical application 5–8 years, and on the price of the additional plant components and their life.

Operating costs The operating costs are composed of labor cost, the maintenance of the plant and energy costs. The labor costs are general directly proportional to the size of the plant and usually calculated as a certain percentage of the investment related costs. The energy required in an electrodialysis process is an additive of two terms: (1) the electrical energy to transfer the ionic components from one solution through membranes into another solution and (2) the energy required to pump the solutions through the electrodialysis unit. Depending on various process parameters, particularly the feed-solution concentration, either one of the two terms may be dominating, thus determining the overall energy costs. The energy consumption due to electrode reactions can generally be neglected since more than 200 cell pairs are placed between the two electrodes in a modern electrodialysis stack. The energy required for operating the process control devices can be neglected.

The total energy required in electrodialysis for the actual desalination process is given by the current passing through the electrodialysis stack multiplied with the total voltage drop encountered between the electrodes:

$$E_{des} = I_{st}U_{st}\,t = I_{st}N_{cell}U_{cell}\,t = I^2 N_{cell}\bar{R}t \tag{5.28}$$

Here, E_{des} is the energy consumed in a stack for the transfer of ions from a feed to a concentrate solution, I_{st} is the current passing through the stack, U_{st} and U_{cell} are the voltage applied across the stack, that is, between the electrodes, and across a cell pair; t is the time of operation.

The total current through the stack is given by Equation 5.21 and the average resistance is given by Equation 5.26. Combination of the two Equations and multiplication by the number of cell pairs in the stack gives the desalination energy:

$$E_{des} = N_{cell}\,\bar{R}_{cell}\,I^2 t = \frac{N_{cell}t}{A}\left[\frac{\Delta \ln \dfrac{C^{fd}}{C^{fc}}\dfrac{C^{c}}{C^{d}}}{\Lambda(C^{fd}-C^{d})} + r^{am} + r^{cm}\right]\left[\frac{Q^{d}_{cell}F(C^{fd}_{s}-C^{d}_{s})}{\xi}\right]^2 \tag{5.29}$$

The specific desalination energy, that is, the energy used per unit product volume is given by:

$$E_{de,spc} = \frac{N_{cell}\,\bar{R}_{cell}\,I^2 t}{V_{pro}} = \frac{N_{cell}t}{AV_{pro}}\left[\frac{\Delta \ln \dfrac{C^{fd}}{C^{fc}}\dfrac{C^{c}}{C^{d}}}{\Lambda(C^{fd}-C^{d})} + r^{am} + r^{cm}\right]\left[\frac{Q^{d}_{cell}F(C^{fd}_{s}-C^{d}_{s})}{\xi}\right]^2 \tag{5.30}$$

Here, E_{des} and $E_{des,spc}$ are the desalination energy and the specific desalination, I is the total current, t is the time of operation; C^{fd} and C^{fc} are the equivalent concentrations of the diluate and the concentrate at the cell inlet, C^{d} and C^{c} are the concentrations of the diluate and the concentrate at the cell outlet, Λ is the equivalent conductivity of the salt solution, r^{am} and r^{cm} are the area resistances of the anion- and cation-exchange membrane, Δ is the cell thickness, ξ is the current utilization, and Q^{d}_{cell} is the diluate

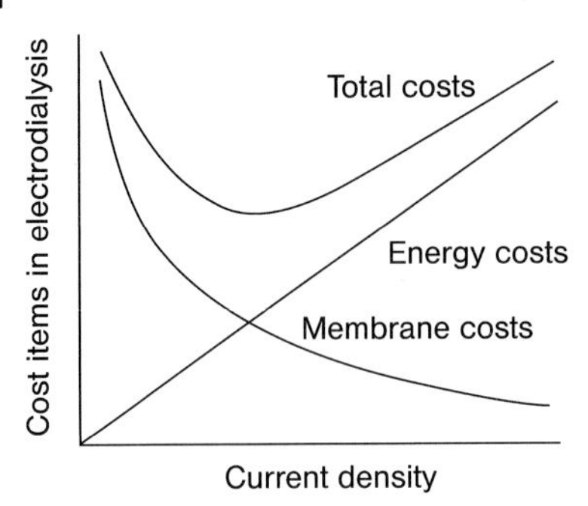

Figure 5.9 Schematic diagram illustrating the various cost items in electrodialysis as a function of the applied current density.

flow rate in a cell, $\bar{R}$ is the average resistance of a cell pair, A is the cell pair area, N_{cell} is the number of cell pairs in a stack, and V_{pro} is a volume product water.

Equation 5.30 shows that the energy dissipation due to the resistance of the solutions and membranes is increasing with the current density, since the electrical energy for a given resistance is proportional to the square of the current, whereas the salt transfer is directly proportional to the current. Hence, the power necessary for the production of a given amount of product increases with the current density. The higher the current density the more power is needed to maintain a given production rate. However, the higher the current density the lower is the required membrane area for a given capacity installation, as illustrated in Figure 5.9, which shows the total costs of desalination and the membrane area and current density related costs as a function of the current density. Figure 5.9 shows that at a certain current density and in the installed membrane area the total desalination costs reach a minimum that must be experimentally determined for a given stack design and feed, diluate, and concentrate. However, the upper limit for the current density of any given installation is determined by the limiting current density that should not be exceeded.

The operation of an electrodialysis unit requires one or more pumps to circulate the diluate, the concentrate, and the electrode rinse solution through the stack. The energy required for pumping these solutions is determined by the volumes of the solutions to be pumped and the pressure drop. It can be expressed by:

$$E_{p,spe} = \frac{E_p}{Q^d t} = k_{eff} \frac{(Q^d \Delta p^d + Q^c \Delta p^c + Q^e \Delta p^e)}{Q^d} \tag{5.31}$$

Here, $E_{p,spec}$ is the total energy for pumping the diluate, the concentrate, and the electrode rinse solution through the stack per unit diluate water, k_{eff} is an efficiency term for the pumps, Q^d, Q^c, and Q^e are the volume flow rates of the diluate, the concentrate, and the electrode rinse solution through the stack.

The energy consumption due to the pressure loss in the electrode rinse solution can be neglected in most practical applications because the volume of the electrode rinse solution is very small compared to the volumes of the diluate and concentrate.

The pressure losses in the various cells are determined by the solution flow velocities and the cell design. The energy requirements for circulating the solution

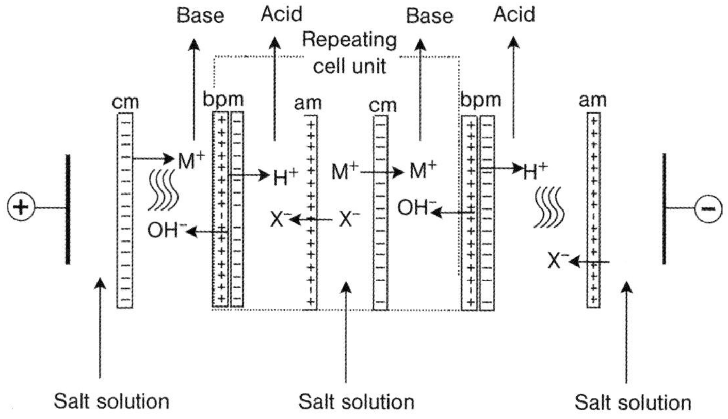

Figure 5.10 Schematic diagram illustrating the acid and base production from the corresponding salt by electrodialysis with bipolar membranes.

through the system may become a significant or even dominant portion of the total energy consumption for solutions with rather low salt concentration.

5.4.2
Electrodialysis with Bipolar Membranes

Electrodialysis with bipolar membranes has gained increasing attention as an efficient tool for the production of acids and bases from the corresponding salts. This process is economically attractive and has many potential applications [24, 25]. A typical arrangement of an electrodialysis stack with bipolar membranes is illustrated in Figure 5.10, which shows the production of an acid and a base from the corresponding salt in a repeating cell unit that consists of three individual cells containing the salt solution, the acid and the base, and three membranes, that is, a cation-exchange, an anion-exchange, and a bipolar membrane. In industrial-size stacks 50–100 repeating cell units may be placed between two electrodes.

The key element in electrodialysis with bipolar membranes is the bipolar membrane. Its function is illustrated in Figure 5.11(a), which shows a bipolar membrane consisting of an anion- and a cation-exchange layer arranged in parallel between two electrodes.

If a potential difference is established between the electrodes, all charged components will be removed from the interphase between the two ion-exchange layers. If only water is left in the solution between the membranes, further transport of electrical charges can be accomplished only by protons and hydroxyl ions that are in a bipolar membrane are regenerated due to the water dissociation taking place in a very thin, that is, 4–5-nm thick transition region between the cation- and anion-exchange layers as shown in Figure 5.11(b). The water dissociation equilibrium is given by:

$$2H_2O \Leftrightarrow H_3O^+ + OH^-$$

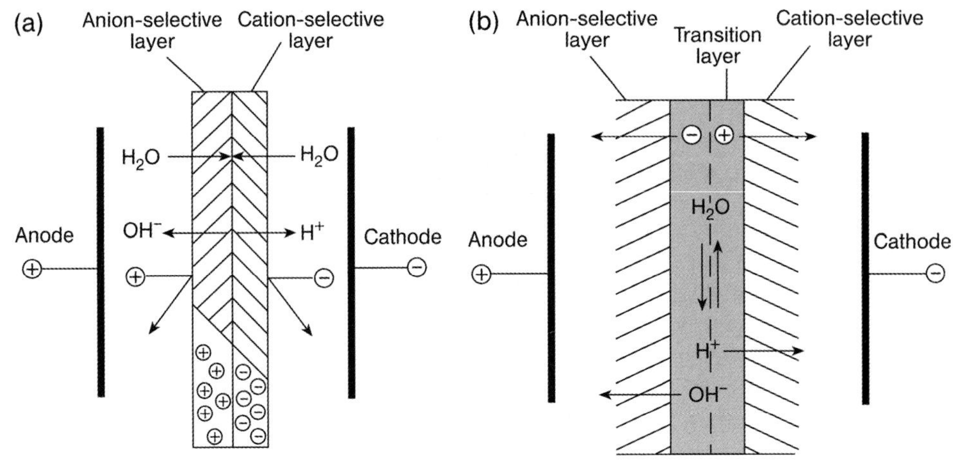

Figure 5.11 Schematic diagram illustrating the function of a bipolar membrane showing (a) a bipolar membrane and (b) the 4–5-nm thick transition region at the interphase of the two cation- and anion-exchange layers.

The energy required for the water dissociation can be calculated from the Nernst equation for a concentration chain between solutions of different pH values. It is given by:

$$\Delta G = F\Delta\varphi = 2.3\ RT\Delta \mathrm{pH} \tag{5.32}$$

Here, ΔG is the Gibbs free energy, F is the Faraday constant, the R is the gas constant, T is the absolute temperature and ΔpH and $\Delta\varphi$ are the pH value and the voltage difference between the two solutions separated by the bipolar membrane. For $1\ \mathrm{mol\,L^{-1}}$ acid and base solutions in the two phases separated by the bipolar membrane ΔG is 0.022 [$\mathrm{kWh\,mol^{-1}}$] and $\Delta\varphi$ is 0.828 [V] at 25 °C.

The transport rate of H^+ and OH^- ions from the transition region into the outer phases cannot exceed the rate of their generation. However, the generation rate of H^+ and OH^- ions in a bipolar membrane is drastically increased compared to the rate obtained in water due to a catalytic reaction [26, 27]. Therefore, very high production rates of acids and bases can be achieved in bipolar membranes.

5.4.2.1 Electrodialysis with Bipolar Membrane System and Process Design

The design of an electrodialysis process with bipolar membranes is closely related to that of a conventional electrodialysis desalination process.

Stack design in bipolar membrane electrodialysis The key component is the stack which in general has a sheet-flow spacer arrangement. The main difference between an electrodialysis desalination stack and a stack with bipolar membranes used for the production of acids and bases is the manifold for the distribution of the different flow streams. As indicated in the schematic diagram in Figure 5.10 a repeating cell unit in a stack with bipolar membranes is composed of a bipolar membrane and a cation- and an anion-exchange membrane and three flow streams in between, that is, a salt

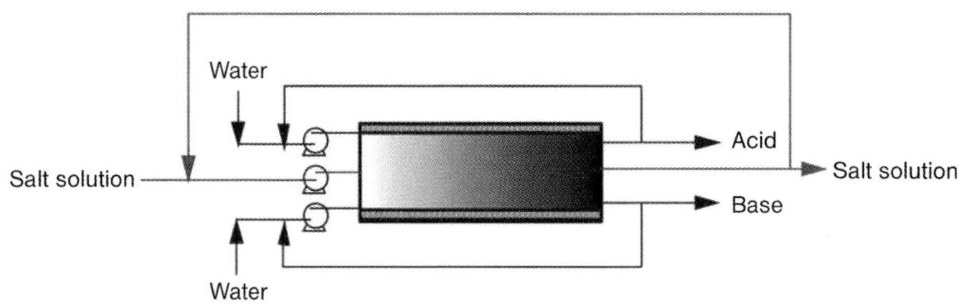

Figure 5.12 Schematic diagram indicating the production of acids and bases from the corresponding salt in a stack with feed and bleed operation - See below for new figure.

solution, a base and an acid flow stream. Since in most practical applications high acid and base concentrations are requested the stack is usually operated in a feed and bleed concept as shown Figure 5.12

Because of the relatively high concentrations of the acid and base as well as the salt solution the limiting current density is in general no problem and a bipolar membrane stack can generally be operated at very high current densities compared to an electrodialysis stack operated in desalination. However, membrane scaling due to precipitation of multivalent ions such as calcium or heavy-metal ions is a severe problem in the base-containing flow stream and must be removed from the feed stream prior to the electrodialysis process with a bipolar membrane.

Problems in the practical application of bipolar membrane electrodialysis In addition to the precipitation of multivalent ions in the base containing flow stream and the stability of the ions in strong acids and bases a serious problem is the contamination of the products by salt ions that permeate the bipolar membrane. In particular, when high concentrations of acids and bases are required the salt contamination is generally high [28] as illustrated in Figure 5.13 that illustrates the conversion of

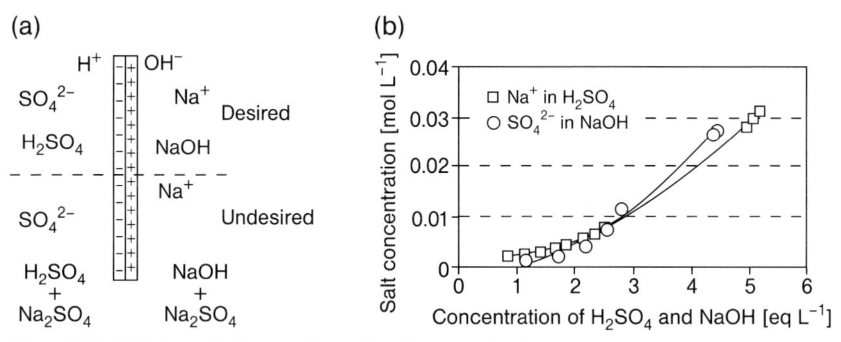

Figure 5.13 (a) Schematic diagram illustrating the contamination of acids and bases by salt due to the incomplete permselectivity of the bipolar membrane for salt ions; (b) experimentally determined salt concentration as a function of the acid and base concentration.

Na_2SO_4 into H_2SO_4 and NaOH by electrodialytic water dissociation. Figure 5.13(a) shows the ion transport in the bipolar membranes.

What is desired is a flux of H^+ and OH^- ions from the transition region due to the applied voltage into the outer phases. However, there is also an undesired transport of Na^+ and SO_4^{2-} ions through the bipolar membrane due to the incomplete permselectivities of the ion-exchange layers of the bipolar membrane. Since the permselectivity of the ion-exchange layers of the bipolar membrane decreases with increasing acid and base concentration due to the Donnan exclusion effect the contamination of the products is increasing with their concentration as demonstrated in Figure 5.13(b) that shows experimentally determined salt concentrations in the acid and base as a function of their concentration.

The salt leakage through the bipolar membrane also effects the current utilization to some extent. However, the current utilization is mainly determined by the properties of the anion-exchange membrane, which has very poor retention for protons due to the tunneling mechanism of the proton transport as illustrated in the schematic drawing of Figure 5.14(a) that shows the undesirable transport of protons through the anion-exchange membrane. The same is true for the hydroxide ions that can permeate the cation-exchange membrane. The net result is that H^+ and OH^- ions generated in the bipolar membrane neutralize each other and thus reduce the current utilization. The fluxes of the protons and hydroxide ions depend on their concentration. At high acid and bases concentrations the current utilization can reach uneconomically low values of less than 30% as indicated in Figure 5.14(b) that shows experimentally determined current utilization as a function of the acid and base concentration.

5.4.2.2 Electrodialysis with Bipolar Membrane Process Costs

The determination of the costs for the production of acids and bases from the corresponding salts follows the same general procedure as applied for the costs in electrodialysis desalination. The contributions to the overall costs are the investment-related cost and the operating costs.

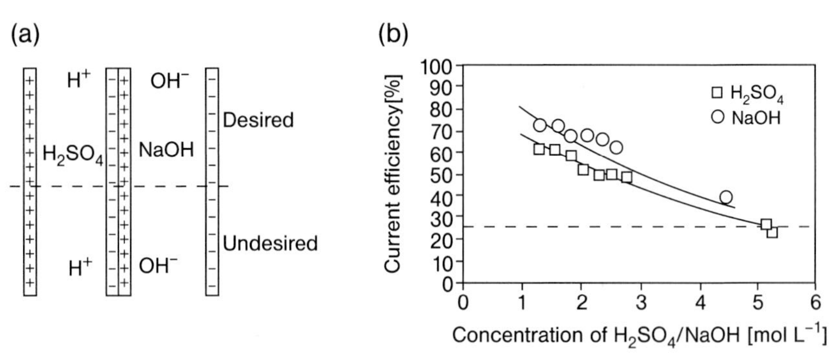

Figure 5.14 (a) Schematic diagram illustrating the decrease of the current utilization during the acid and base production due to the poor acid blocking capability of the anion-exchange membrane; (b) experimentally determined current utilization as function of the acid and base concentration.

Investment costs in electrodialysis with bipolar membranes Investment costs include nondepreciable items such as land and depreciable items such as the electrodialysis stacks, pumps, electrical equipment, and monitoring and control devices. The investment costs are determined mainly by the required membrane area for a certain plant capacity. The required membrane area for a given capacity plant can be calculated from the current density in a stack that is in electrodialysis with a bipolar membrane not limited by concentration-polarization effects. The required membrane area for a given plant capacity is given by:

$$A = \frac{Q_p F\, C_p}{i\,\xi} \tag{5.33}$$

Here, A is the required membrane area, i is the current density, Q_p is the product volume flow, F is the Faraday constant, ξ is the current utilization and C_p is the concentration of the product.

The required membrane area A refers actually to a unit cell area that contains a bipolar membrane, and a cation- and an anion-exchange membrane. Since in strong acids and bases the useful life of the bipolar membrane as well as the anion-exchange membrane is rather limited, the stack-related investment costs are dominating the total investment costs.

Operating cost in electrodialysis with bipolar membranes The operating costs in electrodialysis with bipolar membranes are strongly determined by the energy requirements that are composed of the energy required for the water dissociation in the bipolar membrane and the energy necessary to transfer the salt ions from the feed solution, and protons and hydroxide ions from the transition region of the bipolar membrane into the acid and base solutions. The energy consumption due to the pumping of the solutions through the stack can generally be neglected.

The total energy for the production of an acid and a base from the corresponding salt is as in electrodialysis desalination that has been discussed earlier given the total current passing through the stack and the voltage drop across the stack. The total energy required in electrodialytic water dissociation in a practical process is given by the current passing through the stack multiplied with the total voltage drop encountered between the electrodes.

$$E_{pro} = I \Delta U\, t \tag{5.34}$$

Here, E_{pro} is the energy consumed in a stack for the production of an acid and a base, I is the current passing through a stack or a series of stacks, ΔU is the voltage applied across the stack, that is, between the electrodes, and t is the time of operation.

The current passing through the stack can be derived by rearranging Equation 5.33. It is:

$$I = Ai = \frac{Q_p F\, C_p^{in} - C_p^{out}}{\xi} \tag{5.35}$$

Here, A is a cell unit area. i is current density, I is the current, Q_p is the flow rate of the product, C_p is the concentration of the product, F is the Faraday constant, ξ the current utilization and the superscripts in and out refer to the in- and outlet of the stack.

The voltage drop across the stack is the result of the electrical resistance of the membranes, that is, that of the cation- and anion-exchange membranes and the bipolar membranes and the resistances of the acid, the base- and the salt-containing flow streams in the stack. In addition to the voltage drop required to overcome the various electrical resistances of the stack additional voltage drop is required to provide the energy for the water dissociation which is given by Equation 5.32. Assuming that the three cells of a cell unit in the stack have the same geometry and flow conditions the total energy consumption in an electrodialysis stack is given by:

$$E_{\text{pro}} = N_{\text{cell}} A_{\text{cell}} \left(\frac{\Delta}{\sum_i \Lambda_i \bar{C}_i} + r^{am} + r^{\text{cm}} + r^{\text{bm}} + \frac{N_{\text{cell}} A_{\text{cell}} \xi \, 2.3 \, RT \, \Delta\text{pH}}{Q_{\text{p}} \left(C_{\text{p}}^{\text{out}} - C_{\text{p}}^{\text{in}} \right) F^2} \right) \times \left(\frac{Q_{\text{p}} F \left(C_{\text{p}}^{\text{out}} - C_{\text{p}}^{\text{in}} \right)}{N_{\text{cell}} A_{cell} \xi} \right)^2 t \tag{5.36}$$

Here, E_{pro} is the energy for the production of a certain amount of acid and base, I is the current passing through the stack, N_{cell} is the number of cell units in a stack, A_{cell} is the cell unit area, C and $\bar{C}$ are the concentration and the average concentration in a cell, Δ is the thickness of the individual cells, and Λ is the equivalent conductivity, r is the area resistance, ξ is the current utilization, R is the gas constant, T the absolute temperature, F the Faraday constant, and ΔpH is the difference in the pH value between the acid and base, the subscript p refers to product and the subscript i refers to salt, acid and base, The superscripts am, cm, and bm refer to the cation-exchange, the anion-exchange, and the bipolar membrane, the superscript out and in refer to cell outlet and inlet, Q is the total flow of the acid or base through the stack and t is the time.

The term $Q_p F(C_p^{out} - C_p^{in})/A_{\text{cell}} N_{\text{st}} \xi$ is identical to the current density. This means that for a given stack design the acid and base production energy E_{pro} is proportional to the i^2.

The average concentrations of the acid, the base, and the salt in the bulk solutions are the integral average of the solutions given by:

$$\bar{C}_i = \sum_i \frac{\ln\left(\frac{C_i^{\text{out}}}{C_i^{in}} \right)}{C_i^{out} - C_i^{in}} \tag{5.37}$$

The total costs of the electrodialytic water dissociation with bipolar membranes are the sum of fixed charges associated with the amortization of the plant investment costs and of the operating costs which include energy and maintenance costs and all pre- and post-treatment procedures. The total costs are a function of the membrane properties, of the feed-solution composition, the required acid and base concentrations, and several process and equipment design parameters such as stack construction and operating current density.

5.4.3
Continuous Electrodeionization

Continuous electrodeionization is widely used today for the preparation of high-quality deionized water for the preparation of ultrapure water in the electronic industry or in analytical laboratories. The process is described in some detail in the patent literature and company brochures [29]. There are also some variations of the basic design as far as the distribution of the ion-exchange resin is concerned. In some cases the diluate cell is filled with a mixed bed ion-exchange resin, in other cases the cation- and anion-exchange resins are placed in series in the cell. More recently, bipolar membranes are also being used in the process.

5.4.3.1 System Components and Process Design Aspects

The process design and the different hardware components needed in electrodeionization are very similar to those used in conventional electrodialysis. The main difference is the stack construction. In a continuous electrodeionization stack the diluate cells and sometimes also the concentrate cells are filled with an ion-exchange resin. The different concepts used for the distribution of the cat- and anions in the cell are illustrated in Figures 5.15(a) and (b). In the conventional electrodeionization process the diluate cell is filled with a mixed-bed ion-exchange resin with a ration of cation- to anion-exchange resin being close to 1 as shown in Figure 5.15(a). The mixed-bed ion-exchange resin in the diluate cells of the stack removes the ions of a feed solution. Due to an applied electrical field the ions migrate through the ion-exchange bed towards the adjacent concentrate cells and highly deionized water is obtained as a product. The ion-exchange resin increases the conductivity in the diluate cells to such an extent that the stack resistance is significantly lower and the limiting current density higher than in a conventional electrodialysis stack. Compared to the deionization by a conventional mixed-bed ion-exchange resin the continuous electrodeionization has the advantage that no chemicals are needed for the regeneration of the ion-exchange resins, which is time consuming, labor intensive, and generates a salt-containing wastewater.

But the continuous electrodeionization using a stack with mixed-bed ion-exchange resins in the diluate has also disadvantages. The most important one is the poor removal of weak acids and bases such as boric or silicic acid [30]. Much better removal of weakly dissociated electrolytes can be obtained in a system in which the cation- and anion-exchange resins are placed in a stack in separate beds with a bipolar membrane placed in between, as illustrated in Figure 5.15(b), which shows a diluate cell filled with a cation-exchange resin facing towards the cathode separated by a bipolar membrane from a diluate cell facing the anode. A cation-exchange membrane, a cation-exchange resin, a bipolar membrane, an anion-exchange resin, an anion-exchange resin, and a concentrate cell form a repeating unit between two electrodes.

The main difference between the electrodeionization system with the mixed-bed ion-exchange resins and the system with separate beds is that in mixed-bed electrodeionization systems anions and cations are simultaneously removed from the feed

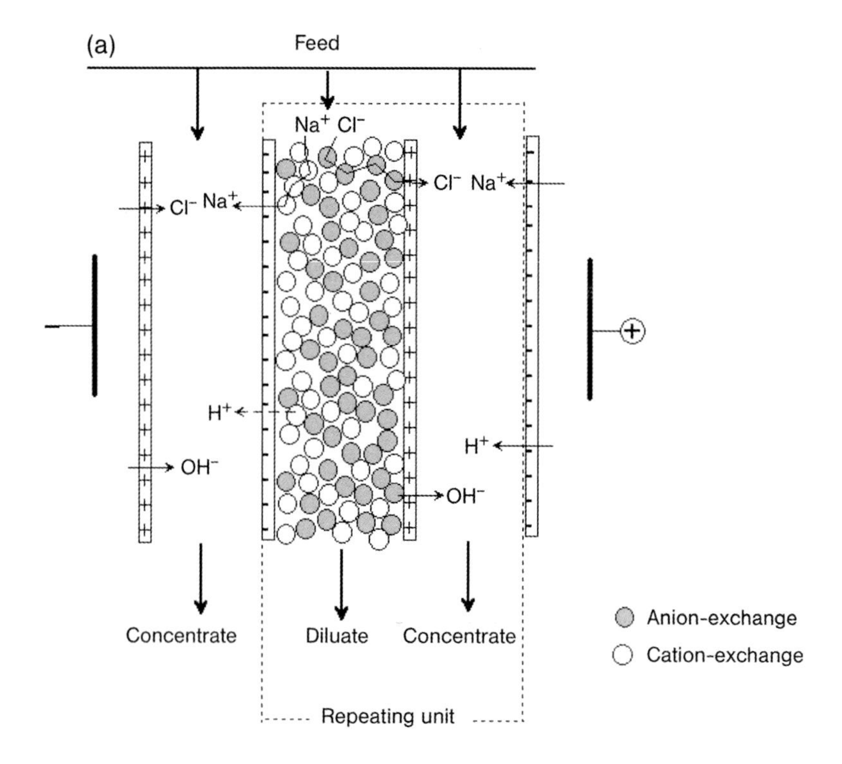

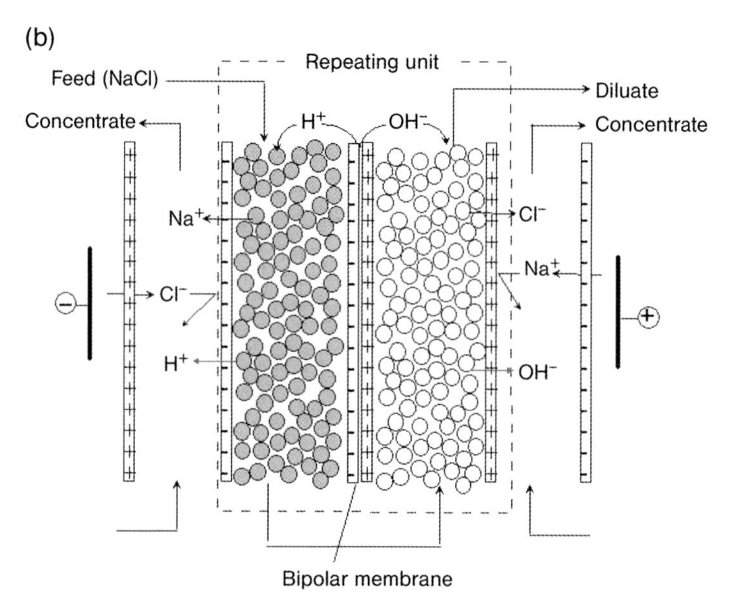

Figure 5.15 Schematic drawing illustrating different stack concepts used in continuous electrodeionization, (a) conventional stack with diluate cells filled with a mixed-bed ion-exchange resin, (b) stack with cation-exchange and anion-exchange resins in different diluate cells and regeneration of the ion-exchange resins by H^+ and OH^- ions generated in a bipolar membrane.

while the solution leaving the diluate cell is neutral. In the electrodeionization system with separate ion-exchange beds and bipolar membranes the cations will first be exchanged by the protons generated in the bipolar membrane with the result that the solution leaving the cation-exchange bed is acidic. This solution is then passed through the cell with the anion-exchange resin where the anions are exchanged by the OH^- ions generated in the bipolar membrane and the solution is neutralized, and at the exit of the anion-exchange-filled cell the solution is also neutral. Both the mixed and the separate bed ion-exchange continuous electrodeionization systems are widely used today on a large industrial scale.

5.4.3.2 Operational Problems in Practical Application of Electrodeionization

In addition to the problems of removing weak acids or bases in the electrodeionization system the mixed bed ion-exchange resin there are problems of uneven flow distribution in the ion-exchange resin beds that lead to poor utilization of the ion-exchange resins. The fouling of the ion-exchange resins by organic components such as humic acids, and bacterial growth on the surface of the resin is a problem that requires a very thorough pretreatment of the feed solution to guarantee a long-term stability of the system. The effect of the cell geometry, that is, the ratio of its length to width and thickness has been studied extensively and is described in various patents.

5.4.4 Other Electromembrane Separation Processes

In addition to the processes discussed so far there are two more electromembrane separation processes in which the driving force is not an externally applied electrical potential but a concentration gradient. The processes are referred to as diffusion dialysis and Donnan dialysis. Diffusion dialysis is utilizing anion- or cation-exchange membranes only to separate acids and bases from mixtures with salts. Donnan dialysis can be used to exchange ions between to solutions separated by an ion-exchange membrane. Both processes have so far gained only limited practical relevance [4] and will not be discussed in this chapter.

List of Symbols

Roman Letters

a	constant [–]
a	*activity* [mol m^{-3}]
A	area [m^2]
C	concentration [mol m^{-3}]
$\bar{C}$	average concentration [mol m^{-3}]
D	diffusion coefficient [m^2 s^{-1}]

E	energy [A V s]
F	Faraday constant [A s eq^{-1}]
G	Gibbs free energy [J]
i	current density [A m^{-2}]
I	current [A]
i_{lim}	limiting current density [A m^{-2}]
J	flux [mol m^{-2} s^{-1}]
J_e	flux of electrical charges [A m^{-2}]
k	coefficient [various]
L	coefficient [mol^2 N^{-1} m^{-2} s^{-1}]
l	length [m]
n	number [–]
N	number [–]
p	pressure [Pa]
q	area [m^2]
Q	volume flow rate [m^3 s^{-1}]
r	area resistance [Ω m^2]
R	electrical resistance [Ω]
$\bar{R}$	average electrical resistance [Ω]
R	gas constant [J mol^{-1} K^{-1}]
S	conductivity [Ω^{-1}]
T	temperature [K]
t	time [s]
T	transport number [–]
U	electrical potential [V]
u	ion mobility [m^2 s^{-1} V^{-1}]
$\bar{V}$	partial molar volume [m^3 mol^{-1}]
V	volume [m^3]
z	directional coordinate [m]
z	charge number [eq mol^{-1}]
Z_b	boundary layer thickness [μ]

Greek Letters

Ψ	membrane permselectivity [–]
Δ	cell thicknes [m]
Δ	difference [–]
$\tilde{\mu}$	electrochemical potential [A V s mol^{-1}]
φ	electrical potential [V]
κ	specific conductivity [Ω^{-1} m^{-1}]
Λ_{eq}	equivalent conductivity [m^2 Ω^{-1} eq^{-1}]
Λ_{mol}	molar conductivity [m2 Ω-1 mol-1]
μ	chemical potential [J mol^{-1}]
ν	stoichiometric coefficient [–]

ρ specific resistance [Ω m]
ξ current utilization [−]

Subscripts

a anion
c cation
cell cell or cell pair
co co ion
eff efficiency
des desalination
e electric charge
fix fixed ion
i component
k component
m membrane
am anion-exchange membrane
cm cation-exchange membrane
p pumping
pro product
s salt
spec specific
st stack
tot total
w water

Superscripts

am anion-exchange membrane
b bulk solution
b constant
bm bipolar membrane
c concentrate
cm cation-exchange membrane
e electrode rinse
d diluate
f feed
fc feed concentrate
fd feed diluate
in inlet
m membrane
out outlet
p product
s solution

References

1 Spiegler, K.S. (1956) Electrochemical operations, in *Ion-Exchange Technology* (eds F.C. Nachod and J. Schubert), Academic Press, New York.

2 Wilson, J.R. (ed.) (1960) *Demineralization by Electrodialysis*, Butterworth Scientific Publications, London.

3 Schaffer, L.H. and Mintz, M.S. (1966) Electrodialysis, in *Principles of Desalination* (ed. K.S. Spiegler), Academic Press, New York, pp. 3–20.

4 Strathmann, H. (2004) *Ion-Exchange Membrane Separation Processes*, Elsevier, Amsterdam, The Netherlands.

5 Liu, K.J., Chlanda, F.P. and Nagasubramanian, K.J. (1977) Use of bipolar membranes for generation of acid and base: An engineering and economic analysis. *Journal of Membrane Science*, **2**, 109.

6 Eisenberg, A. and Yeager, H.L. (1982) *Perfluorinated Ionomer Membranes*, ACS Symposium Series 180, American Chemical Society, Washington, DC.

7 Flett, D.S. (1983) *Ion-Exchange Membranes*, E. Horwood Ltd, Chichester, UK.

8 Bergsma, F. and Kruissink, Ch.A. (1961) Ion-exchange membranes. *Fortschritte der Hochpolymer Forschung*, **21**, 307.

9 Helfferich, F. (1962) *Ion-Exchange*, McGraw-Hill, London.

10 Sata, T. (1986) Recent trends in ion-exchange research. *Pure and Applied Chemistry*, **58**, 1613.

11 Grot, W. (1975) Perfluorinated cation exchange polymers. *Chemie Ingenieur Technik*, **47**, 617.

12 Zschocke, P. and Quellmalz, D. (1985) Novel ion exchange membranes based on an aromatic polyethersulfone. *Journal of Membrane Science*, **22**, 325.

13 Simons, R. (1993) Preparation of a high performance bipolar membrane. *Journal of Membrane Science*, **78**, 13.

14 Wilhelm, F.W. (2000) Bipolar membrane preparation, in *Bipolar Membrane Technology* (ed. A.J.B. Kemperman), Twente University Press, Enschede, The Netherlands.

15 Pourcelly, G. (2002) Conductivity and selectivity of ion exchange membranes: structure-correlations. *Desalination*, **147**, 359.

16 Spiegler, K.S. (1958) Transport processes in ionic membranes. *Transactions of the Faraday Society*, **54**, 1408.

17 Kedem, O. and Katchalsky, A. (1961) A physical interpretation of the phenomenological coefficients of membrane permeability. *The Journal of General Physiology*, **45**, 143.

18 Planck, M. (1890) *Ann Physik u Chem, NF*, **39**, 161.

19 Strathmann, H. (1995) Electrodialysis and related processes, in *Membrane Separation Technology* (eds R.D. Nobel and S.A. Stern), Elsevier, Amsterdam, pp. 213–281.

20 Schaffer, L.H. and Mintz, M.S. (1966) Electrodialysis, in *Principles of Desalination* (ed. K.S. Spiegler), Academic Press, New York, pp. 3–20.

21 Huffmann, E.L. and Lacey, R.E. (1972) Engineering and economic considerations in electromembrane processing, in *Industrial Processing with Membranes* (eds R.E. Lacey and S. Loeb), John Wiley & Sons, New York, pp. 39–55.

22 Katz, W.E. (1979) The electrodialysis reversal (EDR) process. *Desalination*, **28**, 31.

23 Lee, H.J., Safert, F., Strathmann, H. and Moon, S.H. (2002) Designing of an Electrodialysis desalination plant. *Desalination*, **142**, 267.

24 Mani, K.N. (1991) Electrodialysis water splitting technology. *Journal of Membrane Science*, **58**, 117.

25 Kemperman, A.J.B. (ed.) (2000) *Handbook on Bipolar Membrane Technology*, Twente University Press, Endschede, The Netherlands.

26 Strathmann, H., Krol, J.J., Rapp, H.J. and Eigenberger, G. (1997) Limiting current density and water dissociation in bipolar

membranes. *Journal of Membrane Science*, **125**, 123.

27 Mafé, S., Ramirez, P., Alcaraz, A. and Aguilella, V. (2000) Ion transport and water splitting in bipolar membranes, in *Handbook Bipolar Membrane Technology* (ed. A.J.B. Kemperman), Twente University Press, Endschede, The Netherlands.

28 Rapp, H.J. (1965) Die Elektrodialyse mit bipolaren Membranen – Theorie und Anwendung, PhD-Thesis, University of Stuttgart, Germany.

29 Ganzi, G.C. (1988) Electrodeionization for high purity water production, in *New Membrane Materials and Processes for Separation*, **84** (eds K.K. Sirkar and D.R. Lloyd), AIChE Symposium Series, p. 73, American Chemical Society, Washington, DC, USA.

30 Grabowskij, A., Zhang, G., Strathmann, H. and Eigenberger, G. (2006) The production of high purity water by continuous electrodeionization with bipolar membranes. *Journal of Membrane Science*, **281**, 297.

6
Fouling in Membrane Processes

Anthony G. Fane, Tzyy H. Chong, and Pierre Le-Clech

6.1
Introduction

This chapter reviews membrane fouling with particular reference to the pressure-driven liquid-phase membrane processes where the solvent is water. The processes of interest are low-pressure microfiltration (MF) and ultrafiltration (UF) and high-pressure nanofiltration (NF) and reverse osmosis (RO). Fouling presents as a decrease in membrane performance with a loss in solvent permeability and changes to solute transmission. Fouling is caused by deposition of feed components, or growth (as in biofouling and scale formation) onto or into the membrane; it is a widespread and costly problem. The foulant–membrane interaction depends on the nature of the foulant, the membrane and the operating environment. This section provides an overview of fouling and describes various 'generic' fouling mechanisms. It compares the fouling profiles of constant pressure vs constant flux and crossflow vs. deadend. The concepts of critical and sustainable flux are introduced. Fouling aspects of low-pressure microporous membranes and high-pressure 'nonporous' membranes are described in Sections 6.2 and 6.3, respectively.

6.1.1
Characteristics of Fouling

The basic relationship between flux and driving force is given in Equation 6.1 (Table 6.1). When fouling occurs an additional resistance, R_F, is imposed and in some cases (with NF and RO) it may increase $\Delta\Pi$ in Equation 6.1 (see Sections 6.1.3 and 6.3.4). Increasing R_F and/or $\Delta\Pi$ causes a flux decline at constant ΔP (transmembrane pressure, TMP) (Figure 6.1(a)) or causes TMP to rise at constant flux (Figure 6.1(b)). The flux–time profile (Figure 6.1(a)) can be misleading. For membrane M1, with initial flux J_i an increment of resistance ΔR_F reduces flux to 0.5 J_i, then for similar ΔR_F to 0.33 J_i, 0.25 J_i, and so on so flux decline 'appears' to be slowing down. A clearer picture emerges by calculating the changes in ΔR_F. Figure 6.1(a) also

Membrane Operations. Innovative Separations and Transformations. Edited by Enrico Drioli and Lidietta Giorno

ISBN: 978-3-527-32038-7

Table 6.1 Fouling and polarization relationships

Description	Relationship	Equation
Flux-driving force resistances	$J = \dfrac{\Delta P - \Delta \Pi}{\mu(R_m + R_F)}$	(6.1)
Membrane resistance	$R_m = f\{N_{pore}, d_{pore}, \varepsilon_m, L_{pore}\}$	(6.2)
Cake resistance	$R_c = m\alpha_c$	(6.3)
Specific cake resistance	$\alpha_c = \dfrac{180(1-\varepsilon_c)}{(\rho_p d_p^2 \varepsilon_c^3)}$	(6.4)
CP film model (for condition of complete retention of solute)	$J = k \ln\left(\dfrac{C_w}{C_b}\right)$	(6.5)
	$\dfrac{C_w}{C_b} = \exp\left(\dfrac{J}{k}\right)$	(6.6)

illustrates that the low flux membrane M2, seems to decline less dramatically for the same ΔR_F. An intrinsically 'highflux' membrane appears more sensitive to fouling than a 'low flux' membrane. To make a fair comparison the membranes need to be tested at similar fluxes (see Section 6.1.4) and fouling quantified as changes in ΔR_F or TMP (at fixed flux). Figure 6.1(b) shows a steady fouling (i) and an example with a sudden TMP jump (ii) , which is characteristic of MBRs (see Section 6.2.4.2).

The other detrimental effect of fouling is that it changes the separation properties of the membrane. For microporous membranes that transmit some species and retain others the effect of fouling is usually to increase retention of partially transmitted species. This is because fouling leads to pore closure or blockage (see Section 6.1.3), making the membrane 'tighter.' In some cases, such as water

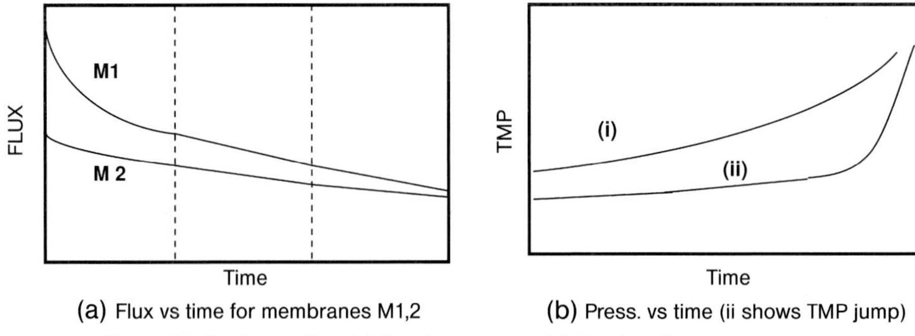

Figure 6.1 Fouling profiles. (a) Constant pressure (b) Constant flux.

treatment and membrane bioreactors, this can be beneficial, giving a greater removal of virus or organic species. In other cases, such as the food and biotechnology industries, this 'tightening' can be a problem if membrane fractionation of species is desired. For 'nonporous' membranes the fouling is a surface layer or cake (see Section 6.1.3) that increases polarization of retained and partially retained species at the membrane surface. The partially retained species then tend to show a higher concentration in the permeate, so observed retention decreases due to fouling. This is opposite to the trend for microporous membranes.

6.1.2
Causes of Fouling

Most dissolved or suspended species have the potential to foul membranes. Fouling interactions could be physicochemical adsorption from solution, precipitation of sparingly soluble salts, growth of biofilms and deposition of suspended matter onto and into the membrane. Table 6.2 lists generic foulants and foulant control by pretreatment or feed-adjustment strategies. Examples of the various forms of fouling are given in Sections 6.2 and 6.3. The factors that encourage or exacerbate fouling are inadequate pretreatment, inadequate fluid management (hydrodynamic environment), excessive flux and unsuitable membrane properties.

Pretreatment. Most feed streams are mixtures with varying characteristics. In many cases there are foulants present that can be minimized by pretreatment. For example, seawater fed to RO desalination plant usually contains turbidity and micro-organisms, which can be partially mitigated by prefiltration (media or membranes) and

Table 6.2 Foulants and foulant control.

Generic foulant	Type of fouling	Membrane	Control of foulant
Inorg. Ions Insoluble salts (Ca etc.)	Scale	NF RO	Concentration, pH, antiscalant
Inorg. Ions Calcium and so on	NOM binding	MF/UF NF	Concentration, pH, coagulants
Organics			
NOM, humics	Cake & biofilms	NF, RO	Adsorb, biotreatment, MF/UF + coagulants
NOM, humics	Cake & pore fouling	MF, UF	Coagulant
Protein (food)	Cake & pore fouling	UF	pH
Protein & polysaccharide	Cake & pore fouling	MF, UF MBR	MBR bioprocess control
Particulates			
Colloids(<1 μm)	Cake	NF, RO	Coagulant + media filter or + MF/UF
Colloids(<1 μm)	Cake & pore fouling	MF, UF	Coagulant
Biological solids	Biofilm	NF, RO	Chlorination, nutrient removal, coagulant + MF/UF

chlorination (followed by dechlorination) respectively. If scale formers are present they can be suppressed by addition of antiscalants. MBR membranes handling municipal wastewater are susceptible to damage and this can be avoided by fine-screen pretreatment. In some cases, such as water treatment, the foulants can be controlled by chemical pretreatment so that filtration is of large floc, rather than macrosolutes or colloids. Where pretreatment is inadequate fouling usually results and leads to frequent cleaning or system upgrade.

*Fluid Management.*Concentration polarization (CP) at the membrane surface is a result of the separation of feed and permeate (mainly water) at the interface. Fouling is a consequence of CP. The concentration of species at the surface (C_w) depends on the imposed flux (J) and the boundary layer mass-transfer coefficient (k) as indicated by the CP film model (Equations 6.5 and 6.6 in Table 6.1). Tangential crossflow is used to limit CP, and does so by increasing k. If fluid management is inadequate for the feed serious CP can lead to fouling. The reasons for inadequate fluid management include badly selected operating conditions, loss of feed flow due to permeate removal or flow maldistribution due to poorly designed membrane module or blockages. In deadend flow (Section 6.1.5), which is applied to dilute feeds, there is no crossflow but it is still important to ensure homogeneous flow distribution and flux to avoid localized fouling.

Flux. The film model (Equation 6.6) illustrates that increasing flux has an exponential effect on CP. If we accept that fouling is a consequence of CP the impact of excessive flux is obvious. As a result 'high flux' membranes tend to be short lived and foul unless improved fluid management is able to enhance k. Selection of the appropriate flux and crossflow velocity is a trade-off between capital and operating costs (see cost of fouling below).

Membranes. The interactions between the membrane and the potential foulants influence the degree of fouling. For example, if the pore size of a microfilter and particulate foulants overlap the particles can enter the membrane and block pores (see Section 6.1.3), and this can cause irreversible fouling. If the surface charge of the membrane and the dominant feed species are opposite it is likely that fouling will occur. Similarly, hydrophobic membranes are prone to fouling by hydrophobic components, including proteins and lipids. A useful rule of thumb is that membranes that are smooth, hydrophilic, of low net charge and narrow pore-size distribution are less susceptible to fouling. However, there are many exceptions to this rule and membrane selection may involve comparison trials.

The cost of fouling has at least 4 components, represented by,

$$c_{\text{Fouling}} = c_{\text{Cleaning}} + c_{\text{Power extra}} + c_{\text{Production loss}} + c_{\text{Membrane replace}} \tag{6.7}$$

c_{Cleaning} includes cost of chemicals, disposal and labor. $c_{\text{Power extra}}$ represents the effect of increased operating pressure. $c_{\text{Production loss}}$ includes the effect of flux decline on throughput as well as down time for cleaning. In respect of $c_{\text{Membrane replace}}$ it should be noted that cleaning events may be the harshest environment for the membrane and replacement tends to increase with cleaning frequency. Fouling costs are application specific but could range from a few % to 10s of % of processing costs.

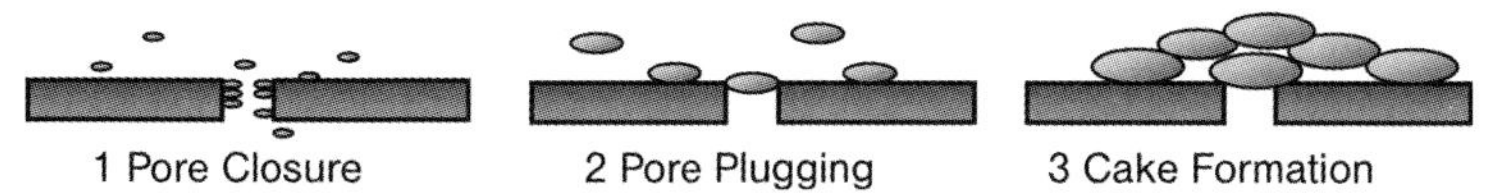

Figure 6.2 Membrane-fouling mechanisms.

6.1.3 Fouling Mechanisms and Theory

Figure 6.2 depicts 3 ways in which microporous membranes can foul; (a) pores can suffer closure or restriction, (b) pores or porosity can be blocked or plugged and (c) a surface cake or layer can cover the membrane. All three mechanisms could apply, probably in sequence (a) then (b) followed by (c). Nonporous membranes are fouled by cake or surface layers (c).

It may be possible to do a membrane autopsy to identify the foulant(s) and fouling mechanism. For microporous membranes the blocking law analysis [1], which uses permeate volume (V) vs. time (t) data, can supplement the observations. The generalized relationship at constant pressure and in dead-end filtration mode gives,

$$\frac{\mathrm{d}^2 t}{\mathrm{d}V^2} = K\left(\frac{\mathrm{d}t}{\mathrm{d}V}\right)^n \tag{6.8}$$

where $n = 0$ for cake filtration, $n = 1.5$ for pore closure ('standard blocking') and $n = 2$ for pore plugging (complete blocking). Linearized forms of the blocking laws have been used to investigate fouling [2].

Some useful relationships describing fouling are given in Table 6.1 with the basic relationship in Equation 6.1. Fouling can increase membrane resistance R_{m} (Equation 6.2), particularly for microporous membranes where loss of pores (N_{pore}) due to plugging and reduction in pore size (d_{pore}) due to restriction may cause 'irreversible' fouling. If fouling is 'irreversible' it implies a resistance not readily removed by cleaning. Fouling due to cake formation adds resistance, R_{F}, (Equation 6.3) and this depends on the foulant load (m) and specific cake resistance (α_{c}) that is increased as particle size (d_{p}) and cake voidage (ε_{c}) decrease (Equation 6.4). Fouling can also increase $\Delta\Pi$ and diminish the driving force. This is known as cake-enhanced osmotic pressure (CEOP) and occurs when retained solutes, such as salts in RO, have hindered backdiffusion in the cake layer so that CP increases and local osmotic pressure rises. CEOP is a feature of fouling in NF and RO and is discussed in Section 6.3.

6.1.4 Critical and Sustainable Flux

The concept of critical flux (J_{CRIT}) was introduced by Field *et al.* [3] and is based on the notion that foulants experience convection and back-transport mechanisms and that there is a flux below which the net transport to the membrane, and the fouling, is negligible. As the back transport depends on particle size and crossflow conditions the J_{CRIT} is species and operation dependent. It is a useful concept as it highlights the role of flux as a driver of fouling. However, in many practical cases J_{CRIT} is very low or difficult to identify. The sustainable flux is an alternative concept and refers to the flux below which the rate of fouling is economically acceptable. A recent review [4] examines these two concepts.

6.1.5 Fouling and Operating Mode

The pressure-driven membrane processes can be operated at fixed pressure (FP) or fixed flux (FF), and FP tends to be lab and small scale and FF is large-scale commercial. Fouling for FP shows as a flux decline and for FF as TMP rise (Figure 6.1(b)). The fouling kinetics differ since FP becomes 'self-limiting' as flux-driven fouling slows down, whereas for FF it is 'self-accelerating' as foulants steadily accumulate and concentration polarization accelerates. These differences mean that extrapolation of FP trends to FF requires caution.

As noted in Section 6.1.2, in most applications the control of CP, and fouling, dictates the use of crossflow. However, for dilute feeds and low-pressure membranes it has been accepted that batch cycles of deadend operation with solids accumulation removed by periodic backwash requires potentially lower energy. Usually, deadend is at FF and the TMP cycles from a minimum to maximum or over a specified cycle time during the batch. If fouling occurs it is evident through a steady rise in TMP_{min} or R_m. Occasional chemical cleaning may restore R_m.

6.2 Low-Pressure Processes

In the context of this review low-pressure membranes include the microporous MF and UF processes. Applications of MF include beverage clarification, cell harvesting, wastewater treatment by membrane bioreactors (MBRs), water treatment and pretreatment prior to RO. UF covers similar applications plus protein concentration (food and dairy) and other macrosolute separations. Crossflow, supplied by pumping, stirring or two-phase flow, is generally applied, except when very low solids content permits deadend filtration with frequent backwash, as in water treatment, RO pretreatment and clarification processes. The liquid streams to be treated by membrane processes usually contain a complex mixture of particulate, colloidal and soluble materials. Before considering two complex feeds (activated sludge and surface water) in Section 6.2.4, general trends observed during the filtration of specific compounds are discussed.

6.2.1 Particulate Fouling

Examples of particles to be filtered through low-pressure membranes include casein micelles from milk, lattices from paint, biomass flocs from activated sludge, bacteria from cell broth and suspended solids from water or wastewater. In the early stages of the filtration, the particles with sizes much smaller than the pores can enter and block the pores (Figure 6.2). Particles larger than the pores of the membrane are totally rejected and may be partially transported back to the bulk. However, the formation of a cake layer on the membrane surface gradually occurs as the filtration proceeds, provided the local flux exceeds the 'critical flux' of the particle (see Section 6.1.4). A high concentration of solids in the feed to be treated can be responsible for an increase in viscosity and a decrease in the translational diffusion coefficient (and the resulting Brownian backdiffusion), both leading to higher concentration polarization and fouling propensity [5]. Brownian diffusion is more applicable to fine colloids (Section 6.2.2) and decreases with particle size, being low for particles larger than about 0.5 μm. However other mechanisms have been proposed for the back transport of larger particles in crossflow microporous membrane processes for both laminar and turbulent conditions. These mechanisms include shear-induced diffusion (migration of interacting retained particles in the direction of decreasing particle concentration), inertial lift (motion of particles across a nonuniform shear field to an equilibrium position away from the channel wall) and flowing cake [5]. The key aspect of these 'particulate' back-transport models is the prediction of more significant fouling in the cases of small particles and at lower crossflow velocities.

Because of their generally reversible nature, fouling by large particles is usually efficiently removed by physical methods such as membrane relaxation (filtration is paused) and backwashing (permeate is pumped in the reverse direction through the membrane). Furthermore, many studies have reported the efficient and optimized use of aeration for fouling limitation in submerged low-pressure membrane processes. Finally, improved pretreatment of the feed by coarse filtration or by coagulation to increase particle size can also be considered to reduce the fouling potential of the particle-based compounds.

6.2.2 Colloidal and Macrosolute Fouling

There is a general consensus that colloidal solutions are composed of small particles whose size could range from 1 nm to 1 μm. Colloids experience double-layer interactions. For example, when silica particles (0.14 μm) were used as model colloids, the ionic strength of the feed strongly influenced the fouling characteristics by increasing the cake packing density, leading to a lower efficiency of the backwash process used to remove the fouling layer [6]. McDonogh *et al.* [7] showed experimentally and by modeling that the limiting fluxes and cake resistances of colloidal suspensions in UF varied significantly with ionic environment as this alters particle–particle interaction through the zeta potential. More discussion of colloidal species is given in Section 6.3.1.

With the majority of MF/UF processes based on the filtration of aqueous solutions, the unavoidable presence of dissolved solutes (salts, organic macromolecules of various nature and sizes) in the feed strongly affects the membrane performances. They can be directly responsible for membrane fouling through mechanisms like adsorption of macromolecular substances or precipitation and deposition of inorganic salts on the membrane surface [5]. Due to their changing nature and large diversity, the naturally occurring polymers (e.g., proteins, carbohydrates and humic substances) present significant complexity to the fouling mechanisms (see case studies in Section 6.2.4), and their filtration through MF/UF have been extensively studied in the literature [8, 9]. Interaction with ions, such as calcium, can also exacerbate the fouling by these macrosolutes. Overall, when considering the effect of dissolved material on membrane fouling, the osmotic pressure of the dissolved compounds has to be taken into consideration, along with the interactions they may have with the particles and colloids in the suspension.

6.2.3 Biofouling and Biofilms

During the early stages of filtration of biomass suspensions, attachment of soluble microbial products (SMP) on the membrane surface through adhesive forces can be observed. This phenomenon has also been reported during passive adsorption when no filtration occurs. The initial conditioning film participates in the reduction of the hydraulic performance, and facilitates further cohesive attachment of colloids and/or particles (including bacteria) on the SMP-covered membrane. The newly immobilized bacteria are then able to grow and colonize the membrane surface, forming a biofilm layer [10]. As permeation continues, the soluble compounds in the feed are transferred through the biofilm structure, providing a constant supply of nutrients and dissolved oxygen (when available) to the growing biofilm. The biofilm formation on the membrane surface highlights the significant role played by the strong interactions existing between particulates and macromolecular components, which can be observed in any multicomponent feed. Biofouling in RO is discussed in Section 6.3.2.

6.2.4 Case Studies

6.2.4.1 Water Treatment and Membrane Pretreatment

As more potable-water treatment plants rely on NF/RO technologies, pretreatment systems able to efficiently and reliably remove suspended solids from the feed water are a critical feature of the overall plant design. A similar requirement pertains to water reclamation plant (used water to high-quality/indirect potable water). Conventional pretreatment processes include coagulation, followed by sedimentation and/or sand filtration, and more recently membrane technology like MF/UF. Indeed, the number of plants based on the use of MF/UF for this application tends to rise as low-pressure membranes remain the only technology able to remove pathogens and bacteria (and decrease biofouling downstream) more effectively than other conventional processes without the need for chemical addition [11]. These applications of low-pressure membranes are characterized by low solids feeds and the trend to use deadend filtration with backwash (Section 6.1.5).

The presence of NOM in surface water and its effect on fouling of low-pressure membranes has been discussed in numerous publications (see recent review [12]). The complex and labile nature of these compounds remains a challenge for researchers to fully understand their fouling mechanisms. As expected, most of the parameters defining the filtration process (membrane and feed characteristics and operating conditions) can influence the propensity of NOM fouling. Within this long list, ionic strength, pH of the feed and presence of divalent cations, such as calcium, have been listed as the main factors affecting the degree of NOM fouling. Due to the wide size distribution of NOM species, fouling mechanisms include membrane adsorption, leading to pore closure and restriction, and cake formation.

For water-treatment applications, MF or UF membranes are configured in submerged or contained (pressurized) systems and usually operated in deadend mode. Cleaning involves hydraulic backwashing (for example, 1 min every 30 min [11] often accompanied by air scouring. During the filtration of a water source containing traces of bacteria, a direct relationship between EPS levels in the feed and the required frequency of backwashing was observed [13]. This study also highlighted that the characteristics of the colloidal fouling were dependent on the value of the applied pressure, which determined maximum flux, rather than the mode of deposition (i.e., constant flux or constant TMP operation).

Through advanced characterization of the temporal changes in the fouling nature observed with submerged MF used for surface water, Yamamura and coworkers [11] proposed a detailed fouling mechanism where the membrane pores are first covered and narrowed by large biopolymer species, followed by humic substances and divalent cations that further block the narrowed pores. After that a cake layer builds up on the membrane surface. The size of the particles to be filtered plays a significant role in the type of fouling obtained in the deadend mode: small particles at around 0.1 μm creating a more compact cake with higher specific resistance (Table 6.1, Equation 6.4) compared to those obtained with larger particles up to 1 μm [14]. As for any membrane process, the applied flux has a crucial role, and has been found to specifically determine the backwashing frequency in dilute feed deadend MF applications. Based on filtration theory the cycle time from low to high TMP is proportional to $(1/\text{flux})^2$ [15]. Since energy use depends on backwash frequency and membrane area depends on imposed flux, there is a trade-off between operating and capital costs and an optimal flux can be obtained [15].

6.2.4.2 Membrane Bioreactor (MBR)

The idea of placing a microporous membrane in direct contact with activated sludge may have been considered pioneering and risky 40 years ago, but the recent widespread application of the MBR has since been proven sustainable, when proper operating conditions are applied. The complex and labile nature of the microbial population present in the MBR process presents new challenges for membrane operators. Activated sludge is composed of suspended solids (large biological flocs, individual micro-organisms and inert particles), colloids and soluble materials (dissolved matter from the wastewater and soluble microbial products (SMP) excreted from biomass activities). A comprehensive review of MBR fouling is available [16].

Over recent years, the methods used to characterize the biomass have been diversified, allowing an improved understanding of the interaction between the compounds present in the activated sludge and the membrane. However, recent efforts to relate the EPS and SMP fractions (generally given in terms of protein and carbohydrate) to the MBR fouling propensity have not been universally successful in explaining the fouling mechanisms in MBRs [16]. Fouling mechanisms in the MBR are likely to include all 3 mechanisms in Figure 6.2 as well as biofouling. A 3 stage history for TMP rise, involving initial membrane fouling by adsorption and pore closure followed by a period of slow ('sustainable') rise and finally a TMP jump (Figure 6.1(b)), has been described [17].

Many membrane suppliers have developed filtration products specifically designed for MBR applications. Low-cost polymeric hydrophilic microporous membranes used in submerged configurations are generally suggested; their pore sizes range from 0.4 μm to 40 kDa [18]. While the large-pore MBRs rely on the formation of a fouling layer to produce high product quality, the intrinsic retention of UF-based systems are not filtration-time dependent and show good performances from the early stage of the filtration. The MBR operating conditions (SRT, HRT) are related to the quantity and quality of the wastewater to be treated and have a strong influence on the nature of the activated sludge [18]. Whilst their direct effect on the fouling propensity is still unclear there are indications that very short or very long SRT are more fouling prone. Air scour is universally used to control fouling in MBRs and the effect of air sparging on membranes has been reviewed in detail [19]. It has been frequently reported that once a certain air flow rate is exceeded, no further significant fouling limitation is observed and the rate of aeration is generally optimized.

Membrane-cleaning strategies are numerous and generally remain proprietary information. Physical cleaning by relaxation or backwashing is used on a frequent basis but the efficiency tends to decrease with filtration time. As irreversible fouling accumulates on the surface, chemical cleanings of various intensities (i.e., cleaner concentration used) can be applied on a weekly to yearly basis [20].

6.3
High-Pressure Processes

The processes of interest are NF and RO where the membranes are either 'nanoporous' or essentially nonporous. In these processes the fouling is a surface layer, the effects of which may be exacerbated by the high retention of solutes by the membrane. Operation is with crossflow and in industry fixed flux is commonly used. This section considers particulate fouling, biofouling and scale formation and then discusses the implications of 'cake enhanced' concentration polarization on fouling outcomes.

6.3.1
Particulate and Colloidal Fouling

Particulate fouling in RO is most likely to be colloidal due to the formation of a colloidal deposit layer on the membrane surface. Examples of colloidal particles include clay, iron oxide, silica, macromolecules (protein, polysaccharides and natural organic matters), and biocolloids (bacteria and viruses). These colloids are usually in the size range of 1–1000 nm (or 0.001–1 μm) and have an upper bound of 1–10 μm, but there is no strict margin [21]. In practice, the particulate fouling tendency of feed waters for RO is characterized by fouling indices, namely the silt density index (SDI) and the modified fouling index (MFI) [22, 23]. These methods are limited because they are based on the retention of colloids by MF/UF membranes under test conditions that differ from the actual RO process. Importantly these indices only predict the hydraulic resistance of the particulate cake layer, but do not measure the cake-enhanced osmotic pressure effect (refer to Section 6.3.4), which can be experienced in RO and NF.

Colloidal fouling involves the transport of the foulant from the bulk fluid to the membrane surface, followed by the particle-attachment process. In a crossflow RO system, particles are convected to and retained on the membrane surface due to the permeation flux, J (perpendicular to the surface) while the crossflow (tangential to the surface) induces particles to be back transported from the membrane surface and into the bulk solution due to the concentration gradient. A particle-polarization layer is formed on the membrane surface that is similar to the concentration-polarization (CP) layer of solute (dissolved ions) in RO. As noted above, the back-transport mechanisms in RO operation include Brownian diffusion (BD) and shear-induced diffusion (SID) [5]. Generally, for submicrometer size particles (typically $< 0.1\,\mu m$), BD is important and $J_{BD} \propto 1/a_p^{2/3}$ whereas SID applies to micrometer-sized particles (typically $> 0.2\,\mu m$) and $J_{SID} \propto a_p^{4/3}$. When a particle is in the vicinity of the membrane/solution interface, colloid–membrane interactions could determine the attachment of the particle onto the membrane surface. The interactions may be attractive (e.g., van der Waals or hydrophobic attraction) or repulsive (e.g., electrical or steric repulsion), which are best described by the classical and extended Derjaguin–Landau–Verwey–Overbeek (DLVO) theory for colloid stability [21]. In general, the nature of the interaction is greatly dependent on the nature of the particle, membrane surface and solution chemistry such as particle size, zeta potential, ionic strength of the solution, and membrane surface roughness [24, 25].

Most RO applications are in a high or raised salinity environment (seawater, brackish water and reclaimed water) where the charge interactions are greatly suppressed. Therefore, the controlling phenomena in membrane fouling are mainly the particle transport step that is a function of convection (permeation drag) and the back-transport mechanisms BD and SID. The concept of critical flux (Section 6.1.4) was introduced from observations in MF and UF [3], but intuitively it should also apply in a RO system. For colloidal fouling the onset and build up of a deposit layer is dependent on the net flux, which is the difference between the operating flux, J, and the critical flux, J_{crit}, or $(J - J_{crit})$ [26].

In RO, a colloidal deposit on the membrane introduces an additional resistance, R_F, and could also cause cake-enhanced concentration osmotic pressure (CEOP) [24]. The CEOP phenomenon is discussed in Section 6.3.4. Large-scale RO plants tend to be operated at a fixed production rate, requiring a fixed average flux.

To compensate for fouling (R_F and CEOP) it is necessary to increase transmembrane pressure (Figure 6.1(b)). The constant flux strategy has important implications. Firstly, if $J > J_{crit}$ of foulant species, that species will continue to deposit. Secondly, as CP is (exponentially) flux-driven (see Equation 6.6 in Table 6.1) it will rise due to CEOP in a self-accelerating fashion. This is in contrast to a fixed-pressure strategy where the flux declines, net convection of foulant drops and CEOP become self-limiting.

6.3.2 Biofouling

Biofouling in RO is a problem of formation of an unwanted biofilm [27]. A biofilm is defined as, 'a surface accumulation, which is not necessarily uniform in time or space, which comprises cells immobilized at a substratum and frequently embedded in an organic polymer matrix of microbial origin' [28]. Biofouling has long been recognized as one of the most problematic types of fouling in the RO process. Even after a 99.9–99.99% removal of bacteria by the use of microfiltration as the pretreatment step, biofouling in RO cannot be eliminated as it only requires a few initial colonies on the membrane surface to eventually form a mature biofilm [27]. In fact, it has been suggested that the majority of the bacterial population involved in biofilm formation are viable but non culturable (VBNC) and are about 0.2 μm in size, making them difficult to remove [29]. In order to survive, these recalcitrant bacteria adhere to the membrane surface, resuscitate (convert from a nonculturable state to a culturable form), multiply and secrete extracellular polymeric substances (EPS), and eventually form a mature biofilm. EPS provides sorption sites for water, inorganic and organic solutes and particles and therefore could induce other types of fouling such as particulate fouling.

In terms of biofilm formation the major difference between RO and other systems is the presence of permeation flux and the complex hydrodynamic conditions, for example, the use of feed spacers in the flow channel. The challenge is to establish a relationship between biofilm formation and flux-mass transfer in a RO system. In general, the biofouling process can be divided into five stages: (1) the formation of a conditioning film, (2) bacteria transport and adhesion, (3) biofilm development and accumulation, and (4) biofilm detachment. The first step in biofilm formation is the adsorption of macromolecules (e.g., humic substances, lipopolysaccharides, or other products of microbial turnover) onto the membrane surface, which is aided by the CP effect. The thin-film organic layer is known as the conditioning film. These adsorbed macromolecules can mask the original surface properties of the membrane and facilitate the attachment of micro-organisms onto the surface. Since most bacteria have an average size of about 1 μm, bacteria cell transport and adhesion is often treated equivalently to particle transport and attachment. Experimental studies that relate the initial stage of bacteria deposition onto the membrane surface with the transport models, for example, the shear-induced diffusion model, and the surface interaction forces, for example, DLVO and acid–base interactions have been demonstrated using direct observation methods [30, 31]. However, this may oversimplify the transport and adhesion process as the actual bacteria cells are semisolid (some have an overall irregular shape due to the flagella) and are surrounded by EPS. Once attached, bacteria cells may grow and proliferate into microcolonies, excreting EPS, colonizing free surface areas, and forming a mature biofilm. Since the attached bacteria multiply at the expense of nutrients, so biofilm growth could be significantly accelerated at high CP level (under high J or low k conditions), which in turn controls the amount of nutrients at the membrane wall [32]. Detachment is an interfacial transfer process, which transfers cells from the biofilm back to the bulk liquid. It was observed that biofilm detachment increases with both fluid shear stress and biofilm mass [33]. Other factors such as the availability of nutrients can also determine the detachment of the biofilm [34]. If the nutrients in the biofilm are consumed, the situation will be unfavorable for the growth of micro-organisms. In natural-water systems, the biofilm layer is limited by the balance between diffusion of nutrient and the rate of consumption of nutrient [35]. In RO, due to the presence of CP, the nutrient supply may be increased in the microenvironment near the membrane surface.

The buildup of biofilm on the membrane surface means an additional resistance to solvent flow as well as the possibility of enhancement of CP level by the biofilm, which is similar to the case of colloidal fouling [32, 36]. In general, the diffusivity is linked to the tortuosity factor of the biofilm [37]. Hence, it is likely that the backdiffusion of solutes in the biofilm on RO is hindered. The enhanced CP is important for two reasons. Firstly, the elevated concentration of solutes at the membrane wall means an increase in the osmotic pressure (CEOP) and hence a loss in the effective TMP. Secondly, the nutrient level is also enhanced and this will further accelerate the growth of the biofilm [32, 36]. So, biofouling in RO becomes an interplay between CP and biofilm development.

6.3.3 Scale Formation

Due to the retention of solutes by an RO membrane, the concentration of sparingly soluble mineral salts such as calcium carbonate, calcium sulfate, calcium phosphate, barium sulfate and so on can exceed the saturation level and cause scaling of the membrane surface. There are two pathways for membrane-scale formation, namely surface (heterogeneous) crystallization and bulk (homogeneous) crystallization [38]. In surface crystallization, the concentration of sparingly soluble salt at the membrane surface exceeds the solubility limit due to the CP effect. As a result, the membrane surface is blocked by the lateral growth of scale deposit, which greatly reduces the effective area for permeation [39]. Thus, for fixed-flux operation, the local flux increases in order to achieve the same average flux, so the local CP level is greatly enhanced due to the exponential relationship of flux and CP (Equation 6.6 in Table 6.1). This means scaling can be exacerbated due to the higher degree of supersaturation. Alternatively, crystal particles form in the bulk phase through homogeneous crystallization when concentrations of salts surpass the saturation level due to high recovery of feed waters. The crystal particles are then transported by

Table 6.3 Crystallization and scaling relationships.

Description	Relationship	Equation
Supersaturation ratio	$S = \frac{C_W}{C_S}$	(6.9)
Induction time for nuclei formation	$\log(t_{ind}) \propto \left(\frac{\sigma_n^3}{T^3(\log S)^2}\right)$	(6.10)
Rate of crystal growth	$\frac{dm_{cry}}{dt} = k_{cry}(C_W - C_S)^n$	(6.11)

the permeation flux and eventually form a porous cake layer on the membrane surface, which can be characterized by an increase in the hydraulic resistance [40], similar to particulate fouling. Furthermore, it should be noted that both scaling mechanisms could occur simultaneously in a RO system. This would be most likely if cake-enhanced concentration polarization was caused by a porous cake layer. CEOP effects could also contribute to fouling.

Membrane-scaling phenomena can be well explained by the concept of crystallization; the relevant equations are given in Table 6.3. The supersaturation ratio is defined in Equation 6.9, where C_w and C_s are the wall concentration (taking into account the effect of CP) and solubility limit of salt, respectively. So when $S > 1$, the salt has a potential to form a scale deposit. However, according to the classical nucleation theory, it takes time for the generation of nuclei, which is a precursor to crystal growth [41]. This time is called the induction time, t_{ind}, and it is related to the supersaturation ratio, as shown in Equation 6.10, where T is the temperature and σ_n is the surface energy. After the induction period, once stable nuclei have been formed, they begin to grow into crystals of finite size. The rate of crystal growth is commonly represented by Equation 6.11, where k_{cry} and n are the rate constant and order of reaction, respectively.

One of the strategies to control scaling is through control of the supersaturation level of the salts [42] and RO systems are operated at a recovery that limits the increase in the bulk concentration of salts. The concentration at the membrane wall can be manipulated by the ratio of J/k of the RO system (Equation 6.6). Other methods include, (1) the injection of acid into the feed stream to reduce the dissolved carbonate ions in order to control scaling due to calcium carbonate, (2) shifting the ion species to precipitate as a more soluble form of salts, for example, calcium forms more soluble complexes with EDTA than calcium carbonate, (3) altering the scale morphology by forming a rather thin deposit layer with less tenacity [43], which could have a lower hydraulic resistance, (4) retarding the crystallization process through extending the induction period of nucleation [44] and adding antiscalants.

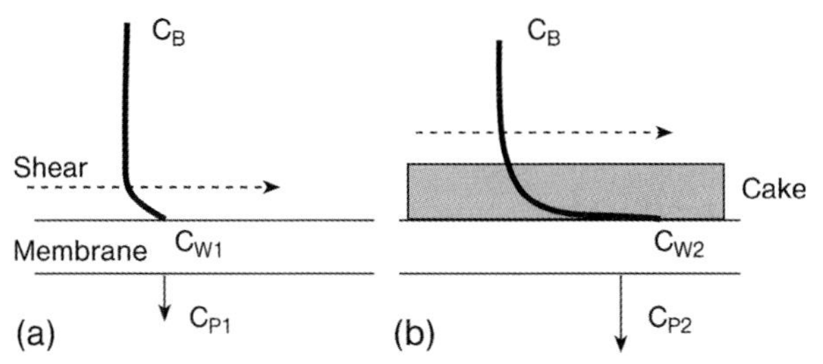

Figure 6.3 (a) CP for clean membrane (b) CECP for fouled membrane.

6.3.4
Cake-Enhanced Osmotic Pressure

As mentioned earlier, the CEOP phenomenon arises due to cake-enhanced concentration polarization CECP. Backdiffusion of retained solutes is hindered because the solutes now need to diffuse through the tortuous paths within the cake layer. The solutes in this 'unstirred' deposit layer are not exposed to crossflow and the concentration and osmotic pressure at the membrane surface are greatly enhanced. The concept is depicted in Figure 6.3 and the relevant equations are given in Table 6.4. Figure 6.3 also explains how fouling in RO (and NF) can decrease solute retention (see Section 6.1.1). The CECP phenomenon causes C_w to rise from C_{w1} to C_{w2} and this can raise permeate concentration C_p from C_{p1} to C_{p2}. Thus fouling in RO reduces both water permeability and permeate quality.

Equation 6.12 relates CP to flux (J) and the effective mass-transfer coefficient (k_{eff}) for retained solute. For a clean membrane $k_{eff} = k$, which is the boundary layer mass-transfer coefficient from Sherwood correlations [45, 46]. However, for a membrane

Table 6.4 Cake-enhanced CP relationships.

Description	Relationship	Equation
Concentration polarization	$\frac{C_W}{C_b} = \exp\left(\frac{J}{k_{eff}}\right)$	(6.12)
Effective mass transfer	$\frac{1}{k_{eff}} = \frac{1}{k} + \frac{1}{k_c}$	(6.13)
Cake mass transfer	$k_c = \frac{D_c}{\delta_c}$	(6.14)
Diffusion in cake	$D_c = \frac{D\varepsilon_c}{\tau_c}$	(6.15)
Cake tortuosity	$\tau_c \approx (1 - \ln \varepsilon_c^2)$	(6.16)

with a cake layer the backdiffusion is hindered. Equation 6.13 shows how k_{eff} is related to the unfouled surface k and the mass transfer in the cake (k_c). The cake mass transfer (Equation 6.14) is given by the diffusion in the cake, D_c, and the cake height, δ_c, where cake diffusion is the free solution diffusion modified by cake voidage, ε_c, and tortuosity, τ_c. An empirical relationship [47] relates tortuosity and voidage. For a typical value of voidage of 0.3 the value of D_c is only about 10% of D. For such a cake with $\delta_c = 20\,\mu m$ the k_c value would be about 20% of the typical k values, so for fouling conditions k tends to be $\gg k_c$ and $k_{eff} \rightarrow k_c$.

Thus, k_{eff} decreases as the fouling layer height δ_c grows, and hence CP increases. It should be noted that for a linear increase in fouling layer thickness, CP increases exponentially due to the form of Equation 6.12. For constant-flux operation of large-scale RO the fouling continues and the required pump pressure (Equation 6.1, Table 6.1) will rise. In practice the pump delivery pressure will have an upper limit (ΔP_{max}) and once this is reached the flux will not be sustainable. As flux declines, net convection of foulant drops and concentration and CEOP become self-limiting. A pseudosteady state could be achieved with ΔP at ΔP_{max} and δ_c constant. However further flux decline would tend to occur due to cake consolidation, accumulation of other foulants (NOM, fine colloids) within the cake, biofilm development or scale formation.

The relative contribution to performance loss (TMP rise at constant flux) of resistance R_F or CEOP depends on the particle size. For particles $>0.5\,\mu m$ the resistance is relatively small (see effect of particle size in Equation 6.4) and CEOP due to cake height, δ_c, is the major effect [48]. It is also observed that biofilms can contribute substantial CEOP effects as well as resistance, and in a recent biofouling study more than 50% of the required TMP rise was due to CEOP effects [32].

6.4 Conclusions

Fouling is a major factor in the application of membranes. For both low-pressure and high-pressure membranes the degree of fouling is a complex function of feed characteristics, membrane properties and operating conditions. However, much is now known about fouling and how it can be controlled. The key to low-fouling operation involves effective pretreatment of feed, careful selection of the membrane and good hydrodynamics within the module, as well as an appropriate flux.

References

1. Hermia, J. (1982) *Transactions of the Institution of Chemical Engineers*, **60**, 183–187.
2. Blanpain, P. and Lalande, M. (1997) *Filtration & Separation*, **34**, 1065–1069.
3. Field, R.W., Wu, D., Howell, J.A. and Gupta, B.B. (1995) *Journal of Membrane Science*, **100**, 259–272.
4. Bacchin, P., Aimar, P. and Field, R.W. (2006) *Journal of Membrane Science*, **281**, 42–69.

5 Belfort, G., Davis, R.H. and Zydney, A.L. (1994) *Journal of Membrane Science*, **96**, 1–58.

6 Hong, S., Krishna, P., Hobbs, C., Kim, D. and Cho, J. (2005) *Desalination*, **173**, 257–268.

7 McDonogh, R.M., Fane, A.G. and Fell, C.J.D. (1989) *Journal of Membrane Science*, **43**, 69–85.

8 Chan, R., Chen, V. and Bucknall, M.P. (2004) *Biotechnology and Bioengineering*, **85**, 190–201.

9 Lee, N., Amy, G., Croue, J.-P. and Buisson, H. (2005) *Journal of Membrane Science*, **261**, 7–16.

10 McDonogh, R., Schaule, G. and Flemming, H.-C. (1994) *Journal of Membrane Science*, **87**, 199–217.

11 Yamamura, H., Chae, S., Kimura, K. and Watanabe, Y. (2007) *Water Research*, **41**, 3812–3822.

12 Fane, A.G., Wei, X. and Wang, R. (2006) *Chapter 7*, in *Interface Science in Drinking Water Treatment* (eds G. Newcombe and D. Dixon), Academic Press.

13 Chellam, S. and Xu, W. (2006) *Journal of Colloid and Interface Science*, **301**, 248–257.

14 Van der Bruggen, B., Kim, J.H., DiGiano, F.A., Geens, J. and Vandecasteele, C. (2004) *Separation and Purification Technology*, **36**, 203–213.

15 Parameshwaran, K., Fane, A.G., Cho, B.D. and Kim, K.J. (2001) *Water Research*, **35**, 4349–4358.

16 Le-Clech, P., Chen, V. and Fane, T.A.G. (2006) *Journal of Membrane Science*, **284**, 17–53.

17 Zhang, J., Chua, H.C., Zhou, J. and Fane, A.G. (2006) *Journal of Membrane Science*, **284**, 54–66.

18 Judd, S. (2006) *The MBR book: Principles and Applications of Membrane Bioreactors in Water and Wastewater Treatment*, Elsevier, Great Britain.

19 Cui, Z.F., Chang, S. and Fane, A.G. (2003) *Journal of Membrane Science*, **221**, 1–35.

20 Le-Clech, P., Fane, A., Leslie, G. and Childress, A. (2005) *Filtration & Separation*, **42**, 20–23.

21 Elimelech, M., Gregory, J., Jia, X. and Williams, R.A. (1995) *Particle Deposition and Aggregation: Measurement, Modelling and Simulation*, Butterworth-Heinemann, Woburn, Massachusetts.

22 Boerlage, S.F.E., Kennedy, M.D., Dickson, M.R., El-Hodali, D.E.Y. and Schippers, J.C. (2002) *Journal of Membrane Science*, **197**, 1–21.

23 Schippers, J.C. and Verdouw, J. (1980) *Desalination*, **32**, 137–148.

24 Hoek, E.M.V., Bhattacharjee, S. and Elimelech, M. (2003) *Langmuir*, **19**, 4836–4847.

25 Brant, J.A. and Childress, A.E. (2004) *Journal of Membrane Science*, **241**, 235–248.

26 Schwinge, J., Neal, P.R., Wiley, D.E. and Fane, A.G. (2002) *Desalination*, **146**, 203–208.

27 Flemming, H.C. (1997) *Experimental Thermal and Fluid Science*, **14**, 382–391.

28 Characklis, W.G. and Marshall, K.C. (1990) *Biofilms: A Basis for an Interdisciplinary Approach in Biofilms* (eds W.G. Characklis and K.C. Marshall), John Wiley & Sons Inc., New York, pp. 3–15.

29 Winters, H. (2005) International Desalination Association World Congress 2005 IDA: Singapore, pp. SP05–200.

30 Boerlage, S.F.E., Kennedy, M.D., Aniye, M.P., Abogrean, E., Tarawneh, Z.S. and Schippers, J.C. (2003) *Journal of Membrane Science*, **211**, 271–289.

31 Kang, S.T., Subramani, A., Hoek, E.M.V., Deshusses, M.A. and Matsumoto, M.R. (2004) *Journal of Membrane Science*, **244**, 151–165.

32 Chong, T.H., Wong, F.S. and Fane, A.G. (2008) *Jounal of Membrane Science*, **325**, 840–850.

33 Sharma, P.K., Gibcus, M.J., van der Mei, H.C. and Busscher, H.J. (2005) *Applied and Environmental Microbiology*, **71**, 3668–3673.

34 Hunt, S.M., Werner, E.M., Huang, B., Hamilton, M.A. and Stewart, P.S. (2004) *Applied and Environmental Microbiology*, **70**, 7418–7425.

35 Wood, B.D. and Whitaker, S. (1998) *Chemical Engineering Science*, **53**, 397–425.

36 Herzberg, M. and Elimelech, M. (2007) *Journal of Membrane Science*, **295**, 11–20.

37 Melo, L.F. (2005) *Water Science and Technology*, **52**, 77–84.

38 Lee, S. and Lee, C.H. (2005) *Water Science and Technology*, **51**, 267–275.

39 Borden, J., Gilron, J. and Hasson, D. (1987) *Desalination*, **66**, 257–269.

40 Pervov, A.G. (1991) *Desalination*, **83**, 77–118.

41 Mullin, J.W. (1993) *Crystallization*, Butterworth-Heinemann, Oxford.

42 van de Lisdonk, C.A.C., Rietman, B.M., Heijman, S.G.J., Sterk, G.R. and Schippers, J.C. (2001) *Desalination*, **138**, 259–270.

43 Tzotzi, C., Pahiadaki, T., Yiantsios, S.G., Karabelas, A.J. and Andritsos, N. (2007) *Journal of Membrane Science*, **296**, 171–184.

44 Hasson, D., Drak, A. and Semiat, R. (2003) *Desalination*, **157**, 193–207.

45 Da Costa, A.R., Fane, A.G., Fell, C.J.D. and Franken, A.C.M. (1991) *Journal of Membrane Science*, **62**, 275–291.

46 Schock, G. and Miquel, A. (1987) *Desalination*, **64**, 339–352.

47 Boudreau, B.P. (1996) *Geochimica et Cosmochimica Acta*, **60**, 3139–3142.

48 Zhang, Y.P., Chong, T.H., Fane, A.G., Law, A., Coster, H.G.L. and Winters, H. (2008) *Desalination*, **220**, 371–379.

7
Energy and Environmental Issues and Impacts of Membranes in Industry

William J. Koros, Adam Kratochvil, Shu Shu, and Shabbir Husain

7.1
Introduction

Short-term economics favor rapid deployment and guaranteed reliability, so energy efficiency and environmental sustainability are often secondary considerations in implementation of new processes. Options with lower energy efficiency and higher environmental impact may be favored over membrane processes, which tend to be less familiar to design engineers and may even require some development time and risk. Moreover, while government regulations can encourage adoption of environmentally beneficial approaches such as membranes, global regulations are difficult to implement. Nevertheless, over the longer run, practitioners can seize a competitive advantage by moving forward ahead of regulations to define the technological landscape. In the early 1970s, Japanese automotive innovations in fuel efficiency enabled a very strong position to be captured in automotive production three decades later. A similar early stage opportunity exists now with regard to large-scale separations processes for production of commodity chemicals, fuels, and water. In many ways, the water-purification sector is more advanced than the other two sectors, and understanding why the broader nonaqueous separation sector has lagged behind is useful. Such understanding provides a framework to efficiently extend the advantages of the membrane platform across the spectrum of separations pertinent to commodity production. While some of this information has already been reported, this chapter provides updated information and significant expansion on the future of membrane separations [1].

Linkage exists between separation energy efficiency and long-term environmental sustainability, and some facts help to clarify this connection. By United Nations estimates, the world currently has 6.7 billion global inhabitants; and, only 1.2 billion people live in 'more-developed countries'[1] such as North America, western Europe, and so on, while 5.5 billion reside in 'less-developed countries' [2]. Estimates suggest

1) A highly industrialized country characterized by significant technological development, high per capita income, and low population growth rates. Examples of such countries include the United States, Canada, Japan, and many countries in Europe.

Membrane Operations. Innovative Separations and Transformations. Edited by Enrico Drioli and Lidietta Giorno

ISBN: 978-3-527-32038-7

that 9.2 billion inhabitants will occupy our planet by 2050, and the ratio 9.2/1.2 = 7.7 provides an upper bound estimate of the increased energy use relative to 2007 to provide an equivalent standard of living *to all global inhabitants by 2050* [2]. Asia is currently classified as 'less developed', but this region has explosively growing economies with visions of joining the ranks of 'more-developed' countries. Moreover, the UN estimates that this region will comprise a population of 5.3 billion by 2050 – over half of the world's inhabitants! A more conservative estimate of the likely world energy consumption in 2050 might be the factor of $(5.3 + 1.2)/1.2 = 5.4$, or roughly $5\times$, and this factor will be used for discussion purposes in this chapter to estimate industrial-sector use.

Raising the standard of living for such a massive group requires huge increases in commodities, including clean water, nonpolluting fuels, and chemicals. From the standpoint of separation processes, providing commodities to the 5.3 billion additional inhabitants in 'more-developed countries' by 2050 is truly a 'grand challenge.' A useful benchmark to guide thinking with regard to separation issues is the well-studied US case where the industrial sector is responsible for 33% of total energy consumption. Over 40% of the energy consumption in the massive chemical and refining and petrochemical industry is consumed by separation processes [3]. Using the scaling $0.4 \times 0.33 = 0.132$, it is reasonable to estimate that 13.2% of total energy consumption can be associated with separation operations. The implications of a 'business as usual' scaling to accommodate a projected $5\times$ increase in global commodities would correspond to a 66% increase in current energy consumption that is associated with *all aspects of the global economy in 2007.* Since the bulk of such energy will come from hydrocarbon sources for many years, this energy burden would bring with it a similar increase in CO_2 emissions and present a major hurdle to worldwide economic sustainability. Fortunately, membranes offer a viable option to address the separations part of this grand challenge. To be effective, however, membrane technology must be introduced *prior to installation of energy inefficient thermally intensive processes.* Indeed, if energy-inefficient process are installed, their long (30–50 year) useful lives will require regulatory intervention to force their replacement.

To support the latter claim, consider production of fresh water by desalting brackish and sea water brines. Currently, there are desalting facilities worldwide with the capacity of 9 billion gallons of water per day, and roughly half is membrane-based and half is thermally driven (e.g., multi-effect and flash evaporators). Since the advent of modern reverse-osmosis (RO) desalting technology, almost all new desalting capacity is based on membranes. Nevertheless, despite more than a $10\times$ higher energy efficiency of membranes, which will be shown in a later section, the old thermal plants remain in use. The same situation can be expected in the chemical and petrochemical industry if scale up to handle the $5\times$ capacity expansion is done by conventional thermally intensive approaches. This reality places added urgency on the need for expedited development of membrane-based processes that expand beyond traditional aqueous purification of brines and micro- or ultrafiltration of aqueous feeds.

A brief review of experience with aqueous feeds will be provided to identify lessons learned that can help expedite expansion of the membrane platform to nonaqueous feeds. The subsequent discussion focuses on large-scale examples where significant reductions in energy consumption (*and hence* CO_2 *emissions*) appear feasible by replacing energy-intensive approaches with membrane processes. Although many net driving forces can be imposed on each penetrant between upstream and downstream membrane faces, transmembrane pressure differences are the most common and are the focus of this discussion.

Most practical membrane processes are continuous steady-state operations with a feed, permeate, and nonpermeate stream. Since membrane processes involve separation of a permeated component A from a second, rejected component B, a measure of separation efficiency is useful. Due to the diversity of applications, many different measures of separation efficiency are used in the various membrane subareas. Probably the easiest to use measure is the so-called 'separation factor,' given in Equation 7.1, which shows the relative enrichment of component A vs. B due to the membrane process [4]:

$$\mathrm{SF} = \frac{\left(\dfrac{\text{Composition of A downstream}}{\text{Composition of B downstream}}\right)}{\left(\dfrac{\text{Composition of A upstream}}{\text{Composition of B upstream}}\right)} \tag{7.1}$$

Since the SF is a 'ratio of ratios,' any measure of composition (mole fraction, mass fraction, concentration, etc.) can be used in Equation 7.1 as long as one consistently uses the same measure for both upstream and downstream phases in contact with the membrane. Locally within a module, the ratio of compositions leaving the downstream face of a membrane equals the ratio of the transmembrane fluxes of A vs. B. Local fluxes of each component are determined by relative transmembrane driving forces and resistances acting on each component. The ratio of the feed compositions in the denominator provides a measure of the ratio of the respective driving forces for the case of a negligible downstream pressure. This form normalizes the SF to provide a measure of efficiency that is ideally independent of the feed composition.

For a given driving force, minimization of the membrane resistance requires the smallest possible effective membrane thickness, ℓ. The ability to minimize ℓ without introducing defects relies upon 'micromorphology control,' and this topic impacts virtually all membrane applications.

7.2 Hydrodynamic Sieving (MF and UF) Separations

Microfiltration (MF) and ultrafiltration (UF) involve contacting the upstream face of a porous membrane with a feed stream containing particles or macromolecules (B) suspended in a low molecular weight fluid (A). The pores are simply larger in MF membranes than for UF membranes. In either case, a transmembrane pressure difference motivates the suspending fluid (usually water) to pass through physically observable *permanent pores* in the membrane. The fluid flow drags suspended particles and macrosolutes to the surface of the membrane where they are rejected due to their excessive size relative to the membrane pores. This simple process concentrates particles or macromolecules in the upstream nonpermeate stream and produces essentially pure low molecular weight permeate downstream (SF $\rightarrow \infty$) if the pore-size distribution prevents any 'B' from passage across the membrane.

Removal of a bulk liquid often represents the major energy cost for processing suspended particles and macromolecular solutes including paints, foods, and myriad waste-recycle streams. For dilute and semidilute feeds (<15 vol%), both MF and UF enable large energy savings compared to evaporation approaches [5, 6]. Despite pumping expenses to drive permeation and minimize accumulation of a rejected component at the membrane surface, energy costs typically range between 0.15–5.0 (kw h)/m^3 of water removed [5, 6]. Generation of electricity using high-pressure steam gives typical efficiencies of 33% or less [7], thereby increasing the 'thermal equivalent' energy cost for the membrane option. Using a median value of 2.5 (kw h)/m^3 for such MF and UF membrane processes and accounting for 33% efficiency of steam-generated electricity, a value of (2.5/0.33) = 7.6 (kw h)/m^3 results. Despite this 'penalty,' the membrane option offers roughly a *10-fold* savings over competitive thermal removal [$\sim$73 (kw h)/m^3] by flash evaporation [8]. Even if thermal energy input is needed in a final finishing step, using membranes in primary concentration steps can provide large overall processing cost savings.

7.3 Fractionation of Low Molecular Weight Mixtures (NF, D, RO, GS)

As the size of both the permeated and rejected components become less than 20 Å, as shown in Figure 7.1, hydrodynamic sieving forces are no longer adequate to perform the subtle size and shape discrimination required. Indeed, the progressions from nanofiltration (NF) $\rightarrow$ reverse osmosis (RO) $\rightarrow$ gas separation (GS) processes represent increasingly more challenging discrimination between entities that often differ by 3–5 Å for nanofiltration or reverse osmosis, down to only fractions of an angstrom for gases. In all of these cases, *intermolecular forces* become dominant determinants of the resistance acting on each penetrant.

For such micromolecularly selective processes, an additional 'partitioning' phenomenon must also be considered in the flux expression to enable describing the

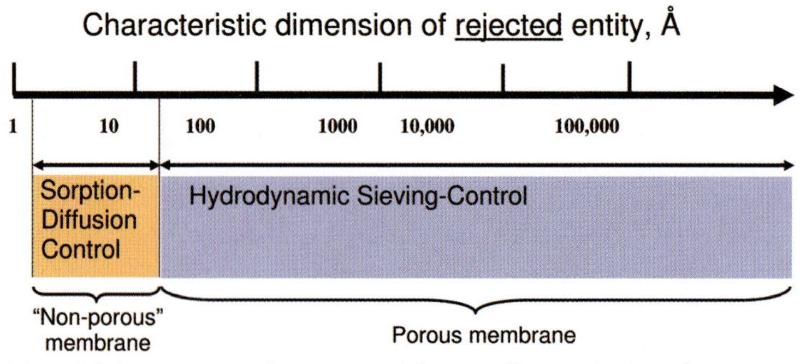

Figure 7.1 Size spectrum of permeate and the controlling mechanism of transport.

process conveniently in terms of external phase conditions. In this case, the partitioning phenomenon can be accommodated as a factor contributing to transport using a 'partition coefficient' typically defined as:

$$K_i = \frac{[\text{Composition of component } i \text{ in membrane}]}{[\text{Composition of component } i \text{ in external phase}]} \tag{7.2}$$

Since K_i is expressed as a ratio, any consistent measure of composition in the membrane and external phases may be used in Equation 7.2. When $K_i > 1$, the membrane acts as a 'concentrator' that attracts component i from the external phase and makes it available at the membrane surface for transmembrane movement. Intermolecular forces of solvation and mixing that are responsible for the partitioning process may be entropic as well as enthalpic in origin. The balance of these forces acting between the membrane and external phase can cause either a higher or lower concentration of a given solute inside the membrane relative to the external phase. If the tendency to enter the membrane is negligible, the partition coefficient approaches zero, that is, $K_i \rightarrow 0$.

The synergistic action of the size-discriminating and partitioning phenomena permits adjustment of the relative compositions of different small molecules or ions in streams contacting the upstream and downstream faces of a membrane. For a given penetrant pair, the ratio of the effective resistance acting on B vs. that acting on A in the membrane specifies the membrane-specific ability to separate this A–B pair. Since the thickness factor, ℓ, cancels, the key ratio of resistances acting on component B vs. A is comprised of a product of partitioning and mobility ratio factors. For most membranes, the mobility ratio can be approximated as D_A/D_B, the ratio of the average diffusion coefficients for component A vs. B within the membrane phase. In this common case, therefore the effective ideal membrane selectivity, α_{AB}, is given by:

$$\alpha_{AB} = \frac{[D_A]}{[D_B]}\frac{[K_A]}{[K_B]} \tag{7.3}$$

Equation 7.3 notes that one can tune *both* 'mobility selectivity', D_A/D_B, and 'partitioning selectivity', K_A/K_B to develop advanced materials for every small molecule separation [9]. This strategy can be applied to virtually any type of membrane material ranging from gels to crystalline zeolites, metals, glasses, or polymers. Moreover, *hybrid materials* comprised of combinations of more than one such material (e.g., a zeolite dispersed in a polymer) allow limitations associated with any specific pure component material type to be overcome. For instance, intrinsic rigidity responsible for outstanding mobility selectivity in zeolites also causes brittleness and difficulties in their high-speed processing. Polymers are processable, but lack the rigidity to perform fine mobility selectivity. Mixtures of zeolites and polymers or molecular-sieve carbons and polymers are now being investigated to create highly selective hybrid materials amenable to economical high-speed fabrication [10, 11]. As will be discussed in more detail, such materials are likely to be increasingly important for dealing with a broad range of future applications.

7.4 Reverse Osmosis – The Prototype Large-Scale Success

As noted earlier, reverse-osmosis (RO) purification of water was the first large-scale commercially viable membrane fractionation of *low molecular weight* liquid mixtures. Like all of the cases involving low molecular weight fractionation, RO purification of potable water from brine relies upon 'partitioning selectivity' and 'mobility selectivity' contributions from Equation 7.3. Optimization of the membrane material and structures for this application took place over a period of more than two decades, and membranes are now rapidly displacing thermal desalting [8, 12]. By understanding how and why RO has displaced distillation in this large-scale application, one can see how to help expand the energy-efficient membrane paradigm more broadly.

Both gas separation and RO require high feed pressures to achieve useful fluxes. Nevertheless, the *utility* of having a *high-pressure* nonpermeate stream leaving the module differs greatly for GS vs. RO cases. For gases, the energy used to compress feed streams is valuable in subsequent processing and product storage. On the other hand, for liquids, after the RO separation is completed 'excess pressure' in the nonpermeate is not needed. Reclaiming this energy is now standard procedure in state-of-the-art RO systems. 'Pervaporation' is a variant of reverse osmosis that uses a low-pressure liquid feed with a *vapor permeate* under vacuum. Effectively, pervaporation involves permeation and evaporation of a portion of the feed, thereby requiring significant thermal energy input [5]. While overcoming the need for high-pressure feed and nonpermeate energy recovery, pervaporation still requires the input of considerable thermal energy. Because reverse osmosis eliminates this thermal inefficiency associated with pervaporation, it became the favored process for water desalination.

The 'effective' driving force for reverse-osmosis permeation of water is proportional to the difference in applied transmembrane pressure, ΔP, and the transmembrane osmotic pressure, $\Delta\Pi$, viz, $(\Delta P - \Delta\Pi)$ [13]. For 50% recovery of feed entering with 34 000 ppm of total dissolved salts in seawater, the stream leaving the module has a very large osmotic pressure. This osmotic pressure must be overcome to produce the last increment of potable water product leaving the module. As noted earlier, providing a large transmembrane ΔP *without* paying an excessive energy cost is commonplace in state-of-the-art reverse-osmosis operations with compact energy recovery turbines [12]. In principle, this practice of recovery of unused energy in compressed nonpermeate streams should be transferable to organic systems as well; however, materials of construction and seals require development for compatibility with organic vs. aqueous feeds.

A state-of-the-art RO seawater system processes *50 million gallons per day* with 50% feedwater recovery as potable water product using a 940-psi (~65 bar) feed pressure [12]. These high pressures and flows are now routinely accommodated economically with compact vessels and high productivity membranes. An optimized thermal distillation plant with the same feedwater requires 1014 Btu/gal [78.5 (kw h)/m^3] of water produced [8], while the state-of-the-art seawater RO system has an energy cost of only 2.2 (kw h)/m^3 [8, 12]. Using the current paradigm of steam cycle generation of electricity with an efficiency of only 33%, the effective 'thermal equivalent' energy cost for the membrane process is (2.2/0.33) $(\text{kw h})/\text{m}^3 = 6.7\ (\text{kw h})/\text{m}^3$. Again, even with such a 'penalty' factor, the membrane option is over *10-fold more efficient* than the thermal approach.

The well-known thermodynamic *inefficiency* in generation of electricity using high-pressure steam can be linked to the unfortunate widespread acceptance of the inefficiency of doing thermally driven separations. Specifically, generation of electricity using high-pressure steam produces excess low-pressure steam: this fact is often used to justify continuation of inefficient separation processes driven by this excess low-value steam [14]. In fact, discussions of thermal separation efficiencies are sometimes based on the efficiency of an ideal heat engine operating between the reboiler and condenser temperatures. Such an approach overlooks the intrinsic limitations of *all* thermally driven processes and perpetuates the unnecessary linkage between thermal energy conversion processes and separation processes. More discussion of issues related to power generation and membrane roles in reducing environmental impact will be offered later.

It is well known that electrochemical oxidation of a fuel to extract power can theoretically be performed in a fuel cell much more efficiently than is possible via a heat cycle. For example, a H_2/O_2 fuel cell reaction at 25 °C has an ideal efficiency of 100%, as compared to 30–33% in standard steam cycles. Current fuel cells still require improvement, and rarely exceed 50–60% efficiency; however, this already surpasses the 33% efficiency for standard steam systems [15]. Realistically, however, scaling such devices to hundreds of megawatt size presents challenges.

Even without the ideal efficiency of a fuel cell, combined-cycle integrated gasification processes are providing significant improvements with efficiencies nearly as high as 50% [16, 17]. Using 50% as the efficiency limit, such a unit coupled to a reverse-osmosis unit would show an improvement of $73/(2.2/0.50) > 16$-fold better than the thermal separation alternative! Whether one considers the already achievable 10-fold reduction with a conventional coal-fired steam turbine or the 16-fold reduction achievable by eventually coupling this membrane process with a high-efficiency integrated gassifier, the numbers are impressive. These numbers also give a vision of a much more energy efficient future if the membrane platform is extended to *nonaqueous applications.*

7.5
Energy-Efficiency Increases – A Look to the Future

The following cases consider advantages in energy savings that extend beyond aqueous filtration and reverse-osmosis applications noted above. Although not yet offering a full factor of 10 savings, this is the same path toward dominance that aqueous separations followed. Almost a decade ago, the concept of a 'disruptive technology' was introduced to describe new approaches that were radically different from the incumbent leading technology in a field. Such technology was also noted in some cases to even perform less well than the incumbent leader but with optimization ultimately improved to become a major player, if not *the* dominant player in a field. In this context, membrane processes are a *potentially* disruptive technology.

7.5.1
Success Stories Built on Existing Membrane Materials and Formation Technology

Valuable savings are possible even using available gas and vapor-separation membrane units, while aggressively pursuing development of nonaqueous RO and its larger energy payoffs over the next decade. Vapor-separation processes are operationally similar to gas-separation units but often use a moderate vacuum downstream, depending upon the vapor pressure of the components at the feed temperature.

A number of applications have been suggested for the removal of organic vapors from gas streams. These include monomer recovery from storage-tank vent streams in the production of polyolefins (e.g., polyvinyl chloride, polyethylene, polypropylene) which will be discussed in a later section [18], removal of natural-gas liquids from fuel gas for gas engines and turbines [19] and removal of solvents from air [20–22]. These applications utilize the high condensability of vapors to achieve high separation efficiencies between condensable and noncondensable components. Unlike separations involving permanent gases, where diffusion selectivity is the dominant factor, these membranes rely upon a so-called 'reverse selective' process based on very high sorption selectivities to achieve separation. As an example, butadiene-acrylonitrile rubber was cited as having a selectivity of around 100 000 for benzene over air [22].

Required selectivities for viablility of the separation with vapors typically lies at selectivities around 100 to 200 to minimize the gas component in the vacuum section [22]. Higher selectivities significantly above this range provide only marginal improvement; since the economics of the separation are driven by the value of the condensable component recovered and reduction in VOC emissions. Using a rubbery membrane in combination with a flash unit and condenser, recovery of up to 500–1000 lb/h of monomer and processing solvent from polymer storage bin purge waste gas streams has been reported with savings of $1 million/year/purge bin [18].

Beyond organic vapor capture applications, more standard membranes involving natural gas represent a large and attractive market for gas-separation membranes. The SACROC installation was one of the first major applications of gas-separation membranes in large-scale separations. This application deals with removal of CO_2 from natural gas associated with crude oil. The Kelly–Snyder field was discovered in 1948 with an estimated size of 2.1 billion bbl of oil. The initial reservoir was produced using water flooding that was later replaced by carbon dioxide injection by the field operator, Chevron. A Benfield (hot promoted potassium carbonate) process and amine scrubbing were employed to remove the CO_2 from the associated gas in the initial stages of operations, prior to the development of membranes. An eventual increase in CO_2 content of the associated gas stream from 0.5 to 40 mole% necessitated an expansion of the CO_2 handling capacity. Membrane units were considered to be ideal due to their modularity that allowed easy scaling, thereby foregoing major capital costs associated with expanding the amine and carbonate units before the added capacity was needed. This staged increase in capacity is another particularly attractive aspect of membrane-based processes. Cellulose acetate hollow-fiber membranes were provided by Cynara who also operated the membrane plant, and initial testing of the membranes was carried out at the Chevron facility to increase confidence in the new technology. A membrane lifespan of 5 years was reported in the presence of adequate pretreatment [23]. Since the initial deployment of membranes, the unit has been expanded from the original 70 MMscfd to 600 MMscfd in 2006 with a gas feed of 87 mole% CO_2 [24, 25]. In many ways, this case delivers on the potential of membranes to be expandable to large applications beyond simple aqueous feeds.

Membrane technology for natural-gas separations is gaining broad acceptance and a number of major membrane-separation plants have come into operation in recent years. These include the Cakerawala production platform (CKP) that processes 700 MMScfd with a 37% CO_2 feed and a plant in Qadirpur, Pakistan that processes 500 MMscfd of 6.5 mole% CO_2 feed to 2% CO_2 pipeline specification [26]. Current plans are being made to double production at the CKP facility [27].

In these applications, membranes offer the unique ability to configure compact systems to perform the desired separation. In some cases, membrane modules placed in series or in parallel enable debottlenecking, while in other cases such units can improve the overall efficiency of the separation. Depending on the needs of the separation, the following figures illustrate possible configurations for membrane separations. Figure 7.2 shows a simple two-stage membrane-separation process where the nonpermeate of the second stage is recycled to the feed of the first stage. A more complex configuration is presented in Figure 7.3, where both the permeate and nonpermeate streams go through a two-stage membrane process in order to achieve a higher purity of both products.

An example of the compact nature and modularity of membrane units is shown in Figure 7.4 where the two membrane units in the foreground replaced the amine-absorption system in the background for removal of CO_2 from natural gas. This application was mentioned earlier with regard to SACROC and subsequent larger offshore applications where space is at a premium.

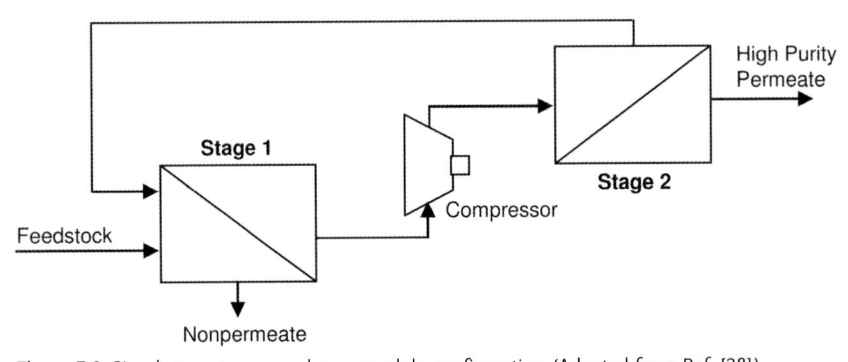

Figure 7.2 Simple two-stage membrane module configuration (Adapted from Ref. [28]).

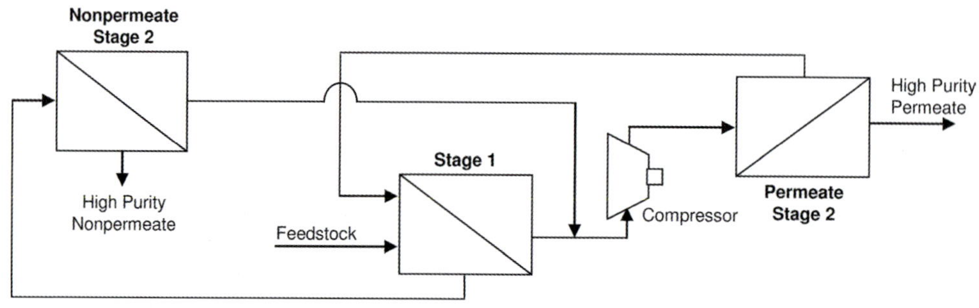

Figure 7.3 Dual two-stage membrane module configuration (Adapted from Ref. [28]).

Hydrogen purification represents another ideal fit for membrane-based separations in many cases. Because of its small molecular size relative to other gases, combined high fluxes and selectivities are often possible, since the diffusional selectivity in Equation 7.3 greatly favors hydrogen, even when the sorption selectivity does not. Initial deployment was carried out in ammonia plants to recover and recycle hydrogen from the product stream. This easy separation has led to a saturation of the ammonia market, with almost all units employing membranes [29]. A second more challenging, and even larger, market exists in the recovery of hydrogen in refinery processes. Increasing use of heavy and sour crude oils require ever larger quantities of hydrogen for oil upgrading to adjust the carbon:hydrogen ratio for lower carbon fuel. The heavier oils are cracked down ideally to pentanes and higher but also result

Figure 7.4 Air Liquide/Medal membrane unit replacement of amine scrubbing towers (used with permission).

in the formation of light hydrocarbons such as methane, ethane, and propane [30]. These gases act as inerts and reduce the vapor pressure of hydrogen in the reactor and must be periodically purged. Typically, 4 moles of hydrogen are lost for every mole of light hydrocarbon removed [4]. Initial deployment of membranes in hydrogen separations was driven by their exceptional payback and modular design, which allowed for their inclusion in existing refinery process lines with little modification. Replacing high-pressure purging and gas absorbers in the hydrocracking process discussed above by using membranes can decrease hydrogen losses by up to 16-fold. Depending on the process requirements of product purity, hydrogen recovery, and product pressure, the economics can justify the use of membrane technology versus traditional adsorption, cryogenic distillation and pressure swing adsorption. Examples of such comparisons are well covered by Zolandz and Fleming [4] and by Baker [30].

Temperature control of the feed stream is critical to membrane operation in order to prevent condensation of the hydrocarbons as hydrogen is removed. As the dew point is reached, the condensing hydrocarbons can lead to plasticization and membrane failure. While the feed temperature is typically kept at 15–20 °C above the dew point of the retentate stream, process upsets or feed changes can still lead to membrane failure. Although as many as 100 membrane plants have been installed in refineries, the global market remains far from saturation. Membranes with increased resistance to plasticization and higher-temperature operation or the use of improved pretreatment would result in greater confidence on the technology and widespread adoption [30].

7.5.2 Future Opportunities Relying Upon Developmental Membrane Materials and Formation Technology

Besides the above success stories, reconfiguring existing thermally driven processes to produce vapor feeds to membrane units for targeted fractionations of valuable components could be an attractive evolutionary strategy. However, as economical nonaqueous RO capability develops, these processes should phase out the older thermal units in the same rapid evolutionary manner that is currently occurring in the aqueous RO arena. As noted earlier, even rapid evolution takes time (10–15 years), as it did for aqueous systems. Such a process should begin now to avoid further proliferation of additional energy-inefficient separation units to meet expanding capacity needs. The following 'forward looking' cases consider opportunities where membranes could have a large impact; however, new membranes will be needed as opposed to the previous cases where existing membranes are adequate.

7.5.2.1 High-Performance Olefin–Paraffin Separation Membranes

The olefins ethylene and propylene are highly important synthetic chemicals in the petrochemical industry. Large quantities of such chemicals are used as feedstock in the production of polyethylene, polypropylene, and so on [31]. The prime source of lower olefins is the olefin–paraffin mixtures from steam cracking or fluid catalytic cracking in the refining process [32]. Such mixtures are intrinsically difficult to

Table 7.1 Physical properties of ethane, ethylene, propane, and propylene.

	Ethylene	Ethane	Propylene	Propane	Ref.
Boiling point (°C)	−103.9	−88.6	−48	−42.2	[33]
Lennard-Jones parameter, σ_{LJ} (Å)	3.7	4.1	4.0	4.3	[34]

separate due to the similar physical properties of the saturated/unsaturated hydrocarbons, as shown in Table 7.1 [33, 34].

Another important factor that distinguishes this separation is that it is not environmentally or economically feasible to simply return a rejected stream to the environment, as in a typical aqueous RO process where the brine can be returned to the ocean. The federal regulations mandate that CO_2 emissions from refineries and chemical plants be reduced to low levels; therefore, facilities can no longer afford to dispose of waste hydrocarbon streams in their flare systems. Pure streams from polyolefin reactors and vents from polymer-storage facilities, which were once flared, must be redirected to recovery systems. To reduce the economic penalty of environmental compliance, these paraffin and olefin mixtures must be recovered and recycled. In other words, two products must be made, a useful fuel and a useful chemical product, hence more process engineering is required in order to achieve such an objective.

A US DOE report estimated that 1.2×10^{14} BTU/year are used for olefin/paraffin separations [35]. The conventional technology to separate olefin/paraffin mixtures is cryogenic distillation, as illustrated in Figure 7.5. The separation is performed at elevated pressures in traditional trayed fractionators. C2 and C3 distillation columns are often up to 300 feet tall and typically contain over 200 trays. Although the separation to achieve chemical grade purity can be accomplished in a single tower [36]. Purifying ethylene/propylene to polymer grade requires a significant increase in the number of fractionating trays or the reflux ratio or both [36]. The large capital expense and energy cost have created the incentive to seek alternative technology for this olefin/paraffin separation.

Membranes offer excellent potential as an alternative for traditional distillation technology. A significant amount of research has shown the potential of membranes in the olefin/paraffin separation arena, which will be briefly discussed later in this section. Capacity expansions of existing thermally driven separation units are ideal ways to introduce membranes more broadly into large-scale use while minimizing risks and building familiarity with this relatively new technology. This approach leverages existing investments without the need to build entirely new thermally driven separation units. Within an existing integrated plant, valuable compounds in a vapor feed stream currently sent to another thermally driven separation unit could be *membrane-fractionated* into higher value products with minimal expense and significant energy savings. Figure 7.6 demonstrates an example of possible implementation of membrane units with C2 and C3 splitters. The introduction of membranes will lead, depending on the separation characteristics of the membrane material, to a significant reduction of the stream fed to the splitter. A possible reduction in capacity of the splitter column might be of the highest interest because the

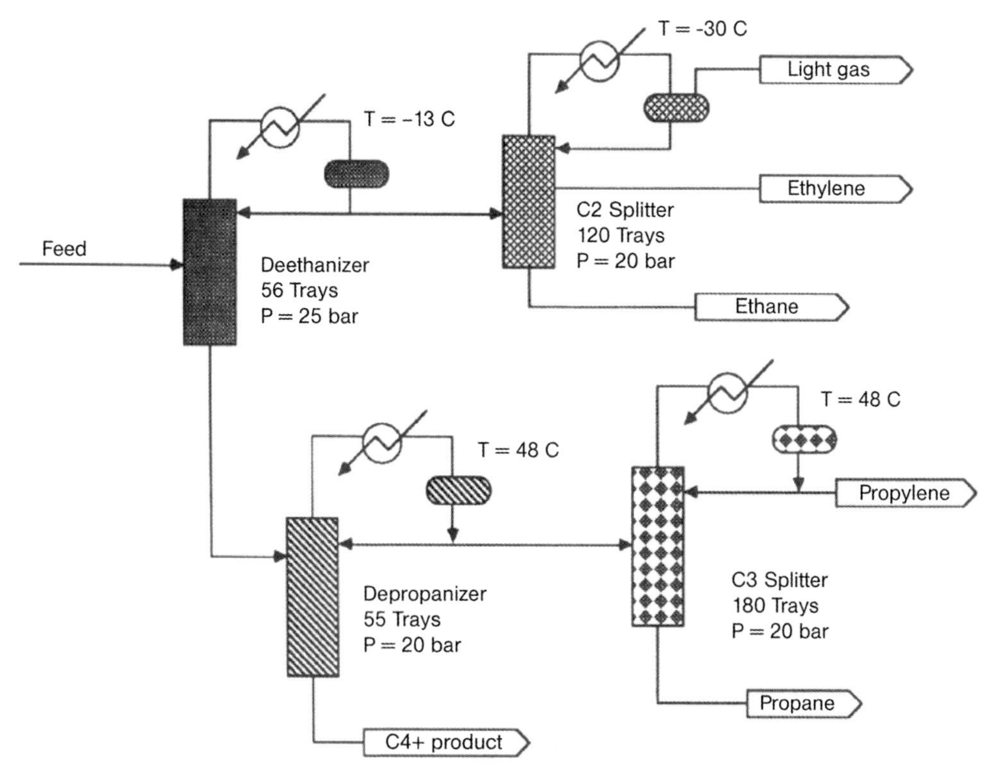

Figure 7.5 Conventional cryogenic distillation process for an olefin/paraffin process (Adapted from Ref. [35]).

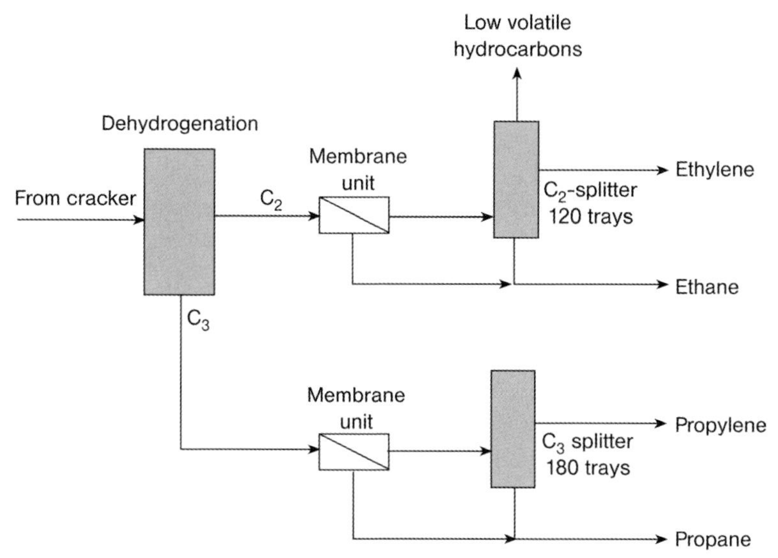

Figure 7.6 Conventional separation of C2 and C3 mixtures integrated with possible membrane units (Adapted from Ref. [37]).

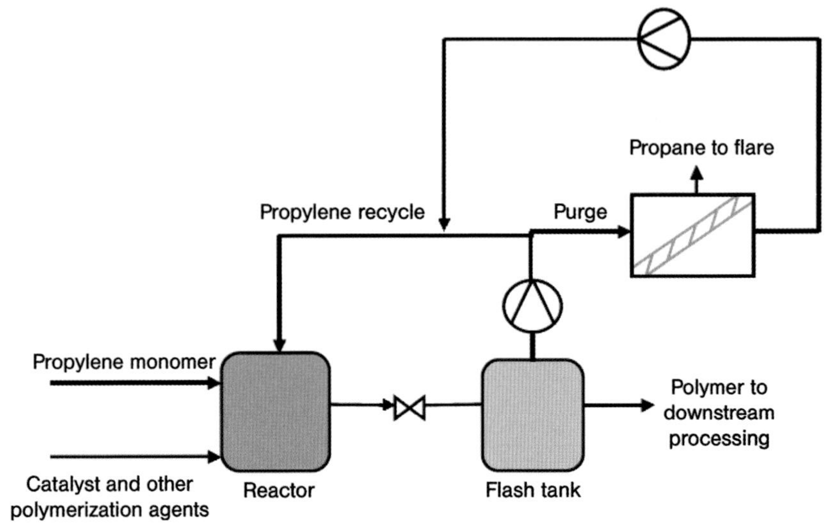

Figure 7.7 Use of a membrane unit to recover and recycle propylene to the polymerization reactor (Adapted from Ref. [38]).

olefin/paraffin separation train is more than half of the total capital cost of an olefin plant. Another example suggested by Baker is illustrated in Figure 7.7 [30]. The integration of a membrane unit into a polypropylene plant could potentially recover previously wasted propylene monomer. This membrane would remove propane, which enters as an impurity in the feed, to allow recycling of the monomer without the potentially hazardous buildup of propane.

Consider, for instance, the 515 Btu/lb (0.151 kw h/lb) reboiler energy is required for the propylene/propane separation using cryogenic distillation [3]. With a typical 50/50 feed and recovery of a 99.5% propylene product, this corresponds to roughly 0.302 kw h/lb propylene product. A recent patent on a vapor permeation membrane cites an energy cost of roughly 0.050 kw h/lb propylene product for this separation with a membrane having intrinsic properties similar to those currently reported in the literature [28]. As in the water RO case, accounting for the current paradigm of steam-cycle-generated electricity with a typical efficiency of only 33% gives the effective 'thermal equivalent' energy cost of $(0.05/0.33)\text{Btu/gal} = 0.151\ \text{Btu/gal}$ – still greatly superior to the thermal option. Moreover, as in the RO example, integrating such a process with a 50% efficient fuel cell or combined-cycle gasification process shows an improvement of $0.302/(0.05/0.5) >$ threefold better than the thermal alternative! First-generation membranes have been reported with properties that suggest this separation can be achieved, so this type of application is likely to develop over the next few years [39].

The first-generation membranes investigated include polymeric membranes and polymer/silver salt composite membranes. Polymers such as cellulose acetate, polysulfone, PDMS, and polyethylene show very poor separation-performance

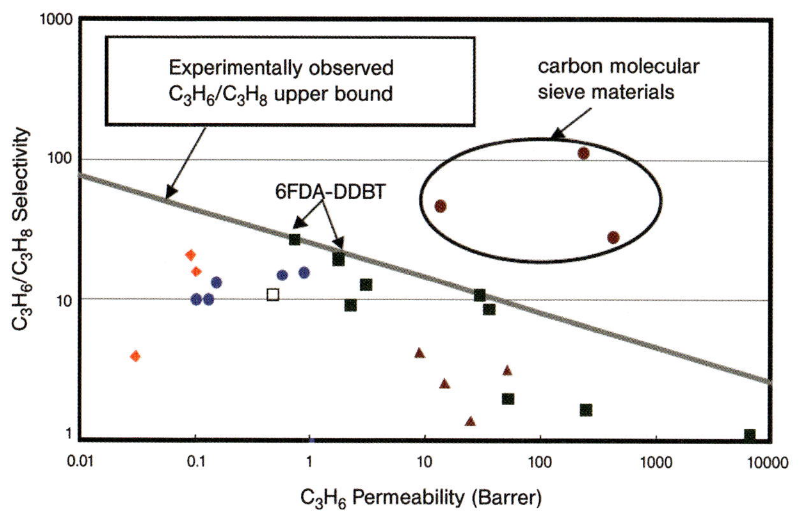

Figure 7.8 C_3H_6/C_3H_8 experimental upper bound based on pure gas permeation data over the range 1–4 atm feed pressure. □ = 100 °C, ■ = 50 °C, ● = 35 °C, ▲ = 30 °C, ◆ = 26 °C (Adapted from Ref. [38]).

stability for olefin/paraffin separation, with selectivities often below 3 [40–42]. Polyimides seem to be the most promising membrane material for this particular separation, yet the performance of polyimides was discovered to be limited by an upper-bound trade-off curve displayed in Figure 7.8 [38]. The permeability and selectivity are in reverse proportion to one another while the commercially attractive region corresponds to the upper right corner of the plot.

Another category of studies focuses on facilitated transport using ion-exchange membranes that contain silver ions as a complexing agent [43–45]. Membranes with facilitated transport properties show very good selectivity and relatively high permeability coefficients for the olefins, but the separation process typically has to be carried out under saturated water vapor to ensure the transport of silver/olefin complexes through the membrane, which adds considerable complications. Moreover, the stability of the silver carrier poses another challenge for industrial application of this technology.

In addition to the polymer and facilitated transport membranes, novel materials are being proposed and investigated to achieve membranes with economically attractive properties. Carbon molecular sieve (CMS) membranes prepared by pyrolysis of polyimides displayed much better performance for olefin/paraffin separation than the precursor membranes [39, 46, 47]. Results obtained with CMS membranes indicated properties well beyond the upper-bond trade-off curve, as shown in Figure 7.8. Nonetheless, this class of materials is very expensive to fabricate at the present time. An easy, reliable, and more economical way to form asymmetric CMS hollow fibers needs to be addressed from a practical viewpoint.

The preceding discussions illustrate that membranes have shown great potential as an alternative for olefin/paraffin separation, yet the performance of current membranes is insufficient for commercial deployment of this technology. Advanced material development is highly desired to improve the membrane properties and reduce cost. Another possible approach involves hybrid membranes with zeolites or CMS incorporated in a continuous polymer phase. More discussion in this regard will be covered later in this chapter.

7.5.2.2 Coal Gasification with CO_2 Capture for Sequestration

Membranes can contribute significantly to new concepts in more energy-efficient and low CO_2 emission power generation, and the following section explores some of these cases as alternatives to conventional amine-absorption-based thermally driven processes.

A state-of-the-art gasifier with integrated gasification combined cycle (IGCC) power plant, shown in Figure 7.9, enables the efficient use of coal for power generation. CO_2 is typically captured following gasification and a water gas shift reaction and prior to syngas combustion in the gas turbine in an IGCC power plant. The water gas shift reactor converts nearly all CO produced during gasification to CO_2. Therefore, the CO_2 concentration in the syngas leaving the shift reactor is typically in the range of 15–60% (dry basis) with total gas pressures ranging from 300 to 400 psia [16, 17, 48, 49]. This precombustion CO_2 capture of the pressurized syngas is typically less costly than postcombustion CO_2 capture, which requires treatment of large volumes of gas near atmospheric pressure.

State-of-the-art precombustion CO_2 capture technology in an IGCC plant employs amine-absorption treatment of the syngas; however, even when optimized, this

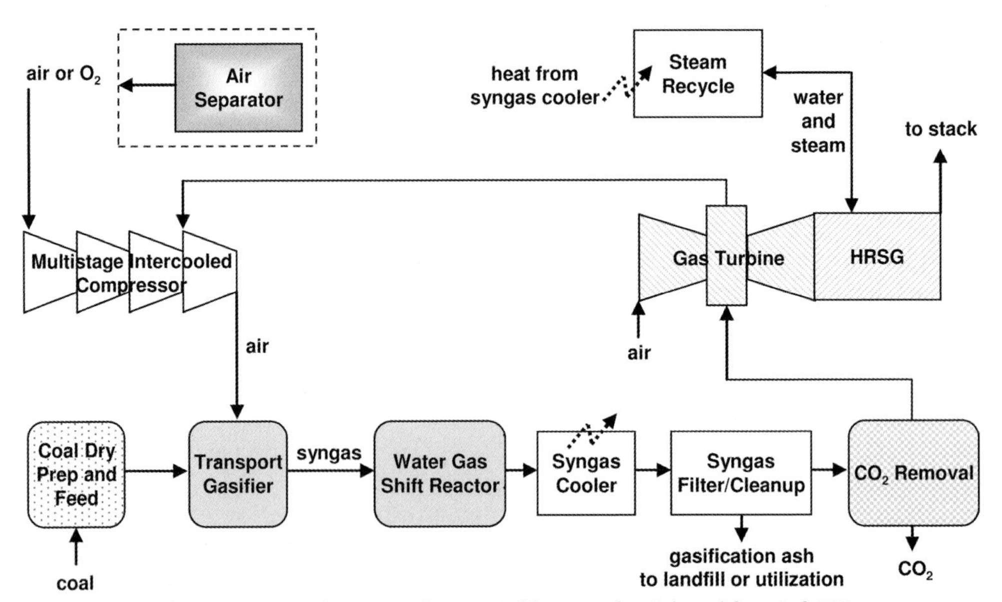

Figure 7.9 IGCC schematic with air or O_2 blown gasifier (Adapted from Ref. [48]).

treatment adds a great deal of cost to the process. In fact, depending on whether an air-blown or oxygen-blown gasifier is used, gasification that utilizes amine absorption for CO_2 capture has 71–89% higher capital costs than without CO_2 capture [48]. Unfortunately, a single-stage polymer membrane unit that performs a similar separation following oxygen-blown gasification has capital costs 105% higher than without CO_2 capture [16], however, as will be discussed next, membrane separation units have lower operational and environmental costs that can offset this higher current capital cost. Incorporation of carbon-capture technology also reduces the overall efficiency of the power plant since the process requires energy to operate. The overall efficiency loss for a membrane capture process in an IGCC is comparable, 9.9–13.5% drop in lower heating value (LHV), to that of the amine-capture process described above, 8.1–15% drop (LHV) [16, 49–51]. Ultimately, when considering different carbon-capture technologies, it is often best to compare the cost per ton of CO_2 avoided, which takes into account the costs associated with the equipment and operation as well as the amount of CO_2 removed from emissions. The amine-absorption capture technology for the oxygen-blown gasification process has an estimated cost of CO_2 avoided of \$48.3/ton; whereas, membrane capture technology for the same gasification process has an estimated cost of CO_2 avoided of only \$41–47/ton [16, 17, 49].

While the initial capital investment of the precombustion membrane-separation process may be slightly more expensive, there are many aspects of membranes along with the lower cost of CO_2 avoided that make them more advantageous than the traditional amine-absorption process. The state-of-the-art amine-treatment approach described in the above air blown IGCC work [16] is similar to the approach that has been implemented by Dakota Gasification at the Great Plains Synfuels Plant [52]. This process utilizes air as the oxidant with a sub-bituminous Powder Ridge Basin (PRB) coal, and the amine system operates with the following complex flow sheet shown in Figure 7.10 [16]. The many units and piping are shown to illustrate the process complexity. The feed to the amine scrubbing system is pretreated with high-efficiency microfiltration membranes, referred to as 'particle-capture devices', to eliminate particulates. Also, the temperature of the stream has been reduced from 550 °F (287 °C) to only 100 °F (38 °C) by efficient steam-cycle condensate and cooling water [16]. Despite the effectiveness of the amine-based system in Figure 7.10, the amine approach adds considerable complexity to the final power system. In addition, while the amine solvent used in this process (MDEA) is considered the most stable and efficient solvent for a high CO_2 concentration stream, it still degrades in the presence of oxygen or when subjected to high temperatures, as found in the regeneration boiler, and can lead to corrosion of the equipment [53].

The pretreatments, described above, that deliver a particulate-free stream at 38 °C to the amine system provide a ready-made feed for processing via membrane modules. This feed can be used with simple and efficient membranes, new structured sorbents, membrane + structured sorbent hybrid systems or more advanced super H_2 selective membranes. These membrane systems can simplify and condense the flow sheet in Figure 7.10, thereby enabling a more compact plant with less piping and associated maintenance concerns.

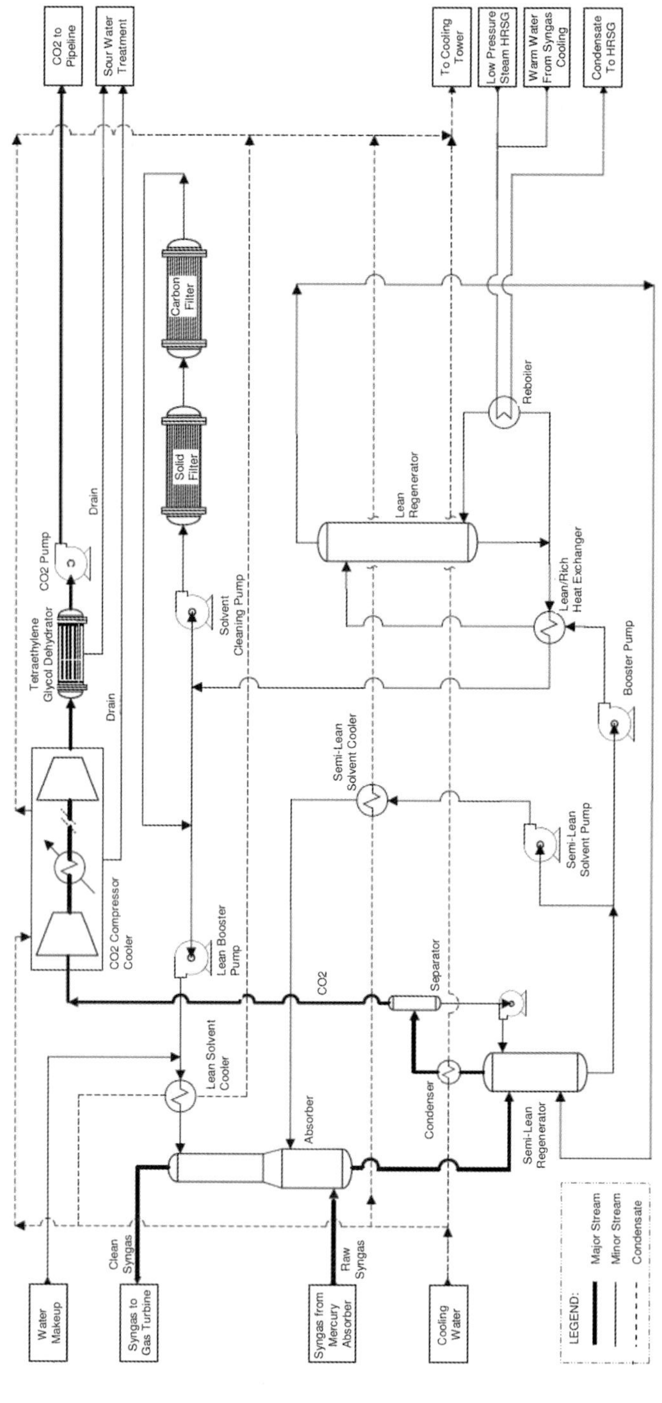

Figure 7.10 Amine process for CO_2 capture [16].

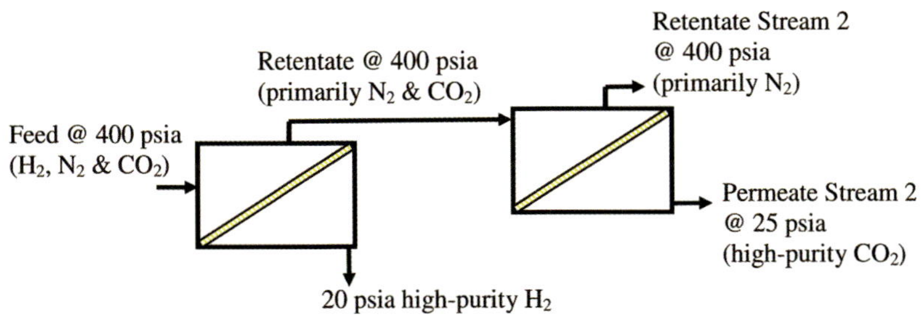

Figure 7.11 Dual-membrane separation utilizing a highly selective metal or inorganic membrane for H_2 purification and a conventional polymer membrane for the CO_2/N_2 separation.

The following example illustrates the potential of membrane-separation processes for precombustion carbon capture in an IGCC. This approach avoids using an expensive air-separation unit (asu) or a difficult-to-implement high-temperature mixed-ion conducting membrane process; however, it still enables capture of CO_2 at purities suitable for commercial use or sequestration.

Figure 7.11 shows a simplified flow sheet of two membrane separation units where the first stage comprises of a zeolite, palladium, or zeolite-ceramic highly selective H_2 membrane and the second stage comprises a conventional polymer membrane having the ability to reject N_2 vs. CO_2. There currently exist high-performance polymer membranes capable of performing the required separation in the second membrane stage. Existing polymer membranes have selectivities for CO_2 vs. N_2 as high as 60 at 35 °C [9, 54–56]. Moreover, one should be able to tailor the properties of various polymer families used to form membranes (e.g., polyimides, polysulfones) to enable tuning of the CO_2/N_2 permselectivity by adding groups with a favorable interaction with CO_2. It is fully expected, therefore, that an economical polymer with a N_2/CO_2 selectivity above 30–40 could be achieved for a feed at roughly 400 psia. In addition, this second-stage membrane would greatly benefit from the high CO_2 driving force at 400 psia after H_2 is removed as a permeate in the first stage. For this unusual application, the more condensable nature of CO_2 and its smaller size, relative to N_2, makes both factors in Equation 7.3 favor CO_2 vs. N_2. This favorable fact enables the desired high permselectivity of CO_2 relative to N_2. This situation will result in the vast majority of the CO_2 permeating through the membrane to the lower-pressure side, while most of the N_2 would be rejected at the high-pressure side. The high-pressure retentate stream will have low levels of CO_2 and H_2 and comprise a large flow of mostly inert N_2. This stream could also be expanded in a gas turbine to claim considerable useful work and be used for other purposes as well. The CO_2 could then be compressed for transport and storage without the added volume, and ultimately cost, of nitrogen. In fact, the energy captured by expansion of the nitrogen retentate would help provide some of this required compression energy for the CO_2.

The main challenge of the first separation involves development of a viable membrane. An economical highly H_2 selective membrane with the ability to reject both N_2 and CO_2 is required for this stage, and such a membrane does not yet exist. Polymer–zeolite or ceramic–zeolite hybrid membranes may provide the required selectivity; however, a palladium, palladium alloy or pure zeolite membrane may be needed to achieve this very difficult goal. This factor certainly presents a major hurdle to application of this dual-membrane system, since the cost of either the metal or zeolite membranes will be considerable. If this hurdle can be overcome, the H_2 permeate stream could then be mixed with air and used in a standard combustion cycle where N_2 will moderate the combustion to avoid excessive temperatures that would damage the turbine or require exotic materials of construction. A potential advantage is that the first-stage process could operate efficiently at elevated temperature allowing cooling to be deferred until after H_2 removal. This first-stage membrane will require considerably more development time to implement than the second-stage membrane; however, it has the potential to be a revolutionary purely membrane-based technology for H_2 production and CO_2 removal.

7.6
Key Hurdles to Overcome for Broadly Expanding the Membrane-Separation Platform

The previous examples for large-scale gas and vapor separations noted above illustrate that much more advanced but still economical membranes are required to better expand the membrane platform. Three related hurdles exist to broadly extending existing membrane separation successes to other low molecular weight organic compounds: these hurdles are the lack of economical materials, membranes, and module fabrication methods. This is a serious situation that must be addressed with integrated programs that seek to develop high-efficiency module formation, high-speed processing, micromorphology control, and advanced materials for membrane implementation. Indeed, large osmotic pressures, higher temperatures, and more aggressive organic feeds in these systems will require even more robust membranes and modules than are currently available for water feed streams. For instance, at 25 °C an osmotic pressure of roughly 79 atm (1161 psi) must be overcome to cause forward flux of propylene from a 75/25 molar mixture to produce a 95/5 molar downstream mixture of propylene and propane. Such pressures can even now be contained within a compact membrane vessel, and some gas-separation modules already operate with higher feed pressures. Such a liquid RO system would even further increase the energy savings below those cited for the propylene/propane vapor-separation case mentioned above. Similarly, a low-cost palladium or pure zeolite membrane that allows only passage of H_2 could enable the first-stage membrane for hydrogen purification in the gasification example noted above.

Much of the technology for gas- and vapor-separation materials, membranes, and modules that are now emerging as large-scale units were derived from work supported on reverse osmosis in the early 1960s [57]. This program by the Office of Saline Water (OSW) targeted energy-efficient processes based upon the promising but unproven (*at the time*) membranes for aqueous separation. These early membranes were, at that time, in a similar state to those for current organic systems, and many problems had to be overcome [13, 57]. While sharing some aspects with aqueous feeds, nonaqueous feeds present new challenges that must be attacked

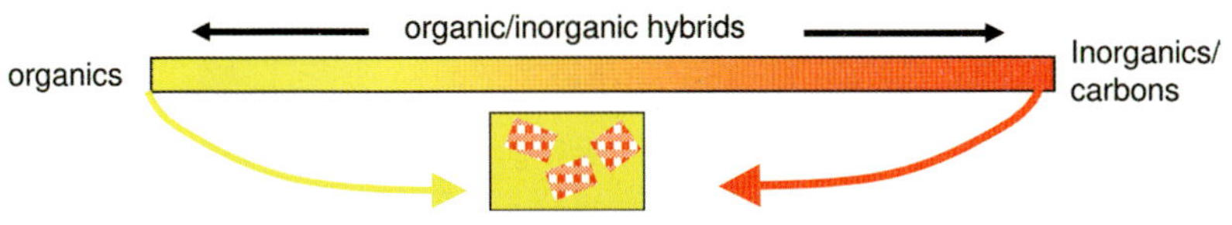

Figure 7.12 Advanced materials spectrum, including not only the extremes of organic polymers and inorganic or carbon materials, but also hybrids of these materials to provide property and processing advantages.

holistically within the framework of the previously stated programs for membrane implementation. A program similar to the OSW initiative, but aimed at organic feeds and high-pressure gases and vapors, would be a positive step in this direction and an investment in the future. Such a sea-changing undertaking probably requires government initiation, as it did in the visionary OSW case.

Despite demanding requirements for selectivity and robustness in this next generation of applications, membranes and modules *must retain their attractive cost advantages*. Realistically, therefore, any program to introduce truly new high-performance membranes should incorporate hybrid materials within its enabling vision. A complete picture of membrane materials includes the spectrum ranging from purely inorganics and carbons to purely organic polymers shown in Figure 7.12.

Current work has really only explored the two extreme ends of this spectrum, plus a few hybrids containing 10–15 vol% inorganic or carbon-dispersed phases in a polymer continuous phase. For future demanding applications, it is likely that the optimum position in the materials spectrum in Figure 7.12 may be even past the 'midpoint' in hybrid composition. Indeed, very high percentages of inorganic or carbon solids, compatibly bound within an appropriate polymer matrix, could be the preferred membrane material of the future for many applications. Such hybrids have the potential to provide the selectivity and strength of inorganics and carbons and the processability and flexibility of polymers.

While silane-treated zeolites dispersed in a polymer matrix have been reported to possess excellent performance in dense films [58], the performance has been difficult to replicate in asymmetric hollow-fiber membranes. Recently, considerable success has been achieved in approaching the major hurdle of zeolite/polymer interface by a novel route [59–63], which overcomes the limitation of silane-coupling agents observed in the phase-separating environment of asymmetric membrane formation. The approach, using an acid halide and a Grignard reagent, modifies the surface of zeolite particles to increase surface roughness and has been successfully employed to modify the surface of two small-pore zeolites, SSZ-13 and zeolite A. These modified zeolites were found to form strong adhesion with high glass transition polymers such as Ultem 1000 polyetherimide and Matrimid 5218 polyimide in dense film and asymmetric hollow-fiber membranes, thereby providing superior gas-separation performance. Figure 7.13 displays a scanning electron micrograph of a dual-layer hollow fiber composed of an Ultem 1000 polymer matrix with 10 wt% Grignard-treated submicrometer zeolite A. The inset focuses on the skin region of the fiber showing the homogeneous dispersion of the modified zeolite particles and the excellent adhesion with the polymer matrix.

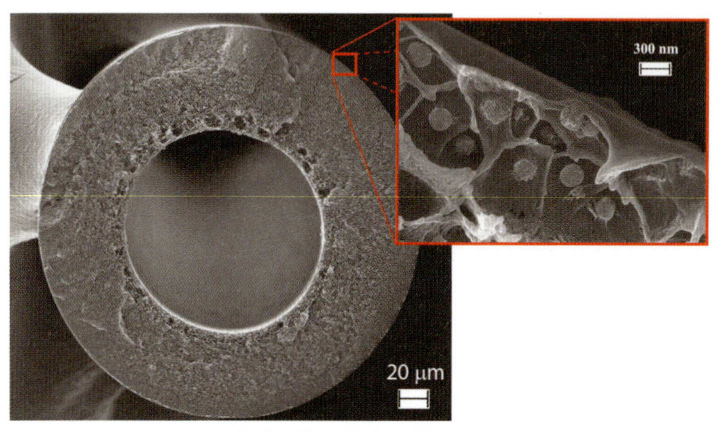

Figure 7.13 SEM images of a dual-layer hollow fiber with zeolite insert. Enlargement reveals good polymer adhesion to the zeolite particles (Adapted from Ref. [62]).

The above laboratory-based successes show the future of this technology is promising, but they need to be supplemented with novel modification and processing techniques that can be scaled for high production levels. These hybrid membranes show considerable potential, however, they still require extensive research before implementation. Indeed if the number of patents filed recently is any indication, hybrid membranes are attracting industrial attention [64–81].

7.7 Some Concluding Thoughts

The above illustrations show that major, even revolutionary, energy savings are possible relative to competitive, thermally driven options by introducing membrane processes for separations. Nevertheless, the discussion also clarifies the need for a large-scale integrated systematic approach to greatly broaden the economical application of membranes to more aggressive feed streams. This information highlights the need for modeling and analysis that starts at megascale plant systems and ranges down to the molecular scale where most separations ultimately occur. Materials science is a critical component; however, technologies to engineer supermolecular membrane morphologies and economical modules are equally critical to build such an expanded platform.

In addition to its central role in advanced separation devices considered here, aspects of membrane technology indirectly impact fuel cells, advanced batteries used in hybrid vehicles, and low-cost flexible solar-energy cells. Applying all of these related energy-saving devices across the various sectors of society mentioned in the introduction of this chapter would motivate rational change toward energy efficiency. The special opportunities for synergistic combination of fuel cells and membrane-separation technologies should be vigorously pursued to break the unnecessary current linkage between inefficacies in thermal energy-conversion processes and separation processes. In order to move toward 'green' energy processes, a high-profile effort is underway to promote the introduction of fuel cells, advanced batteries, and solar cells; however, much less aggressive action is apparent to promote energy-efficient separations. A concerted effort focused on developing the membrane platform beyond its current state to enable rapid replacement of energy-inefficient separation processes is greatly needed. It is extremely important for developing countries with fewer installed thermal processes to make investments in these more-efficient approaches, and for more-developed countries to phase out these thermal processes. Ultimately, an economy based on thermal dinosaurs stands to be the biggest loser – *natural selection works*!

Acknowledgments

Support is acknowledged from the Georgia Research Alliance and the DOE, under grant DE-FG03-95ER14 538, and GRP Investigator grant KUS-I1-011-21.

References

1 Koros, W.J. (2004) Evolving beyond the thermal age of separation processes: membranes can lead the way. *AIChE Journal*, **50** (10), 2326.

2 United Nations (2007) Department of Economic and Social Affairs, Population Division. World Population Prospects: The 2006 Revision, Highlights, Working Paper No. SA/P/WP.202.

3 Humphrey, J.L. and Keller, G.E. (1997) Energy Considerations, in *Separation Process Technology*, McGraw-Hill, New York, NY, Chapter 6.

4 Ho, W.S. and Sirkar, K.K. (1992) Overview, in *Membrane Handbook*, Van Nostrand Reinhold, New York, NY, Chapter 1.

5 Eykamp, W. (1997) Membrne Separation Processes, in *Perry's Chemical Engineers' Handbook*, 7th edn, McGraw Hill, New York, NY, Chapter 22.

6 Blume, I.(May 2004) Norit Ultrafitration as Pretreatment for RO for Wastewater Reuse: The Sulaibiya Project. Presentation at Advanced Membrane Technology II Conference, Irsee, Germany, Engineering Conferences International, organizer N.N. Li.

7 Smith, J.M., van Ness, H.C. and Abbot, M.M. (2001) Production of Power from Heat, in *Introduction to Chemical Engineering Thermodynamics*, 6th edn, McGraw Hill, New York, NY, Chapter 8.

8 Pankratz, T. and Tonner, J. (2003) *Desalination.com: An Environmental Primer*, Lone Oak Publishing, Houston, TX.

9 Koros, W.J., Fleming, G.K., Jordan, S.M., Kim, T.H. and Hoehn, H.H. (1988) Polymeric membrane materials for solution-diffusion based permeation separations. *Progress in Polymer Science*, **13**, 339–401.

10 Ekiner, O.M. and Kulkarni, S.S.(December 2003) Process for Making Mixed Matrix Hollow Fiber Membranes for Gas Separation, US Patent 6,663,805.

11 Vu, D., Koros, W.J., Mahajan, R. and Miller, S.J. (January 2003) Gas Separation Using Mixed Matrix Membranes. US Patent 6,503,295.

12 Marks, D.H., Balaban, M., Falagan, B.A., Jacangelo, J.G., Jones, K.L., Koros, W.J., Letey, J., Pankratz, T.M., Sakaji, R.H., Turner, C.D. and Wilf, M. (2004) *Review of*

the Desalination and Water Purification Technology Roadmap, National Research Council, National Academies Press, Washington, DC.

13 Merten, U. (1966) Transport Properties of Osmotic Membranes, in *Desalination by Reverse Osmosis*, M.I.T. Press, Cambridge, MA, Chapter 2.

14 Steinmeyer, D. (1997) Energy Conservation in Separation Processes, in *Encyclopedia of Separation Technology*, John Wiley & Sons, Ltd, New York, NY, pp. 749–759.

15 Brandon, N.P., Skinner, S. and Steele, B.C.H. (2003) Recent advances in materials for fuel cells. *Annual Reviews of Material Science*, **33**, 183–213.

16 Kaldis, S.P., Skodras, G. and Sakellaropoulos, G.P. (2004) Energy and capital cost analysis of CO_2 capture in coal IGCC processes via gas separation membranes. *Fuel Processing Technology*, **85** (5), 337–346.

17 Metz, B., Davidson, O., de Coninck, H.C., Loos, M. and Meyer, L.A. (2005) *IPCC Special Report on Carbon Dioxide Capture and Storage*, Cambridge University Press, Cambridge, United Kingdom, and New York, NY, USA.

18 Gottschlich, D. and Jacobs, M.L. (1998) *Monomer Recovery Process*, Membrane Technology and Research Inc., USA, p. 14.

19 Lokhandwala, K.A. and Jacobs, M.L. (2000) Membranes for fuel gas conditioning. *Hydrocarbon Engineering*, **5** (5), 81–82, 84.

20 Strathman, H., Bell, C.M. and Kimmerle, K. (1986) Development of synthetic membranes for gas and vapor separation. *Pure and Applied Chemistry*, **58** (12), 1663–1668.

21 Peinemann, K.V., Mohr, J.M. and Baker, R.W. (1986) The separation of organic vapors from air. *AIChE Symposium Series*, **82** (250), 19–26.

22 Baker, R.W., Yoshioka, N., Mohr, J.M. and Kahn, A.J. (1987) Separation of organic vapors from air. *Journal of Membrane Science*, **31** (2–3), 259–271.

23 Juyal, U. (2002) CO_2 membrane Technology Matures. *Oil & Gas Journal*, **100** (15), 46.

24 Parro, D. (1984) Membrane carbon dioxide separation proves out at Sacroc tertiary recovery project. *Oil & Gas Journal*, **82** (39), 85–86, 88.

25 Marquez, J. Jr and Brantana, M. (2006) High-capacity gas membrane elements reduce weight, cost. *Oil & Gas Journal*, **104** (28), 51–54.

26 UOP (29 March 2004) Expanded OGDCL Membrane Plant Meets UOP's Revamp Guarantees, Press Release.

27 Callison, A. and Davidson, G. (2007) Offshore processing plant uses membranes for CO_2 removal. *Oil & Gas Journal*, **105** (20), 56–58, 60, 62, 64–65.

28 Collings, C.W., Huff, G.A. and Bartels, J.V. (January 2004) Separation Processes Using Solid Perm-Selective Membranes in Multiple Groups for Simultaneous Recovery of Specified Products from a Fluid Mixture. US Patent Appl. Publ. 20040004040 A1.

29 Nunes, S.P. and Peinemann, K.V. (2006) *Membrane Technology: In the Chemical Industry*, Wiley-VCH, Weinheim, Germany.

30 Baker, R.W. (2002) Future Directions of Membrane Gas Separation Technology. *Industrial & Engineering Chemistry Research*, **41** (6), 1393–1411.

31 Gottschlich, D.E. and Roberts, D.L. (1990) Energy Minimization of Separation Processes using Conventional/Membrane Hybrid Systems, Department of Energy, Report No. DE-AC-07-76ID01570.

32 (June 2001) Ethylene/Propylene, Nexant Chem Systems Report.

33 Perry, R.H. and Green, D. (1984) *Perry's Chemical Engineers Handbook*, 6th edn, McGraw-Hill, New York.

34 Hirschfelder, J.H., Curtiss, C.F. and Bird, R.B. (1964) *Molecular Theory of Gases and Liquids*, John Wiley & Sons, Ltd, New York.

35 Eldridge, R.B. (1993) Olefin paraffin separation technology - a review. *Industrial*

& *Engineering Chemistry Research*, **32**, 2208–2212.

36 Humphrey, J.L., Seibert, A.F. and Koort, R.A. (1994) Separation Technologies Advances and Opportunities, Department of Energy, Report No. DE-AC-07-90ID12920.

37 Staudt-Bickle, C. and Koros, W.J. (2000) Olefin/paraffin gas separations with 6FDA-based polyimide membranes. *Journal of Membrane Science*, **17**, 205–214.

38 Burns, R.L. (2002). Investigation of Poly (pyrrolone-imide) Materials for the Olefin/Paraffin Separation, PhD Dissertation, The University of Texas at Austin.

39 Yoshino, M., Nakamura, S., Kita, H., Okamoto, K.-i., Tanihara, N. and Kusuki, Y. (2003) Olefin/paraffin separation performance of asymmetric hollow fiber membrane of 6FDA/BPDA-DDBT copolyimide. *Journal of Membrane Science*, **212**, 13–27.

40 Ito, A. and Hwang, S.T. (1989) Permeation of propane and propylene through cellulosic polymer membranes. *Journal of Applied Polymer Science*, **38**, 483.

41 Tanaka, K., Taguchi, A., Hao, J., Kita, H. and Okamoto, K. (1996) Permeation and separation properties of polyimide membranes to olefins and paraffins. *Journal of Membrane Science*, **121**, 197.

42 Shimazu, A., Ikeda, K. and Hachisuka, H. (1998) Method of selectively separating unsaturated hydrocarbon, US Patent 5,749,943.

43 Ho, W.S. and Dalrymple, D.C. (1994) Facilitated transport of olefins in Ag+ containing polymer membranes. *Journal of Membrane Science*, **91**, 13.

44 Yang, J.S. and Hsiue, G.H. (1997) Selective olefin permeation through Ag(I) contained silicone rubber-graft-poly(acrylic acid) membranes. *Journal of Membrane Science*, **126**, 139.

45 Yamaguchi, T., Baertsch, C., Koval, C.A., Noble, R.D. and Bowman, C.N. (1996) Olefin separation using silver impregnated ion-exchange membranes and silver salt/polymer blend membranes. *Journal of Membrane Science*, **117**, 151.

46 Suda, H. and Baraya, K. (1997) Alkene/alkane permselectivities of a carbon molecular sieve membrane. *Chemical Communications*, **93**.

47 Hayashi, J., Mizuta, H., Yamamoto, M., Kusakabe, K. and Morooka, S. (1996) Separation of ethane/ethylene and propane/propylene systems with a carbonizad BPDA-pp'ODA polyimide membrana. *Industrial & Engineering Chemistry Research*, **35**, 4176.

48 Bonsu, A.K., Eiland, J.D., Gardner, B.F., Powell, C.A., Rogers, L.H., Booras, G.S., Breault, R.W. and Salazar, N. (2006) Impact of CO_2 Capture on Transport Gasifier IGCC Power Plant. International Technical Conference on Coal Utilization and Fuel Systems, Clearwater, FL, May 21–25.

49 Grainger, D. and Hägg, M.-B. (2007) Techno-economic evaluation of a PVAm CO_2-selective membrane in an IGCC power plant with CO_2 capture. *Fuel*, **81** (8), 14–24.

50 Davison, J., Bressan, L. and Domenichini, R. (October 2003) *Coal Power Plants with CO_2 Capture: The IGCC Option*, Gasification Technologies, San Francisco, California.

51 Kanniche, M. and Bouallou, C. (2007) CO_2 capture study in advanced integrated gasification combined cycle. *Applied Thermal Engineering*, **27**, 2693–2702.

52 Perry, M. and Eliason, D. CO_2 Recovery and Sequestration at Dakota Gasification Company. (2004) Gasification Technologies Conference, San Francisco, California.

53 Korens, N., Simbeck, D. and Wilhelm, D. (2002) Process Screening Analysis of Alternative Gas Treating and Sulfur Removal for Gasification, Revised Final Report, SFA Pacific, Inc.

54 Bixler, H.J. and Sweeting, O.J. (1971) *The Science & Technology of Polymer Films*, vol. **II**, John Wiley & Sons, Ltd, New York.

55 Koros, W.J., Coleman, M.R. and Walker, D.R.B. (1992) Controlled permeability polymer membranes. *Annual Review of Materials Science*, **22**, 47–89.

56 Powell, C.E. and Qiao, G.G. (2006) Polymeric CO_2/N_2 gas separation membranes for the capture of CO_2 from power plant flue gases. *Journal of Membrane Science*, **279**, 1–49.

57 Lonsdale, H.K. (1982) The Growth of Membrane Technology. *Journal of Membrane Science*, **10**, 81–181.

58 Mahajan, R. and Koros, W.J. (2002) Mixed matrix membrane materials with glassy polymers. Part 2. *Polymer Engineering and Science*, **42** (7), 1432–1441.

59 Shu, S., Husain, S. and Koros, W.J. (2007) Formation of nanostructured zeolite particle surfaces via a halide/Grignard route. *Chemistry of Materials*, **19** (16), 4000–4006.

60 Husain, S. and Koros, W.J. (2007) Mixed matrix hollow fiber membranes made with modified HSSZ-13 zeolite in polyetherimide polymer matrix for gas separation. *Journal of Membrane Science*, **288** (1–2), 195–207.

61 Shu, S., Husain, S. and Koros, W.J. (2007) A general strategy for adhesion enhancement in polymeric composites by formation of nanostructured particle surfaces. *Journal of Physical Chemistry C*, **111** (2), 652–657.

62 Husain, S. (2006). Mixed Matrix Dual Layer Hollow Fiber Membranes for Natural Gas Separation, in Chemical & Biomolecular Engineering, PhD Dissertation, Georgia Institute of Technology.

63 Shu, S. (2007) Engineering the Performance of Mixed Matrix Membranes for Gas Separations, in Chemical & Biomolecular Engineering, PhD Dissertation, Georgia Institute of Technology.

64 Koros, W.J. *et al.* (2002) Mixed matrix membranes and manufacture and purification of methane from a gas stream, (USA). Application: US. 2002056369.

65 Koros, W.J. *et al.* (2002) Mixed matrix membranes with pyrolyzed carbon sieve particles and methods of making and using the same in synthesis gas purification, (Chevron USA Inc., USA; The University of Texas System). Application: WO. 2002024310.

66 Kulkarni, S.S., Ekiner, O.M. and Hasse, D.J. (2005) Method for forming of mixed matrix composite permselective fluid sepn. membrane using washed mol. sieve particles, (USA). Application: US 20050230305.

67 Kulkarni, S.S. and Hasse, D.J. (2005) Molecular sieve-containing polyimide mixed matrix membranes, (USA). Application: US. 20050268782.

68 Kulkarni, S.S. *et al.* (2006) Making mixed matrix membranes using electrostatically stabilized suspensions, membranes for fluid separation, and separating a fluid from fluid mixture, (USA). Application: US 20060117949.

69 Koros, W.J. *et al.* (January 2003) Gas separations using mixed matrix membranes. US Patent 6,503,295.

70 Miller, S.J. *et al.* (December 2002) Purification of p-xylene using composite mixed matrix membranes. US Patent 6,500,233.

71 Koros F W.J. *et al.* (July 2003) Mixed matrix membranes and methods for making the same. US Patent 6,585,802.

72 Hasse, D.J. *et al.* (September 2003) Mixed matrix membranes incorporating chabazite type molecular sieves. US Patent 6,626,980.

73 Ekiner, O.M. and Kulkarni, S.S.(December 2003) Process for making hollow fiber mixed matrix membranes. US Patent 6,663,805.

74 Kulkarni, S.S. *et al.* (January 2003) Gas separation membrane with organosilicon-treated molecular sieve. US Patent 6,508,860.

75 Koros, W.J. *et al.* (June 2004) Crosslinked and crosslinkable hollow fiber mixed matrix membrane and method of making same. US Patent 6,755,900.

76 Kulprathipanja, S. and Charoenphol, J. (April 2004) Mixed matrix membrane for separation of gases. US Patent 6,726,744.

77 Guiver, M.D. *et al.* (August 2003) Composite gas separation membranes. US Patent 6,605,140.
78 Marand, E. *et al.* (September 2006) Mixed matrix membranes. US Patent 7,109,140.
79 Miller, S.J. *et al.* (November 2006) Mixed matrix membranes with low silica-to-alumina ratio molecular sieves and methods for making and using the membranes. US Patent 7,138,006.
80 Miller, S.J. *et al.* (January 2007) Mixed matrix membranes with small pore molecular sieves and methods for making and using the membranes. US Patent 7,166,146.
81 Miller, S.J. and Yuen, L.-T.(September 2007) Mixed matrix membrane with super water washed silica containing molecular sieves and methods for making and using the same. US Patent 7,268,094.

8
Membrane Gas-Separation: Applications

Richard W. Baker

8.1
Industry Background

Gas separation with membranes is now a US$300–400 million dollar-per-year industry, and the industry is growing at a double-digit rate [1–3]. In the early 1980s, the first successful membrane-based industrial gas-separation plants were built by Permea (now a division of Air Products). These plants recovered hydrogen from ammonia reactor purge gas or purified hydrogen recycle streams in refinery hydrocrackers. However, the industry really took off a few years later when Medal, Generon, Ube, and Permea developed membranes to separate nitrogen from air. Nitrogen production now represents half of the membrane gas-separation equipment business. The average nitrogen separation unit cost is small, usually in the range of US$10 000–100 000, but several thousand are made each year. Another major application of gas-separation membranes is the separation of carbon dioxide from natural gas. In contrast to nitrogen units, carbon-dioxide separation plants are often very large, and cost from US$20 to 50 million each. More than twenty of these large plants have now been installed and this application continues to grow rapidly. A table describing the current membrane gas-separation industry is given below (Table 8.1).

8.2
Current Membrane Gas-Separation Technology

All gas-separation membranes have an anisotropic structure with a thin, dense selective layer facing the high-pressure feed gas. The selective layer is supported on a much thicker microporous support layer that provides mechanical strength. The chemical structure determines the permeability of the selective layer.[1] The selective-

1) Permeability is the general term used to describe the rate at which a gas will pass through a material. Permeability is most commonly measured in Barrer, defined as 1×10^{-10} cm^3 cm^3 (STP) cm/cm^2 s cmHg.

Membrane Operations. Innovative Separations and Transformations. Edited by Enrico Drioli and Lidietta Giorno

ISBN: 978-3-527-32038-7

Table 8.1 Characteristics of the current membrane gas separations industry – 2008.

Separation	Principal producers	Membrane/ modules	Market size (US$million/y)
Nitrogen from air Water from air	Permea	Polysulfone/capillary fiber	150
	Medal	Polyimide/capillary fiber	
	Dow, Generon	Polyimide/capillary fiber	
	Ube	Polyimide/capillary fiber	
	Aquilo	Polyphenylene oxide/capillary fiber	
Carbon dioxide from natural gas	Cynara	Cellulose triacetate/ fine fiber	100
	Medal	Polyaramide/fine fiber	
	Grace, Separex	Cellulose acetate/spirals	
	MTR	Perfluoro polymers/spirals	
Refining: H_2/CH_4 Ammonia plants: H_2/N_2, Ar Syngas: H_2/CO	Permea	Polysulfone/fine fiber	75
	Medal	Polyaramide/fine fiber	
	MTR/Ube	Polyimide/fine fiber	
C_{3+} Hydrocarbons/ nitrogen	MTR	Silicone rubber/spirals	30
	Borsig	Silicone rubber/ plate-and-frame	
Everything else: Nitrogen/ natural gas, Helium/ natural gas, H_2S/natural gas, CO_2/H_2, Miscellaneous petrochemicals	—	Many	20

layer thickness determines membrane permeance.[2] When the selective layer is made from a low-permeability material, such as an amorphous glassy polymer, the selective layer is made as thin as possible, typically between 0.2 to 1.0 micrometers. When the selective layer is made from a high-permeability material, such as a rubbery polymer, the selective layer has a thickness of 0.5–5.0 micrometers. Thinner rubbery membranes could be made, but the formation of concentration gradients at the membrane surface puts a limit on the maximum permeance that can be used practically [4, 5].

8.2.1
Membrane Types and Module Configurations

Membranes can be made as flat sheets in long rolls or in the form of thin, hollow tubes. Production of current gas-separation membrane modules is divided approxi-

2) Permeance is permeability divided by the thickness of the membrane material. Permeance is expressed in terms of gas permeation units (gpu), defined as $1 \times 1 \times 10^{-6}$ cm^3 (STP)/cm^2 s cmHg.

mately evenly between those based on hollow fine fiber membranes (hydrogen separation, carbon dioxide from natural gas), capillary membranes (nitrogen from air) and flat-sheet membranes (carbon dioxide from natural gas, hydrocarbon/nitrogen, methane separations).

8.2.1.1 Hollow Fine Fiber Membranes and Modules

Hollow fine fiber membranes are extremely fine polymeric tubes 50–200 micrometers in diameter. The selective layer is on the outside surface of the fibers, facing the high-pressure gas. A hollow-fiber membrane module will normally contain tens of thousands of parallel fibers potted at both ends in epoxy tube sheets. Depending on the module design, both tube sheets can be open, or as shown in Figure 8.1, one fiber end can be blocked and one open. The high-pressure feed gas flows past the membrane surface. A portion of the feed gas permeates the membrane and enters the bore of the fiber and is removed from the open end of the tube sheet. Fiber diameters are small because the fibers must support very large pressure differences feed-to-permeate (shell-to-bore).

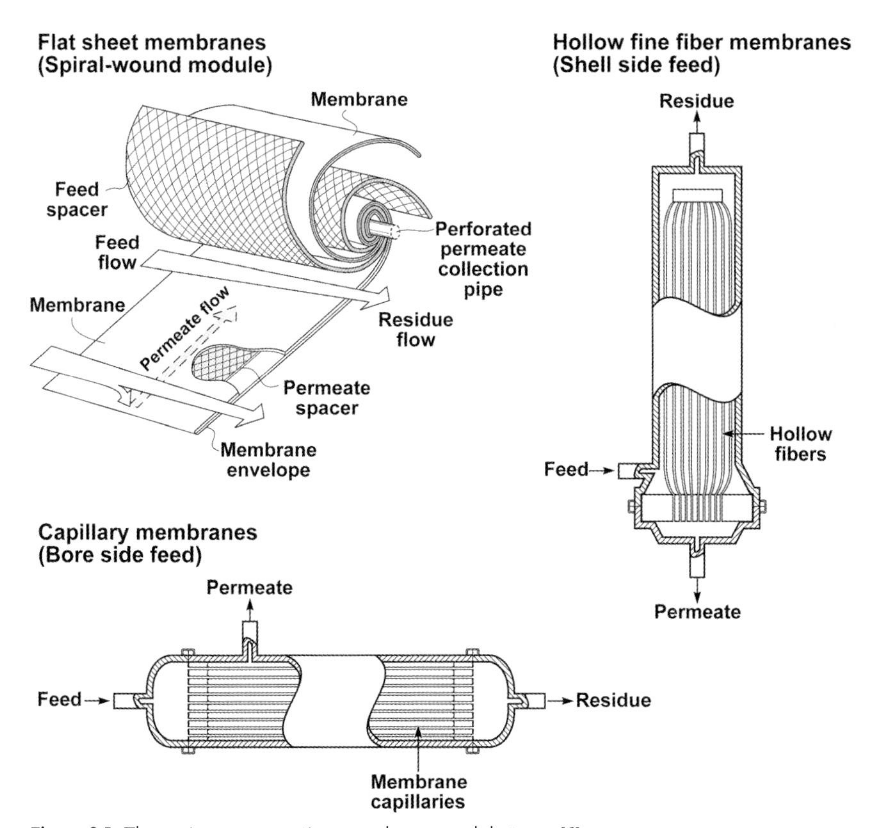

Figure 8.1 The main gas-separation membrane module types [6].

8.2.1.2 Capillary Fiber Membranes and Modules

Capillary fibers are produced using similar equipment to hollow fine fibers, but have a larger diameter, typically 200–400 micrometers, and the selective layer is formed on the inside surface of the fibers. In a capillary fiber module, the feed gas flows through the bore of the fibers, as shown in Figure 8.1. A portion of the feed permeates the membrane and passes to the shell side of the modules, and is removed as permeate. The pressure difference feed-to-permeate (bore-to-shell) that capillary fibers can support is limited and typically does not exceed 10–15 bar (1 bar = 100 kPa). Higher pressures may rupture fibers, and even a single defective fiber can seriously degrade the separation capability of the module.

Capillary membrane modules are not as inexpensive or compact as hollow fine fiber modules, but are still very economical. Their principal drawback is the limited pressure differential the fibers can support, typically not more than 10 to 15 bar. This limitation means capillary modules cannot be used at the high pressures needed for hydrogen or natural-gas processing applications. However, capillary modules are ideally suited to lower-pressure separations, such as nitrogen from air or air dehydration. In these applications, capillary modules have essentially the entire market.

8.2.1.3 Flat-Sheet Membranes and Spiral-Wound Modules

Flat-sheet membranes are made in continuous rolls 500–5000 m long. Sheets of membrane 1–2 m long are cut and folded and then packaged as spiral-wound module envelopes. A single module may contain as many as thirty membrane envelopes. Currently, the industry standard spiral-wound module is 8 inches (1 inch = 2.54 cm) in diameter and about 35–40 inches long; it contains 20–40 m^2 of membrane.

Each membrane/module type has advantages and disadvantages [2, 7]. Hollow fine fibers are generally the cheapest on a per-square-meter basis, but it is harder to make very thin selective membrane layers in hollow-fiber form than in flat-sheet form. This means the permeances of hollow fibers are usually lower than flat-sheet membranes made from the same material. Also, hollow fine fiber modules require more pretreatment of the feed to remove particulates, oil mist and other fouling components than is usually required by capillary or spiral-wound modules. These factors offset some of the cost advantage of the hollow fine fiber design.

The investment in time and equipment to develop a new membrane material in a high-performance hollow fine fiber or capillary form is far larger than that required to develop flat-sheet membranes, and many materials cannot be formed into fiber modules at all. For this reason, flat-sheet membranes, formed into spiral-wound modules, are used in many niche applications which cannot support the development costs associated with fiber modules. Spiral-wound modules are also competitive in the natural-gas processing area, where their general robustness is an asset.

8.2.2 Module Size

In the early days of gas separation, the average membrane module was just a few inches in diameter. In recent years, the trend has been to obtain economies of scale by

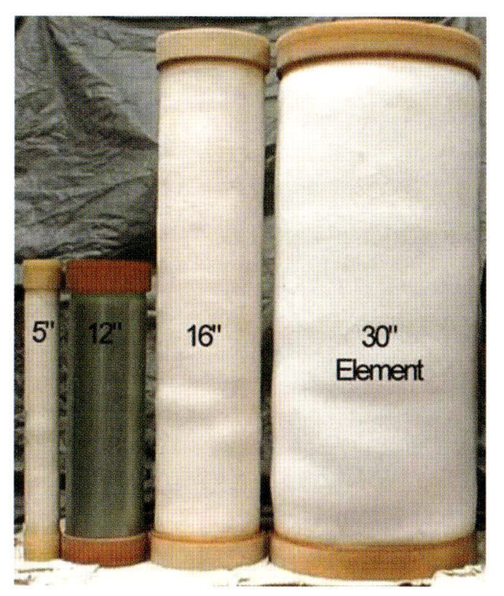

Figure 8.2 The expanding diameter of Cynara hollow-fiber membrane modules, from the first 5-inch modules of the 1980s to the 30-inch diameter behemoths now being introduced (Photo used courtesy of NATCO Group, Inc.) [9].

developing ever larger modules. Figure 8.2 shows the evolution of cellulose triacetate fine fiber membrane modules made by Cynara (a division of NATCO) for natural-gas treatment. Spiral-wound modules are also increasing in size, from the current 8-inch diameter module to 12-inch or larger modules. The driver for these changes is the high cost of gas-separation skids. Gas-separation systems require membrane modules contained in high-pressure, code-stamped vessels. The cost of the vessels, frames, and associated pipes and valves can be several times the cost of the membrane modules. Considerable savings are obtained by packaging larger membrane modules into fewer vessels, or housing multiple modules within a single large pressure vessel [8].

8.3 Applications of Gas-Separation Membranes

8.3.1 Nitrogen from Air

The largest current application of gas-separation membranes is separation of nitrogen (N_2) from air. Capillary modules formed into bore-side feed modules are used almost exclusively in this application [10, 11]. The feed air is compressed to 6–10 bar and pumped through the membrane capillaries. Oxygen (O_2) permeates the membrane preferentially, leaving an oxygen-depleted, nitrogen-rich residue stream. The first membranes used for this application were based on poly(4-methyl-1-pentene) and ethyl cellulose, and had O_2/N_2 selectivities of about 4. Because of the modest

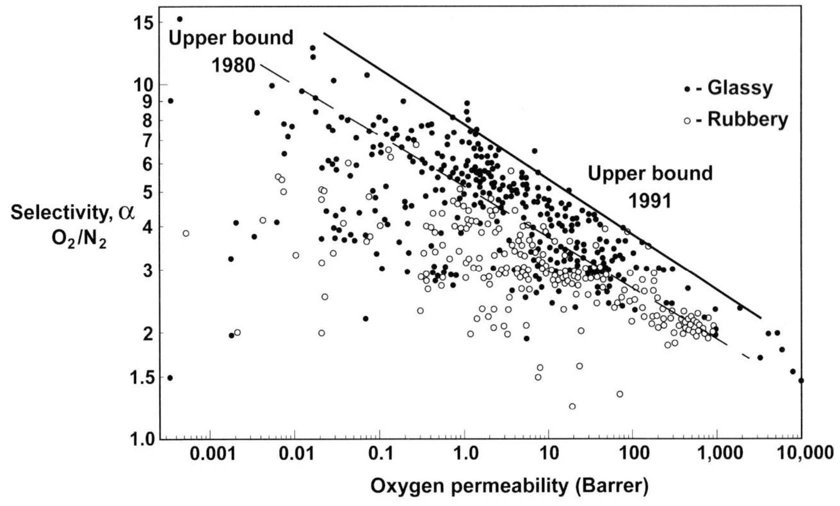

Figure 8.3 Oxygen/nitrogen selectivity as a function of oxygen permeability. The upper-bound line represents the point above which no better membranes are known [12]. This line shows the trade-off relationship between membrane permeability and selectivity.

selectivity, a significant fraction of the nitrogen in the feed air was lost with the oxygen permeate. Within a few years, improved materials with O_2/N_2 selectivities of 6 to 8 were introduced. Units incorporating these membranes recovered a much higher fraction of the feed-air nitrogen. Very little change in membrane materials has taken place in the last 15 years, despite ongoing research manifested by the flood of publications describing materials with improved properties. The problem is the flux/selectivity trade-off relationship, illustrated by the Robeson plot.

The Robeson plot shown in Figure 8.3 was created in 1991 [12]. The plot shows the O_2/N_2 selectivity and oxygen permeability of every membrane material reported at that time. Since 1991, other materials have been reported, but the position of the upper bound line has not moved significantly. Many high-selectivity materials are known, but higher selectivity is always obtained at the expense of an exponential reduction in membrane permeability. Using a high-selectivity membrane means that a better separation is obtained, and so the size of the compressor required to produce a unit of product nitrogen decreases. However, this decrease in the cost of the compressor is offset by an increase in the cost of the extra membrane area needed because of the lower membrane permeability. Figure 8.4 illustrates the trade-off between compressor horsepower and membrane area for various membrane units producing the same 100 standard cubic feet per minute (scfm) (1 scf = 0.0286 N m^3) of 99% nitrogen. The base case is taken to be a membrane with an O_2/N_2 selectivity of 6 and an oxygen permeance of 8 gpu (a permeability of 0.8 Barrer and a membrane thickness of 0.1 micrometer). There is a significant decrease in compressor horsepower as the membrane selectivity changes from 4 to 6, but thereafter, the improvement is small. However, the membrane area required to produce the same amount of

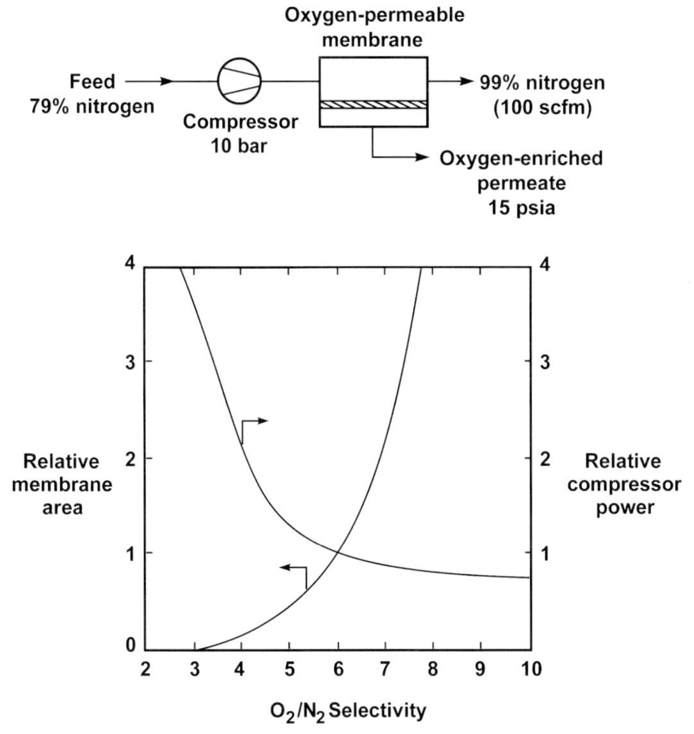

Figure 8.4 The compression power used and membrane area required for nitrogen membrane production as a function of membrane selectivity. The membrane permeability used for each selectivity is taken from the Robeson upper-bound trade-off line shown in Figure 8.3. All numbers are shown relative to a membrane with a selectivity of 6 and an oxygen permeability of 0.8 Barrer.

product nitrogen increases sharply from a selectivity of 4 to 6 and even more sharply at selectivities above 6. Barring an unexpected breakthrough, today's membranes with a selectivity of 6 to 8 are likely to continue as the industry standard.

8.3.2 Air Drying

Capillary membrane modules very similar to those used for nitrogen production are also used to produce dry air. The water molecule is smaller and more condensable than oxygen and nitrogen, so many membrane materials are available with water/air selectivities of several hundred.

In air-drying applications, it is important to operate the modules in a counterflow mode, usually with a small sweep flow from the residue gas. Some calculations illustrating the importance of counterflow and counterflow/sweep operation are shown in Figure 8.5.

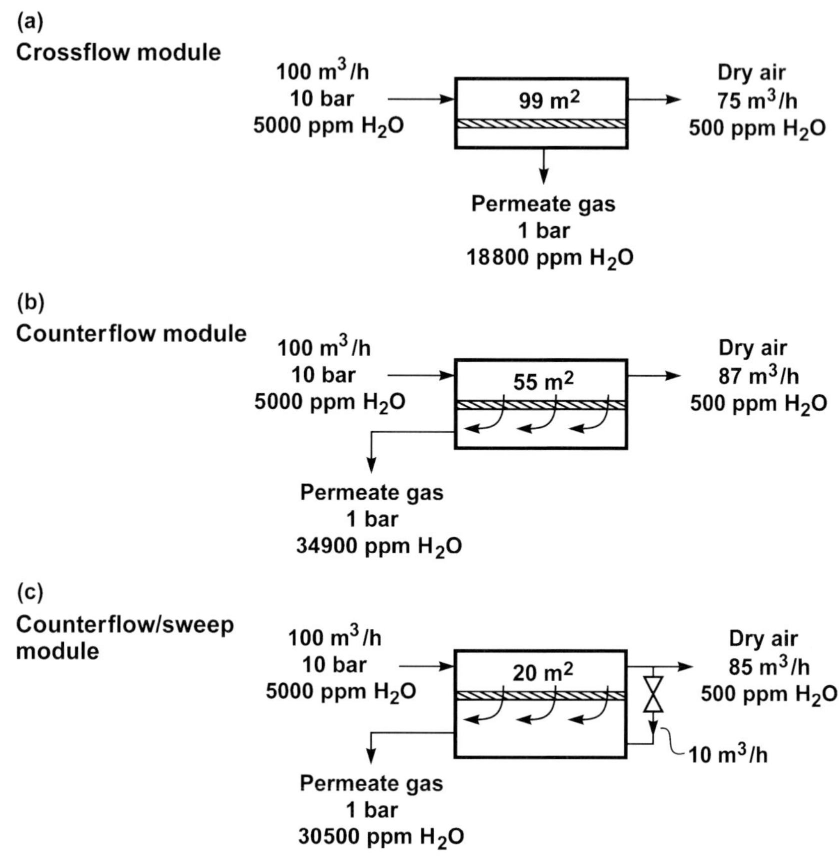

Figure 8.5 Comparison of (a) crossflow, (b) counterflow and (c) counterflow sweep module performance for the separation of water vapor from air. Membrane water/air selectivity = 100, water permeance = 1000 gpu.

In the crossflow module illustrated in Figure 8.5(a), the pooled permeate stream has a water concentration of 1.88%. The counterflow module illustrated in Figure 8.5(b) performs substantially better, providing a pooled permeate stream with a concentration of 3.49%. Not only does the counterflow module perform the separation twice as well, it also requires only about half the membrane area. This improvement is achieved because the gas permeating the membrane at the residue end of the module contains much less water than the gas permeating the membrane at the feed end of the module. Permeate counterflow dilutes the permeate gas at the feed end of the module with low-concentration permeate gas from the residue end of the module. This increases the water concentration driving force across the membrane and so increases the water flux.

In the case of the counterflow/sweep membrane module illustrated in Figure 8.5(c), a portion of the dried residue gas stream is expanded across a valve and used as the permeate-side sweep gas. The separation obtained depends on how much gas is used as a sweep. In the calculation illustrated, 10% of the residue gas is used as a sweep, and the result is dramatic. The concentration of water vapor in the permeate gas is 3.05%, almost the same as for the counterflow module shown in Figure 8.5(b), but the membrane area required to perform the separation is one-third of the counterflow case. *Mixing residue gas with the permeate gas improves the separation!* The cause of this paradoxical result is discussed in a number of papers by Cussler *et al.* [13], and is illustrated graphically in Figure 8.6.

Figure 8.6(a) shows the concentration of water vapor on the feed and permeate sides of the membrane module in the case of a simple counterflow module. On the high-pressure side of the module, the water-vapor concentration in the feed gas drops from 5000 ppm to about 1500 ppm halfway through the module and to 500 ppm at the residue end. The graph directly below the module drawing shows the theoretical maximum concentration of water vapor on the permeate side of the membrane. The actual calculated permeate-side concentration is also shown. The difference between these two lines is a measure of the driving force for water-vapor transport across the membrane. At the feed end of the module, this difference is about 15 000 ppm, but at the permeate end the difference is only about 500 ppm.

Figure 8.6(b) shows an equivalent figure for a counterflow module in which 10% of the residue gas containing 500 ppm water vapor at 10 bar is expanded to 1 bar and introduced as a sweep gas. The water-vapor concentration in the permeate gas at the end of the membrane then falls from 4500 to 500 ppm, producing a dramatic increase in water-vapor permeation through the membrane at the residue end of the module. The result is a two-thirds reduction in the size of the module required for the separation.

Counterflow modules are always more efficient than crossflow modules, but the advantage is most noticeable when the membrane selectivity is much higher than the pressure ratio across the membrane and a significant fraction of the most permeable component is being removed from the feed gas. This is the case for air-dehydration membrane modules, so counterflow capillary modules are almost always used. With most other gas-separation applications, the advantage offered by counterflow designs does not offset the extra cost of making the counterflow type of module, so they are not widely used.

8.3.3 Hydrogen Separation

Hydrogen (H_2) is a highly permeable gas; several glassy polymeric materials are known with good hydrogen permeabilities and H_2/CH_4 and H_2/N_2 selectivities of more than 50. In early applications, membranes made from these materials were used to recover hydrogen from various reactor purge streams [14]. Two typical processes are shown in Figure 8.7. The first involves the separation of hydrogen from nitrogen, methane, and argon. In ammonia reactors, nitrogen from air and hydrogen from a methane reformer are reacted at high pressure to produce ammonia. The ammonia product is removed by cooling and condensation, leaving unreacted gas that is recycled to the reactor. Methane and argon that enter the reactor with the feed streams build up in this reactor loop, gradually degrading the performance of

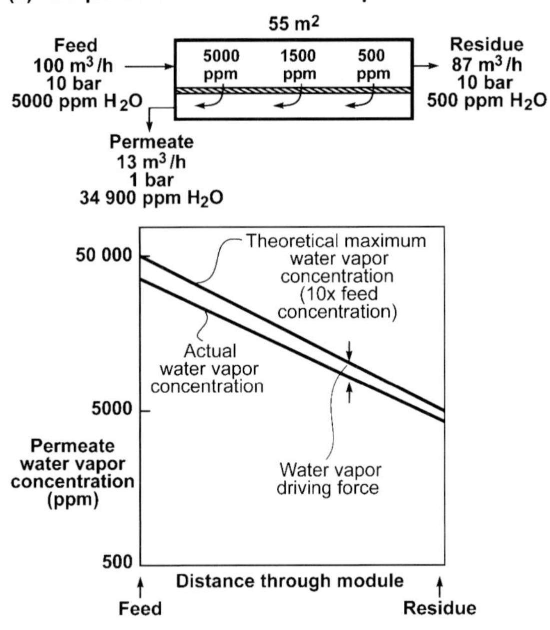

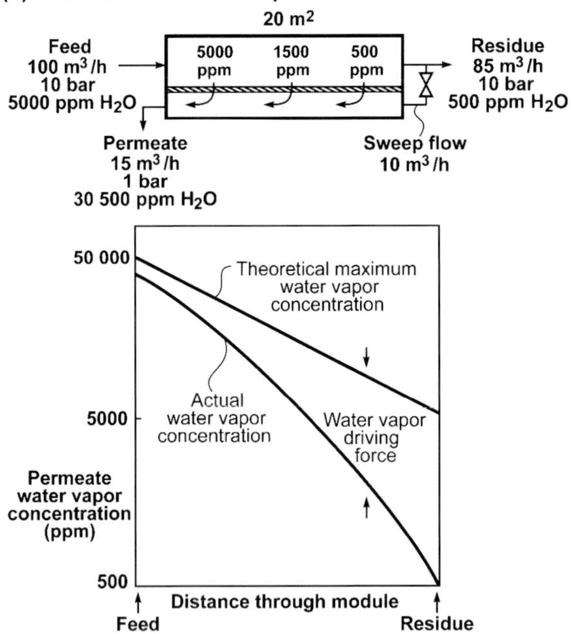

Figure 8.6 The effect of a small permeate-side, counterflow sweep on the water-vapor concentration on the permeate side of a membrane. In this example calculation, use of a sweep reduces the membrane area by two-thirds [6].

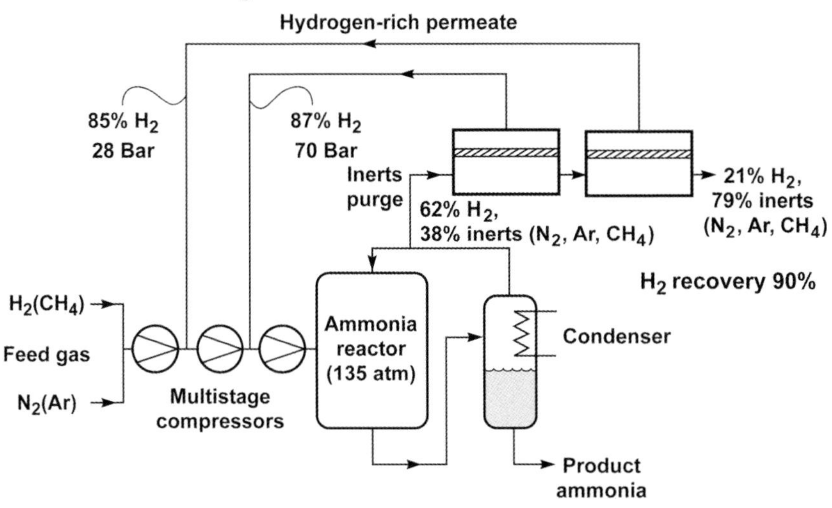

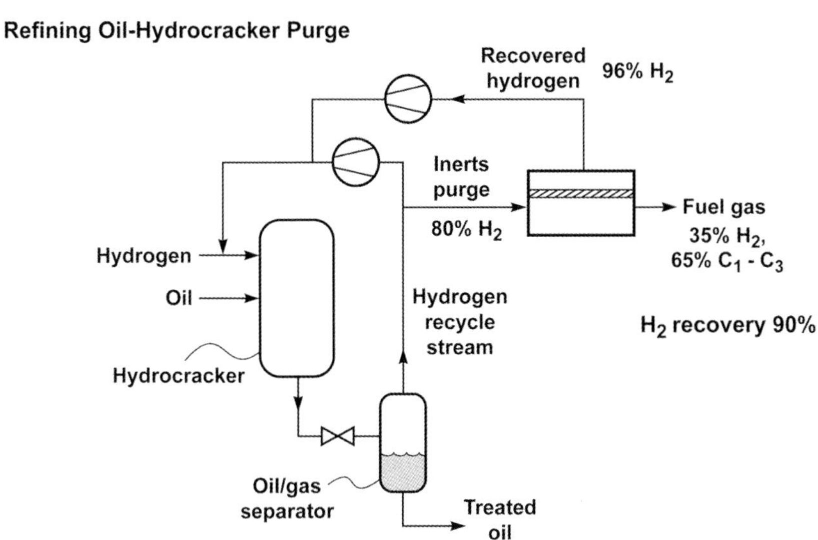

Figure 8.7 Membrane systems to recover and recycle hydrogen lost with the reactor inert-gas purge stream [6].

the reactor. To control the concentration of these inerts, a portion of the recycle loop is purged. About 4 moles of hydrogen are lost with every mole of inert gas purged. Recovery of this hydrogen is well worthwhile and easily accomplished by installing a membrane unit on the purge gas. Ninety per cent hydrogen recovery is usually obtained.

Recovery of hydrogen from the recycle streams of refinery hydrocrackers is a similar application [15, 16]. In these units, heavy oil is treated with hydrogen to crack C_8–C_{12} hydrocarbons into C_4–C_6 molecules, and an inert-gas purge is used to remove

Figure 8.8 Photograph of a Permea hydrogen recovery unit installed at an ammonia plant. The hollow fine fiber modules are mounted vertically [6].

methane, ethane, and propane produced as by-products. A hydrogen-permeable membrane is used to recover the hydrogen content of this purge gas. Again, 90% hydrogen recovery is obtained.

The competitive technologies for these separations are cryogenic condensation and fractionation, or pressure-swing adsorption. The gas flows are usually too small to make cryogenic technology applicable and the pressures involved are above the normal operating range of pressure-swing adsorption. These reasons, together with the simple flow scheme, easy operation, and relatively small footprint of membrane units have made them the standard technology in these processes. Most plants use hollow fine fiber membrane modules from Permea, Ube or Medal. A photograph of a Permea unit installed at an ammonia plant is shown in Figure 8.8. Because the gas being treated is hydrogen at high pressure, thick-walled vessels and special metallurgy are required, together with expensive controls and valves. The cost of these components far exceeds the cost of the membrane modules.

8.3.4 Natural-Gas Treatment

Removal of impurities from natural gas is, by volume of gas to be treated, the largest gas-separation application [1, 17]. About 150 trillion scf of natural gas are produced each year worldwide. All of this gas requires some treatment before it can be used. So far, membranes have captured only 5% of this market, but the membrane share is growing; currently, this is the fastest growing segment of the membrane

Table 8.2 Composition specifications for natural gas delivery to the US national pipeline grid [17].

Component	Specification	Range in US well compositions	
		% of total US gas	Component content
CO_2	<2%	72%	<1%
		18%	1–3%
		7%	3–10%
		3%	>10%
		100%	
Water	<120 ppm	—	800–1200 ppm
H_2S	<4 ppm	76%	<4 ppm
		11%	4–1000 ppm
		4%	1000–10000 ppm
		8%	>10000 ppm
		100%	
C_{3+} Content	950–1050 Btu/scf; dew point: $<-20\,°C$	—	—
Total Inert Gases (N_2, He)	<4%	14%	≥4%
		86%	<4%

gas-separation industry. Raw natural gas varies substantially in composition from source to source [18]. Methane is always the major component, typically 75–90% of the total, but natural gas also contains significant amounts of ethane, some propane and butane, and 1–3% of other higher hydrocarbons. In addition, the gas contains undesirable impurities: water, carbon dioxide, nitrogen, and hydrogen sulfide. Although the composition of raw gas varies widely, the composition of gas delivered to commercial pipeline grids is tightly controlled. Typical US natural-gas pipeline specifications are shown in Table 8.2.

8.3.4.1 Carbon-Dioxide Separation

At present, the largest membrane application in natural-gas processing is carbon dioxide (CO_2) removal. The traditional carbon-dioxide removal technology is amine absorption. Amine plants are able to reduce the carbon-dioxide concentration to less than 1%. Generally, 2–4% of the gas processed is used as fuel for the amine plant or is lost with the removed carbon dioxide. Amine plants are relatively large, complex operations with an absorber and stripping tower and the need to heat and cool large volumes of recirculating fluid. Corrosion caused by amine-degradation products is a critical maintenance issue, and careful, well-monitored operating procedures are required to control the amine chemistry. Membrane plants require significantly less operator attention and smaller units often operate unattended. For these reasons,

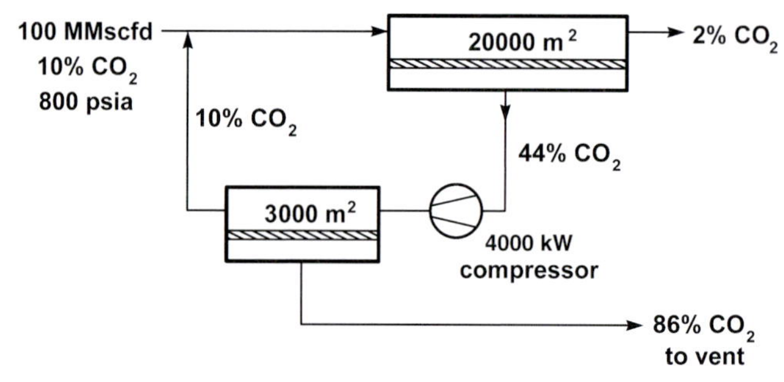

Figure 8.9 Block diagram of a two-stage membrane system to process 100 million scfd of natural gas. Reproduced with permission from *Ind. Eng. Chem. Res.* 2008, 47(7), 2109–2121. Copyright 2008 American Chemical Society [17].

they are favored for use in remote locations, especially on offshore platforms where their lower weight and smaller footprint are additional attractive features. Both hollow fine fiber membrane modules (cellulose triacetate from Cynara) and spiral-wound membrane modules (cellulose acetate from Grace and Separex, perfluoro membranes from MTR) are used.

A block diagram of a typical 100 million scfd gas processing plant is shown in Figure 8.9. A plant of this size costs about US$20 million, depending on location and overall complexity. More than 30 plants of this size or larger have been installed. A photograph of the Kandanwari, Pakistan, plant installed by UOP is shown in Figure 8.10. This plant treats 500 million scfd of gas and is currently

Figure 8.10 Photograph of the 500-million scfd CO_2 removal plant installed by UOP at Kandanwari, Pakistan. The UOP Separex system reduces CO_2 content of a natural-gas stream from 6.5 to 2% CO_2. Photo used courtesy of UOP, LLC.

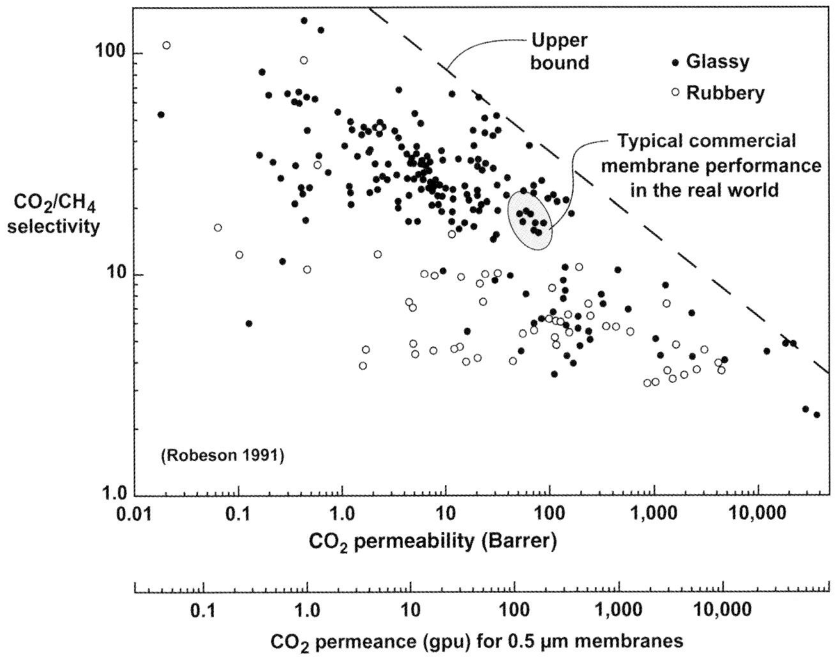

Figure 8.11 Robeson plot of CO_2/CH_4 selectivity versus membrane permeability and permeance [12]. The points shown are based on low-pressure, pure-gas measurements. The performance of commercial membranes when used to separate carbon dioxide from high-pressure natural gas is shown on the same figure for comparison.

the world's largest membrane gas-separation unit. Even larger plants are on the drawing board.

The Kandanwari plant and almost all of the large carbon-dioxide separation plants installed to date have used cellulose acetate or triacetate membranes with a CO_2/CH_4 selectivity in operation of 10 to 20. This fact may seem strange, since membrane materials with selectivities of 50 or more and higher permeabilities than cellulose acetate are routinely reported. Figure 8.11 shows a Robeson plot for carbon-dioxide/methane separations. The position of today's cellulose-acetate membranes is shown on the plot. The commercial membranes in use have half the reported selectivity of the best upper-bound materials. This difference reflects the difference between selectivity estimated from the ratio of pure methane and carbon-dioxide permeability measurements and the 'real world' selectivity measured with high-pressure gas mixtures containing plasticizing components, including not only carbon dioxide, but also water, heavy hydrocarbons, and aromatics [19, 20]. Developing membranes and processes that are able to operate under real-world conditions is where the bulk of industry research is focused.

8.3.4.2 Separation of Heavy Hydrocarbons

Rubbery polymers, most commonly silicone rubber, are used to separate heavy hydrocarbons from natural gas. The traditional technology for this separation is cooling and condensation, but membrane units have found a use in small applications where simplicity and ease of operation are needed. For example, at remote locations, raw untreated natural gas is used as fuel for field compressor engines or power-generating turbines. This gas is often produced in association with oil, so it can contain high levels of heavy hydrocarbons and aromatics that cause coking and preignition when the gas is used as engine fuel. The design of a simple membrane unit to treat such a gas by preferentially permeating the heavy components is shown in Figure 8.12. The feed to the unit is a slip stream from the compressed pipeline gas. The clean residue, stripped of the heavy hydrocarbons, is used as compressor engine fuel, and the heavy hydrocarbons are recycled to the suction side of the field compressor [21].

8.3.4.3 Nitrogen Separation from High-Nitrogen Gas

A second application of rubbery membranes in natural-gas processing is separation of nitrogen from high-nitrogen gas. Pipeline gas must normally contain less than 4% nitrogen, but the pipeline operator will often accept high-nitrogen gas if sufficient low-nitrogen gas is available to dilute the off-spec gas. When dilution is not possible, cryogenic, adsorption, or membrane treatment of the gas is required [22–24].

Methane is about three times more permeable than nitrogen through silicone rubber membranes, so these membranes can be used to perform a separation. Because the membrane selectivity is low, multistep or multistage systems must be used. The design of a two-step nitrogen separation plant installed in a Sacramento River Delta gas field in California is shown in Figure 8.13 [17]. The feed gas contains 16% nitrogen. The heating value of the gas is 900 Btu/scf ($1\ \text{Btu} = 1.0550 \times 10^3$ joules). The pipeline accepts gas for dilution with low-nitrogen gas if the heating value is raised to 990 Btu and the nitrogen content reduced to about 9%. To reach this target, the feed gas, at a pressure of 980 psia, is passed through three sets of modules in series. The permeate from the front set of modules is preferentially enriched in methane, ethane, and the C_{3+} hydrocarbons, and the nitrogen content is reduced to 9% nitrogen. These changes raise the heating value of the gas to 990 Btu/scf. This gas is compressed and sent to the pipeline. The residue gas (containing 22% nitrogen) is sent to a second membrane step where it is concentrated to 60% nitrogen. The permeate from the second step contains 18% nitrogen and is recycled to mix with the feed gas. The residue gas from the second unit is then sent to a final small module unit to be fractionated. The permeate gas from the final unit contains 40% nitrogen and is used as fuel for the compressor engines. The final residue contains 65–70% nitrogen, and is essentially stripped of all C_{3+} hydrocarbons. This gas is vented. The unit recovers about 96% of the heating value of the feed gas in the product stream, about 2% of the gas is used as compressor fuel and another 2% is lost with the nitrogen vent.

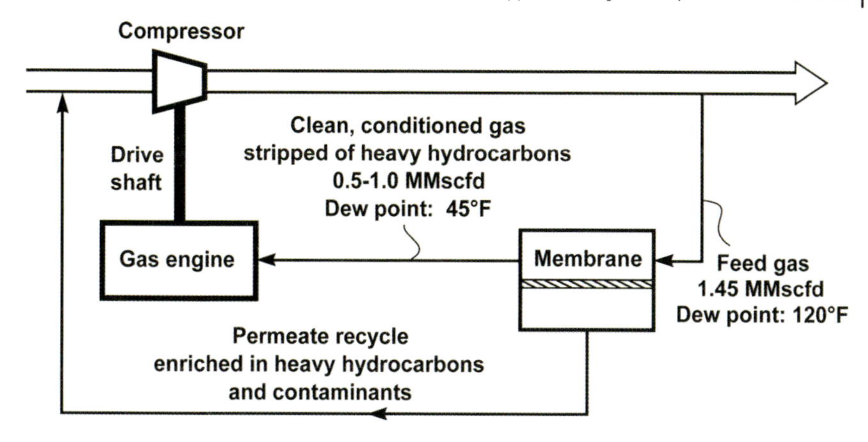

Figure 8.12 Block diagram and photograph of a membrane fuel-gas conditioning unit (FGCU) used for a field gas compressor engine (the unit uses silicone rubber membranes in spiral-wound modules). The membrane modules are contained in the horizontal pressure vessels. The unit produces 0.5–1.0 MMscfd of clean gas. Reproduced with permission from *Ind. Eng. Chem. Res.* 2008, 47(7), 2109–2121. Copyright 2008 American Chemical Society [17].

8.3.5
Vapor/Gas Separations in Petrochemical Operations

In the separation of vapor/gas mixtures, rubbery polymers, such as silicone rubber, can be used to permeate the more condensable vapor components, or glassy polymers can be used to permeate the smaller gases. Although glassy, gas-permeable membranes have been proposed for a few applications, most installed plants use

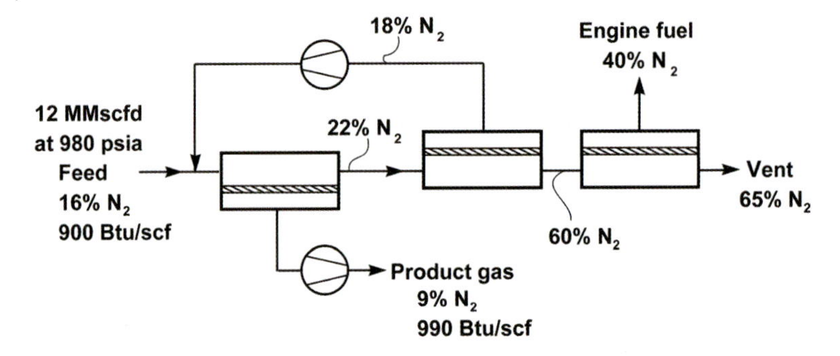

Figure 8.13 Flow diagram and photograph of a 12 MMscfd membrane nitrogen removal plant installed on a high-nitrogen gas well in the Sacramento River Delta region of California. Reproduced with permission from *Ind. Eng. Chem. Res.* 2008, 47 (7), 2109–2121. Copyright 2008 American Chemical Society [17].

rubbery vapor-permeable membranes, often in conjunction with a second process such as condensation [25–27]. The first plants were used in the early 1990s to treat gasoline terminal vent gases or chlorofluorocarbon (CFC) vapor vents from industrial refrigeration plants. Membranes are now used to recover hydrocarbons and processing solvents from petrochemical plant purge streams. Some of these streams are large, and discharge vapors with a recovery value of US$2–5 million/y.

One of the most successful petrochemical applications is illustrated in Figure 8.14: treatment of resin-degassing vent gas in a polyolefin plant [28]. In these plants, olefin monomer, catalyst, solvents, and other coreactants are fed at high pressure into a polymerization reactor. The polymer product (resin) is removed from the reactor and separated from excess monomer in a flash-separation step. The recovered monomer is recycled to the reactor. Residual monomer is removed from the resin powder by

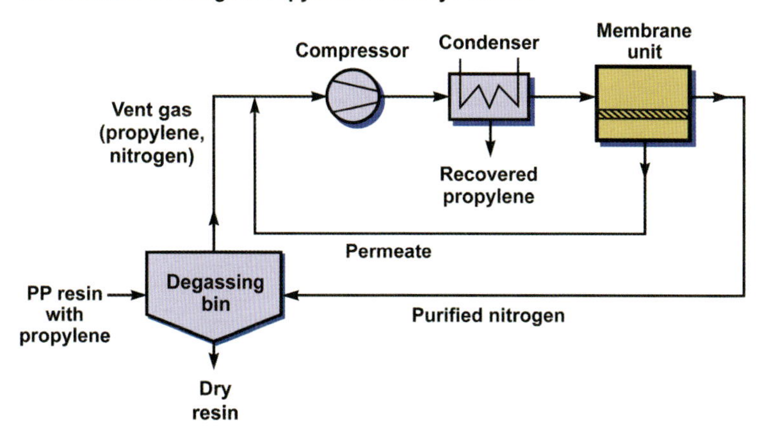

Figure 8.14 Flow diagram showing the use of hydrocarbon-permeable membranes to recover unreacted monomers from a polyolefin plant resin degassing unit. The photograph is of a system installed by MTR in Qatar in 2007.

stripping with nitrogen in a fluidized bed. The composition of the resulting degassing vent stream varies greatly, but it usually contains 20–30% mixed hydrocarbon monomers in nitrogen. The monomer content of the gas represents about 1% of the hydrocarbon feedstock entering the plant. This amount might seem small, but because polyolefin plants are so large, the recovery value of the stream can be significant. About 40 such plants are in use worldwide.

Degassing a vent stream with a membrane system is shown in Figure 8.14. The compressed vent gas is sent to a condenser, where a portion of the hydrocarbon vapors is removed as a liquid. The uncondensed hydrocarbons and nitrogen are separated in the membrane unit, which produces a hydrocarbon-enriched permeate and a purified nitrogen stream (>98% nitrogen). The nitrogen stream is recycled to the resin degasser. The hydrocarbon-enriched permeate is returned to the front of the compressor for hydrocarbon recovery; the hydrocarbon liquid stream from the condenser is upgraded in the monomer purification section of the plant and then

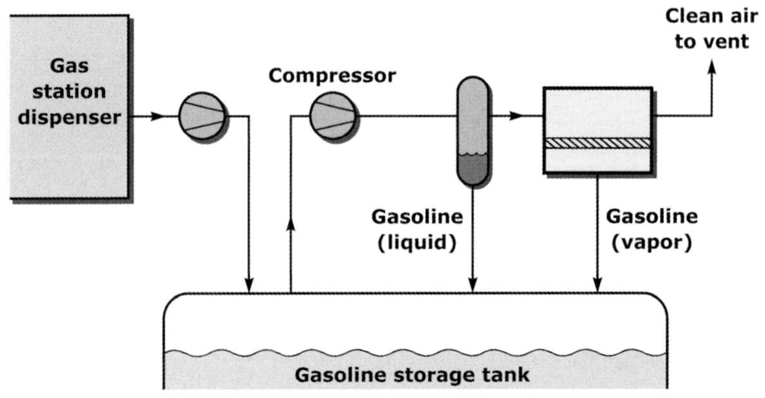

Figure 8.15 Flow schematic of the OPW Vaporsaver unit to minimize gasoline vapor emissions. More than 1000 of these systems have been installed at United States retail gasoline stations.

recycled to the reactor. Similar compression–condensation-membrane separation systems have been installed to separate hydrocarbon/nitrogen mixtures produced by a wide variety of petrochemical processes. More than 20 large systems have also been installed at gasoline terminals to separate gasoline vapors from vent streams produced during gasoline loading operations [29].

At the other end of the scale, more than 1000 small systems have been installed at gasoline stations to minimize the release of hydrocarbon vapors to the atmosphere. These systems use a small pump to draw air and vapors from the gasoline dispensing nozzle (see Figure 8.15). For every liter of gasoline dispensed from the pump, as much as two liters of air and gasoline vapor are returned to the storage tank. Build-up of air in the tank leads to atmospheric releases of gasoline vapor-laden air from the tank head space. Systems fitted with membranes to recover gasoline vapors and return them to the storage tank reduce hydrocarbon emission by 95–99%.

8.4 Future Applications

The applications described above cover the bulk of the current industrial membrane gas-separation business. A number of applications in various stages of development that could become commercial in the next few years are described briefly below.

8.4.1 CO_2/N_2 Separations

Worldwide, approximately 5000 coal-based electric power plants release a total of 10 billion metric tons of carbon dioxide into the atmosphere each year. Separation of this carbon dioxide from power-plant flue gas and sequestration as liquid carbon dioxide

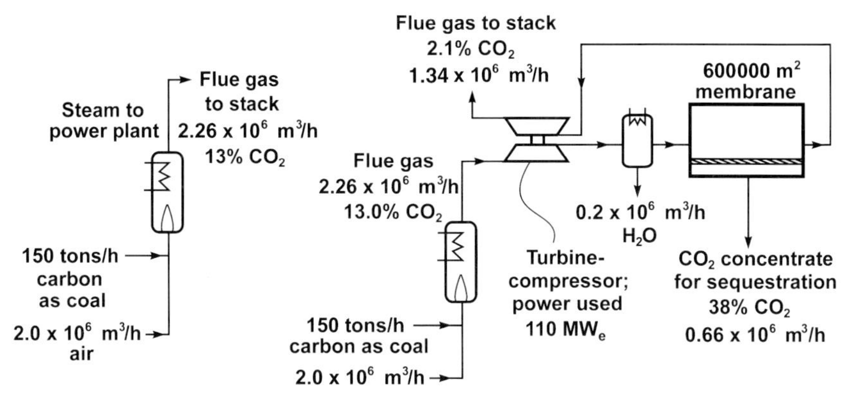

Figure 8.16 Block diagrams illustrating the use of CO_2-selective membranes to separate CO_2 from power-plant flue gas. These calculations are based on a power plant producing 600 MW_e of power. The membrane permeances and size of the membrane units are based on membranes with a CO_2 permeance of 1000 gpu and a CO_2/N_2 selectivity of 40.

into deep aquifers/salt domes is a target of research programs around the world [30]. The cost of carbon-dioxide capture from flue gas depends on the type of power plant producing the gas, the fuel input source (coal, oil, or natural gas), and the capture technology (absorption, adsorption, chemical scrubbing, or membranes) [30, 31]. Currently, carbon-dioxide capture with amine absorption seems to be the leading candidate technology – although membrane processes have been suggested [32].

The use of selective membranes to separate carbon dioxide from flue gas is illustrated in Figure 8.16. Figure 8.16(a) shows a simplified flow diagram of a conventional power plant. For ease of calculation, the fuel input is assumed to be 150 tons/h of carbon as medium-volatility coal. Combustion of this amount of fuel with an excess of air would generate $2.26 \times 10^6\,m^3/h$ of flue gas containing 13% carbon dioxide. This hypothetical plant would produce approximately 600 MW_e of electric power (at 10 000 Btu heat/kW power).

Figure 8.16(b) shows a single-stage membrane process for treating the flue gas. The process cuts carbon dioxide emissions by 90%. In this process, the flue gas is compressed and cooled, which removes most of the water vapor. The gas, which contains about 13% carbon dioxide, then passes across the surface of a CO_2-permeable membrane, producing a permeate containing 38% carbon dioxide and a pressurized residue stream containing 2.1% carbon dioxide. The residue stream is expanded through the turbine compressor, which reduces the power consumption of this unit by more than 50%. The net energy consumption of the turbine compressor is about 110 MW_e, or 19% of the electric power produced. A further 15% is required to concentrate, compress, and condense the carbon dioxide in the low-pressure permeate stream, to produce pure, high-pressure, supercritical carbon dioxide for pipeline sequestration. So approximately 35% of the power plant's electricity is used to separate and sequester the carbon dioxide produced. The membrane plant is also very large. The design shown in Figure 8.16 uses membranes with very high permeance and selectivity, but still requires 600 000 m^2 of membrane. Very low cost membranes and membrane modules are needed to make this process viable.

Innovative process designs are being developed to reduce the size of the membrane unit and the energy needed to separate, condense and inject the carbon dioxide. It seems possible to reduce the energy consumption of the membrane process to about 20–25% of the power plant output. If this work is successful and these membrane plants are built, this application will dwarf all other gas-separation membrane processes.

8.4.2 CO_2/H_2 Separations

The production of hydrogen from coal is expected to become an important aspect of future energy supply [33]. A block flow diagram of one of the proposed coal-to-hydrogen production processes is shown in Figure 8.17. The process starts with a gasification step, in which coal is reacted with oxygen and steam at high temperature and pressure to produce hot syngas, containing mainly hydrogen and carbon monoxide. This gas is scrubbed of tars, particulates and sulfur compounds and cooled to 400–450 °C. The gas then passes through two catalytic shift reactors, at 400 and 250 °C, respectively, to convert carbon monoxide to carbon dioxide and hydrogen via the reaction (8.1)

$$CO + H_2O \rightleftarrows CO_2 + H_2 \tag{8.1}$$

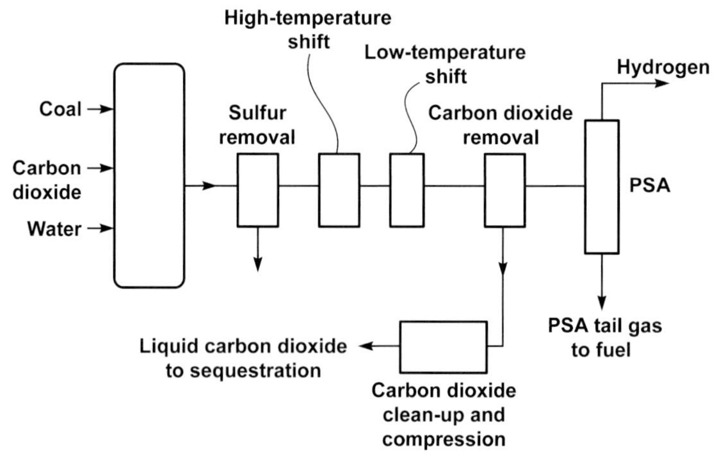

Figure 8.17 Block diagram of a coal-to-hydrogen production plant. Many variants of the process have been proposed.

The gas is then cooled to 30–50 °C and the carbon dioxide is removed by amine absorption or other processes. The remaining impurities – carbon monoxide, methane, nitrogen, argon – are removed in a final pressure-swing adsorption (PSA) step to yield >99.5% pure hydrogen. One of the main problems with this process is that the carbon dioxide is removed by the amine unit as a low-pressure gas. This gas must be compressed to 80 bar to be pipelined for sequestration. This compression step alone requires massive compressors and uses 4–5% of the total power output of the plant. The amine treatment step itself uses even more energy, so the total energy consumption is 15% of the power produced by the plant.

One possible improvement suggested by many authors is the use of hydrogen-permeable membranes instead of an amine unit to remove carbon dioxide [30]. These membranes separate the hydrogen as a low-pressure permeate gas, and the carbon dioxide is left as a high-pressure residue gas. Condensation of the carbon dioxide to liquid carbon dioxide is then much less costly. A number of zeolite [34], ceramic and metal membranes [35, 36] have the required selectivity, but are expensive to produce, so the plant's capital cost may be high. Some polymerics have selectivities in the range of 5–15 and could also be used. Unfortunately, permeabilities are relatively low [37, 38]. An alternative approach is to use membranes that preferentially permeate the carbon dioxide and retain the hydrogen. A number of polar polyethylene oxide-related polymers [39, 40] have been found with CO_2/H_2 selectivities of up to 10 and good permeabilities. Development of this type of membrane has promise.

8.4.3 Water/Ethanol Separations

A pervaporation plant for the dehydration of bioethanol was in operation in 1982. In the succeeding two decades, many small systems were installed, but the technology never really got off the ground [41]. A number of developments are about to change this. First, since about 2003, a very large bioethanol industry has grown up in the United States and Brazil. About 7 billion gallons (1 fluid gal = 3.785 liters) of ethanol were produced from corn in the USA in 2007, and a further 6 billion gallons were produced from sugar in Brazil. Worldwide production in 2007 was about 14 billion gallons. Production may reach 30 billion gallons by 2012, especially if cellulose-to-ethanol technology (now at the demonstration stage) becomes commercial.

The current ethanol dehydration technology – two-stage distillation followed by a molecular-sieve dryer, as shown in Figure 8.18(a) – uses approximately 16 000–20 000 Btu of energy/gal of ethanol produced. This is about 20% of the energy value of the ethanol produced. There is a considerable interest in membrane technology that would be lower in cost and less energy intensive.

Most of the current industrial development efforts are focused on processes that separate water from the overhead ethanol/water vapor of the distillation column, replacing the molecular sieve drier as shown in Figure 8.18(b). The overhead vapor mixture is sent to a water-permeable membrane, producing a dry ethanol residue and a low-pressure permeate enriched in water, which is recycled to the column. Another option, shown in Figure 8.18(c), is to use the membrane-separation step to replace

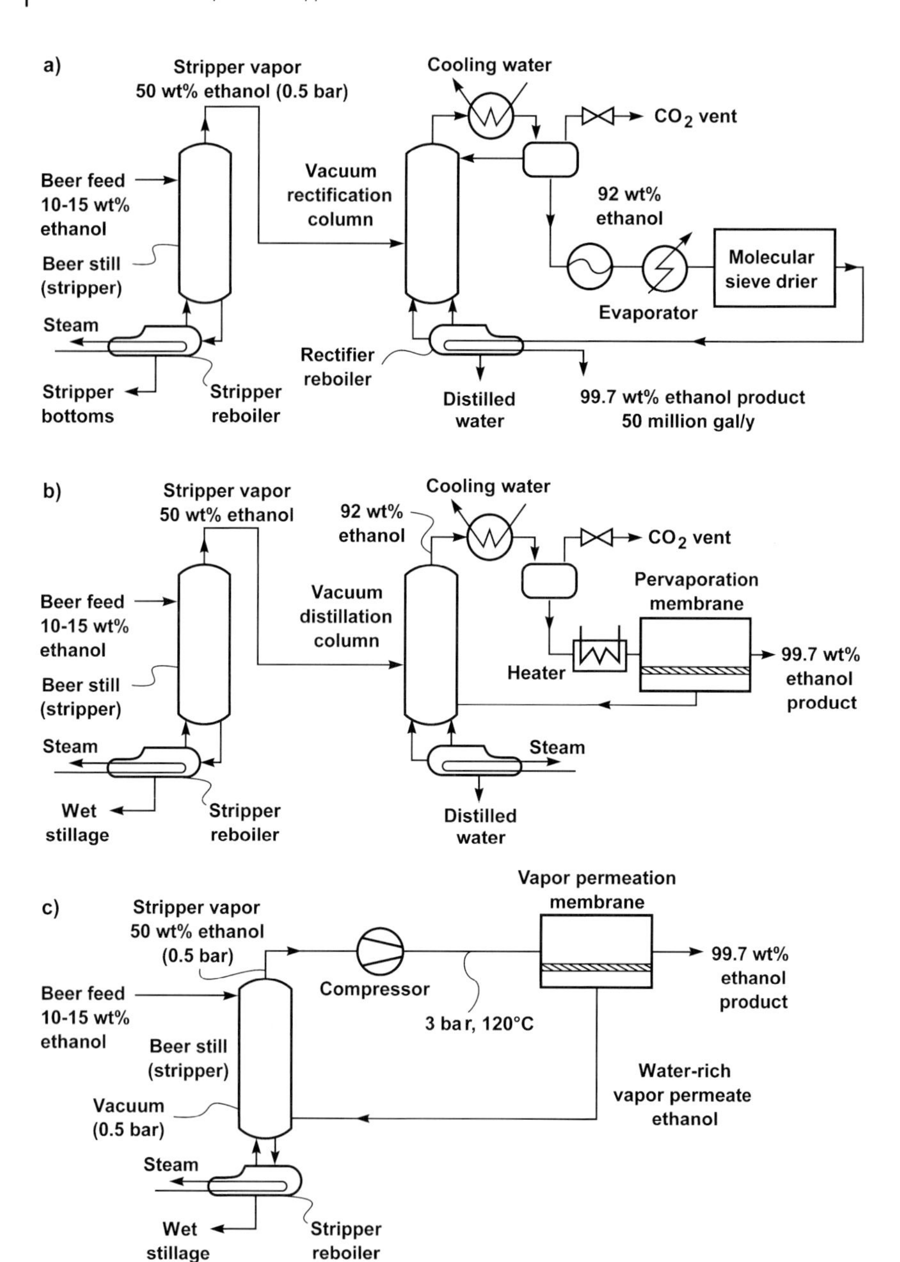

Figure 8.18 Flow schemes of the separation train of a 50-million gallon/y bioethanol plant. Current technology is illustrated in (a). Pervaporation membranes can be used to replace the molecular-sieve drier of the plant (b) or vapor-permeation membranes can be used to replace the rectifier column and molecular-sieve units (c).

both the second distillation column and the molecular-sieve drier. Depending on the operating conditions of the distillation column, the membranes for these options must separate ethanol/water liquid or vapor mixtures at temperatures of 100–120 °C. Ceramic membranes can perform this separation, but are likely to be expensive; polymeric membranes cost much less, but are often unstable in the presence of hot ethanol/water mixtures [42]. The academic literature is flooded with papers describing the separation of ethanol-water mixtures at 50–60 °C. Research of this type has no industrial relevance for this application.

8.4.4 Separation of Organic Vapor Mixtures

The only vapor/vapor mixture currently separated on an industrial scale is ethanol/water. These components have sufficiently different properties that membranes can be found that retain good selectivities, even at high temperatures. Finding membranes that can separate hydrocarbon mixtures has proven much more difficult. The most widely studied vapor mixture is propylene/propane. In the United States, over 34 billion pounds (1 billion pounds = 0.446 million metric tons) of propylene were produced in 2004. All of this propylene had to be separated from propane by distillation. These compounds have very similar boiling points, and the separation requires a very pure propylene overhead (99.5 wt%) and a relatively pure (95 wt%) propane bottoms stream. This means very large columns with 180–240 trays are needed. The distillation column reflux ratios are also as high as 15 to 35, which leads to very high energy costs.

The problem with use of polymeric membranes in this application is plasticization, leading to much lower selectivities with gas mixtures than the simple ratio of pure-gas permeabilities would suggest. For this type of separation, a Robeson plot based on the ratio of pure-gas permeabilities has no predictive value. Although membranes with pure-gas propylene/propane selectivities of 20 or more have been reported [43, 44], only a handful of membranes have been able to achieve selectivities of 5 to 10 under realistic operating conditions, and these membranes have low permeances of 10 gpu or less for the fast component (propylene). This may be one of the few gas-separation applications where ceramic or carbon membranes have an industrial future.

8.5 Summary/Conclusion

The membrane gas-separation industry has come a long way from its starting point in 1980. Plant operators at that time were only ready to consider a membrane solution to their problems if the economics were too good to ignore and if conventional technology could not do the job. Today, membrane gas-separation plants are commonplace and are recognized as being reliable, efficient, and cost effective. As a consequence, the industry is growing quickly and prospects for future growth look good as membranes compete successfully against absorption, adsorption, and cryogenic technology. However, the real future for membranes will be in new applications. If CO_2 sequestration is ever used, it will require many very large plants separating CO_2/H_2 and CO_2/N_2 mixtures; these are great applications for membrane separations. Separation of water from ethanol is also likely to be adopted by industry in the next few years. And finally, there is the separation of propylene/propane mixtures. Ten years ago, I thought development of membranes to separate these mixtures was just around the corner. I was wrong. Useful membranes for this separation remain an unsolved problem, but hope springs eternal, and several laboratories are still pursuing projects to make these membranes. The first thirty years have set the stage and have overcome resistance to the use of membranes in gas separation; the next thirty years will certainly extend their use to larger and more varied applications.

References

1 Spillman, R.W. (1989) Economics of gas separation by membranes. *Chemical Engineering Progress*, **85**, 41.

2 Koros, W.J. and Fleming, G.K. (1993) Membrane-based gas separation. *Journal of Membrane Science*, **83**, 1.

3 Baker, R.W. (2002) Future directions of membrane gas-separation technology. *Industrial & Engineering Chemistry Research*, **41**, 1393.

4 Lüdtke, O., Behling, R.-D. and Ohlrogge, K. (1998) Concentration polarization in gas permeation. *Journal of Membrane Science*, **146**, 145–157.

5 He, G., Mi, Y., Yue, P.L. and Chen, G. (1999) Theoretical study on concentration polarization in gas-separation membrane processes. *Journal of Membrane Science*, **153**, 243–258.

6 Baker, R.W. (2004) *Membrane Technology and Applications*, John Wiley & Sons Ltd, Chichester, UK.

7 Zolandz, R.R. and Fleming, G.K. (1992) Design of Gas Permeation Systems, in *Membrane Handbook* (eds W.S. Ho and K.K. Sirkar), Van Nostrand Reinhold, New York.

8 Kaschemekat, J., Fulton, D. and Wynn, N. (2008) Gas-separation Membrane Module Assembly, U.S. Patent 7,404,843.

9 Blizzard, G., Parro, D. and Hornback, K. (2005) CO_2 separation membranes a critical part of the mallat CO_2 removal facility. Proceedings of the Laurance Reid Gas Conditioning Conference, University of Oklahoma.

10 Prasad, R., Shaner, R.L. and Doshi, K.J. (1994) Comparison of Membranes with Other Gas-separation technologies, in *Polymeric Gas-separation membranes* (eds D.R. Paul and Y.P. Yampol'skii), CRC Press, Boca Raton, FL, pp. 531–614.

11 Prasad, R., Notaro, F. and Thompson, D.R. (1994) Evolution of membranes in commercial air separation. *Journal of Membrane Science*, **94**, 225.

12 Robeson, L.M. (1991) Correlation of separation factor versus permeability for polymeric membranes. *Journal of Membrane Science*, **62**, 165.

13 Wang, K.L., McCray, S.H., Newbold, D.N. and Cussler, E.L. (1992) Hollow fiber air drying. *Journal of Membrane Science*, **72**, 231.

14 Henis, J.M.S. (1994) Commercial and Practical Aspects of Gas-separation membranes, in *Polymeric Gas-separation membranes* (eds D.R. Paul and Y.P. Yampol'skii), CRC Press, Boca Raton, FL, pp. 441–512.

15 Bollinger, W.A., MacLean, D.L. and Narayan, R.S. (1982) Separation systems

for oil refining and production. *Chemical Engineering Progress*, **78**, 27.
16 MacLean, D.L., Bollinger, W.A., King, D.E. and Narayan, R.S. (1986) Gas Separation Design with Membranes, in *Recent Developments in Separation Science* (eds N.N. Li and J.M. Calo), CRC Press, Boca Raton, FL, p. 9.
17 Baker, R.W. and Lokhandwala, K. (2008) Natural-gas processing with membranes: an overview. *Industrial & Engineering Chemistry Research*, **47** (7), 2109–2121.
18 Hugman, R.H., Springer, P.S. and Vidas, E.H. (1990) Chemical Composition in Discovered and Undiscovered Natural Gas in the Lower-48 United States, Report No. GRI-90/0248, Energy and Environmental Analysis, Inc., Arlington, VA.
19 Sanders, E.S. (1988) Penetrant-induced plasticization and gas permeation in glassy polymers. *Journal of Membrane Science*, **37** (1), 63–80.
20 Visser, T., Koops, G.H. and Wessling, M. (2005) On the subtle balance between competitive sorption and plasticization effects in asymmetric hollow fiber gas-separation membranes. *Journal of Membrane Science*, **252**, 265–277.
21 Jariwala, A., Lokhandwala, K. and Baker, R.W. (2006) Only raw sour gas for engine fuel? Proven membrane process cleans gas for engines. Proceedings of the Laurance Reid Gas Conditioning Conference, University of Oklahoma.
22 Tannehill, C.C. (1999) Nitrogen Removal Requirement from Natural Gas, Topical Report, Report No. GRI-99/0080, Gas Research Institute, Chicago, IL.
23 Mitariten, M. (2004) Economic nitrogen removal. *Hydrocarbon Engineering*, **9** (7), 53–57.
24 Baker, R.W., Lokhandwala, K.A., Pinnau, I. and Segelke, S. (1997) Methane/Nitrogen Separation Process, U.S. Patent 5,669,958.
25 Wijmans, J.G.Process for Removing Condensable Components from Gas Streams, U.S. Patents 5,199,962, 1993, and 5,089,033, 1992.
26 Baker, R.W. and Wijmans, J.G. (1994) Membrane Separation of Organic Vapors from Gas Streams, in *Polymeric Gas Separation Membranes* (eds D.R. Paul and Y.P. Yampol'skii), CRC Press, Boca Raton, FL, pp. 353–397.
27 Ohlrogge, K., Wind, J. and Belling, R.D. (1995) Off gas purification by means of membrane vapor separation systems. *Separation Science and Technology*, **30**, 1625.
28 Baker, R.W. and Jacobs, M. (1996) Improved monomer recovery from polyolefin resin degassing. *Hydrocarbon Processing*, **75**(3), 49.
29 Ohlrogge, K. and Stürken, K.S. (2001) The Separation of Organic Vapors from Gas Streams by Membranes, in *Membrane Technology in the Chemical Industry* (eds S.P. Nunes and K.-V. Peinemann), Wiley-VCH, Weinheim, Germany.
30 Hendriks, C. (1994) *Carbon-dioxide removal from Coal-Fired Power Plants*, 1st edn, Kluwer Academic Publishers, Boston, MA.
31 (2005) *Carbon Capture and Sequestration Systems Analysis Guidelines*, U.S. Department of Energy, National Energy Technology Laboratory.
32 Favre, E. (2007) Carbon dioxide recovery from post-combustion processes: can gas permeation membranes compete with absorption? *Journal of Membrane Science*, **294**, 50.
33 U.S. DOE Office of Fossil Energy (2006) Hydrogen from Coal RD&D Plan, http://www.fossil.energy.gov/programs/fuels/publications/programplans/2005/Hydrogen_From_Coal_RDD_Program_Plan_Sept.pdf.
34 Bredesen, R., Jordal, K. and Bolland, O. (2004) High-temperature membranes in power generation with CO_2 capture. *Journal of Membrane Science*, **43**, 1129–1158.
35 Ma, Y., Mardilovich, I.P. and Engwall, E.E. (2003) Thin composite palladium and palladium/alloy membranes for hydrogen separation. *Annals of the New York Academy of Sciences*, **984**, 346–360.

36 Dolan, M.D., Dave, N.C., Ilyushechkin, A.Y., Morpeth, L.D. and McLennan, K.G. (2006) Composition and operation of hydrogen-selective amorphous alloy membranes. *Journal of Membrane Science*, **285**, 30–55.

37 Young, J.S., Long, G.S. and Espinoza, B. (2006) Cross-linked Polybenzimidazole Membrane for Gas Separation, U.S. Patent 6,997,971.

38 Pesiri, D.R., Jorgensen, B.J. and Dye, R.C. (2003) Thermal optimization of polybenzimidazole meniscus membranes for the separation of hydrogen, methane, and carbon dioxide. *Journal of Membrane Science*, **218**, 11–18.

39 Lin, H. and Freeman, B.D. (2005) Materials selection guidelines for membranes that remove CO_2 from gas mixtures. *Journal of Molecular Structure*, **739**, 57–74.

40 Lin, H., Van Wagner, E., Freeman, B.D., Toy, L.G. and Gupta, R.P. (2006) Plasticization-enhanced H_2 purification using polymeric membranes. *Science*, **311**, 639–642.

41 Wynn, N. (2001) Pervaporation comes of age. *Chemical Engineering Progress*, **97**, 66.

42 Vane, L.M. and Alvarez, F.R. (2008) Membrane-assisted vapor stripping: energy-efficient hybrid distillation vapor permeation processes for alcohol-water separation. *Journal of Chemical Technology and Biotechnology*, **83**(9), 1275–1287.

43 Staudet-Bickel, C. and Koros, W.J. (2000) Olefin/paraffin gas separations with 6-FDA-based polyimide membranes. *Journal of Membrane Science*, **170**, 205.

44 Shimadzu, A., Miyazaki, T., Maeda, M. and Ikeda, K. (2000) Relationship between the chemical structure and the solubility diffusivity and permselectivity of propylene and propane in 6-FDA-based polyimides. *Journal of Polymer Science Part B-Polymer Physics*, **38**, 2525.

9
CO_2 Capture with Membrane Systems

Rune Bredesen, Izumi Kumakiri, and Thijs Peters

9.1
Introduction

9.1.1
CO_2 and Greenhouse-Gas Problem

Economical growth and well fare are directly linked to energy consumption. The world's energy needs are currently mainly provided by combustion of fossil fuels ($\sim$85%), making CO_2 the most important anthropogenic greenhouse gas (GHG) [1]. The UN Intergovernmental Panel on Climate Change (IPCC) and the International Energy Agency (IEA) have recently published reports [2, 3] that forecast a substantial increase in energy demand and GHG emissions in the coming years. A Reference Scenario used by IEA anticipates that GHG emissions increase by 57% between 2005 and 2030 [3]. Only strong political measures can stabilize the emission in one or two decades, and reduce it in longer terms, thereby limiting the expected dramatic increase in average global temperature. The GHG emissions can only be reduced by parallel actions to improve energy efficiency, changing to renewable and nonfossil-fuel-based energy sources, and through broad deployment of carbon capture and storage (CCS) technology. CCS technology is already in use in sweetening of natural gas [2, 4], for example, the Statoil Sleipner natural-gas production installations in the North Sea capture and store nearly 1 Mtonne CO_2/year [4]. CCS technology is most cost effective at large-scale point emissions, like power plants and some large industries. Currently, no full-scale power plant exists with CCS technology, but erection of several large-scale demonstration plants have been announced [5–7]. In the short-term perspective, demonstration at large scale is technologically valuable, however, for broad deployment of CCS technology the cost of CO_2 capture, representing 60–80% of the total CCS cost depending on fuel and process, must become economical on commercial terms. The capture cost is therefore the most critical issue to be solved.

Membrane Operations. Innovative Separations and Transformations. Edited by Enrico Drioli and Lidietta Giorno

ISBN: 978-3-527-32038-7

To comply with IPCC recommendations, substantial GHG emission reduction from small- and medium-scale sources is also necessary [8]. A change from distributed use of fossil fuel to large-scale production, including CO_2 capture, and distribution of hydrogen and electricity is a plausible route to such reduction. A relevant question is whether capture and storage from small- and medium-scale sources would be economically and practically feasible. A recent IEA report [9] suggests that distributed collection through pipelines from medium-size sources in industrialized areas (North West of England was used as a case study) could be an alternative, assuming that safety issues are handled satisfactorily. The report concludes that emission sources of the order of 5 ktonne CO_2/yr only constitutes 4% of the total emission, and that this CO_2 is considerably more expensive to collect compared to the emission from large (emission >1 Mtonne CO_2/yr, 73% of total) and medium-size sources (emission > 45 ktonne CO_2/yr, 23% of total). The medium-size sources are typically energy-intensive industries, like iron and steel, glass, cement, pulp and paper, bulk chemicals production, and refinery sites. Although this chapter mainly focuses on membranes for CO_2 mitigation in power generation, we believe that future developments and implementation of cost-effective sustainable technology in other sectors will involve extensive use of energy-lean membrane technology.

9.1.2
CO_2 Capture Processes and Technologies

Capture of CO_2 from fossil-fuel-based large-scale power generation sites is commonly described along three processes routes (Figure 9.1(a)–(c)), (i) postcombustion capture, (ii) precombustion decarbonization, and (iii) oxy-fuel (combustion of fuel in oxygen without the presence of nitrogen). The figure does not include separation steps for gas cleaning in the three process routes (e.g., removal of sulfur, mercury, particles), as this falls outside the scope of this chapter.

The key separation processes envisaged for the three capture routes, indicated as membrane separation processes, are listed in Table 9.1.

(i) In postcombustion processes, CO_2 is separated at ambient pressure from the exhaust gas after combustion of fossil fuel in air. Depending on the fuel and the process design, the concentration of CO_2 in the exhaust gas may typically vary from 30 to 5 vol.%, with N_2 being the main component. Separation is usually considered at close to ambient temperature, but it could in principle be performed at higher temperatures during cooling of the exhaust gas.

(ii) In precombustion decarbonization processes, carbon is separated from the fuel before combustion. This is typically done firstly by converting the fuel to synthesis gas ($CO + H_2$), and secondly, by transformation of the heating value of CO to H_2 by the water gas shift (WGS) reaction. The synthesis gas may be produced by gasification of, for example, coal, or reforming of hydrocarbons. The produced CO_2 and H_2 are separated, a process being facilitated by the high CO_2 concentration and pressure. Elimination of carbon as solid, for example, in

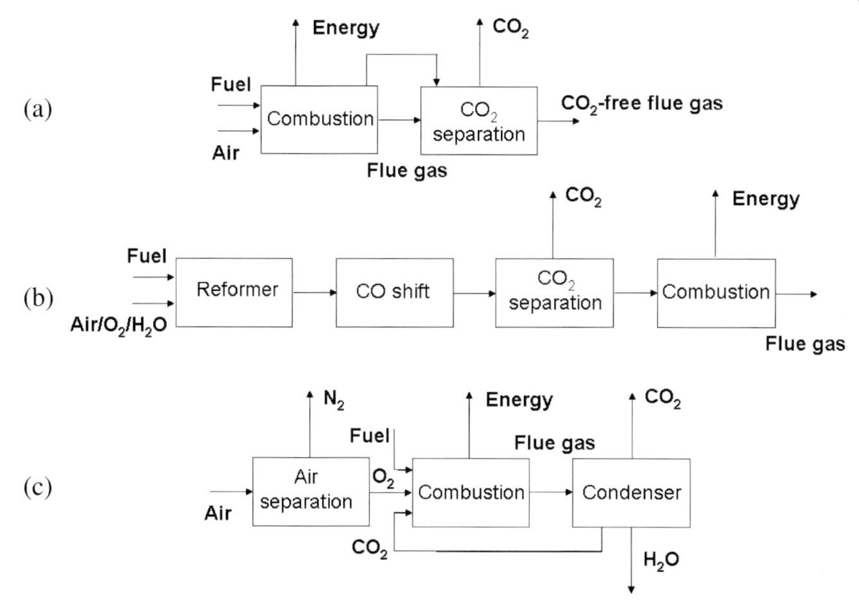

Figure 9.1 Basic schematic diagram of (a) postcombustion capture, (b) precombustion decarbonization, and (c) oxy-fuel processes.

plasma or catalytic cracking of hydrocarbons to produce H_2, may also be categorized as precombustion processes.

(iii) In oxy-fuel processes, combustion of fossil fuel is carried out using oxygen instead of air, thus avoiding N_2 dilution of the exhaust gas. To reduce the high temperature generated, CO_2 from the cooled-down exhaust gas is recycled and used as a dilutant in the combustion [10]. For solid oxide fuel cells (SOFC) based on oxygen-ion-conducting electrolyte membranes, an overstoichiometric amount of air flow can be used on the cathode side to control temperature.

Current commercial CO_2 capture methods [11] include absorption (chemical and physical), adsorption, cryogenic processes, and polymeric membrane technologies. CO_2 capture technologies have not yet been integrated in any large-scale fossil-fuel-based power generation plant [12]. In the coming years, there is a definite need to establish CO_2 capture experience on a large scale, to improve existing technologies and to develop alternative technologies and processes, to lower the cost and energy consumption related to CO_2 capture. To illustrate the magnitude of separation cost of existing technologies, Table 9.2 gives values for 4 types of power plants.

The cost of the separation units constitutes ~20–40% of the total capital cost of the power plants with CO_2 capture. Traditional absorbers and distillation systems are energy consuming and therefore reduce net efficiency of the power plant significantly. The efficiency penalty is also negatively affected by the increasing price of fossil fuels, thus making energy-lean technologies more competitive. In the following

Table 9.1 Typical (membrane) separation processes for CO_2 mitigation.

Membrane process	Main components, feed	Main components, permeate	Relevance
O_2 separation	Compressed air	O_2	Oxygen for gasification of fossil fuels (coal, heavy oil, biomass)
			Oxygen for oxy-fuel combustion
			Oxygen for combustion of remaining fuel from various process streams
	SOFC cathode air stream	$CO_2 + H_2O$	Combustion of remaining fuel in SOFC anode exhaust gas
	Air (compression not necessary)	$CO + H_2$	Partial oxidation of natural gas in precombustion
CO_2 separation	Flue gas $N_2/H_2O/CO_2$	CO_2	Postcombustion capture
	Flue gas H_2O/CO_2	CO_2	Recycling of CO_2 in oxy-fuel process
	Synthesis gas $CO/H_2/H_2O/CO_2$	CO_2	Precombustion decarbonization (WGS)
	Natural gas + steam	CO_2	Precombustion decarbonization (MSR + WGS)
	Natural gas CH_4/CO_2	CO_2	Natural gas sweetening
H_2 separation	Synthesis gas $CO/H_2/H_2O/CO_2$	H_2	Precombustion decarbonization (WGS), H_2 production for fuel cells
	Natural gas + steam	H_2	Precombustion decarbonization (MSR + WGS), H_2 production for fuel cells, and so on.
	Anode exhaust gas SOFC containing unconverted fuel	$H_2O + CO_2$	Combustion of SOFC anode off-gas with cathode air stream (SOFC-GT concept).
		H_2	Recycling of fuel or for PEM, and so on.

Table 9.2 Capital cost and % cost of CO_2 separation unit for different power plants with CO_2 capture, after [13].

Plant type	Capital cost ($/kW)	Separation unit	Capital cost (% total)
NGCC (GTCC)	916	*Postcombustion*	
		Amine chemical absorption	24
PC	1962	*Postcombustion*	
		Amine chemical absorption	18
IGCC	1831	*Precombustion*	
		Air separation (O_2 production)	18
		WGS/selexol physical absorption	13
PC	2417	*Oxy-fuel*	
		Air separation (O_2 production)	32
		CO_2 distillation	7

NGCC/GTCC, Natural gas combined cycle (often termed gas-turbine combined cycle); PC, Pulverized-coal-fired power plant; IGCC, Integrated gasification combined cycle, Oxy-fuel (PC boiler) plant. Flue-gas desulfurization and air particulate control is included in the total cost, but not in the separation unit cost.

we will first continue looking at integration of membrane processes in energy systems with CO_2 capture, and then highlight relevant membrane properties and critical issues related to this technology.

9.2 Membrane Processes in Energy Systems with CO_2 Capture

9.2.1 Processes Including Oxygen-Separation Membranes

In precombustion CO_2 capture schemes, oxygen-separation membranes can be used for synthesis gas production, either by partial oxidation of natural gas, or in gasification of fossil fuel (coal, oil or biomass) [14, 15]. In partial oxidation, dense ceramic oxygen-separation membrane can provide oxygen directly to high pressure natural gas at $\sim$800–1000 °C in a membrane reactor. A technical and economical evaluation [14] of reforming processes integrating oxygen-ion-conducting membranes has shown that an oxygen flux exceeding $\sim$10 mL/cm^2 min produces a competitive technology (Figure 9.2). Such flux values have been reported for existing membranes) [16].

For IGCC coal-based processes, where pure oxygen is used for gasification to produce synthesis gas, a significant cost reduction may appear by changing the production method from cryogenic to membrane technology (Figure 9.3) [15]. This reduction has been estimated to be 35% in capital cost and 37% in power consumption [17], for a 438-MW IGCC plant.

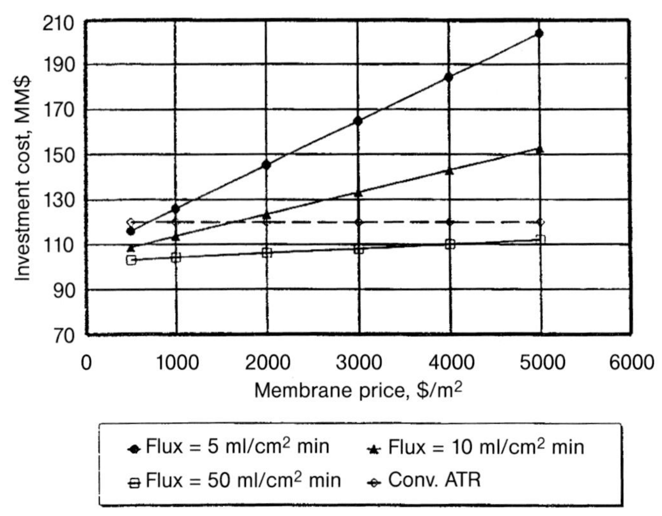

Figure 9.2 Sensitivity of investment cost to membrane price and O_2 flux in autothermal reforming, after [14]. Cost employing conventional cryogenic air separation technology shown by broken line.

Another IGCC process-simulation study, employing steam as the membrane permeate-side sweep gas, reported that the net efficiency of the plant is critically dependent on the applied sweep condition [18]. Increasing steam sweep flow rate and pressure reduce efficiency significantly, illustrating that the operational conditions of the membrane unit can not be optimized isolated from a total cycle analysis. Conditions that often improve the membrane separation, such as high pressure on the feed side and applying sweep or vacuum on the permeate side, may add to investment cost and energy consumption. Design of power-generation cycles

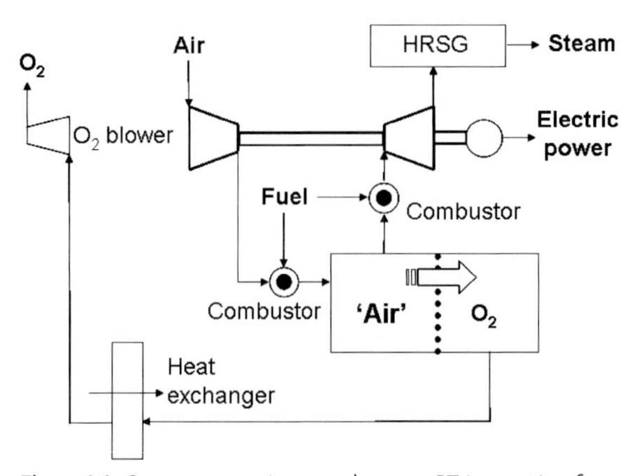

Figure 9.3 Oxygen-separation membrane – GT integration for production of oxygen, after [15].

commonly starts from assumptions related to key units, like the gas turbine, which subsequently defines the borderlines for the other unit operations. Membrane integration with minimum cost and energy efficiency penalty has consequences for the operation conditions, and therefore, for the choice of membrane materials. Separating the membrane process from the fuel-gas stream necessitates few preventive actions to avoid poisoning from contaminants [19]. In the oxy-fuel process, CO_2 from the combustion is recycled and could be used as the sweep gas for an integrated membrane unit to produce a stream of $O_2 + CO_2$ for the combustor. Fuel cleanup and possible use of more stable, but less permeable membranes could add extra cost, which in this case must be considered in a total analysis. The AZEP oxy-fuel concept [20] utilizes hot exhaust gas from the natural-gas-fed combustor as sweep gas. In addition, heat is transferred across the membrane to the air stream, before the hot depleted air is expanded in the turbine to generate electricity, see Figure 9.4. The stability of the membrane limits the operation to temperatures significantly lower than the inlet temperature of modern gas turbines (>1400 °C), which reduces the cycle efficiency. To circumvent this limitation a combustion chamber can be introduced to heat the depleted air before entering the turbine expander. This reduces the CO_2 capture rate from 100 to 85%, however, the net efficiency (LHV) increases from 49.6 to 53.4%, compared to 57.9% for the reference 400-MW GTCC plant without CO_2 capture. The penalty of ~4.5% for CO_2 capture represents one of the lower values reported for thermally produced electricity, which demonstrates the gain achievable if the membrane unit could operate at higher temperature. Advantageous cost efficiency for oxygen membrane integration has also recently been shown by IEA for CO_2 capture from 50-MW boilers [8]. Oxy-fuel

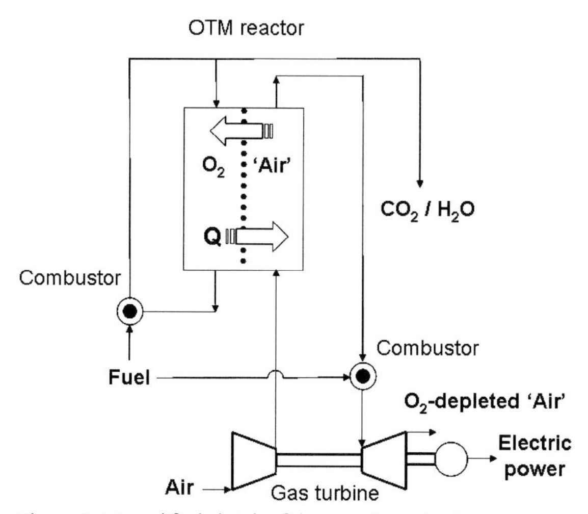

Figure 9.4 Simplified sketch of the mixed conducting oxygen-separation membrane reactor part in the AZEP concept, after [20]. Second combustor placed before the turbine improves efficiency, but also increases CO_2 emission.

combustion employing membrane technology, and postcombustion capture with amine scrubbing, gave cost of CO_2 avoidance of ~22 and 70 €/tCO_2, respectively. In the analysis, the capital cost for the membrane unit was estimated to 6.8 M€ (assuming 1500 \$/m^2), anticipating an oxygen permeation rate of 1 g/m^2 s (4.2 mL(STP)/cm^2 min). The competitiveness of the membrane system can be explained by the lower energy penalty, and lower capital and operating costs. The potentially lower cost of oxygen membrane systems may open other medium- to small-scale applications such as the combustion of SOFC anode off-gas [21]. The efficiency of an integrated SOFC-GT process, for combined electrochemical and thermal electricity production, can reach more than 65% (LHV), and deliver an exhaust stream containing $CO_2 + H_2O$ only [21].

9.2.2
Precombustion Decarbonization Processes Including Hydrogen and Carbon Dioxide Membrane Separation

Large-scale production of hydrogen for fuel and chemicals starts from fossil fuels, typically by methane steam reforming (MSR) and WGS processes.

(1) Methane Steam Reforming	$CH_4 + H_2O = CO + 3H_2$	$\Delta H^0_{288} = 206\ kJ/mol$
(2) Water gas shift	$CO + H_2O = CO_2 + H_2$	$\Delta H^0_{288} = -41\ kJ/mol$
(3) Total reaction	$CH_4 + 2H_2O = CO_2 + 4H_2$	$\Delta H^0_{288} = 165\ kJ/mol$

The strongly endothermic equilibrium-limited steam reforming reaction is carried out at high temperature ~850–900 °C to reach high conversion. Water gas shift is a weakly exothermic equilibrium-limited reaction favored at low temperature. Since the yield for both reactions is limited, conversion will be governed by removing either H_2 or CO_2 from the reactor. Thus, steam reforming could be performed at lower and water gas shift at higher temperature, respectively, in a membrane reactor without compromising yield, see Figure 9.5. Heat is required to sustain the endothermic steam reforming reaction, which can either be supplied to the reactor externally by heaters, or internally by partial oxidation with air or oxygen. As shown in a recent study, spending some produced hydrogen as fuel for external burners, rather than natural gas, is an efficient solution in routes that include CO_2 capture [22].

For hydrogen-selective membranes, H_2 will be obtained at lower partial pressure, and not all hydrogen can be transferred to the permeate side for subsequent use. If hydrogen is used as fuel for a gas turbine, the gas pressure has to be high, for example, 18–20 bars, which could mean that expensive compression of H_2 is necessary at the permeate side. Alternatively, higher pressure applied at the feed side enables direct production of high pressure H_2 at the permeate side [23]. For steam reforming, which is not favored by high pressure, the hydrogen flux still increases with total pressure due to an overall increase in the partial pressure. By avoiding compression, cycle efficiencies above 50% (LHV) including CO_2 capture are reached [24, 25]. Combustion of the remaining hydrogen and unconverted fuel in the

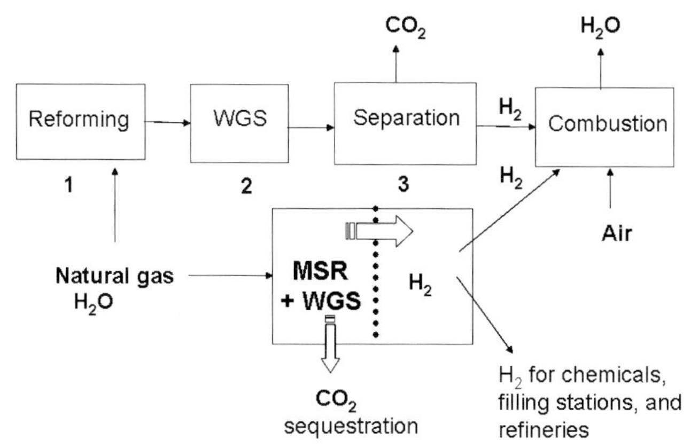

Figure 9.5 Hydrogen-selective membrane combining steps 1 + 2 + 3 in precombustion decarbonization. Starting from synthesis gas, for example, after gasification of coal, step 2 + 3 may be combined in the membrane unit. After WGS, the membrane unit may perform step 3, only.

retentate stream with pure oxygen will reduce the need for complete conversion of the fuel [26], and is a parameter to consider in a total cost and efficiency analysis. An elegant process to (i) provide heat to the reforming of natural gas, (ii) generate *in-situ* sweep gas to the WGS membrane reactor, and (iii) produce a pressurized stream of $N_2 + H_2O + H_2$ fuel for the GT is demonstrated in the hydrogen membrane reformer (HMR) concept, see Figure 9.6 [27]. This highly integrated, high temperature membrane operation (1000–700 °C), gives only a 5% reduction in cycle efficiency when CO_2 capture and compression to 150 bar are included.

For production of low-pressure H_2, Middleton and coworkers have estimated the cost of CO_2 avoidance in production of 230 000 Nm^3/h of H_2 at 1.5 bar from natural gas and coal using thin Pd membranes [28]. The total installed cost for the process gave 30% reduction relative to the baseline cost for amine scrubbing of the flue gas, while the cost of CO_2 avoided was 33% lower. The cost of CO_2 removal, given in Figure 9.7 versus membrane permeability and specific membrane cost for precombustion routes starting from natural gas reforming, illustrates that more expensive

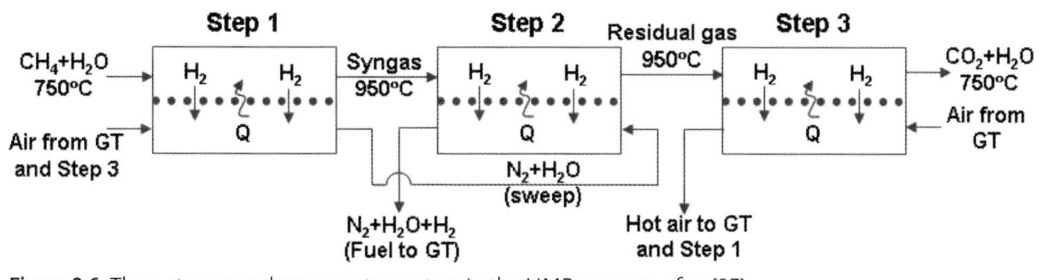

Figure 9.6 Three-stage membrane reactor system in the HMR concept, after [27].

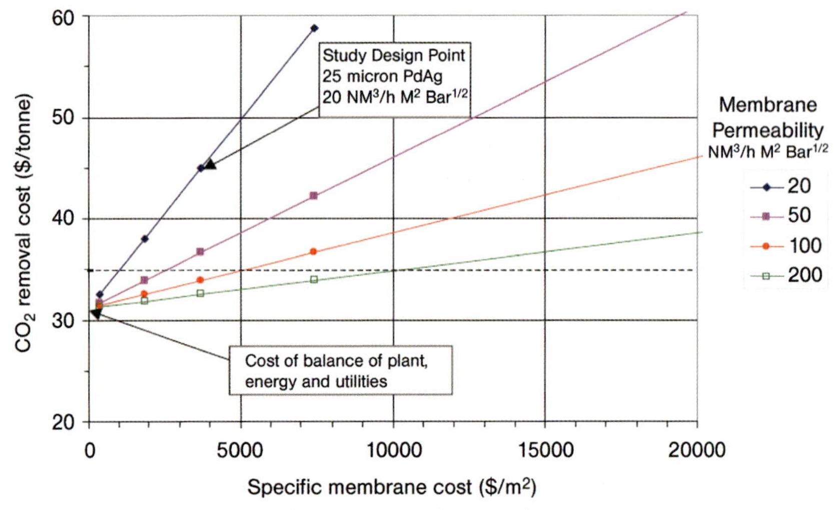

Figure 9.7 Cost of CO_2 capture for membrane reforming, after [28].

materials can be considered provided they offer increased performance. The results also demonstrated that the Pd metal cost is insignificant when the thickness reaches approximately 5 μm. The applied permeability–thickness relation versus cost of CO_2 avoidance is indicative in terms of allowable cost, and also holds for other types of hydrogen-selective membranes.

Extended use of hydrogen in small- and medium-size applications is expected to grow as a consequence of CO_2-mitigation actions and transition to the hydrogen economy. Distributed hydrogen production via MSR + WGS for refueling stations, employing Pd-based hydrogen-selective membranes, has recently been addressed in a techno-economic assessment [29]. The authors investigated production systems in the range 0.2–10 MW membrane reactors and found capture cost of 14 $/t$CO_2$. Furthermore, for thin Pd alloy membranes (<10 μm), the cost of local CO_2 separation (a 2 MW unit), collection in a grid of pipelines and sequestration is less than regional (40 MW) and comparable to centralized hydrogen production including CO_2 sequestration, see Figure 9.8. The main cost of local production is the first pipeline branch with low capacity, however, the cost of separation is significantly lower than for conventional MSR. Tokyo gas has operated their Pd-based membrane reformer producing 40 Nm^3/h hydrogen for more than 3000 h. The hydrogen purity obtained is 99.999%, with an energy efficiency of 70–76% [30].

In the case of CO_2-selective membranes in precombustion processes, the fuel heating value remains at the high-pressure retentate side (H_2 and unconverted fuel) in the WGS separation process. The power-cycle efficiency for natural-gas-fuelled GTCC including CO_2-selective membranes in the WGS reactor appears less pressure dependent compared to hydrogen-selective membranes, due to the lower amount of CO_2 produced in the MSR + WGS reactions [24]. A relative simple simulation for precombustion capture made (based on CO_2/H_2 selectivity of 50) [24], suggests that

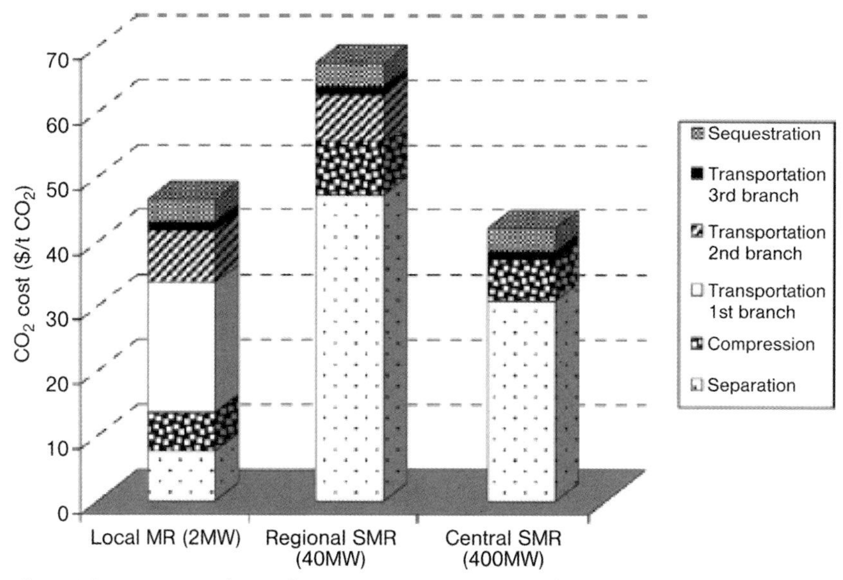

Figure 9.8 Disaggregated cost of CO_2 separation, transport and sequestration, after [29].

integration of CO_2-selective membranes has approximately the same efficiency penalty as H_2 selective membranes for typical GT inlet pressures (18–25 bar).

9.2.3
Postcombustion Capture Processes with Membrane Separation

The low pressure and CO_2 concentration in postcombustion flue gas streams is a demanding challenge for membrane technology. Simulations employing different membrane selectivities CO_2/N_2 (50–200) and flue-gas compositions (10, 20, and 30% CO_2) show that high CO_2 recovery (>80%) is difficult to achieve for a CO_2 concentration below 20% [31]. Compared to conventional amine absorption capture, which requires about 4–6 GJ/tCO_2, the study shows that membranes may potentially reduce this amount to 0.5 GJ/tCO_2 for streams containing ~20% CO_2 or more [31]. The target membrane selectivity required in this case is around 60 (CO_2/N_2). This has already been achieved by the most promising membrane materials [32]. The concentration of CO_2 in flue gases originating from cement production lies between 15–30% by volume, which is higher than in flue gases from power production (3–15% by volume), which could warrant CO_2 removal [2]. Also in the iron and steel industry, high concentration of both CO_2 and CO (~20 vol.%) in the blast-furnace gas could render possible CO_2 removal, and return of CO-rich gas to the furnace [2]. Based on CO_2 removal for a conventional blast furnace, it is concluded that a significant reduction of the CO_2 removal costs to ~17 €/tCO_2 can be obtained by employing membrane technology [33]. As a consequence, CO_2 removal from blast furnaces is comparable with CO_2 removal in IGCC plants.

9.3 Properties of Membranes for Hydrogen, Oxygen, and Carbon Dioxide Separation

In this section, we will treat in more detail some critical properties related to membranes that are considered for use in CO_2 capture processes. As demonstrated in the previous section, the manner of membrane integration in the different processes has major implications for the operating working conditions. For a certain membrane type, the operational window is defined by the expected performance and stability. The flux and selectivity are mainly assessed on basis of short-term studies in model gases. The experience with real gases and long-term studies is limited, which is reflected in the assessments of critical issues given below.

9.3.1 Membranes for Oxygen Separation in Precombustion Decarbonization and Oxy-Fuel Processes

9.3.1.1 Flux and Separation

Ceramic membranes for oxygen separation can be divided into electrolyte type, in which only oxygen ions are mobile, and mixed-conducting types, in which both electrons and oxygen ions are mobile. The latter type can be used for pressure-driven oxygen separation, while electrolytes can be used in SOFCs for electricity production, or in the reverse mode, for oxygen pumping applying an electrical voltage. Various materials show high oxygen ion conductivity, but currently, oxides from fluorite- (MO_2) and perovskite- (ABO_3) related families appear as being most promising [34, 35]. The similarity in materials for SOFC and oxygen membranes is also reflected in the common temperature of operation, being mainly between 800–1000 °C [36]. Integration of these technologies as previously discussed is therefore possible in efficient CO_2 handling. Flux reported in the open literature rarely exceeds values of 10 mL/cm^2 min, and is usually obtained at fairly low absolute oxygen pressure differential [37]. It is evident from the literature that many mixed conductors have slow surface kinetics and flux in thin membranes can be limited by this effect [38].

Ambitious industrial efforts to develop oxygen-separation membranes have resulted in large progress in membrane materials, and membrane and module design. Monoliths and flat structures allow high membrane packing density and can reduce the cost of production including sealing. For instance, modules with contact area of $>500\,m^2/m^3$ have been produced with a checkerboard pattern of channels, and extrapolation to the AZEP process conditions expected to give an oxygen production rate of around 37 mol $O_2/(m^3\,s)$, or 15 MW/m^3 power density [20]. According to the AZEP developers, these values correspond to targets set and confirm the feasibility of the concept. Air Products (US) has adapted two flat membrane concepts for synthesis gas and oxygen production. A pilot unit producing 5 tonne O_2/day has been developed relying on stacked wafer-like membranes connected to a center tube collecting the oxygen [39]. Further development assumes commercial capacities (500–2000 tonne O_2/day) in 2012 [17].

9.3.1.2 Stability Issues

The demanding operational conditions, often combining high temperature, large gradients in total and partial pressures, and the presence of reactive-gas components challenge membrane stability. Reaction with CO_2 is a problem for several good mixed-conducting perovskites (ABO_3) that contain basic alkaline-earth elements on the A site. Thermodynamical stability of carbonates and oxy-carbonates is governed by lower temperatures, thus, reaction might occur during unexpected shut-downs even if avoided at the operational temperature. Acidic sulfur-containing gases (H_2S, SO_2) easily reacts with the same membrane components, and need therefore to be removed before the separation unit. The material designer must also consider possible evaporation of membrane components. The high temperature in combination with steam can lead to increased evaporation by metal-hydroxy components. Kinetic demixing seems to be an unavoidable phenomena originating from difference in diffusivity of the metal components in thermodynamic potential gradients [40]. The effect may lead to decomposition of the membrane, even if the membrane is thermodynamically stable in the whole oxygen partial pressure range of operation [41]. This long-term effect increases for thin membranes operating under large gradients, as encountered in synthesis gas production. Various changes in composition and morphology are often observed for membranes operated under such conditions, but this is probably due to a combination of degradation processes. The combination of high temperature and mechanical pressure also induces creep in the material, which has been studied in some oxygen-separation membranes [42]. Creep may also be a mechanism to reduce stress resulting from differences in thermal expansion of different components and chemical expansion due to reduction in oxygen content in the lattice. Chemical expansion is particularly severe in perovskites with B cations of Co and Fe. For some typical perovskites, Sirman has tabulated relative effects of various cations on A and B sites on essential membrane properties, such as oxygen ion and electronic conductivity, oxygen surface exchange rate, thermal and chemical expansion coefficients, CO_2 tolerance, and resistance to creep [39]. In high-flux (La-alkaline earth)(Co, Fe)$O_{3-\delta}$ perovskites, for example, addition of elements as Mn, Cr, Ti can improve operational stability.

9.3.2 Membranes for Hydrogen Separation in Precombustion Decarbonization

Hydrogen-separation membranes include both dense and porous types covering a temperature regime from ambient to $\sim$1000 °C. We may conveniently distinguish between different types based on the thermal operational window as this is decisive for potential applications. For low-temperature polymer-based membranes, which utilize differences in solubility and diffusivity as the separation mechanism, current research is aimed at exploiting these properties [43]. In absolute terms, both glassy and rubbery polymer membranes have moderate fluxes and selectivity. Although the use of crosslinked polymers has improved the performance, the complexity of implementing this approach on the large industrial scale must be solved before they find widespread use [43]. If separation/flux combination exceeding the upper bound in the well-known Robeson plot is required [44], alternative membrane types are to be considered [43].

9.3.2.1 Microporous Membranes

Flux and separation Microporous inorganic membranes for gas separation mainly include microporous carbon, silica-based or related materials, and zeolite types. In the temperature regime ~100–300 °C, depending on material and operation condition, the presence of adsorbing components, like H_2O and CO_2, will hinder hydrogen diffusion leading to flux reduction [45]. These membranes should preferentially work at sufficiently high temperature, free from surface adsorption, and with selectivity given by size exclusion. In practice, defects and a distribution in pore size result in limited selectivity dependent on molecular size. For zeolite and zeolite-like membranes, where the zeolite pore size can be controlled accurately, intercrystalline diffusion paths are difficult to fully eliminate, which results in moderate separation factors [46]. High-quality microporous membranes show permeance in the range of 10^{-7}–10^{-6} mol/m^2 s Pa [47–50]. Amorphous silica membranes, probably the most studied and advanced microporous membrane for hydrogen separation, have a thickness in the range ~20–70 nm [47, 51]. Thus, a further reduction in thickness to increase permeance, still maintaining a low defect concentration, appears as a considerable challenge [47]. A promising approach is a stage-wise sol-gel and chemical vapor deposition (CVD) synthesis process where the silica membrane obtained combines high selectivity ($H_2/N_2 = 2300$) and good permeance (6.43×10^{-7} mol/m^2 s Pa) [52]. Generally, high selectivity is desirable but the necessity varies with the application. For current proton exchange membrane (PEM) fuel-cell applications even low CO contents in the hydrogen must be avoided due to poisoning of the anode catalyst. For combustion in, for example, gas turbines, heaters and boilers the presence of some unconverted fuel, steam and CO_2 is not critical and selectivity requirements are less. For these latter applications, high flux is most important, which can be increased in microporous membranes by sacrificing selectivity. Microporous C, Si–O–C, Si–O–N materials [49], with varying content of oxygen, have also been investigated, but currently these fall in the same flux/selectivity range as silica membranes. Hydrogen fluxes in zeolite membranes are generally about 5–10 times lower than for sol-gel silica membranes due to the thicker zeolite layer needed to obtain defect-low membranes [53].

Stability issues Microporous silica membranes produced by traditional sol-gel methods are not stable in the presence of steam [51]. Different approaches have been investigated to improve the hydrothermal stability ranging from metal doping [54], inclusion of Si–O–C bonds in the structure to increase hydrophobicity and reduce hydroxyl formation [49], to changing to compositions mainly consisting of Si–C, Si–N and Si–C–N [55–57]. Promising results have been obtained, but the authors are not aware of results demonstrating steam stability in typical high-pressure WGS or MSR conditions. On the other hand, the stability towards other WGS components (CO_2/CO/CH_4) and H_2S is expected to be high; an advantageous property of this membrane type. Pure-carbon membranes, however, have limited stability in some gases (CH_4, H_2, CO_2, O_2) at relevant temperatures [58], and appear less feasible for MSR and WGS processes [59]. Generally, zeolite membranes are expected to have good thermal stability, but under hydrothermal conditions the stability appears limited due to the dissolution of aluminum from the zeolite framework. Improving the hydrothermal stability seems possible, for example, by low aluminum content zeolite or titanosilicate membranes.

9.3.2.2 Dense Metal Membranes

Flux and separation Dense inorganic membranes for hydrogen separation include metal, ceramic, and cermet (metal + ceramic) types [43, 60, 61]. The metal membranes can be divided into two main groups, palladium based, and those containing Group IVB and VB metals. In addition some other metals (e.g., Ni) and amorphous phases are investigated [61]. At present, Pd-based composite membranes can be made thinner than refractory-alloy-based membranes, which in terms of flux compensates for the higher permeability of the latter. For highly selective ~2-μm thick Pd-23w%Ag composite membranes, a H_2 flux reaching ~1200–1500 mL/cm^2 min depending on pre-treatment at 25 bar differential pressure has been reported [62, 63], a value that corresponds to a permeance of 6.4×10^{-3}–1.5×10^{-2} mol/m^2 s Pa$^{0.5}$. The permeance is considerably reduced (5–10 times) in WGS conditions, particularly due to CO surface poisoning [62]. The refractory metals need a catalyst on the surface to enhance the kinetics of the surface reaction, and a layer of Pd or Pd-alloy is commonly applied for this purpose. The Pd layer also serves to protect the reactive refractory metal from corrosion as these easily form oxides, carbides, and nitrides. The amount of Pd coating necessary to obtain fairly stable performance [6] is marginally less than that used in state-of-the-art Pd-based composite membranes [6]. Hydrogen flux reaching 423 mL/cm^2 min has been reported in H_2/He feed for Pd-coated refractory metal membranes at 34 bar hydrogen differential pressure [64]. Hydrogen fluxes up to 150 mL/cm^2 min were achieved in WGS mixtures at pressures up to 31 bar [64].

9.3.2.3 Stability Issues

Interdiffusion between the refractory metal or porous metal support and Pd layer reduces performance and long-term stability [60, 65]. To reduce the problem, barrier layers of, for example, TiN [65], oxides [66–68] or porous Pd–Ag [69, 70] are coated on the metal support. Investigations of Pd-based membranes in continuous sulfur-free operation have demonstrated long-term stability [30, 71–73]. Thermal cycling is more demanding due to differences in thermal and chemical (due to hydrogen dissolution) expansion between the Pd layer and support structure. The thermal expansion coefficient (TEC) mismatch $((TEC_{Pd\text{-}layer} - TEC_{substrate})/TEC_{substrate})$ is high, >30%, for refractory metals and porous ceramic supports [64]. The expansion and contractions in refractory metals and Pd alloys induce stress that leads to deformation, wrinkles and possible detachment from the support layer [62]. Less interfacial stress is generated for thin Pd layers [74, 75], particularly on porous steel supports that have closer TEC values. Hydrogen embrittlement in metals due to hydrogen dissolution is also a concern, but can be avoided by control of operation conditions and appropriate alloying. Furthermore, metal supports are also prone to creep at lower temperatures than ceramics. This could limit the total pressure differential across the membrane in MSR and WGS applications.

Many fossil fuels contain sulfur components, which react with the Pd/Pd-alloy leading to flux reduction by surface blocking, or even complete disintegration of the membrane. Investigations of some Pd–Cu [76–82] and Pd–Au alloy [83] membranes have shown improved chemical stability towards H_2S, but reports about performance in real industrial gases are meagre. Sulfur resistance appears to correlate with the Pd–Cu crystalline structure, which is determined by the operating temperature and alloy composition [79]. Failure seems to depend on H_2S concentration, and not exposure time. For 125-μm thick $Pd_{70}Cu_{30}$ membranes, stable operation at 1173 K in the presence of H_2S-to-H_2 ratios as high as 0.0011 (~1100 ppm H_2S-in-H_2) appears possible [84]. Under certain conditions, carbon deposition in the membrane can also occur affecting the stability [85, 86]. The many stress-generating effects, and reactive components the membrane is subjected to probably cause the commonly observed microstructural changes in thin Pd-based membranes [81, 87, 88]. Further optimization of the performance requires better understanding of these features. Alternative cermet membranes, where an interconnected Pd-based phase is confined to the pores of the ceramic support may possibly offer some stability advantage, though clear evidence is lacking [89].

9.3.2.4 Dense Ceramic Membranes

Flux and separation Relatively high hydrogen permeability is found in many oxides, particularly those with soft lattices containing large basic metal ions [90, 91]. The reason is that oxygen ions move temporarily close together during vibration, allowing protons to jump from one oxygen to the next. More seldom is the combination of high mixed protonic and electronic conductivity required for pressure-driven hydrogen-separation membranes. The possibility of a non-negligible contribution of neutral-hydrogen diffusion has been suggested, but further studies are needed to verify this effect [90]. Known mixed proton and electron conductor membranes require temperatures higher than 600–800 °C to reach appreciable permeability. Recent publications list conductivity and some flux data for several common membrane materials [90]. The maximum flux reported, as far as the author know, are in the range 15–20 mL/cm^2 min [92, 93]. This is in the same range as for ceramic mixed conducting oxygen-separation membranes. The addition of an electron-conducting second phase to good proton conductors, to increase the ambipolar conductivity has been reported [92, 94]. For example, by nickel addition, flux through 266-μm $Ba(Zr_{0.1}Ce_{0.7}Y_{0.2})O_3$ membranes reached nearly 1 mL/cm^2 min at 900 °C in pure H_2. Bulk diffusion appeared rate limiting, thus lowering thickness may give an interesting flux [94].

Stability issues The many stability issues discussed for ceramic oxygen-ion conductors apply also to proton conductors. Reactions with acidic gas components and water are of similar concern for these oxides containing large often basic (Ba, Sr) elements [95]. Other issues such as kinetic demixing, creep and strength have also equal importance, but the authors are not aware of problems related to chemical expansion in this type of membranes.

9.3.3 Membranes for CO_2 Separation in Precombustion Decarbonization

Recent developments demonstrate possibilities for inorganic CO_2 selective membranes. Microporous membranes with strong CO_2 adsorption show CO_2 selectivity if other gas species are hindered in accessing the pores. For instance, at intermediate temperatures, limited CO_2 selectivity to N_2 (to about 400 °C) and H_2 (to about 200 °C) is reported for MFI zeolite membranes [96]. Also, at high pressure (10–15 bars) CO_2 selectivity has been demonstrated in MFI membranes (CO_2/N_2 separation factor ~ 13) with promising CO_2 permeance of 2.7×10^{-7} mol/m^2 s Pa, though these results were obtained at 25 °C [97]. A new interesting membrane type, with the potential of high-temperature operation, is the dual-phase membrane, which consists of an interconnected molten carbonate phase in a porous support [98, 99]. The electrical current loop, set up by the transport of CO_2 as carbonate ions, is closed by electrical transport in the solid supporting phase. Therefore, oxygen ion conducting or metals have been used as supports to facilitate the countercurrent. It has been shown that enhanced flux is obtained in the presence of oxygen on the feed side, implying that carbonate ions are the actual carrier, and not just dissolved CO_2 gas. The first few results reported show CO_2/CH_4 selectivity of 5, and a permeance in the order of 1×10^{-8} mol/m^2 s Pa, at 500–600 °C [99, 100].

9.3.4 CO_2 Separation in Postcombustion Capture

9.3.4.1 CO_2 Separation Membranes

CO_2 capture by polymeric membranes from low-pressure flue gas was early considered. The suggested necessary combination of permeability and selectivity (50 Barrer, $CO_2/N_2 = 200$), however, can not be reached with existing commercial membranes [101]. Postcombustion capture has the great disadvantage compared to precombustion (20–30% CO_2 at 20–50 bars) that separation is from low-pressure flue gas with low CO_2 concentration. If the CO_2 concentration of the flue gas could be increased, for instance by combustion in oxygen-enriched air, polymeric membranes may represent an alternative to amine scrubbing. Furthermore, different modifications such as mixing inorganic nanoparticles with the polymer have given enhanced membrane selectivity by increasing the solubility and the diffusivity of CO_2 [102]. Dendrimer liquid membranes are also reported to have high CO_2/N_2 selectivities over 1000 with 1×10^{-9} m^3/m^2 s Pa CO_2 permeance [103]. However, this immobilized liquid membrane may have insufficient tolerance to handle the large pressure differences required, though recent promise has been reported [104]. These composite PAMAM dendrimer membranes are currently under evaluation, and

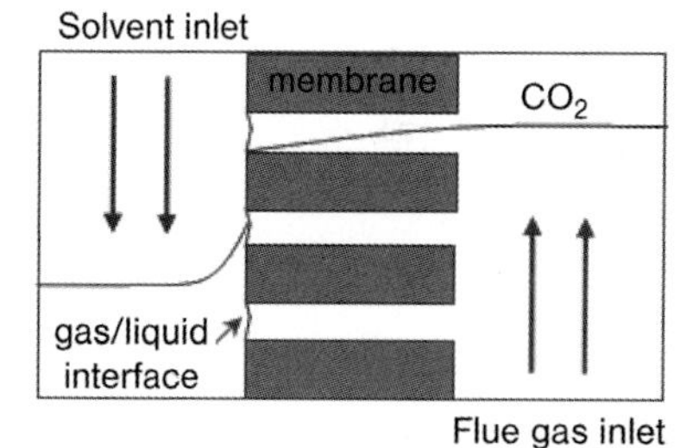

Figure 9.9 Schematic principle of the nonwetted mode of a membrane contactor.

preliminary results indicate CO_2/N_2 selectivity over 200 with $5 \times 10^{-10}\,m^3/m^2\,s\,Pa$ CO_2 permeance [105].

9.3.4.2 Membrane Contactors for CO_2 Capture

Over the past 20 years, membrane contactors, a technology based on the combination of membrane separation and chemical absorption, have been evaluated for CO_2 capture applications [106]. The nonwetting porous membrane is generally not selective, but solely acts as a barrier between the flue gas and the liquid adsorbent, see Figure 9.9 [106]. Separation is determined by the reaction of one component (typically CO_2 or H_2S) in the gas mixture with the absorbent in the liquid.

The energy-consuming regeneration of the amine solution to isolate CO_2 determines to a large degree the energy required for the CO_2 capture [107]. Currently, new and more energy-efficient absorbents are under development, which will benefit the membrane contactor technology [108]. In industry, Kvaerner Oil & Gas and W.L. Gore & Associates GmbH demonstrated the membrane contactor technology in a pilot plant at Statoil's gas processing plant in Kårstø on the west coast of Norway [109].

9.4 Challenges in Membrane Operation

9.4.1 Diffusion Limitation in Gas-Phase and Membrane Support

In recent decades, membrane developers have focused on developing skills to prepare thin selective membrane layers. This effort has resulted in some membranes with both high flux and selectivity. Typically the flux is either determined by the thickness of the selective layer, or (slow) surface kinetics. Strategies to circumvent these limitations are usually to decrease thickness and increase the surface area and/or catalytic properties. These commonly encountered cases are typical for membranes with low to medium permeability. For highly permeable membranes, however, the gas-phase diffusion in the support or in the bulk gas may become rate limiting. In this case, the design of the membrane structure is highly important as illustrated in the two following examples.

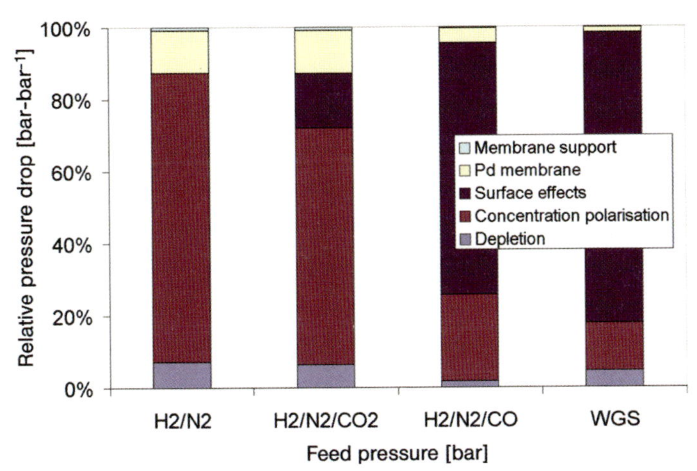

Figure 9.10 Hydrogen pressure drop due to depletion, concentration polarization, surface effects, transport in the palladium membrane and porous support, compared to the total hydrogen partial pressure drop. (a) $H_2:N_2 = 50:50$; (b) $H_2:N_2:CO_2 = 50:25:25$; (c) $H_2:N_2:CO = 50:25:5$; (d) $H_2:CO_2:H_2O:CO:CH_4 = 60:19:16:4:1$. $P_{feed} = 20$ bar, $T = 400\,°C$. Depletion means the lowering of H_2 bulk gas concentration due to H_2 removal along the tube length. Data after [62].

(i) Highly permeable 1–3-μm thick Pd-23 wt% hydrogen-selective membranes supported on 0.48-mm thick porous stainless steel tubes with 2-μm pore size have been reported with a pure H_2 permeance of 6.4×10^{-3} mol/m^2 s Pa$^{0.5}$ [62]. Operation of these membranes in gas mixtures, for example, $H_2 + N_2$ (N_2 assumed to behave as an inert) suggests that the hydrogen flux is mainly limited by a gas-phase diffusion limitation at the feed side. A hydrogen-depleted concentration-polarization layer is built up, reducing the efficient partial pressure of hydrogen, and thereby also the gradient in pressure sustaining the flux. Figure 9.10 illustrates the estimated partial pressure drop for sustaining the flux by three major processes; gas diffusion to the membrane feed surface, transport through the Pd–Ag 23 wt.%, and transport through the porous steel support, respectively. The gas-phase limitations imply a need for improving membrane and module design, and an optimization of feed-flow conditions to reduce the thickness of the hydrogen-depleted layer. The example using a Pd-based membrane may also be used to illustrate the problem of surface reaction rate limitation. Adsorption of other gas molecules on the surface hinders H_2 incorporation, and therefore reduces flux. The effect is particularly strong for CO, which is illustrated in Figure 9.10. A comparison with the high-flux situation with only an inert molecule present shows that the importance of the gas-phase diffusion limitation is drastically reduced. This illustrates that the operation of highly permeable Pd-based membranes (or Group IV and V membranes with Pd-catalyst layer) in WGS conditions is strongly limited by a combination of surface effects and gas-phase diffusion limitations. Thus, if the surface effect could be

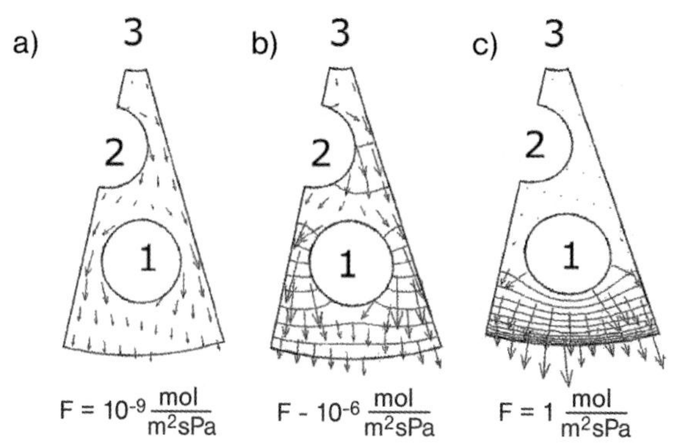

Figure 9.11 2D pressure profiles (isobars) and 2D flux profiles (arrows) as a function of permeance (*F*) of microporous silica membranes on the inside wall of multichannel supports. Very low *F* (a), value for the state-of-the-art (b), and very high *F* (c), after [110].

reduced, the expected flux increase will be limited by gas-phase diffusion. These design and operational implications set by gas-phase diffusion limitations are not limited to the case of Pd-based membranes, but to all highly permeable gas-separation membranes.

(ii) For highly permeable separation layers the resistance of the support structure must be considered. Hollow fibers, multichannel and honeycomb elements all have high surface area per volume, and as such represent possible membrane designs for cheap large-scale gas-separation systems. While hollow fibers and honeycomb structures divide the feed and permeate streams by similar separating membrane walls, the distance from the feed to the permeate side in common multichannel elements varies considerable depending on channel position. The flux per area efficiency of the multichannel element thus depends on the net contribution from all channels, that is, the resistance of the support must be insignificant compared to the resistance of the selective layer [110]. This is illustrated in Figure 9.11 where the contributions from channel 3 (inner channel) and 2 diminish as the permeability of the membrane layer on the inside channel wall increases. The permeance of highly permeable membranes is typically found in the region $10^{-6}\,\mathrm{mol/m^2\,s\,Pa}$, where the resistance from commercially available multichannel supports influence the efficiency.

9.4.2
Membrane Module Design and Catalyst Integration

The mechanical properties of the membrane are essential in operation and module design. For instance, hollow carbon fibers fabricated by pyrolysis of polymers are seemingly too brittle for practical applications [111]. Ceramic capillaries prepared by extrusion are much stronger, but appear limited in maximum length due to vibrations that might occur during operation. To improve the resistance to mechanical stress one faces the dilemma of increasing the wall thickness and/or reducing porosity and/or pore size; all strategies that will increase support resistance. The optimal capillary design is therefore a compromise between sufficiently mechanical strength and permeance. Monolithic membranes with thin walls offer probably the most stable and efficient design, the latter not only due to the high surface area, but also because of the small comparative sealing area. Several designs have been suggested and recently summarized by Carolan [112]. It should, however, be noted that monoliths that provides possibility of crossflow of two separate gas streams give complex manifold systems that may not be easy to fabricate [20]. Ideally, the channels should be made sufficiently small to reduce the gas-phase diffusion limitation, but at the same time not generate too high a crossflow pressure drop. The flat design, which in different forms has been widely investigated in planar SOFC stacks, also provides a means to create high membrane surface area and narrow gas-flow paths. Stacks of flat membranes using spacers to control the distance between membrane plates have been developed by Air Products and partners [113] for their ceramic oxygen-separation membranes in synthesis gas production. The wafer-like design operates with the membrane in mechanical compression between an outer porous support and an internal core of microchannels that distributes the gas evenly. The low-pressure stream is confined to the internal channels, while the pressurized gas is streaming between the wafers. The planar designs reported for synthesis gas and oxygen production [39, 112] limit the extent of necessary metal–ceramic seals for integration in steel housings, but require high-temperature ceramic–ceramic seals to connect the wafers in the stack. Extensive ceramic–ceramic sealing is also demonstrated for hollow fibers, which can be bundled together and sealed to ceramic end sealings [51].

The two main reactions, discussed in this chapter, MSR and WGS, require catalyst and operational control of mass and heat flow. For the highly endothermic MSR reaction, heat is traditionally provided by natural-gas burners, which would require additional systems for CO_2 capture. More elegant is *in-situ* oxidation, for example, as done in the previously discussed hydrogen membrane reformer (HMR) [27]. Alternatively, by employing an oxygen membrane providing oxygen to the fuel side, partial oxidation (exothermic) and steam reforming can be combined to control temperature [14]. The reactor design for these processes should aim at reducing mechanical stress originating from thermal and chemical expansion mismatch, total pressure and temperature gradients. Close integration of the catalyst in MSR to ensure sufficient heat transfer, could be a complicating factor if the chemical compatibility of catalyst and membranes is not sufficient, or if the membrane and catalyst have different lifetimes. For monolithic structures the compatibility issue appears more important than for tubes or plates where the catalyst can be placed more easily externally to the membrane surface, while still being in close proximity. The WGS reaction is only weakly exothermic providing higher flexibility in catalyst integration. Middleton and coworkers suggested that in the WGS process, the reactor containing the catalyst could be separated from the membrane separation unit, in a three-stage sequential process of reaction and separation [114]. There are several advantageous with this concept, (i) each reactor and separation step can be optimized with respect to sizing, design, and to a certain degree, temperature, (ii) feed flow rate and sweep gas can be optimized in each separation step, (iii) the exchange of membranes and catalyst can be done independently, (iv) problems related to chemical noncompatibility of catalyst and membrane can be eliminated. The downside of the concept is an increase in catalyst volume (33%) and membrane area (29%) compared to a single-stage catalytic membrane reactor process designed for capture of 2 Mtonne/yr CO_2 in the production of hydrogen by authothermal reforming and WGS at Grangemouth refinery in Scotland [114].

9.5 Concluding Remarks

Increasing awareness about environmentally related problems has led to large efforts for developing clean and energy-efficient technology. In this chapter we have given examples demonstrating the many opportunities, offered by emerging membrane technology, to efficiently solve key problems related to GHG emission control. The encouraging involvement of industry and public funding organization ensures faster realization, as well as illustrating the competitiveness of the technology. Sufficient flux and selectivity capacity is reported for several existing membrane systems, though long-term performance verification is less clear. Further R&D efforts are therefore still needed to further verify these critical parameters, and particularly, considerable more attention should be directed to stability issues. Optimization of the membrane operation is though a compromise that includes design of the membrane and module on the one hand, and process integration deciding operation condition on the other. Introduction of O_2-, H_2-, and CO_2-selective membranes in large-scale CO_2-capture processes is still some years into the future. In the coming years membrane development will continue on a broad basis, and novel ways of process integration will evolve that strengthens the future impact of membrane technology in CO_2 mitigation.

Acknowledgments

The support from the Research Council of Norway through the NANOMAT projects Functional Oxides for Energy Technology (Grant No. 15 851/431) and Materials for Hydrogen Technology" (Grant No. 158 516/S10) is gratefully acknowledged.

References

1 Davison, J. (2007) *Energy*, **32**, 1163.
2 IPCC (2005) *IPCC Special Report on Carbon Dioxide Capture and Storage, Prepared by Working Group III of the Intergovernmental Panel on Climate Change* (eds B. Metz, O. Davidson, H.C. de Coninck, M. Loos and

L.A. Meyer), Cambridge University Press, United Kingdom and New York, NY, USA, p. 442.

3 IAE – International Energy Agency (2006) Key World Energy Statistics 2006, IAE - International Energy Aogency.

4 Steeneveldt, R., Berger, B. and Torp, T.A. (2006) *Chemical Engineering Research & Design*, **84**, 739.

5 (2007) European Technology Platform for Zero Emission Fossil Fuel Power Plants http://www.zero-emissionplatform.eu/website/docs/ETP%20ZEP/EU%20Flagship%20Programme%20for%20CCS.pdf

6 (2007) Alstom Environment Control Systems Alstom and Statoil to jointly develop project for chilled ammonia-based CO_2 capture for natural gas in Norway.

7 International Energy Agency (IEA) (2007) Near-term opportunities for carbon dioxide capture and storage. Global Assessments Workshop.

8 IEA Greenhouse Gas R&D Programme (IEA GHG) (2007) CO_2 capture from medium scale combustion installations 2007/7.

9 IEA Greenhouse Gas R&D Programme (IEA GHG) (2007) Distributed collection of CO_2, 2007/12, September 2007.

10 Kather, A., Mieske, K., Hermsdorf, C., Klostermann, M. and Köpke, D. (2006) 8th International Conference on Greenhouse Gas Control Technologies (GHGT-8), Trondheim, Norway.

11 Allam, R.J., Bredesen, R. and Drioli, E. (2003) in *Carbon Dioxide Recovery and Utilization* (ed. M. Aresta), Kluwer, Dordrecht, pp. 55–118.

12 Rubin, E.S., Chen, C. and Rao, A.B. (2007) *Energy Policy*, **35**, 4444.

13 Rubin, E.S., Yeh, S., Antes, M., Berkenpas, M. and Davison, J. (2007) *International Journal of Greenhouse Gas Control*, **1**, 188.

14 Bredesen, R. and Sogge, J. (1996) Presentation at the United Nation Seminar on the Ecological Applications of Innovative Membrane Technology in the Chemical Industry, Cetraro, Calabria, Italy.

15 Stiegel, G.J. (1999) *Membrane Technology*, **110**, 5.

16 Vente, J.F., Haije, W.G. and Rak, Z.S. (2006) *Journal of Membrane Science*, **276**, 178.

17 Armstrong, P.A., Bennet, D.L., Foster, E.P.T. and Stein, V.E. (2005) Presentation at Gasification Technologies 2005, San Francisco, USA.

18 Sander, F., Foeste, S. and Span, R. (2006) (eds R. Bredesen and H. Ræder), Proceedings of the 9th International Conference on Inorganic Membranes, Lillehammer, Norway, pp. 224–228.

19 Trembly, J.P., Gemmen, R.S. and Bayless, D.J. (2007) *Journal of Power Sources*, **163**, 986.

20 Sundkvist, S.G., Julsrud, S., Vigeland, B., Naas, T., Budd, M., Leistner, H. and Winkler, D. (2007) *International Journal of Greenhouse Gas Control*, **1**, 180.

21 Maurstad, O., Bredesen, R., Bolland, A., Kvamsdal, H.M. and Schell, M. (2004) in *7th International Conference on Greenhouse Gas Control Technologies (GHGT-7), Vancouver, Canada* (eds E.S. Rubin, E.W. Keith and C.F. Gilboy), Elsevier Science, Oxford, UK.

22 Feng, W., Ji, P.J. and Tan, T.W. (2007) *AIChE Journal*, **53**, 249.

23 Jordal, K., Bredesen, R., Kvamsdal, H.M. and Bolland, O. (2004) *Energy*, **29**, 1269.

24 Kaggerud, K., Gjerset, M., Mejdell, T., Kumakiri, I., Bolland, A. and Bredesen, R. (2005) *Greenhouse Gas Control Technologies*, **7**, 1857.

25 Kvamsdal, H.M., Jordal, K. and Bolland, O. (2007) *Energy*, **32**, 10.

26 Johannessen, E. and Jordal, K. (2005) *Energy Conversion and Management*, **46**, 1059.

27 Åsen, K., Vigeland, B., Norby, T., Larring, Y. and Mejdell, T. (2004) in *7th International Conference on Greenhouse Gas Control Technologies (GHGT-7), Vancouver, Canada* (eds E.S. Rubin, E.W. Keith and C.F. Gilboy), Elsevier Science, Oxford, UK.

28 Middleton, P., Solgaard-Andersen, H. and Rostrup-Nielsen, H.T. (2002) World

Hydrogen Energy Conference, Montreal, Canada.

29 Sjardin, M., Damen, K.J. and Faaij, A.P.C. (2006) *Energy*, **31**, 2523.

30 Yasuda, I. and Shirasaki, Y. (2007) *Materials Science Forum*, **539–543**, 1403.

31 Bounaceur, R., Lape, N., Roizard, D., Vallieres, C. and Favre, E. (2006) *Energy*, **31**, 2556.

32 Favre, E. (2007) *Journal of Membrane Science*, **294**, 50.

33 Lie, J.A., Vassbotn, T., Hagg, M.B., Grainger, D., Kim, T.J. and Mejdell, T. (2007) *International Journal of Greenhouse Gas Control*, **1**, 309.

34 Steele, B.C.H. (1989) in *High Conductivity Solid Ionic Conductors, Recent Trends and Applications* (ed. T. Takahashi), World Scientific, Singapore, pp. 402–446.

35 Kilner, J.A. (2000) *Solid State Ionics*, **129**, 13.

36 Fontaine, M.-L., Norby, T., Larring, Y., Grande, T. and Bredesen, R. (2008) in *Inorganic Membranes: Synthesis, Characterization and Applications* (eds R. Mallada and M. Menendez), Elsevier, Amsterdam.

37 Vente, J.F., McIntosh, S., Haije, W.G. and Bouwmeester, H.J.M. (2006) *Journal of Solid State Electrochemistry*, **10**, 581.

38 Bouwmeester, H.J.M., Kruidhof, H. and Burggraaf, A.J. (1994) *Solid State Ionics*, **72**, 185.

39 Sirman, J. (2006) in *Nonporous Inorganic Membranes for Chemical Processing* (eds F. Sander and M.V. Mundschau), Wiley-VCH, Weinheim, pp. 165–184.

40 Stølen, S. and Grande, T. (2004) in *Chemical Thermodynamics of Materials* (eds S. Stølen and T. Grande), John Wiley & Sons, Ltd, New York, pp. 1–27.

41 Schmalzried, H. and Laqua, W. (1981) *Oxidation of Metals*, **15**, 339.

42 Majkic, G., Wheeler, L. and Salama, K. (2000) *Acta Materialia*, **48**, 1907.

43 Ockwig, N.W. and Nenoff, T.M. (2007) *Chemical Reviews*, **107**, 4078.

44 Robeson, L.M. (1991) *Journal of Membrane Science*, **62**, 165.

45 Hong, M., Li, S., Falconer, J.L. and Noble, R.D. (2008) *Journal of Membrane Science*, **298**, 182.

46 McLeary, E.E., Jansen, J.C. and Kapteijn, F. (2006) *Microporous and Mesoporous Materials*, **90**, 198.

47 Verweij, H., Lin, Y.S. and Dong, J.H. (2006) *MRS Bulletin*, **31**, 756.

48 Mottern, M.L., Shqau, K., Shi, J.Y., Yu, D. and Verweij, H. (2007) *International Journal of Hydrogen Energy*, **32**, 3713.

49 de Vos, R.M., Maier, W.F. and Verweij, H. (1999) *Journal of Membrane Science*, **158**, 277.

50 Peters, T.A., Fontalvo, J., Vorstman, M.A.G., Benes, N.E., van Dam, R.A., Vroon, Z.A.E.P., van Soest-Vercammen, E.L.J. and Keurentjes, J.T.F. (2005) *Journal of Membrane Science*, **248**, 73.

51 Yoshino, Y., Suzuki, T., Nair, B.N., Taguchi, H. and Itoh, N. (2005) *Journal of Membrane Science*, **267**, 8.

52 Gopalakrishnan, S., Yoshino, Y., Nomura, M., Nair, B.N. and Nakao, S.I. (2007) *Journal of Membrane Science*, **297**, 5.

53 Gu, X., Tang, Z. and Dong, J. (2008) *Microporous and Mesoporous Materials*, **111**, 441.

54 Kanezashi, M. and Asaeda, M. (2006) *Journal of Membrane Science*, **271**, 86.

55 Ciora, R.J., Fayyaz, B., Liu, P.K.T., Suwanmethanond, V., Mallada, R., Sahimi, M. and Tsotsis, T.T. (2004) *Chemical Engineering Science*, **59**, 4957.

56 Iwamoto, Y., Sato, K., Kato, T., Inada, T. and Kubo, Y. (2005) *Journal of the European Ceramic Society*, **25**, 257.

57 Volger, K.W., Hauser, R., Kroke, E., Riedel, R., Ikuhara, Y.H. and Iwamoto, Y. (2006) *Journal of the Ceramic Society of Japan*, **114**, 567.

58 Koresh, J.E. and Soffer, A. (1987) *Separation Science and Technology*, **22**, 973.

59 Lu, G.Q., Diniz da Costa, J.C., Duke, M., Giessler, S., Socolow, R., Williams, R.H. and Kreutz, T. (2007) *Journal of Colloid and Interface Science*, **314**, 589.

60 Dolan, M.D., Dave, N.C., Ilyushechkin, A.Y., Morpeth, L.D. and McLennan, K.G.

(2006) *Journal of Membrane Science*, **285**, 30.

61 Phair, J.W. and Donelson, R. (2006) *Industrial & Engineering Chemistry Research*, **45**, 5657.

62 Peters, T.A., Stange, M., Klette, H. and Bredesen, R. (2008) *Journal of Membrane Science*, **316**, 119.

63 Peters, T.A., Stange, M., Bredesen, R. (2008) Thin Pd-23w%Ag membranes for hydrogen separation, Presentation at the 10th International conference on inorganic membranes. (ICIM-10, August 18–22, 2008), Tokyo, Japan.

64 Mundschau, M.V., Xie, X. and Evenson, C.R. (2006) (eds F. Sander and M.V. Mundschau), Nonporous Inorganic Membranes for Chemical Processing. Wiley-VCH, Weinheim, pp. 107–138.

65 Nam, S.E. and Lee, K.H. (2005) *Industrial & Engineering Chemistry Research*, **44**, 100.

66 Yepes, D., Cornaglia, L.M., Irusta, S. and Lombardo, E.A. (2006) *Journal of Membrane Science*, **274**, 92.

67 Huang, Y. and Dittmeyer, R. (2006) *Journal of Membrane Science*, **282**, 296.

68 Su, C.L., Jin, T., Kuraoka, K., Matsumura, Y. and Yazawa, T. (2005) *Industrial & Engineering Chemistry Research*, **44**, 3053.

69 Ma, Y.H., Akis, B.C., Ayturk, M.E., Guazzone, F., Engwall, E.E. and Mardilovich, I.P. (2004) *Industrial & Engineering Chemistry Research*, **43**, 2936.

70 Ayturk, A.E., Mardilovich, I.P., Engwall, E.E. and Ma, Y.H. (2006) *Journal of Membrane Science*, **285**, 385.

71 Tosti, S., Basile, A., Bettinali, L., Borgognoni, F., Chiaravalloti, F. and Gallucci, F. (2006) *Journal of Membrane Science*, **284**, 393.

72 Matzakos, A. (2006) Presentation at the 9th International Conference on Inorganic Membranes, Lillehammer, Norway.

73 Pex, P.P.A.C., Delft, Y.C.v., Correia, L.A., Veen, H.M.v., Jansen, D. and Dijkstra, J.W. (2004) in *7th International Conference on Greenhouse Gas Control Technologies (GHGT-7), Vancouver, Canada* (eds E.S. Rubin, E.W. Keith and C.F. Gilboy), Elsevier Science, Oxford, UK.

74 Tosti, S., Bettinali, L., Castelli, S., Sarto, F., Scaglione, S. and Violante, V. (2002) *Journal of Membrane Science*, **196**, 241.

75 Guazzone, F., Payzant, E.A., Speakman, S.A. and Ma, Y.H. (2006) *Industrial & Engineering Chemistry Research*, **45**, 8145.

76 Roa, F., Block, M.J. and Way, J.D. (2002) *Desalination*, **147**, 411.

77 Alfonso, D.R., Cugini, A.V. and Sholl, D.S. (2003) *Surface Science*, **546**, 12.

78 Kulprathipanja, A., Alptekin, G.O., Falconer, J.L. and Way, J.D. (2004) *Industrial & Engineering Chemistry Research*, **43**, 4188.

79 Kulprathipanja, A., Alptekin, G.O., Falconer, J.L. and Way, J.D. (2005) *Journal of Membrane Science*, **254**, 49.

80 Grimmer, P.J., Xie, X., Evenson, C.R., Mundschau, M.V. and Wright, H.A. (2006) Presentation at the 23th Annual International Pittsburgh Coal Conference, Pittsburgh, USA.

81 Yuan, L., Goldbach, A. and Xu, H. (2007) *Journal of Physical Chemistry. B*, **111**, 10952.

82 Morreale, B.D., Ciocco, M.V., Howard, B.H., Killmeyer, R.P., Cugini, A.V. and Enick, R.M. (2004) *Journal of Membrane Science*, **241**, 219.

83 Gade, S.K., Keeling, M.K., Steele, D.K., Thoen, P.M., Way, J.D., DeVoss, S. and Alptekin, G.O. (2006) in Proceedings of the 9th International Conference on Inorganic Membranes, Lillehammer, Norway (eds R. Bredesen and H. Ræder), pp. 112–117.

84 Iyoha, O., Enick, R., Killmeyer, R. and Morreale, B. (2007) *Journal of Membrane Science*, **305**, 77.

85 Li, H., Goldbach, A., Li, W. and Xu, H. (2007) *Journal of Membrane Science*, **299**, 130.

86 Gao, H.Y., Lin, Y.S., Li, Y.D. and Zhang, B.Q. (2004) *Industrial & Engineering Chemistry Research*, **43**, 6920.

87 Shu, J., Bongondo, B.E.W., Grandjean, B.P.A. and Kaliaguine, S. (1997) *Journal of Materials Science Letters*, **16**, 294.

88 Mekonnen, W., Arstad, B., Klette, H., Walmsley, J.C., Bredesen, R., Venvik, H. and Holmestad, R., *Journal of Membrane Science*, **310**, 337.

89 Balachandran, U., Lee, T.H., Chen, L., Song, S.J., Picciolo, J.J. and Dorris, S.E. (2006) *Fuel*, **85**, 150.

90 Norby, T. and Haugsrud, R. (2006) in *Nonporous Inorganic Membranes for Chemical Processing* (eds F. Sander and M.V. Mundschau), Wiley-VCH, Weinheim, pp. 1–48.

91 Fontaine, M.L., Larring, Y., Norby, T., Grande, T. and Bredesen, R. (2007) *Annales de Chimie-Science des Materiaux*, **32**, 197.

92 Hamakawa, S., Li, L., Li, A. and Iglesia, E. (2002) *Solid State Ionics*, **148**, 71.

93 Vigeland, B. and Åsen, K. (2006) 8th International Conference on Greenhouse Gas Control Technologies (GHGT-8), Trondheim, Norway.

94 Zuo, C.D., Lee, T.H., Dorris, S.E., Balachandran, U. and Liu, M.L. (2006) *Journal of Power Sources*, **159**, 1291.

95 Kreuer, K.D. (1997) *Solid State Ionics*, **97**, 1.

96 Kumakiri, I., Lecerf, N. and Bredesen, R. (2004) *Transactions of the Materials Research Society of Japan*, **29**, 3271.

97 Sebastian, V., Kumakiri, I., Bredesen, R. and Menendez, M. (2007) *Journal of Membrane Science*, **292**, 92.

98 Chung, S.J., Park, J.H., Li, D., Ida, J.I., Kumakiri, I. and Lin, J.Y.S. (2005) *Industrial & Engineering Chemistry Research*, **44**, 7999.

99 Kawamura, H., Yamaguchi, T., Nair, B.N., Nakagawa, K. and Nakao, S. (2005) *Journal of Chemical Engineering of Japan*, **38**, 322.

100 Yamaguchi, T., Niitsuma, T., Nair, B.N. and Nakagawa, K. (2007) *Journal of Membrane Science*, **294**, 16.

101 Powell, C.E. and Qiao, G.G. (2006) *Journal of Membrane Science*, **279**, 1.

102 Zimmerman, C.M., Singh, A. and Koros, W.J. (1997) *Journal of Membrane Science*, **137**, 145.

103 Kovvali, A.S. and Sirkar, K.K. (2001) *Industrial & Engineering Chemistry Research*, **40**, 2502.

104 Duan, S.H., Kouketsu, T., Kazama, S. and Yamada, K. (2006) *Journal of Membrane Science*, **283**, 2.

105 Kazama, S., Kai, T., Taniguchi, I., Duan, S., Chowdhury, F. and Fujioka, Y. (2007) Presentation at 4th Trondheim Conference on CO_2 Capture, Transport and Storage.

106 Feron, P.H.M., Jansen, A.E. and Klaassen, R. (1992) *Energy Conversion and Management*, **33**, 421.

107 Abu-Zahra, M.R.M., Niederer, J.P.M., Feron, P.H.M. and Versteeg, G. (2006) 8th International Conference on Greenhouse Gas Control Technologies (GHGT-8), Trondheim, Norway.

108 Ma'mun, S., Svendsen, H.F., Hoff, K.A. and Juliussen, O. (2007) *Energy Conversion and Management*, **48**, 251.

109 International Energy Agency (IEA) (2003) Zero Emission Technologies for Fossil Fuels Technology Status Report. Working Party on Fossil Fuels.

110 Zivkovic, T., Benes, N.E. and Bouwmeester, H.J.M. (2004) *Journal of Membrane Science*, **236**, 101.

111 Liang, C., Sha, G. and Guo, S. (1999) *Carbon*, **37**, 1391.

112 Carolan, M. (2006) in *Nonporous Inorganic Membranes for Chemical Processing* (eds F. Sander and M.V. Mundschau), Wiley-VCH, Weinheim, pp. 215–244.

113 Allam, R., White, V., Stein, V., McDonald, C., Ivens, N. and Simmonds, M. (2005) *Carbon Dioxide Capture for Storage in Deep Geologic Formations*, Elsevier Science, Amsterdam, pp. 513–535.

114 Middleton, P., Klette, H., Bredesen, R., Larring, Y., Ræder, H. and Lowe, C. (2004) in *7th International Conference on Greenhouse Gas Control Technologies (GHGT-7), Vancouver, Canada* (eds E.S. Rubin, E.W. Keith and C.F. Gilboy), Elsevier Science, Oxford, UK.

10
Seawater and Brackish-Water Desalination with Membrane Operations

Raphael Semiat and David Hasson

10.1
Introduction: The Need for Water

According to UN reports, between 20–25% of the world's population do not have access to good-quality water. People are dying daily due to illnesses related to poor-quality water. The availability of drinking water is continuously decreasing due to the over-usage of aquifers and traditional water sources. This is due in part to how humankind treats the environment, resulting in the pollution of water resources. This is causing people to concentrate in large cities where they expect a better life. These cities are also starting to suffer from a lack of natural, good-quality water. In many places, people are responsible for getting their own water from a distance, wasting considerable time and effort in fulfilling this important task. In many places farmers are dying of hunger since they lack both the technique and the capability to pump water from a nearby river to irrigate their crops.

Over 98% of water sources on earth are undrinkable due to salt content. Only a fraction of the good-quality water is actually used due to the naturally uneven distribution of the water. The problem of water shortage is not only a problem of proper techniques; it is also a social and educational problem depending in many cases on national and international efforts as well as on technical solutions. We need better techniques to provide good-quality water at a low cost, and we must educate people to make better usage of this cheap, yet very costly, resource.

The aim of this chapter is to deal with some of the best available water production and purification techniques, as well as discuss desalination issues based on membranes. Increasing production at affordable costs is one of humankind's most important objectives.

10.2
Membrane Techniques in Water Treatment

The term 'desalination' has lately started to include diverse treatments to purify different water sources, from slightly polluted water, through wastewater and

Membrane Operations. Innovative Separations and Transformations. Edited by Enrico Drioli and Lidietta Giorno

ISBN: 978-3-527-32038-7

brackish water, up to seawater. Membrane techniques are used in many ways to improve water quality. reverse osmosis (RO) is currently the fastest growing desalination technique in industry, emerging even faster than evaporation techniques. Electrodialysis is used for the treatment of slightly polluted water. Other types of membranes are used in different techniques to remove suspended and dissolved matter from raw waters. The main pretreatment steps before using RO membranes are based on the removal of suspended matter from feed water, sometimes including disinfection substances to kill bacteria, followed by a means to remove organic matter and chlorine compounds by active carbon, acidulation to remove carbonate, and more. The feed water is then pumped to an elevated pressure, high enough to overcome the osmotic pressure of the salt-concentrated solution resulting from the actual product recovery of the feed water. Other techniques based on water evaporation are also used for desalination, yet are not included in the scope of this chapter [1].

Osmotic pressure is a property of a solution containing dissolved matter, such as salts, starch or sugar in water; the latter are similar to materials existing in the roots of most plants. The relatively high concentration enables transferring water from the soil surrounding the root through a membrane at the skin of the root. Applying increased pressure to such a concentrated solution behind the membrane reduces water passage and may stop the flow of water (the pressure level that stops the flow is defined as the osmotic pressure of the solution). Higher pressure applied on the solution side of a synthetic membrane, well above the osmotic pressure, will overpower the solution's properties and transfer water from the concentrated solution through the membrane in a direction opposite to the natural action at the plant root. This is the basis of the reverse-osmosis process: it enables selective water permeation through a membrane from the saline side to the freshwater side [2].

Salts rejected by the membrane stay in the concentrating stream but are continuously disposed from the membrane module by fresh feed to maintain the separation. Continuous removal of the permeate product enables the production of freshwater. RO membrane-building materials are usually polymers, such as cellulose acetates, polyamides or polyimides. The membranes are semipermeable, made of thin 30–200 nanometer thick layers adhering to a thicker porous support layer. Several types exist, such as symmetric, asymmetric, and thin-film composite membranes, depending on the membrane structure. They are usually built as envelopes made of pairs of long sheets separated by spacers, and are spirally wound around the product tube. In some cases, tubular, capillary, and even hollow-fiber membranes are used.

Water passage through reverse-osmosis membranes is based on water dissolution in the membrane walls followed by diffusion to the other side of the membrane. RO membranes are denser membranes, containing almost no holes. The membrane skin, supported by a porous polymeric layer, is responsible for the membrane properties. The solubility of water in the membrane is much higher than the solubility of the salts present in feed water, hence enabling the separation between water and salt ions, which are also relatively large molecules surrounded by water molecules. The integrity of the skin layer is very important for the rejection of salts. Scratches and holes in the skin enhance the passage of salt ions and thus reduce salt rejection. A SEM picture of a RO membrane is presented in Figure 10.1. The skin,

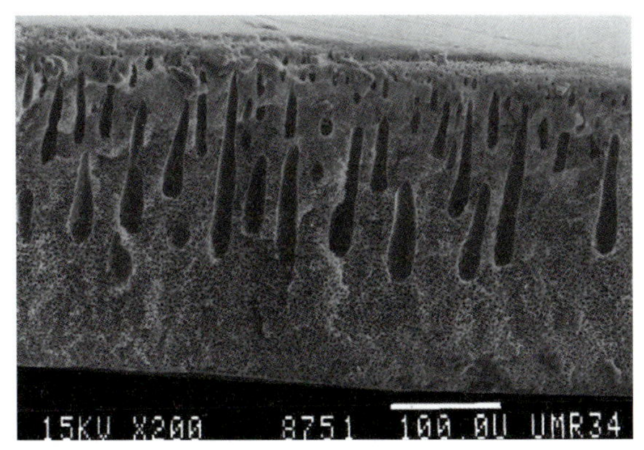

Figure 10.1 The structure of a RO membrane – the thin skin is responsible for membrane flux and rejection properties. Picture taken from Blanco *et al.* [53].

presented at the top of the picture, is responsible for the membrane properties. Membranes are spirally wounded as shown in Figure 10.2 (8″ and 16″ in diameter) and are inserted into pressure vessels.

Ultrafiltration and microfiltration membranes produce high porosities and pore sizes in the range of 30–100 nanometers (UF) and higher (MF), which enable the passage of larger dissolved particles and even some suspended particles. The separation-filtration mechanism is based on molecule/particle sizes. The nanofiltration membrane lies between the UF and RO membranes, combining the properties of both so that the two mechanisms coexist. In addition, the NF membrane may be

Figure 10.2 Spiral-wound RO membranes, 8″ and 16″ in diameter. Cooperation between Nitto Denco/Hydranautics and Graham Tec.

charged electrically, depending on functional groups acting on the membrane surfaces. This charge affects the passage of molecules through the membranes. The membranes may be found in different types of modules, enabling their use in water purification and treatment. Wilbert *et al.* [3] described various treatments available for surface water and other sources.

Nanofiltration membranes are used to remove hardness from drinking water [4, 5]. They may also be used to remove other unwanted dissolved species, even the partial removal of nitrates from ground water. It was recently shown that RO and NF membranes may be backwashed by direct osmotic pressure to clean membrane surfaces, a simple and very beneficial technique [6, 7].

Ultrafiltration and microfiltration can be backwashed occasionally to remove accumulated solids from membranes. UF and MF membranes may be used to remove micrometer-sized and upper suspended particles, namely bacteria, algae, and so on, they can also be used to remove *Guardia* and *Cryptosporidium*, as well as most viruses found in surface water. In fact, the solid layer ('cake') adhering to the membranes in the latter two techniques acts like a dynamic membrane [8, 9], removing smaller particles even at colloidal and virus levels.

The use of MF membranes may be cheaper than sand filtration in the treatment of surface water. The international water company, Ondeo (Lyonnaise des Eaux), uses MF membranes combined with active coal and sedimentation stages to purify polluted Seine River water for drinking purposes [10]. Veolia also uses MF combined with NF to get good-quality water. Many other companies, membrane manufacturers or users are involved in producing clean wastewater, either directly together with a MBR bioreactor [11] or using membranes after they have passed through the bioreactor. In Singapore, wastewater is treated with UF and RO membranes to make NewWater for usage in microelectronic fabrication [12]. Part of the water is mixed with surface water for regular usage. More on wastewater treatment is provided below.

Electrodialysis (ED), or reversible electrodialysis (RED), involves applying a DC electrical field across a membranes stack. Ions are transferred through semipermeable membranes into concentrated streams, leaving behind a diluted salt solution. This was considered a promising technique mainly because of the relative insensitivity of the membranes for fouling, and due to the thermodynamic transfer properties of this technique. Unfortunately, the technique did not succeed in taking its naturally expected position among other processes. It is currently used primarily for brackish-water desalination and water purification [13]. EDR membranes are also used to remove special salts, such as nitrates, from slightly polluted water. Strathmann [14] provides a cost estimate of the ED process.

The use of membranes is infiltrating into the process industry, where improved water quality is needed. Power stations, petrochemical and high-tech production plants are seeking improved water quality and are using different types of membranes to meet their needs. Additional information on different aspects of desalination processes was reported by Semiat [1].

Electrical power is the energy source for RO desalination. A reverse-osmosis desalination plant is presented schematically in Figure 10.3. Electricity is used to

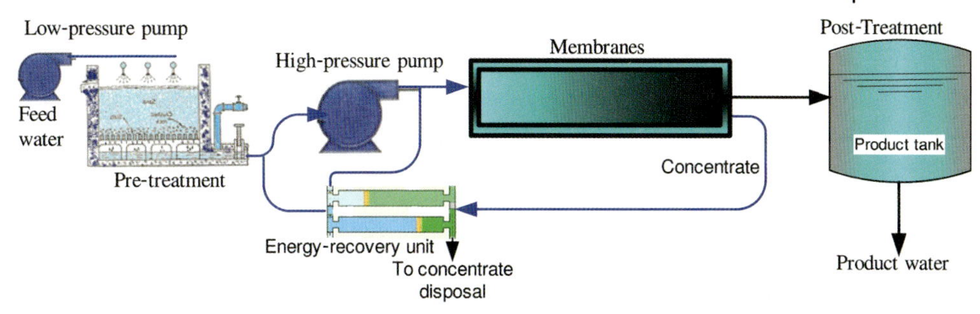

Figure 10.3 Schematic presentation of a reverse-osmosis desalination plant.

pump the water at a relatively high operating pressure. The product penetrates the membrane and exits at a predesigned recovery, defined as the product-to-feed ratio. The high-pressure purged concentrate contains energy that may be recovered using turbines or pressure-exchange devices [15]. The osmotic pressure of seawater, for example, varies from 24 bars to twice as much for concentrate at 50% recovery. Operating pressures therefore vary between 10–25 bars for brackish water and 60–80 bars for seawater in order to allow sufficient permeation at relatively high concentrations of the brine along the pressure vessel. The process takes place at ambient temperatures. Water conversion can increase to 70–95% recovery in the case of contaminated or brackish water, or 35–50% recovery using seawater. The low recovery from seawater is due to the high osmotic pressure of the concentrate leaving the membrane modules, depending on the recovery ratio and the need to operate at higher pressure, where investment increases significantly with operating pressure. Lower water recovery is obtained in relatively closed water bodies, such as the Red Sea or the Persian Gulf, due to higher salt concentrations.

Water temperature influences membrane performance. Flux through a membrane increases with water temperature and is bounded by membrane limitations, yet salt rejection and product quality are reduced with an increase in water temperature. Hot seawater flowing from the cooling system of a large power plant may increase efficiency at the expense of water quality. The quality of the water produced depends on membrane-rejection properties together with the degree of water recovery and system design. Some relatively small molecules, such as carbon dioxide, hydrogen sulfide, silica, and boric acid, may penetrate and reduce water quality. Silica and CO_2 are not a problem; low acidity in the product is preferred as a means for dissolving lime in order to add calcium carbonate to the water produced and reduce water aggressiveness. Secondary or higher membrane stages, aeration or ion exchange may solve other problems.

The boron problem still exists due to the low rejection of boric acid through the membranes, yet several other solutions exist, as described below. Final mixing of the water is advisable in some cases to increase salt concentration slightly. Small organic compounds dissolved in the feed water may also find their way into the water produced. Salt content depends on feed quality (brackish or seawater) and may vary between 50–600 ppm of TDS. A secondary stage may improve quality with only a slight cost increase. This is useful in cases where high recovery from seawater is required or where ultrapure water is needed.

10.3
Reverse-Osmosis Desalination: Process and Costs

Figure 10.3 depicts a schematic flow sheet of a typical desalination plant. Feed pretreatment for the removal of suspended material, bacteria, and organics is carried out by sand filtration followed by media filtration. UF or MF modules are used in modern plants. Residual chlorine, if present, is removed with active carbon filters or by the injection of sodium bisulfate solution. The high-pressure pump used to feed the membrane module may be connected along a single shaft with a motor and a turbine [16, 17] in order to recover the energy content of the pressurized concentrate. Other means, such as independent turbines for secondary stages, may also be used for energy recovery [15]. Concentrate disposal is simple in the case of seawater desalination but more difficult in the case of inland desalination. Measures that can be taken in this case include natural and enhanced evaporation ponds, underground injection, and pipe transport to the sea.

The reverse-osmosis membrane process is considered universally as the most promising technology for brackish and seawater desalination [18]. Potential directions for reducing desalination costs may be deduced by analyzing the cost of the components.

After the investment, energy is the second cost component to consider. The cost of energy was reduced in the design of the Ashkelon plant with the use of a dedicated gas-turbine power station; this power station reduces energy costs because it is insensitive to the common *sine* wave of power consumption curve involving fluctuations in day–night, summer–winter electricity demand. Modern energy conservation devices also reduce energy costs, albeit at the expense of increased capital cost. A trend towards increased investment to replace energy will increase with energy cost.

Figure 10.4 presents an estimated cost breakdown of desalinated water produced in a typical plant. The main component is, of course, the capital and financial cost, comprised of the cost of the main equipment items: feed tanks, pretreatment filtration units, pumps, pressure exchangers and piping, controls, membranes and membranes housing, post-treatment and product tanks.

It is obvious from the data in Figure 10.4 that cost reduction may be examined in two main ways. The first is to reduce energy cost and the second is to reduce investment. Energy cost depends on the market costs of energy, which are currently rising, and on the efficient use of energy in the process. This is explained later. Investment expenses highly depend on the process operating pressure. The improvement of other parameters will not have a great effect since their consumption is relatively low (manpower, chemicals, membrane replacements, etc.). Good practice, namely good pretreatment, will save membrane replacements. Some items, such as membranes or high-pressure pumps, are restricted to the desalination industry and their cost may simply be lowered by market forces. Investment in sophisticated automation and control equipment can

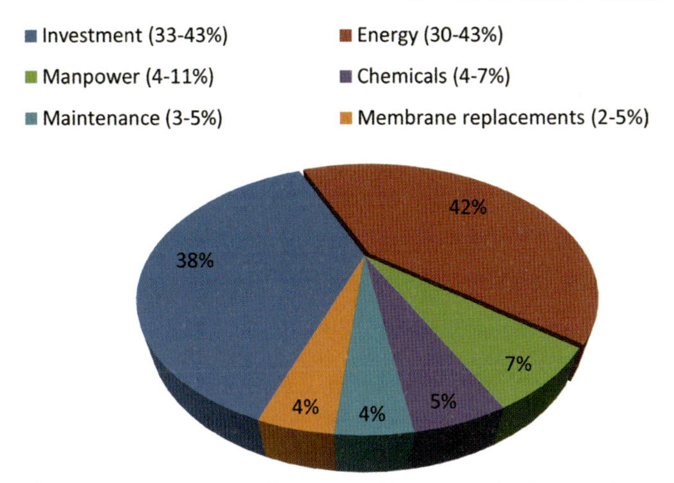

Figure 10.4 Cost estimate of a common RO seawater desalination plant.

reduce water costs by maintaining stable high throughputs and savings in labor costs. As can be seen in Figure 10.4, labor costs are no longer a significant cost item since modern desalination plants can operate largely unattended.

Wilf [19] presents the energy demand components in a two-pass RO desalination plant. Information about the Ashkelon plant costing may be found in Kronenberg [20] and Velter [21]. More information about RO costing history can be found in Glueckstern [22]. Better predictions for the future are problematic due to the current energy crisis.

Compliance with proper operational procedures and following a careful maintenance program can also reduce desalination costs by minimizing the replacement of damaged membranes, reducing the use of cleaning chemicals, and reducing the inventory of membranes and spare parts.

The design of a desalination plant is usually a site-specific task. Pretreatment is the most important local design. It is also envisaged that when operators are insufficiently trained, the design and investment will be based invariably on exaggerated safety factors. Well-trained and experienced operators can increase desalination plant production by identifying and debugging bottlenecks.

The relatively high cost of seawater desalination can be tolerated easily by the urban customer. In some cases, customers in large cities are paying up to three Euros per m^3 of treated water. The monthly cost of water in organized cities is usually lower in comparison to all other utilities. Different industries can usually handle the cost of desalinated water. Some industries need the high quality obtained by desalination; others may reuse and circulate the processed water. The problem is usually in agriculture – simple flood irrigation cannot afford desalination costs. The cost of water in greenhouses is only a small part of the total production cost, allowing for desalination costs. The need for water in agriculture and its cost may be tolerated by ensuring better use of the water. For example, drip irrigation directly to plant roots may save between 50–90% of the water currently used.

10.3.1 Quality of Desalinated Water

The quality of the water produced can be tailored to meet the needs of the consumer. Practically, it is possible to clean water at a low sodium chloride level. This may be mixed with brackish water or allowed to pass through a bed of $CaCO_3$ to dissolve this salt into the water. The quality of water produced also depends on the quality of brackish water available for mixing. The expected quality of desalinated seawater depends more and more on the permissible concentration level. Seawater contains approximately 5 ppm of boron. Due to the insufficient rejection of current membranes, the water product may contain over 1 ppm of boron. Boron is an important component, especially for plant growth. However, for many crops, too high a boron concentration is harmful and can cause a significant reduction in crop yield. Boron may be removed from water by ion exchange, together with secondary and higher RO stages, by increasing the pH of water on the feed side of the membrane and by using EDR. A combination of techniques is also possible [23, 24].

Current demand in Israel requires the production of water containing less than 0.4 ppm of boron in the Ashkelon plant and 0.3 ppm in the future Hadera plant (under design). The reason behind this is related to the recovery of wastewater following the treatment of desalinated water. Boron reaches the wastewater from different sources, which may damage crops irrigated with treated wastewater. In Ashkelon, for example, this demand required using up to four stages of RO membranes to remove the boron, resulting in a significant reduction in salts, to a level below 60 ppm TDS. Thermal processes may produce water containing between 5–50 ppm of TDS, similar in composition to feed seawater with thermal techniques. The boron problem does not exist in evaporation techniques.

The RO product of brackish water may contain between 200–500 ppm of TDS, which is basically NaCl, and a smaller portion of other salts. Some minor constituents, such as boric acid, hydrogen sulfide and CO_2, may also be present in the product depending on the composition of the feed water, but may be removed with adequate pre- or post-treatment. Feed water containing dissolved volatile organic compounds will generate water, unless special care is taken, that is slightly contaminated with the same components. This may be true for RO and evaporation techniques.

The water product is aggressive, tends to corrode iron pipes, and dissolves protective layers containing calcium and other salts on the inner sides of the mains. This may cause a phenomenon called 'red water,' the release of corrosion products by water dissolving the pipes' protective $CaCO_3$ layer. Therefore, water requires post-treatment that usually involves an increase in pH level, the addition of Ca (preferably to a level of about 100 ppm as $CaCO_3$), and alkalinity, namely HCO_3^- (also to a level of about 100 ppm as $CaCO_3$), according to local water regulations or WHO recommendations.

Desalinated water contains a low concentration of salts. Some salts are needed for maintaining a proper balance in bodily functions, so a complementary source of salts is needed for both human and animal diets. Certain agricultural crops may also suffer from a lack of minor constituents when irrigated with desalinated water; others may benefit from this. The addition of magnesium to the desalinated water produced is now being considered for human health and agricultural needs.

10.3.2 Environmental Aspects

Desalination processes may be characterized by their effluents emitted to the environment, the land and atmosphere nearby, and the sea. Desalination is highly dependent on energy, and generally uses fossil energy. All types of air pollution associated with energy production, namely, the emission of NO_x, SO_2, volatile compounds, particulates and CO_2, also exist through the use of electricity produced by conventional power stations or by a dedicated power station. Using gas turbines may increase efficiency and therefore reduce pollutants.

Effluents from desalination plants contain a relatively high concentration of salts and depend on water recovery from the feed brine. In the case of seawater desalination, rejected brine is almost twice as concentrated as the original seawater solution. The concentrate also contains chemicals used in the pretreatment of the feed water. The latter may contain low concentrations of antiscalants, surfactants and acid added to the feed water that reduces pH. Occasional washing solutions or rejected backwash slurries from feed water may be added to this. In small-scale operations, the problem is minor and no serious damage is caused to marine life. In large-scale water production, the problem is more serious. However, the dilution and spreading of effluents may solve this problem. Natural chemicals that do not harm the environment will probably replace part of the added chemicals in the future. Concentrates should be released to the sea a few meters above the sea floor from a few nozzles pointing upward, at an angle between 45–60 degrees. The volume of high concentration will be minimized, no damage will be caused to marine vegetation and small species, and fish will avoid this region.

The more serious problems involve concentrates that are produced in-land in the case of brackish-water desalination or wastewater recovery. The concentrate composition here differs from the seawater composition. In most cases, the solution contains more calcium and magnesium; sometimes other components are involved depending on the composition at the source. The problem is less severe when the solutions are purged into the open sea; in this case, care must be taken to prevent possible salt precipitation from supersaturated concentrates along the discharge pipeline [25]. If the sea is not easily accessible, the concentrate may increase groundwater salinity if allowed to penetrate into the ground. A possible solution to this problem involves zero discharge treatment, which is the evaporative separation between solids and water to enable solids to be stored properly in-land. This solution may be performed using solar evaporation ponds or by forced evaporation with available heat sources. The process is expensive, but the basis for comparison is the cost of transporting brine to the nearest possible authorized area, taking into account the influence of this treatment on product cost. Another approach is adding a crystallizer working at elevated pH to remove the supersaturation of the dissolved salts from the concentrate stream. In most cases, it is then possible to add a second RO stage to produce more water [26].

10.3.3 Energy Issues

Desalination, as a separation process, requires energy. The specific energy for reverse-osmosis desalination has decreased significantly over the past decade and is approaching the theoretical thermodynamic minimum. This was achieved through the development of large pumps having an efficiency as high as 92% equipped with modern, efficient turbines and other energy-recovery devices. The newer devices, known as 'turbochargers,' 'pressure exchangers' or 'work exchangers' (names adopted by different producers), represent efficient ways of recovering the energy content of the high-pressure concentrate. Turbines turn concentrate pressure into the velocity of jets that spin a wheel, which is used either to reduce the power consumption of the motor driving the pump or to boost the pressure of the feed to a second stage. Other methods of exchanging the pressure of the exiting concentrate involve simple devices transferring pressure to the seawater feed. Using the new techniques, a reduction in power consumption of the desalination systems was achieved. For example, processing 3.5% seawater at a recovery rate of 50% requires about 2.7 KWh/m^3 of the water produced by recovering the concentrate energy using turbines. Pressure exchangers can go even lower, to about 2.2 kWh/m^3 of water produced. Since more energy is consumed for feed and concentrate pumping, as well as for the pretreatment stages, overall energy needs are less than 3.7 kWh/m^3 for seawater production. Production from brackish water, wastewater or slightly polluted sources can go as low as 1.7 kWh/m^3 of water produced or less, depending on salts concentration and possible recovery.

The energy cost of an optimized desalination plant is approximately 30–40% of the total water cost. This cost may vary since energy costs may be replaced by equipment investment. For example, fossil energy may be replaced theoretically by different types of solar collectors. The problem is that since solar energy is available only 25% of the time, the investment fraction increases by a factor of four, before taking into account the equipment needed for electricity production. The water cost will be much higher compared to the use of a regular energy source.

The optimization is made during the design of the plant, yet the energy cost may vary significantly during the project's lifetime. For example, during the writing of this chapter, the cost of natural oil was increasing at a significant rate in comparison to its cost at the design stage of the Ashkelon plant. It is difficult to change the optimal design of a plant once it has been built. However, it is possible to minimize losses by designing for more flexible changes in terms of variable energy consumption and equipment costs.

A 100-million-m^3 RO-based seawater desalination plant requires an electrical energy supply of less than 50 MW. A dedicated power station can work at a much higher efficiency than a regular power station for this purpose since it is operated constantly without the known *sine* wave, representing day–night, summer–winter

Table 10.1 RO energy consumption in comparison to other alternatives.

Subject	Fuel			
	Natural gas	Gasoil	Heavy fuel	Coal
Caloric value kcal/kg fuel	9000	10 750	10 000	7700
Caloric Value kWh/kg fuel	10.5	12.5	11.6	9
Electricity production (45% eff.) kWh/kg fuel large power station	4.7	5.6	5.2	4
Electricity production (80% eff.) high-efficiency gas turbine kWh/kg fuel	8.4			
Capacity – seawater desal. (50% recovery) m^3/kg fuel	1.3	1.6	1.5	1.2
80% efficiency	2.4			
Fuel consumption/ton desalinated water kg fuel/m^3	0.7	0.6	0.7	0.9
80% efficiency	0.4			
How many km can I drive with 1 m^3 desalinated water fuel consumption?	2–7	2–6		
How many hours of a single room AC (2.5 kWh) can I operate?	1.4			

changes in consumption. Better efficiency is expected for gas turbines since the high temperature of the gases may also be used. Therefore, the real energy required is lower than for other common uses. Critics among environmentalists often express concern about the energy consumption for water desalination. Water is needed for many people on earth and for supplying the basic needs of the majority of these people. The introduction of desalination may leave less water for the environment. This takes priority over using energy for air-conditioning or running large, energy-consuming cars. A look at Table 10.1 will show the real energy consumption and a comparison to other energy usages. The table shows consumption in terms of different sources of energy, natural gas, gasoil, heavy fuel and coal. This is the best way of comparing energy demand when comparing electricity and fuel. A large-scale power station generates electricity from coal or heavy fuel at around 45% efficiency. If operated effectively, gas turbines may go as high as 80% efficiency. Fuel consumption, in terms of how many kWh is produced from 1 kg of fuel, is presented in the table. Next, follows the calculations of how many cubic meters (tons) of water can be produced from a single kg of fuel by using a large RO plant. From here we can see how much fuel is needed to make a ton of water, the cheapest product on earth. At this point, one can see that this amount of energy can drive us only 2–7 kilometers in our car (depending on its size), or determine how long we can operate our small room AC with the same amount of fuel. Looking at these numbers, the cost of energy for desalination in comparison with other energy usages is rather small.

Environmental concern about the CO_2 'greenhouse' effect associated with the use of hydrocarbon fuel has led to the goal of supplying desalination energy from renewable-energy sources. Renewable-energy sources may soon be compatible and economic for general electricity production. At this stage, they will also be suitable for desalination purposes.

No doubt, greater efforts should be devoted towards exploiting renewable-energy sources. However, the real test of any new energy source is its acceptance for electricity production or other common energy uses. Savings on CO_2 emissions must be made in terms of other energy forms and not regarding the very delicate issue of desalination for freshwater production. The use of nuclear energy, which is currently more expensive than fossil energy, is dangerous in areas where political instabilities prevail. It is also problematic where the technology is not accessible and it is necessary to rely on imported, trained, and sophisticated labor.

A possible method of efficient energy use in a sufficiently large desalination plant involves the design of a hybrid plant consisting of a membrane unit combined with a vapor-compression unit [27] using electrical energy and a multieffect evaporation plant using heat energy. Such an operation is common in the chemical industry. Energy costs could be minimized by coupling the desalination plant with a dedicated power plant generating electricity and waste heat at optimal economic conditions.

The advantage of the day–night, summer–winter electricity production cycle is that desalinated water is produced during the night, involving lower power consumption. The main disadvantage is that the desalination equipment is not used for a large percentage of the time. This is a mistake, since, as in any modern plant, production costs are greater if the equipment is not in full use. An efficient desalination plant should therefore be operated 24-hours-a-day, 365-days-a-year, with exceptions only for maintenance. During this time, a full supply of energy is required at the lowest cost.

10.4
Treatment of Sewage and Polluted Water

A large source of water for reuse may be obtained by reclaiming polluted water. Sources of polluted water emanate from domestic wastewater, industrial waste solutions, agricultural effluent as runoff water, recirculated greenhouse water, and fish pond waste. All these must be treated to a tolerable quality to prevent deterioration of the soil and aquifers, and pollution of lakes, rivers, and the sea. Above all, this is a source of usable water.

The current global trend in dealing with this important problem involves secondary biotreatment. This is followed by the increased use of membranes. MF and UF membranes are capable of almost completely removing suspended matter, including bacteria 6–9 orders of magnitude (6–9 logs of removal), waterborne protozoa, and reducing virus content by two to three orders of magnitude. Parameters affecting virus removal are associated with particulate agglomeration occurring next to the membrane, possible adsorption to the cake layer, virus association in groups, pore-size reduction by the cake, gel layer and fouling build-up on the membrane, and also pH effects. RO or NF membranes can be used later to remove salinity and some

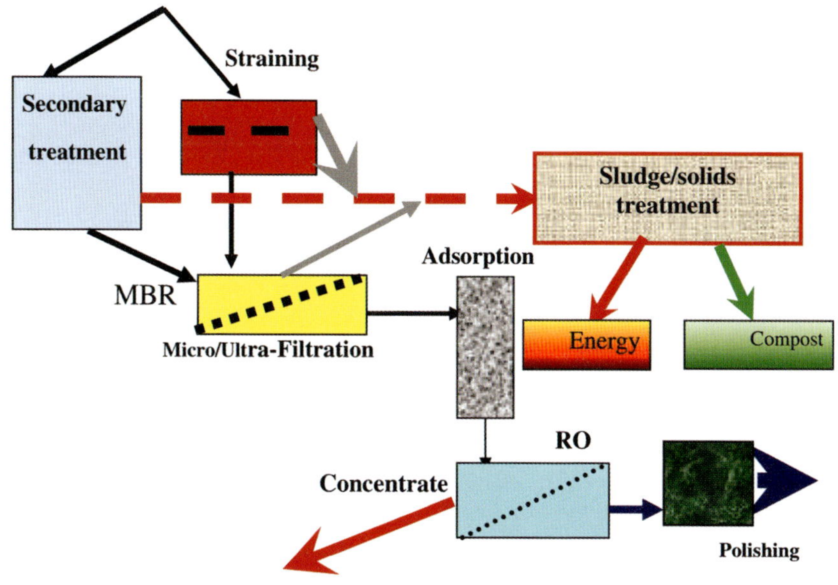

Figure 10.5 Schematic presentation of a wastewater treatment plant.

dissolved organic matter, while reducing TOC, COD, and BOD. Water can be disinfected by UV, ozone, chlorine, and chloramines.

Figure 10.5 illustrates the wastewater-treatment process. It is possible to use strained, presettled wastewater primary effluent or secondary effluent following biological treatment. Proper usage of the biotreatment removes the organic matter and almost all ammonia and phosphates from water solutions. Anaerobic operation reduces nitrates to elementary nitrogen. When membranes are submerged in a bioreactor, it is known as a membrane bioreactor (MBR) [11]. Effluent from the bioreactor may also be treated with external MF or UF membranes, and the concentrate is circulated back to the bioreactor. Water at this stage may be used mainly for irrigation also in some places but as drinking water.

The MBR-treated effluents can be fed directly into the RO/NF system. The final product, following RO/NF, surpasses current (and future) environmental requirements by far, as well as those for unrestricted use in agriculture, aquifer reinfiltration, and eventually evens all municipal uses.

Polishing with AC to remove trace organics may be used as means of reducing RO/NF membrane fouling and deterioration and final polishing following RO/NF product. Water exiting the last membranes requires the addition of calcium and possibly magnesium salts to be accepted by WHO recommendations.

Certain parameters affect membrane fouling: particle nature; particle size and size distribution; membrane type and structure; surface interactions; and the clogging mechanism. An important parameter is the method applied to the filtration technique, namely, crossflow or deadend filtration. The latter requires less pumping energy but tends to clog the membrane faster.

10.4.1
Membrane Bioreactors

Combining UF or MF membrane technologies with biological reactors for the treatment of wastewater in a one-stage process has led to the generation of the MBR concept in which MF or UF membranes have replaced the traditional sedimentation tank. An efficient clarification of the treated wastewater is achieved, membranes can reduce the disinfection practices such as chlorination, and a pathogen-free, tertiary quality effluent is thus obtained [11].

The main advantages of MBR over waste activated sludge processes are:

- Small footprint;
- Complete solid removal;
- Effluent disinfection;
- High loading rate capabilities;
- Low to zero sludge production;
- No bulking problems;
- High oxygen utilization rate.

Other parameters that require attention are:

- Operation with deadend or crossflow;
- Membrane configuration;
- Cleaning of fouled membranes;
- Washing;
- Frequency of backwashing;
- Use of air;
- Use of cleaning solutions.

Different sources of wastewater may contain different materials, so every type of wastewater must be tested in order to choose the right conditions. Harussi *et al.* [28] compared the alternative costs of feed-water desalination for cities to wastewater desalination. The first alternative may be cheaper but the second is necessary in order to be compatible with the environment.

10.4.2
Reclaimed Wastewater Product Quality

Product quality depends heavily on previous treatment stages. Organic matter and ammonia may exist in the product in the cases of insufficient removal. However, low ammonia content may react with chlorine to form chloramines, a long-term disinfector. Chlorination of organic matter may result in halogenated organic matter. Insufficient removal of phosphates in the biotreatment stage will result in a reduced recovery of the RO process if supersaturation is reached [29]. Insufficient de-nitrification, the removal of nitrates, will cause high nitrate concentration in the concentrate, which is a problem with concentrate removal. Boron content increases in the city due to many sources, especially from detergents and from industries using bleaching processes.

The concentration of urban wastewater contaminants is relatively low, so high-recovery membrane processes can generally be used to solve the problem relatively inexpensively. The case is completely different regarding industrial wastewater; each stream having different contaminants must be dealt with independently.

Many treatment facilities at different locations were installed to produce water from wastewater for different uses. In some cases, MF membranes are used directly on strained wastewater to remove suspended particles that are too large for the gap between two membranes [30]. Simple wastewater-treatment facilities in Europe exist along all large rivers. Secondary treated waters flowing into the rivers are again pumped at a distance of about 200 meters downstream, treated with active carbon and UF membranes, disinfected and then distributed to the system. This is wastewater treatment without an RO section due to the low salinity of the water. The process cannot handle dissolved medicines, hormones, drugs, and other contaminants that could be removed with RO or NF membranes. In some cases, NF membranes are used for better treatment of the water. Information on wastewater costing may be found in Adham *et al.* [31].

10.5 Fouling and Prevention

The main problem in membrane usage for water purification is the fouling layer that adheres to the membrane. The source of the fouling layer is the different species existing in the feed water and their increased concentration next to the membrane wall. When water permeates through the membrane, all rejected species accumulate next to the membrane wall, their concentration increases in comparison to the bulk concentration, and the motion away from the membrane is controlled by diffusion to the bulk of flow against the flux of the water flowing to the membrane.

The main types of fouling are:

1. Suspended particles.
2. Salt precipitation due to supersaturation.
3. Dissolved organic matter.
4. Biofouling.

The denser RO and NF membranes may face all of the above problems. UF and MF are open membranes and hence encounter problems associated with salt precipitation. Other important parameters are listed below:

- Clogging by suspended solids, precipitating salts, bacteria cultures, and so on.
- Difficulties in membranc surface cleaning.
- Sensitivity to degradation by organic chemicals.
- Sensitivity to different types of bacteria.
- Deterioration due to free chlorine.
- Permeability of nonpolar substances, including low molecular weight organic substances.

The critical issue for a successful RO plant is pretreatment. Long-term operating experience proves the viability of continuous MF/UF pretreatment of RO for the desalination of a wide variety of water sources. MF/UF has proven to simplify and reduce the costs of traditional pretreatment, comprised of deep-bed media filters combined with chemical treatment. MF/UF produces filtrate of a consistent quality almost irrespective of fluctuations in feed-water quality. In the last five years, RO-membrane improvements, combined with the use of membrane filtration for pretreatment, have halved the cost of advanced treatment and are now more widely used for the reuse of municipal wastewater.

Suspended-particle precipitation is caused by attraction forces with the membrane wall. They may be removed effectively by good pretreatment, either by sand filtration or by using UF/NF membranes. Salt precipitation is caused by the increased concentration of low solubility salts such as calcium salts – carbonate, fluoride, and sulfate (mainly from brackish water), phosphate (domestic wastewater), silica, iron oxides, and other metal salts originating from water feed. Many techniques are available to prevent their precipitation, such as acidulation to remove carbonates or increase the solubility of salts, the removal of special ions by ion exchange, hardness removal prior to feeding the membranes, and so on. This is not a major problem in seawater desalination, however, it is very important in brackish water and recovered wastewater desalination. The recovery level of water produced from a certain source is controlled by the ability to maintain high supersaturation of the salts before precipitation. This is done by adding 'antiscalants' or crystallization inhibitors [32]. The antiscalants are usually medium-length polymers, such as polyphosphates, polyphosphonates, and polyacrylates. Their mechanism is not completely understood, yet they interfere with crystal growth through adsorption on the active sites or by applying electrical charges that prevent the crystals from growing. Techniques were developed to estimate the recovery level while comparing different antiscalants at different concentrations [33–39].

Organic fouling is the basis of biofouling, which is one of the most severe problems seeking a solution [40–43]. Currently, only good pretreatment may limit biofouling.

10.5.1
How to Prevent

Like in medicine, prevention is the key in most cases for healthy membrane life. A good pretreatment allows much longer membrane life at lower cleaning costs. Pretreatment is used mainly for the removal of suspended solids, bacteria, and large organic molecules. This is done mostly by sand filtration, but UF membranes can also be found for this purpose in modern designs. Dosing the water with coagulants is used to agglomerate the small particles for easy separation. dissolved air flotation (DAF) was also introduced as a means for removing both suspended and organic matter [44, 45].

Disinfection of water is used to destroy bacteria. However, RO membranes are sensitive to oxidizers so they must be removed before entering the membranes. Active-carbon beds are sometimes used to remove traces of organic matter together with excess chlorine. It was found in some cases that improper usage brings the bacteria level, after the active carbon, back to the original count before treatment.

Other pretreatment steps involved in the removal of special contaminants include acidulation, ion exchange of some important contaminants, and the removal of H_2S, if it exists.

10.5.2 Membrane Cleaning

Membrane cleaning is an important stage that should be minimized, if possible. Cleaning fouled membranes depends on the type of fouling. A simple wash along the membrane or opposite the flow direction is the easiest type. Low pH is used when salts are precipitated. High pH is effective when silica is precipitated, and includes organic matter or even bacteria. Backwash is the usual treatment of UF/NF membranes, usually involving the dosing of disinfectant. It is impossible, however, to apply high pressure to the permeate side in RO/NF operations since this may damage the membranes. Recent developments have shown that backwash is possible in these membranes based on direct osmosis, the difference between osmotic pressure of the feed side and the permeate side. Short, frequent osmotic backwash may maintain clean membranes and increase their life. Instructions for cleaning methods are given by membrane manufacturers.

10.6 R&D Directions

10.6.1 Impending Water Scarcity

The need for more, better-quality water is increasing all the time. Global warming, either naturally or as a result of excessive use of fossil energy, is causing glaciers to melt and consequently ocean water levels to rise. This alone will cause significant problems associated with current water systems. River levels will rise, flooding their immediate surroundings. An increase in seawater level will change the current balance between seas and shore aquifers, causing a major penetration of seawater into the aquifers. There is no way to accommodate these problems other than by relying on desalinated water and wastewater recovery. Additional R&D work is therefore required in order to continue to reduce the cost of water by implementing improved recovery techniques and ensuring better usage in agriculture.

10.6.2 Better Membranes

The main components of RO desalination costs are energy cost and equipment investment. Itemizing the equipment in use, membranes, pressure vessels, pumps, tubing, and flow devices and energy-recovery units illustrate that there is no special item that is significantly more expensive than the others. However, membranes play the most important role in possible cost reduction. The cost of RO membranes represents about 8% of the overall investment. Membranes may be improved significantly. Permeation may be increased, maintaining similar rejection properties. Increased flux through the membranes will enable a pressure reduction and hence less energy at the same recovery ratio. The pressure reduction may also reduce the costs of the expensive metals used at high pressure in a highly corrosive environment. This will reduce the cost of pumps and flow devices. Larger membrane modules will reduce the plant's footprint.

Other important future membranes properties include improved resistivity to extreme pH enabling better cleaning performance and resistance to oxidizers, organic solvents and particulate fouling. More important properties exist, yet the most important may be the resistance to fouling of the different sources.

Similar properties are needed for other types of membranes in use such as MF, UF, and NF membranes. Improved membranes may be used in other separation processes, not necessarily related to water [46–48].

10.6.3 New Membranes-Based Desalination Processes

Recent developments have brought significant attention to other types of membrane processes reported mainly in research papers: forward osmosis [49, 50], and membrane distillation [51, 52]. Forward osmosis, or direct osmosis, is defined by water passage from a salt solution or a polluted solution through a membrane to a solution containing dissolved matter of higher osmotic pressure. A possible advantage of such a process is that the separation of the water from the higher osmotic pressure solution is easier than the separation through RO. These separations were proposed by using magnetic nanoparticles covered with organic matter, separated by magnetic field, distillation of the dissolved material like in the case of ammonium carbonates, or a possible simpler separation, such as crystallization. It is also important to minimize contaminant traces of the high osmotic-pressure material in product water according to drinking-water regulations. Thermodynamics, however, teaches us that minimum separation energy is dependent on concentration. RO separation is very close to minimal thermodynamic separation energy. So, the objective here is to find a process that consumes less energy than the RO process.

Another new trend is called membrane distillation. This is based on open hydrophobic membranes that enable the passage of water vapor only. The product quality is expected to be better than RO since only water vapor may pass through the membrane. Vapor condensation is allowed on colder surfaces adjacent to the membranes or outside the membrane module, where vapors are pumped out. Another way is to condense the vapor in direct contact with a cold-water stream. The main problem using this technique is the need to evaporate the water. The energy demand for this is around 650 kWh/m^3. This enormous amount of energy may be reduced when energy reuse is possible, in a similar way to the multieffect distillation desalination process. This may reduce the energy demand down to about 60 kWh/m^3 if energy is reused more than 10 times within the desalination plant. More energy is needed for pumping the water and for cooling. Heat-transfer flux through the membrane is low, so a large transfer area is needed. Developers of this technique claim that the high energy demand may be supplied by low-grade, cheap energy yet this claim is also true for the multieffect distillation process. The only possible advantage of the multieffect distillation technique is the possibility of lower volume and low footprints of the plant. However, the design is complicated, as is shown in Figure 10.6, representing the preliminary design of membrane distillation with heat recovery, similar but not identical to the multieffect distillation process. In this design, entering seawater exchanges heat, leaving hot streams of product and concentrate. External heat must be added to the feed water. The water is then allowed to flow in parallel to the membranes where evaporation takes place. Vapor leaving the membrane condenses on the heat-transfer wall and transfers the heat of condensation to the water in the next stage. In this way, the evaporation energy is reused to heat the concentrating solution. Other designs are possible, yet in terms of energy demands, the technique cannot compete with RO.

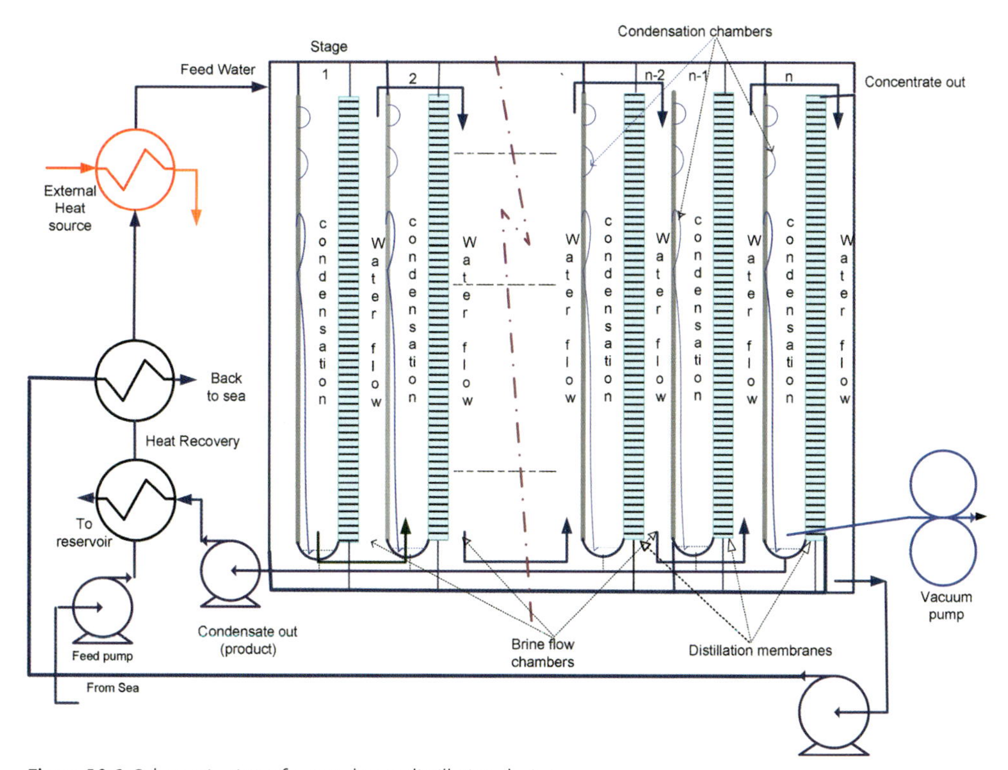

Figure 10.6 Schematic view of a membrane distillation design based on multieffect distillation technology.

10.7 Summary

The use of different types of membranes for water production and purification is presented here. Solutions are available for good-quality water production at affordable costs. Important related aspects such as environmental and energy issues are presented. Future directions are reviewed. Additional research is required in order to improve the processes and reduce the cost of water produced. The subject of water is one of the most important subjects that humankind must solve, together with renewable energy and environmental problems.

References

1 Semiat, R. (2000) Desalination – Present and future, Invited Article for IWRA'21. *Water International*, **25** (1), 54–65.

2 Faller, K.A., Murray, P. and Livingston, A. (1998) Reverse osmosis and nanofiltration manual of water supply practices, *AWWA Manual of Water Supply Practice*, **M46** Amer Water Works Assn., Elbert, Colorado, USA.

3 Wilbert, M.C., Leitz, F., Abart, E., Boegli, B. and Linton, K. (1998) *The Desalting and Water Treatment Membrane Manual: A Guide to Membranes for Municipal Water Treatment*, 2nd edn, Water Treatment Technology Program Report No. 29, Bureau of Reclamation, United States Department of Interior.

4 Bergman, R.A.(Nov. 1995) Florida – A Cost Comparison Update, Membrane Softening vs. Lime Softening. International Desalination and Water Reuse.

5 Hassan, A.M., Al-Sofi, M.A.K., Al-Amoudi, A., Jamaluddin, T.M., Dalvi, A.G.I., Kitner, N.M., Mustafa, G.M. and Al-Tisan, I.A. (1998) A new approach to membranes and thermal seawater desalination processes using nanofiltration membranes. International Desalination and Water Reuse Part 1: May 1998, Part 2: Aug.

6 Sagiv, A. and Semiat, R. (2005) Backwash of RO spiral wound membranes. *Desalination*, **179**, 1–9.

7 Liberman, B. (2004) Methods of direct osmosis membrane cleaning online for high SDI feed after pre-treatment. IDA Workshop, Tempa-San Diego.

8 Altman, M., Hasson, D. and Semiat, R. (1999) Dynamic membranes. *Reviews in Chemical Engineering*, **15**, 1–40.

9 Rumyantsev, M., Shauly, A., Yiantsios, S.G., Hasson, D., Karabelas, A.J. and Semiat, R. (2000) Parameters affecting the properties of dynamic membranes formed by Zr hydroxide colloids. *Desalination*, **131**, 189–200.

10 Baudin, I., Chevalier, M.R., Anselme, C., Cornu, S. and Laine, J.M. (1997) L'Apie and Vigneux Case studies: first months of operation. *Desalination*, **113**, 273–275.

11 Yang, W.B., Cicek, N. and Ilg, J. (2006) State-of-the-art of membrane bioreactors: Worldwide research and commercial applications in North America. *Journal of Membrane Science*, **270**, 201–211.

12 Hai, O.H. and de Ryck, L.(Sept. 2005) Water reuse and Ulu Pandan NEWater project. IDA World Congress on Desalination and Water Reuse, Singapore.

13 Thampy, S.K., Rangarajan, R. and Indusekhar, V.K.(August 1999) 25 years of electrodialysis experience at Central Salt and Marine Chemicals Research Institute Bhavnagar, India. International Desalination and Water Reuse, Vol. 9/2.

14 Strathmann, H. (2004) Assessment of electrodialysis water desalination process costs. Proceedings of the International

Conference on Desalination Costing, Limassol, Cyprus.

15 Voutchkov, N. and Semiat, R. (2008) in *Advanced Membrane Technology and Applications* (eds N.N. Li, W.S. Winston Ho and A.G. Fane), John Wiley & Sons, Inc., New York.

16 Glueckstern, P. and Priel, M. (1999) Experience, capability and plans for erecting large seawater desalination plants. Proceedings of the 2nd Annual IDS Conference, Haifa.

17 Glueckstern, P. and Priel, M. (2000) Desalination of brackish and marginal water sources in Israel: Past present and future. Proceedings of the 3rd Annual IDS Conference, Tel-Aviv.

18 Furukawa, D.H.(Sept. 1997) A review of seawater reverse osmosis. IDA Desalination Seminar, Cairo, Egypt.

19 Wilf, M. (2004) Fundamentals of RO-NF Technology. Proceedings of the International Conference on Desalination Costing, Limassol, Cyprus.

20 Kronenberg, G. (2004) Ashkelon 100 MC/Year BOT Project. Proceedings of the International Conference on Desalination Costing, Limassol, Cyprus.

21 Velter, G. (2004) Ashkelon desalination project operations. Proceedings of the International Conference on Desalination Costing, Limassol, Cyprus.

22 Glueckstern, P. (2004) History of desalination cost estimations. Proceedings of the International Conference on Desalination Costing, Limassol, Cyprus.

23 Glueckstern, P. and Priel, M. (2003) Optimization of boron removal in old and new SWRO systems. *Desalination*, **156**, 219–228.

24 Sagiv, A. and Semiat, R. (2004) Analysis of parameters affecting boron permeation through reverse osmosis membranes. *Journal of Membrane Science*, **243** (1–2), 79–87.

25 Semiat, R., Hasson, D., Zelmanov, G. and Hemo, I. (2004) Threshold scaling limits of RO concentrates flowing in a long waste disposal pipeline. *Water Science and Technology*, **49** (2), 211–219.

26 Lisitzin, D., Hasson, D. and Semiat, R.(June 2006) Membrane crystallizer for increased desalination recovery, ECI – Advanced Membranes Technology III. Membrane Engineering for Process Intensification, Cetraro, Calabria, Italy.

27 Awerbuch, L. (1997) Dual purpose power desalination/hybrid systems/energy and economics. IDA Desalination Seminar, Cairo, Egypt.

28 Harussi, Y., Rom, D., Galil, N. and Semiat, R. (2001) Evaluation of membrane processes to reduce the salinity of reclaimed wastewater. *Desalination*, **137**, 71–89.

29 Greenberg, G., Hasson, D. and Semiat, R. (2005) Limits of RO recovery imposed by calcium phosphate precipitation. *Desalination*, **183**, 273–288.

30 Johnson, W.T., Phelps, R.W. and Beatson, P.J. (Oct. 1997) Wastewater reuse using membranes. IDA World Congress on Desalination and Water Science, Madrid.

31 Adham, S., Kumar, M. and Pearce, W.H. (2004) Development of a model for brackish and reclaimed water membrane desalination costs. Proceedings of the International Conference on Desalination Costing, Limassol, Cyprus.

32 Hasson, D. and Semiat, R. (2006) Scale control in saline and wastewater desalination. *Israel Journal of Chemistry*, **46** (1), 97–104.

33 Hasson, D., Semiat, R., Bramson, D., Busch, M. and Limoni-Relis, B. (1998) Suppression of $CaCO_3$ scale deposition by anti-scalants. *Desalination*, **118**, 285–296.

34 Drak, A., Glucina, K., Busch, M., Hasson, D., Laine, J.M. and Semiat, R. (2000) Laboratory technique for predicting the scaling propensity of RO feed waters. *Desalination*, **132**, 233–242.

35 Hasson, D., Drak, A. and Semiat, R. (2001) Inception of $CaSO_4$ scaling on RO membranes at various water recovery levels. *Desalination*, **139**, 73–81.

36 Hasson, D., Drak, A. and Semiat, R. (2003) Induction times induced in an RO system by antiscalants delaying $CaSO_4$ precipitation. *Desalination*, **157**, 193–207.

37 Semiat, R., Sutzcover, I. and Hasson, D. (2001) Technique for evaluating silica scaling and its inhibition in RO desalting. *Desalination*, **140**, 181–193.

38 Semiat, R., Sutzkover, I. and Hasson, H. (2003) Characterization of the effectiveness of silica anti-scalants. *Desalination*, **159**, 11–19.

39 Semiat, R., Sutzkover, I. and Hasson, D. (2003) Scaling of RO membranes from silica supersaturated solutions. *Desalination*, **157**, 169–191.

40 Ivnitsky, H., Katz, I., Minz, D., Shimoni, E., Chen, Y., Tarchitzky, J., Semiat, R. and Dosoretz, C.G. (2005) Characterization of membrane biofouling in nanofiltration processes of wastewater treatment. *Desalination*, **185**, 255–268.

41 Kang, S., Asatekin, A., Mayes, A.M. and Elimelech, M. (2007) Protein antifouling mechanisms of PAN UF membranes incorporating PAN-g-PEO additive. *Journal of Membrane Science*, **298**, 42–50.

42 Asatekin, A. *et al.* (2006) Antifouling nanofiltration membranes for membrane bioreactors from self-assembling graft copolymers. *Journal of Membrane Science*, **285**, 81–89.

43 Kimura, K., Yamato, N., Yamamura, H. and Watanabe, Y. (2005) Membrane fouling in pilot-scale membrane bioreactors (MBRs) treating municipal wastewater. *Environmental Science & Technology*, **39**, 6293–6299.

44 Huehmer, R. and Henthorne, L. (2006) Advances in RO pretreatment techniques innovations and applications of sea-water and marginal water desalination. IDS 8th Annual Conference Technion, Haifa.

45 Bonnélye, V., Sanz, M.A., Mazounie, P., Vion, P., Del Castillo, J. and Rovel, J.M. (2006) Advances in DAF high rate flotation, innovations and applications of sea-water and marginal water desalination. IDS 8th Annual Conference Technion, Haifa.

46 Akthakul, A., Salinaro, R.F. and Mayes, A.M. (2004) Antifouling polymer membranes with sub-nanometer size selectivity. *Macromolecules*, **37**, 7663–7668.

47 Ulbricht, M. (2006) Advanced functional polymer membranes. *Polymer*, **47**, 2217–2262.

48 Eisen, M. and Semiat, R. (2008) Membranes in Desalination and Water Treatment. To appear in MRS publication, special issue on materials for water.

49 McCutcheon, J.R., McGinnis, R.L. and Elimelech, M. (2005) A novel ammonia-carbon dioxide forward (direct) osmosis desalination process. *Desalination*, **174**, 1–11.

50 Cath, T.Y., Adams, D. and Childress, A.E. (2005) Membrane contactor processes for wastewater reclamation in space II. Combined direct osmosis, osmotic distillation, and membrane distillation for treatment of metabolic wastewater. *Journal of Membrane Science*, **257** (1–2), 111–119.

51 Alklaibi, A.M. and Lior, N. (2005) Membrane-distillation desalination: Status and potential. *Desalination*, **171** (2), 111–131.

52 Gilron, J., Song, L. and Sirkar, K.K. (2007) Design for cascade of crossflow direct contact membrane distillation. *Industrial & Engineering Chemistry Research*, **46**, 2324–2334.

53 Blanco, J.F., Sublet, J., Nguyen, Q.T. and Schaetzel, P. (2006) Formation and morphology studies of different polysulfones-based membranes made by wet phase inversion process. *Journal of Membrane Science*, **283**, 27–37.

54 Asatekin, A., Kang, S., Elimelech, M. and Mayes, A.M. (2007) Anti-fouling ultrafiltration membranes containing polyacrylonitrile-graft-poly(ethylene oxide) comb copolymer additives. *Journal of Membrane Science*, **298**, 136–146.

55 Hasson, D., Drak, A., Komlos, C., Yang, Q. and Semiat, R. (2007) Detection of fouling on RO modules by residence time distribution analyses. *Desalination*, **204** (1–3), 132–144.

56 Hester, J.F. and Mayes, A.M. (2002) Design and performance of foul-resistant poly(vinylidene fluoride) membranes prepared in a single step by surface segregation. *Journal of Membrane Science*, **202**, 119–135.

57 Wang, Y.Q. *et al.* (2005) Remarkable reduction of irreversible fouling and improvement of the permeation properties of poly(ether sulfone) ultrafiltration membranes by blending with pluronic F127. *Langmuir*, **21**, 11856–11862.

11
Developments in Membrane Science for Downstream Processing

João G. Crespo

11.1
Introduction

This chapter discusses the use of membrane processes for recovery, concentration, and purification of biologically active compounds from complex media. This chapter is not organized and written as a review paper aiming at referring all major developments in the use of membranes for *downstream processing* but, rather, it presents the author's perspective about this field, its main constraints and challenges.

Most scientific reviews in this field [1–3] are focused on the recovery and purification of large molecules with biological activity – protein harvesting and protein purification (including purification of monoclonal antibodies), purification of DNA and RNA – but the recovery of small molecules [4] with biological activity ($M < 500$ Da) usually attracts little attention. This chapter will pay attention to both small and large biomolecules and will discuss the critical issues related with their recovery and purification. Membrane bioreactors are excluded from this discussion, although some of the most interesting processes under development involve the integrated concept of bioconversion and product recovery [4–6].

11.1.1
Why Membranes for Downstream Processing?

Membrane separations are regarded as particularly suitable for biotech applications because (1) in general, they can be operated under mild conditions of temperature, pressure, and shear stress, therefore preserving the biological activity of the compounds to be recovered and the properties of the original media/matrices; (2) they do not require any extraction mass agents such as solvents, avoiding product contamination and the need for subsequent purification and (3) a large variety of membrane materials are available: they can be polymers, inorganic matrices or composites. The wide range of possible materials underlines one of the strengths of membrane separations: the possibility of designing and fine tuning the membrane for a specific

Membrane Operations. Innovative Separations and Transformations. Edited by Enrico Drioli and Lidietta Giorno

ISBN: 978-3-527-32038-7

need, through development of materials with adequate structural properties or by suitable modifications of their surface chemistry and topography [7–9].

Additionally, membranes have the unique advantage of allowing the simultaneous contact with two different media, at each membrane side, creating 'compartments' with different properties. Therefore, membranes offer the potential to promote the spatial organization of catalytic compartments and selective barriers. This feature is used with advantage in new concepts of membrane multiphasic (bio)reactors and membrane contactors.

When using porous membranes for filtration processes, they act primarily on the basis of size exclusion, leading to permeate fluxes that are high when compared with other competing processes, due to convective transport through the porous structure of the membrane. Therefore, high throughputs are usually referred to as one advantage of membrane filtration processes, such as microfiltration and ultrafiltration, when compared with other separation processes involving porous media, as happens with chromatographic systems.

11.2 Constraints and Challenges in Downstream Processing

The main constraints and problems associated with the use of membrane processes for downstream processing have been extensively discussed in the literature and the understanding of their nature and mechanisms has driven research towards the development of new solutions.

11.2.1 External Mass-Transport Limitations

Mass-transport limitations are common to all processes involving mass transfer at interfaces, and membranes are not an exception. This problem can be extremely important both for situations where the transport of solvent through the membrane is faster and preferential when compared with the transport of solute(s) – which happens with membrane filtration processes such as microfiltration and ultrafiltration – as well as with processes where the flux of solute(s) is preferential, as happens in organophilic pervaporation. In the first case, the concentration of solute builds up near the membrane interface, while in the second case a depletion of solute occurs. In both situations the performance of the system is affected negatively: (1) solute accumulation leads, ultimately, to a loss of selectivity for solute rejection, promotes conditions for membrane fouling and local increase of osmotic pressure difference, which impacts on solvent flux; (2) solute depletion at the membrane surface diminishes the driving force for solute transport, which impacts on solute flux and, ultimately, on the overall process selectivity towards the transport of that specific solute.

A large number of methods for improving external mass transport in membrane systems have been proposed and evaluated. Several of them may lead to a significant process improvement under defined conditions. Still, these methods – use of corrugated membranes [10], module development using Taylor vortices [11] and Dean vortices [12], vibrating modules [13], use of optimized spacers and static promoters [14] – refer to situations where the energy input is not totally used at the correct location, the membrane surface locals where solute buildup or depletion is occurring, being partially lost within the bulk liquid. New strategies, involving the fabrication of membranes with a specifically designed surface 3D topography have been proposed and developed [15]. Besides their use as nanostructured surfaces for cell differentiation these membranes, fabricated with different techniques such as nanoimprint lithography, open new opportunities for improved mass transport at the membrane scale.

11.2.2 Membrane Fouling

Fouling is the most used word in the membrane literature. Considering the character of most biological media, where proteins and polysaccharides are usually present beside salts and other compounds, fouling is inevitable. Still, the extent of fouling and its more or less reversible character can be minimized and controlled by using an adequate combination of operating and environmental conditions, and a judicious selection of the membrane to be used. The membrane fouling literature is extensive but a few papers are fundamental for the comprehension of this phenomenon [16–18]. Mitigation of membrane fouling has been addressed through different approaches: (1) optimization of the operating and fluid dynamics conditions; (2) tuning of the environmental conditions of the fluid phase; (3) use of membranes with improved properties.

Traditional operation of membrane filtration systems uses pressure differences (typically from 0.5 to 2 bar in ultrafiltration processes) which promote a significant convective transport of media components towards the membrane surface. These transport conditions lead to an increase of solute concentration at the membrane surface and promote solute–solute and solute–membrane interactions that may lead to severe fouling. The understanding of these phenomena supported the concept of critical permeate flux, above which fouling occurs. The critical-flux concept [19–21] explains the success of submerged membrane bioreactors, applied by Kubota and Zenon in wastewater treatment, where the permeate flux is imposed by suction at subcritical or near to critical conditions, assuring low fouling and a long-term operation without the need for membrane cleaning. This improvement is achieved at the expenses of large membrane areas, which became affordable even for wastewater treatment due to the significant decrease of membrane cost. Operation under controlled permeate flux, below or near to critical flux conditions, is a strategy that is also used for downstream processing, namely for recovery of proteins from biological media, as will be discussed later in this chapter.

It is interesting to note that the problem of operation with large pressure differences between the feed/retentate side and the permeate compartment of membrane filtration modules was identified long ago. The concept of operation under low uniform transmembrane pressure (UTMP) was pioneered and first

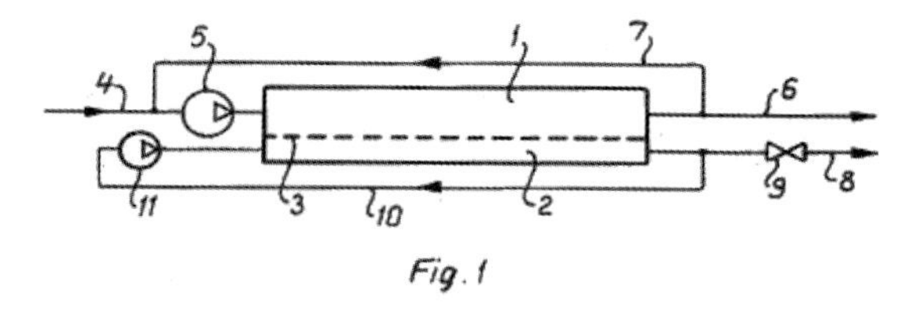

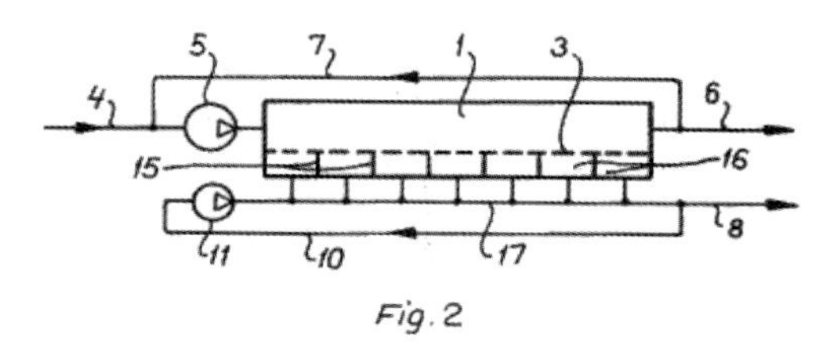

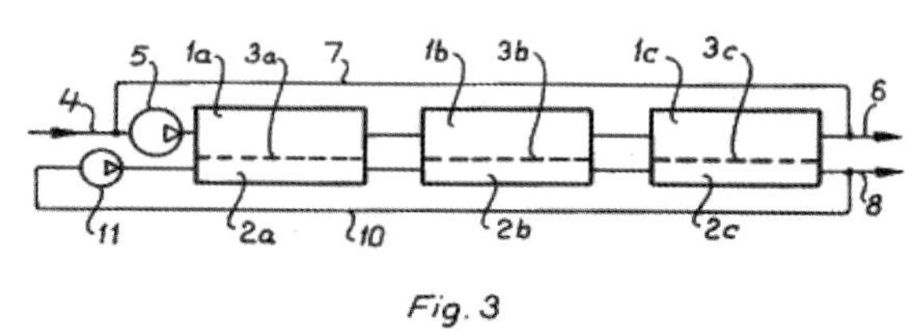

Figure 11.1 Different schemes proposed by Sandblom [22] for permeate recirculation in the shell side of the membrane module in order to assure an approximately constant transmembrane pressure.

patented by Sandblom [22]. UTMP operation makes it possible to benefit from an efficient particle backtransport from the membrane wall at high wall shear rates, while maintaining low TMP in the pressure-dependent regime [23]. Sandblom [22] suggested the use of permeate recirculation in the shell side of the membrane module so that the pressure gradient on the feed/retentate side can be kept approximately constant (Figure 11.1). The TMP can be maintained at a low uniform value throughout the length of the module, independently of the axial velocity in the feed/retentate side. This concept has been revisited recently and considered for protein recovery by membrane processing [23].

Fouling problems have also been tackled through the development of surface-modified membranes, namely by covalent attachment of either quaternary amine or sulfonic groups, in order to provide a desired positive or negative charge. These membranes exhibit lower fouling tendency, depending from the environmental conditions tuned for the media to be processed, and may provide high selectivity by exploring both size and charge effects. A recent, and very exciting area of research, deals with the development of membranes that may reverse the hydrophobic/hydrophilic character of their surface when exposed to an external stimulus. These stimulus-responsive membranes may change their character reversibly when exposed to alternate UV/visible light radiation conditions. This approach opens new perspectives for reversing fouling without the use of any external mass agent [24, 25].

11.2.3 Membrane Selectivity

Selectivity is one of the main issues when assessing the potential use of membranes for downstream processing. Membranes are often regarded as limited in selectivity, when the solutes to be fractionated exhibit close molecular weights and molecular properties. Tuning of the main environmental conditions that characterize the media to be processed may, however, lead to significant improvements in selectivity and overall performance. By controlling the pH and ionic strength of the media to be processed it is possible to maximize the differences between the effective volume of the product to be retained and impurities or other products we aim to permeate [26, 27]. As an example, the effective volume of a charged protein accounts for the presence of a diffuse electrical double layer surrounding the protein. Increasing the protein charge, or reducing the solution ionic strength, increases the effective volume thus reducing protein transmission through the membrane. Optimal performance is typically attained by operating close to the isoelectric point of the lower molecular weight protein, to minimize its rejection, and relatively low salt concentrations (10 mM or lower ionic strength) to maximize electrostatic exclusion of the more retained species (e.g., larger protein).

Electrical-charge effects can be further exploited by using charged membranes (as referred to above) to increase retention of all species with like polarity. It is important to mention that it may be possible to exploit electrostatic interactions even for solutes with similar isoelectrical points, due to different charge–pH profiles for the different species present. The membrane pore-size distribution also affects selectivity by altering the solute sieving coefficients locally. Narrow pore-size distributions, especially for electrically charged membranes, will impact very positively on membrane selectivity and overall performance.

In membrane chromatography (see additional discussion below) microfiltration membranes are modified by attachment of functional ligands, from the inner pore surface throughout the membrane, in order to provide conditions for highly selective binding interactions with target solutes. In many situations, the reason why membrane chromatography fails to reach commercial application is not related with the intrinsic selectivity of the adsorptive membranes but, rather, due to their dynamic binding capacity that hardly competes with bead-based processes for bind/elute applications [1].

11.3 Concentration and Purification of *Small* Bioactive Molecules

Recently, the recovery and purification of small bioactive molecules from complex media has gained a new interest. Small bioactive molecules comprise a large variety

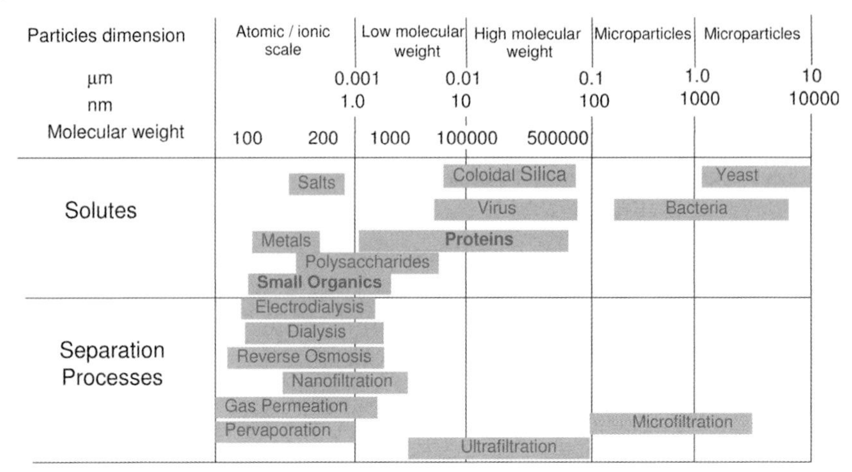

Figure 11.2 Diagram relating the size of different particles/solutes and the corresponding membrane processes.

of compounds with a molecular weight typically below 1 kDa, which includes compounds valuable due to their use as flavors and fragrances, as building blocks or precursors in the fine-chemistry industry, and compounds with antioxidant or anticarcinogenic activity (among other types of desirable biological activity).

These compounds are commonly present in complex fermentation media, or even in natural raw materials and subsidiary streams resulting from the processing of these materials. Their recovery is usually difficult due to their low concentration, often vestigiary, and the complexity of the original matrix where they have to be recovered from. This chapter discusses, and illustrates with recent applications, the use of different membrane processes able to deal with the recovery of small biologically active molecules (see Figure 11.2): electrodialysis, pervaporation, and nanofiltration.

11.3.1 Electrodialysis

The most interesting examples of the use of electrodialysis for recovery of target small molecules are related with the *in-situ* integration of this technique during biotransformations. It is often observed that the final product of a biotransformation process causes inhibition of the biocatalyst involved, even at relatively low concentrations ($<500\,\mathrm{mmol\,L^{-1}}$). To overcome this problem it is necessary to recover the product formed in an integrated process, in order to keep its concentration within a desirable concentration level. Jonsson and coworkers developed and patented a process [28] which allows for the recovery of lactic acid from an active fermentation by integrating *in situ* an electroenhanced dialysis step followed by bipolar electrodialysis where the ionic species are concentrated. This procedure allows for a continuous production and recovery of lactate, while simultaneously controlling the fermentation pH at its optimal level. Another interesting integrated approach has been described by Zelic *et al.* [29], where an electrodialysis step is used for the continuous recovery of pyruvate from an active fermentation process.

The simultaneous separation and recovery of acidic and basic bioactive peptides by employing electrodialysis with ultrafiltration membranes has also been investigated recently [30]. This work aims at demonstrate the feasibility of separating peptides from a beta-lactoglobulin hydrolysate, using an ultrafiltration membrane stacked in an electrodialysis cell, and a study of the effect of pH on the migration of basic/cationic and acid/anionic peptides in the electrodialysis configuration.

11.3.2
Pervaporation

The recovery of flavors and fragrances from diluted aqueous streams may be of industrial interest under different circumstances: (1) recovery of complex aroma profiles and/or target aroma compounds from active biocatalytic processes; (2) recovery of complex aroma profiles and/or target aroma compounds from natural extracts and industrial processes aqueous streams. Pervaporation offers a unique solution for the recovery of complex aroma profiles. An example for the recovery of complex aroma profiles, faithful to their origin, is the recovery of a muscatel aroma from an ongoing wine-must fermentation [31, 32].

Coupling pervaporation to active bioconversion processes is extremely interesting because it allows for continuous removal of target compounds that, otherwise, may simultaneously exert an inhibitory effect over the biocatalyst. Several examples have been discussed in the literature [33, 34] referring to the advantages of integrating bioconversion processes and pervaporation. However, not much has been discussed about the problem of production of noncondensable gases during biological processes (namely carbon dioxide), which permeate the membrane. The presence of noncondensable gases requires an additional energy input in order to keep the downstream pressure at desirable levels and leads to a decreasing energy efficiency of the condensation process.

Although membranes may exhibit a high affinity towards aroma compounds, the high diffusivity of water, even through hydrophobic membranes, limits the degree of selectivity for aroma recovery from diluted aqueous media, such as fermentation media and most natural matrices and process streams. The membrane material of choice for organophilic pervaporation has been polydimethylsiloxane (PDMS), including chemically modified derivates with bulky side groups, introduced in order to reduce water flux (such as polyoctylmethylsiloxane – POMS) and other elastomeric materials such as polyether-polyamide block-copolymers (PEBA), ethylene-propylene-diene monomer (EPDM) elastomers and filler-type membranes [35]. Because one aims at employing selective membranes that are as thin as possible, in order to have a high sorption affinity and minimize the relevance of diffusion selectivity, most organophilic pervaporation membranes are composites consisting of a thin selective active layer and a macroporous support, which assures mechanical stability.

Mass-transfer limitations due to poor hydrodynamic conditions in the feed-side/membrane interface are common in organophilic pervaporation (as referred above).

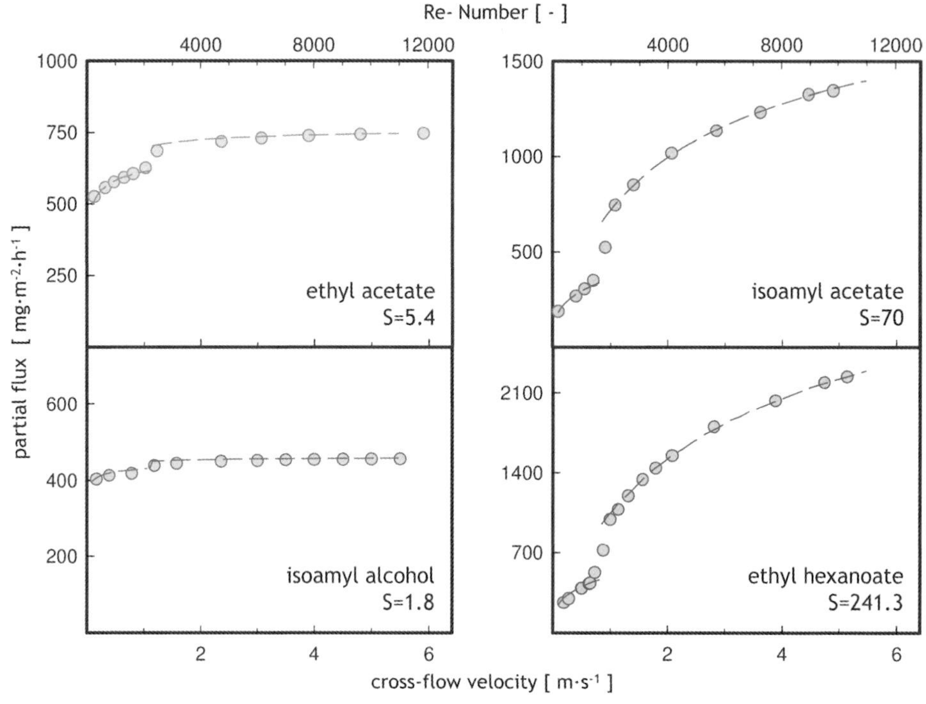

Figure 11.3 Partial fluxes of isoamyl alcohol, ethyl acetate, isoamyl acetate and ethyl hexanoate as a function of their feed crossflow velocity (bottom axis) and Reynolds number (top axis) in a single-channel module, using a POMS-PEI composite membrane. Notice that external mass-transfer limitations are not fully overcome when solutes with a high affinity towards the membrane are processed (Adapted from Ref. 32.)

This effect, usually known as feed-side concentration polarization, may become particularly relevant for solutes with a high sorption affinity towards the membrane, which may lead to its depletion near the membrane interface if external mass-transfer conditions are not sufficiently good to guarantee their fast transport from the bulk feed to the interface [32, 36] (see Figure 11.3). As a consequence of their depletion near the interface the driving force for transport, and the resulting partial fluxes, become lower.

Product recovery and capture can be carried out in a series of condensation stages, at different temperatures, in order to achieve different fractions enriched in target compounds. The temperature of each condenser has to be adjusted according to the downstream pressure in the circuit and the character of the compounds to be separated and recovered [37, 38]. Capture of the target-permeating compounds by condensation remains one of the main problems for competitive use of pervaporation systems, due to the energy costs involved to keep an adequate downstream pressure and to cool down the permeating stream.

New ways of capturing the permeating vapors have to be developed, in order to render this process competitive, enabling continuous operation and reducing energy input. One of the most interesting approaches under development involves the capture of the target aroma compounds by promoting their incorporation (i.e. by solubilization) into a designed system, which will be used as a delivery agent directly in the final food, cosmetic or pharmaceutical product. This is a very effective and elegant way to use the same system to, firstly, capture the aromas from the permeating vapor stream and, secondly, to deliver them into the end product.

Most research on aroma recovery by organophilic pervaporation has been conducted using aqueous aroma model solutions [39, 40], although in recent years a significant interest has been devoted to the recovery of aroma compounds from natural complex streams, such as fruit juices [41, 42], subsidiary streams from the food industry [43] and other natural matrices [44]. The increasing demand for natural aroma compounds for food use, and their market value, opens a world of possibilities for a technique that allows for a benign recovery of these compounds without addition of any chemicals or temperature increase. Considering the strong growth predicted for the low-fat and low-sugar foods and beverages market, the global demand for flavors is expected to grow significantly. Hence, the development of new technologies and delivery systems that improve the use of flavors in food products is likely to be crucial to the future development of this market.

Organophilic pervaporation may play an important role for replacement of evaporative techniques as well as aroma recovery processes based on solvent extraction, in particular when the label 'natural' is considered crucial. The technical challenges discussed above have to be addressed in order to render organophilic pervaporation a competitive process, in particular, the way of capturing target aromas from the permeate stream has to be reassessed in terms of energy consumption and labor intensity.

11.3.3 Nanofiltration

Nanofiltration is a pressure-driven membrane process that lies between ultrafiltration and reverse osmosis (see Figure 11.2) in terms of its ability to reject molecular or ionic species. Usually it is considered that nanofiltration membranes may exhibit nominal cutoffs between 1000 and 200 Da, but this classification should not be regarded as a clear and sharp domain between ultrafiltration (larger cutoffs) and reverse osmosis (smaller cutoffs). The membranes used in nanofiltration – most are polymeric membranes because ceramic nanofiltration membranes are still in their infancy – may be so dense that pores cannot be regarded as such and tighter nanofiltration membranes have to be considered as dense, with the molecular transport taking place through free-volume elements of the polymer. Nanofiltration membranes usually provide for a good retention of small organic molecules and inorganic salts, especially if multivalent ions are involved.

The high rejection of these compounds may lead to significant osmotic pressure differences across the membrane, which decreases the actual driving force for transport. Therefore, in order to assure high solvent fluxes and good solute(s) rejection(s), these membranes should exhibit a high affinity for the solvent (hydrophilic membranes for processing of aqueous streams) and low affinity for the solute(s).

Taking into consideration its rejection characteristics nanofiltration has been proposed for the recovery and fractionation of bioactive molecules with molecular weight lower than 1 kDa from complex media. Recent work describes the successful use of nanofiltration for the recovery of biologically active oligosaccharides from milk [45], using a combination of enzymatic treatment of defatted milk with beta-galactosidase and nanofiltration. It was shown that enzymatic hydrolysis of lactose significantly improves the efficiency and selectivity of this process. The human milk oligosaccharides recovered by this method were shown to inhibit binding of intimin, an adhesion molecule of enteropathogenic *Escherichia coli*, to epithelial cells *in vitro*. No significant difference was found in the oligosaccharide profile between samples prepared by this method and conventional gel-permeation chromatography. The approach developed was also shown to be suitable for the recovery of substantial quantities of tri- and tetra-saccharides from caprine milk.

A similar concept has been proposed and applied to the recovery of bioactive peptides, namely peptides derived from different proteins from bovine colostrum, milk, and cheese whey [46], for human nutrition and promotion of human health. Active peptides can be liberated during gastrointestinal digestion or milk fermentation with proteolytic enzymes. Such peptides may exert a number of physiological effects *in vivo* on the gastrointestinal, cardiovascular, endocrine, immune, nervous, and other body systems. A number of bioactive peptides have been identified in fermented dairy products, and there are already a few commercial dairy products enriched with blood-pressure-reducing milk protein peptides. However, the industrial-scale production of such peptides has been limited by a lack of suitable technologies. Nanofiltration offers the potential for recovering of these peptides and fractionating them according to their molecular weight and charge, excluding higher molecular weight fractions. Size exclusion is not the sole mechanism to be explored in order to achieve a desirable selectivity: membrane–solute electrostatic effects may be explored favorably by playing with the selection of the membrane material and the environmental conditions of the media to be processed.

Recovery of valuable bioactive compounds by nanofiltration, from natural products or streams resulting from the processing of natural products, is also gaining an increasing interest. Recent examples include the production of natural extracts from olive oil subproducts, which are rich in the most potent natural antioxidant compound identified so far (hydroxytyrosol) as well as the production of natural extracts from grape pomace residues, which are rich in a number of high-value compounds.

The first recognized properties of hydroxytyrosol were its ability to prevent the oxidation of the low-density lipoprotein (LDL) [47] and the aggregation of blood platelets [48]. Manna *et al.* [49] proved that this compound is able to protect several cellular human systems from the toxicity induced by reactive oxygen species.

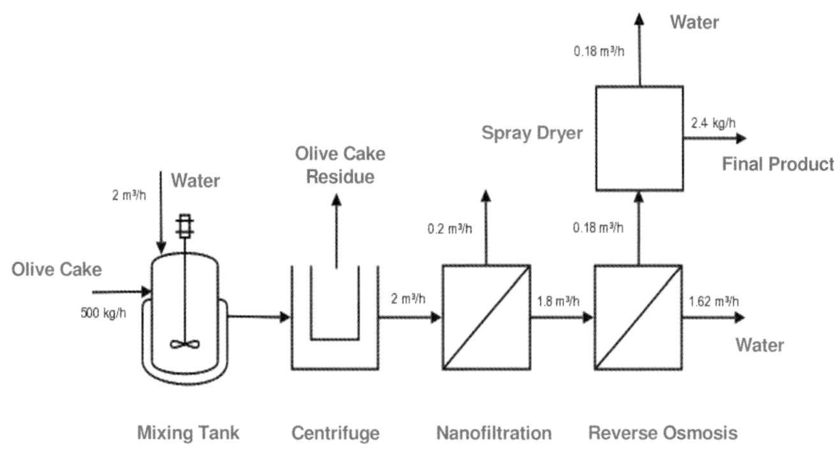

Figure 11.4 Process diagram for the production of hydroxytyrosol-rich natural extracts produced from olive-oil solid residues.

Hydroxytyrosol has been also referred to as a potent agent with anticarcinogenic activity. Grape pomace is rich in a large number of polyphenolic compounds with high antioxidant activity, with resveratrol being probably the most known and referred to in the scientific literature.

In both cases, solid residues are extracted either with water or hydro-alcoholic solutions at mild temperatures (an enzymatic pretreatment may also be added in order to break glycosidic bonds), and the supernatant is processed by nanofiltration (Figure 11.4). This operation allows for recovering a fraction of compounds within a desirable range of molecular weights, while rejecting higher molecular weight compounds (frequently with an undesirable biological activity) such as pesticides and heavy metals, very efficiently excluded by nanofiltration [50].

This approach can be extended to a large number of natural products rich in compounds with desirable biological activity, which include fruits and other plants – for example, the recovery of phytosterols and tocopherol from vegetable oils, as well as natural marine products.

11.4 Concentration and Purification of *Large* Bioactive Molecules

As noted above, most review papers dealing with downstream processing are focused on the recovery and purification of large bioactive molecules, namely proteins. Early biotechnology products were highly active hormones (e.g., insulin, human growth hormone, erythropoietin), thrombolytic agents (e.g., tissue-type plasminogen activator) and clotting factors (e.g., factor VIII). These compounds were typically produced on quite a small scale, considering the annual production requirements ranging from 1 to 10 kg. Recent products are monoclonal antibodies, used for the treatment of breast cancer, B-cell lymphoma and rheumatoid arthritis, among others.

These molecules act stoichiometrically, binding to a particular receptor or cell type, thus requiring much higher dosing levels and batch sizes, in order to satisfy annual production requirements of the order of 1000 kg [1]. As a consequence, process optimization and, in particular, downstream processing is become more relevant in terms of productivity and costs of production.

The excellent review by van Reis and Zydney [1] provides a comprehensive discussion of all major uses of membranes for processing of large molecules. This chapter will be essentially focused on the use of ultrafiltration for the concentration and fractionation of proteins, and the development of membrane chromatography systems.

11.4.1
Ultrafiltration

Ultrafiltration is mostly used for protein concentration and buffer exchange at the industrial scale, replacing size-exclusion chromatography. For protein concentration the selection of membranes with low nominal cutoff values is usually recommended, although, as previously discussed, high rejection and controlled fouling operation may be achieved by using charged membranes together with a correct adjustment of the environmental conditions (pH and ionic strength) of the media to be processed.

Recently, the term 'high-performance tangential flow filtration' (HPTFF) became common to describe the operation of ultrafiltration processes under optimized conditions, with a particular emphasis on protein fractionation [51]. Recent results obtained with HPTFF [1] have shown the potential of this process for the fractionation of monomers from oligomers [51], protein variants differing at only a single amino acid residue [52], and an antigen binding fragment from a similar size impurity [53].

The improved selectivity achieved in HPTFF can only be obtained through the understanding of the phenomena taking place when the solute(s) to be processed interacts with the membrane surface (Figure 11.5). A series of optimized

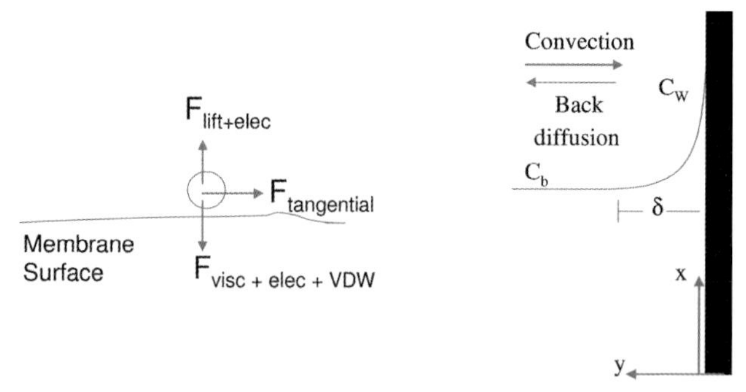

Figure 11.5 Scheme representing the different forces acting on a solute near a membrane surface, during a convective filtration process.

approaches, developed in recent years and partially discussed above in this chapter, can therefore be implemented: (1) operation in the pressure-dependent regime, below a critical flux, employing the concept of low uniform transmembrane pressure (UTMP); (2) adjustment of the environmental conditions of the media to be processed by adequate regulation of its pH and ionic strength; (3) use of electrically charged membranes, if possible with a narrow pore-size distribution, in order to increase the retention of species with like polarity. Overall process optimization is achieved by determining the best compromise between yield and purity, as a function of selectivity and mass throughput [54].

As previously discussed, operation under controlled permeate flux (in opposition to operation under controlled transmembrane pressure) offers a high degree of process control, making it possible to operate under gentle convective transport conditions, which minimize solute transport to the membrane surface keeping osmotic pressure differences low and reducing fouling. Figure 11.6a and b [55]

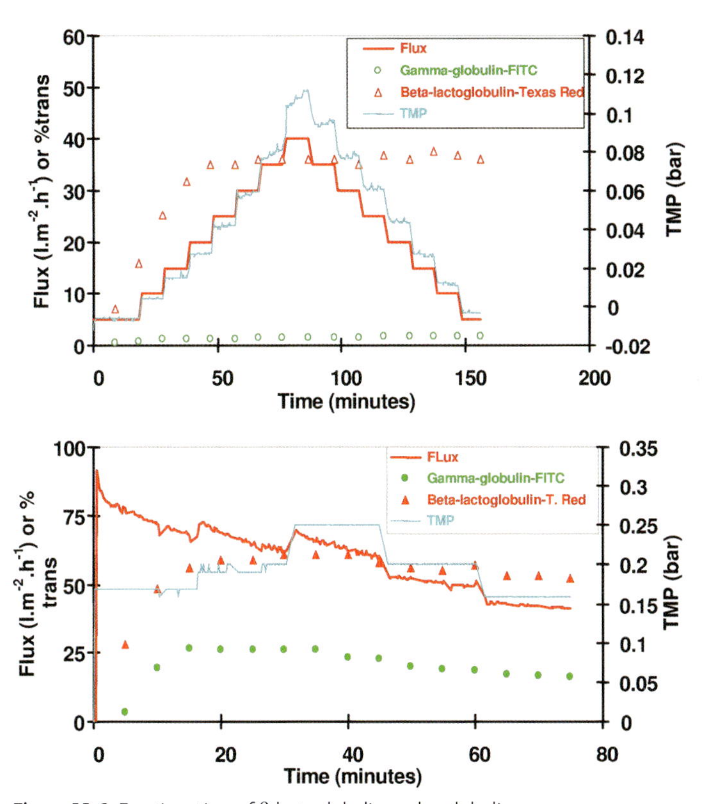

Figure 11.6 Fractionation of β-lactoglobulin and γ-globulin (detection by fluorescence) through a prefouled 50-kDa PES membrane under controlled permeate flux operation (a). Fractionation of β-lactoglobulin and γ-globulin through a prefouled 50-kDa PES membrane under controlled transmembrane pressure operation (b). Adapted from Ref. 55.

compare the fractionation of gamma-globulin (155 kDa) and beta-lactoglobulin (monomer with 18.3 kDa, although it forms dimers and tetramers easily, according to the environmental media conditions) using a polyethersulfone membrane with a nominal cutoff of 50 kDa. As can be seen, for a controlled permeate flux of $25\,l\,m^{-2}\,h^{-1}$ and above this critical value, fouling occurs and the resulting transmembrane pressure increases (Figure 11.6a). Still, rejection of gamma-globulin is as high as 98.2%, while rejection of beta-lactoglobulin is always lower than 63%; operation under these conditions allows for a good fractionation of the two proteins if a cascade procedure is employed, because a good rejection of the larger protein is achieved. In contrast, if a controlled transmembrane-pressure strategy is used (Figure 11.6b), even under low TMP conditions (between 0.15 and 0.25 bar), rejection of the larger protein decreases significantly up to values between 75 and 70% without a significant improvement in the transmission of beta-lactoglobulin. As a consequence, fractionation of the two proteins becomes quite ineffective. Additionally, under controlled-pressure operation, fouling became quite significant and the permeate flux declined dramatically with time. Operation under controlled TMP with low pressure differences, comparable with the ones achieved during controlled flux operation, could be a strategy to follow but in traditional ultrafiltration modules it is not possible to assure such a uniform and low TMP.

Since selectivity is a function of the local permeate flux, and thus the local transmembrane pressure, selectivity can be further improved by maintaining a nearly uniform and low transmembrane pressure throughout the ultrafiltration module. Following the early work from Sandblom, HPTFF technology uses, when required, a cocurrent permeate flow that is accomplished by the addition of a coflow loop and pump on the permeate side.

During permeation, proteins are exposed to different processing conditions, such as shear stress (e.g., when permeating the membrane pores) and interaction with the membrane and the membrane pores' surface, which may induce reversible or irreversible changes in the protein structure. In fact, permeation of protein molecules through membranes with a molecular weight cutoff lower than their mass has been reported (see Figure 11.6 showing permeation of gamma-globulin through a 50-kDa membrane). This permeation may be attributable to pore-size distribution (e.g., the presence of pores with a higher diameter than the nominal cutoff), or to the protein shaping and orientation, which may favor its passage through the membrane pores. However, since distinct conformational states of proteins possess small energetic differences, the hypothesis of protein structural alteration (e.g., molecular elongation) during the permeation process, which facilitates their passage through the membrane pores, has to be considered.

The impact of parameters such as the ratio of protein mass to the membrane cutoff (λ), or the effect of the membrane material during permeation, on their structure/function, is currently unknown and constitutes an essential requisite for the full development and implementation of membrane processes for the selective fractionation of proteins.

All these alterations, which may be induced when a protein molecule interacts with a membrane surface either during convective (filtration) processes or under diffusive

- Intrinsic fluorescence probe with the highest fluorescence quantum yield in proteins, which can be exclusively excited at λ_{exc} of 290 nm
- Fluorescence emission depends on the environment in its vicinity

→ Exposure of tryptophan to environments with different polarities induces shifts of its emission maximum

Increase of polarity => **Red shift**

Decrease of polarity => **Blue shift**

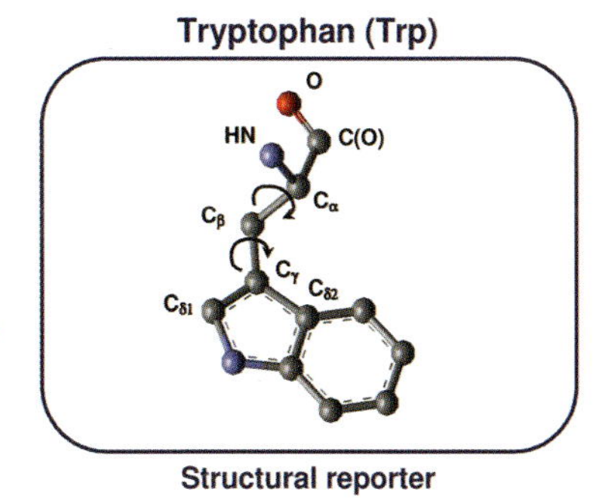

→ Changes in the **environmental polarity** and the presence of a **nearby quencher** induce changes on its fluorescence intensity

Increase of environmental polarity

Proximity of quenchers: aspartate, disulfide or hydrogen bonds

→ Decrease of Trp fluorescence emission

Figure 11.7 Explanatory diagram showing the use of Tryptophan as an intrinsic fluorescence probe for proteins.

conditions (diafiltration), can be reported by intrinsic fluorescence probes such as tryptophan residues and detected by using appropriate natural fluorescence techniques (see explanatory scheme in Figure 11.7).

One of the most interesting features of natural fluorescence results from the fact that the fluorescence response of a given molecule depends very much on their microenvironment. This feature can be used in order to gather information about the structure of complex molecules such as polypeptides and proteins, which may integrate several fluorescent amino acids residues such as tryptophan, tyrosine, and phenylalanine. Among these, tryptophan is the one that exhibits the highest quantum yield, which makes it a good candidate to be used as an intrinsic fluorescence reporter.

The basic concept is the use of the fluorescence response of tryptophan residues, embedded in the polypeptide/protein structure, which is sensitive to changes in its microenvironment. Therefore, if a given protein interacts with a membrane surface and, due to this interaction, changes its tridimensional structure, it can be anticipated that the relative position of the tryptophan residue(s) may be altered. Such changes, may include: (1) a higher exposure of a buried tryptophan to the surrounding solvent (usually water) due to unfolding processes, which leads to contact with a more hydrophilic environment; (2) the opposite process, involving the movement of a tryptophan residue to a more buried, hydrophobic environment; (3) the increase of a tryptophan residue mobility, which will occur if the protein assumes a more unfolded conformation in the region where the tryptophan is located; (4) the change of the relative position of a tryptophan residue towards internal protein quenchers, such as aspartate residues and disulfide bonds, which may occur as a result of processes of folding/unfolding.

In a recent series of papers [56–59], using complementary information acquired by different natural fluorescence techniques, such as steady-state fluorometry, steady-state fluorescence anisotropy and time-decay fluorescence, it was shown that the ratio of protein mass to the membrane cutoff (λ) may determine significant changes on the protein conformation leading to its unfolding under higher pore mouth and intrapore stress conditions. It was also shown that selection of the membrane material can be a key issue, because some materials may exhibit a high affinity for the metal center of enzymes, such as the C isoenzyme of horseradish peroxidase, where the prosthetic heme group partially loses its protoporphyrin ring coordinating iron ion, leading to a deep change of its structure and, ultimately, loss of activity.

The use of these natural fluorescence techniques offers not only the possibility of studying the interaction of proteins with membranes, under convective and diffusive conditions, but also they may be easily extended to studies involving proteins and other porous materials such as chromatography media. The areas of application of these techniques will range from polypeptide and protein fractionation to the monitoring of systems where protein–surface interactions are relevant.

11.4.2 Membrane Chromatography

A large number of membrane materials containing functional ligands has been developed, namely ion-exchange, affinity, reversed-phase, and hydrophobic interaction membranes. Although the binding capacity in membranes tends to be low, the convective flow through the pores reduces mass-transfer resistance compared with bead chromatography. The main benefit of membrane chromatography is associated with the shorter transport times achieved, in comparison with bead chromatography, as the interaction between molecules and active sites at the membrane surface occurs in a convective through pore flow, rather than in a stagnant fluid phase inside the pores of an adsorbent particle. This feature can be particularly important for purification of large biomolecules and viruses that may exhibit significant transport limitations in conventional chromatographic media. Membrane chromatography also offers higher flow rates, lower pressure drops, and shorter processing times than traditional chromatography. Their use is also accompanied by reduced protein unfolding and denaturation, as (discussed above) [3].

Taking these features together, membrane chromatography is emerging as an alternative for flow-through applications, where a rapid processing of high volumetric and dilute feed streams is required [60]. Flow-through applications include also the removal of DNA, viruses, and endotoxins [1].

In addition, it has been shown that membrane adsorbers are competitive in bind-and-elute applications for large solutes such as DNA, RNA, and viruses. Most bead chromatographic media have pore sizes that are too small, which exclude large molecules from entering and binding to specific sites. Under these circumstances, membrane chromatography exhibits a competitive binding capacity for these molecules, such as DNA.

11.5 Future Trends and Challenges

The constraints discussed in this chapter represent new challenges and opportunities for improved membrane materials and technologies for *downstream processing*.

The use of membrane processes for the recovery of small biologically active molecules is expected to grow significantly within the next years, in particular in what refers to the use of nanofiltration for recovery of high added value compounds with impact on human nutrition and promotion of human health. Some of the recovery processes will demand membranes that are more stable in hydro-alcoholic media, or even in organic media compatible with the final product and its use, exhibiting simultaneously a sharp cutoff behavior. The discovery of small molecules with desirable biological properties, present in natural matrices such as plants and marine products, will boost the need for recovery processes regarded as clean and sustainable that allow the use of the label 'natural' in the final product. Membrane processes fit perfectly this demand due to the mild conditions under which they operate.

In what concerns ultrafiltration, it has replaced size-exclusion chromatography in almost all final formulation processes. Charged ultrafiltrafion membranes, in conjunction with optimum operating parameters, as previously discussed, can also be used to enable protein purification with HPTFF. In fact, recent developments in membrane chromatography and HPTFF enable, for the first time, complete purification of proteins using membrane systems [1].

Membrane chromatography has been evolving slowly and its low binding capacity has been the major obstacle. To be competitive for applications in which the product is bounded and then eluted, membranes will need to have a binding capacity equivalent to that of bead chromatography, and similar process time and cost. On the other hand, flow-through chromatography applications that involve the binding of impurities, whereas the products flow through the matrix, are nowadays competitive.

One interesting issue is the development of disposable systems (and disposable, single-use membranes). Disposable systems may become attractive for production processes, because they eliminate the need for development and validation of cleaning cycles. In this case, the development of biodegradable membranes will become an interesting opportunity.

In many cases, the improvement in external mass transfer is achieved at the expense of membrane packing density. This is the case for a number of solutions involving the use of spacers or, for example, the case of rotating modules that promote the formation of Taylor vortices. As mentioned above, membranes with micro- or nanostructured 3D topography offer an extremely elegant solution for improvement of mass transfer by inducing turbulence at the membrane scale, avoiding the spending of energy in the bulk fluid and allowing for the construction of highly packed modules.

In fact, fascinating breakthroughs are expected in the coming years, through the development of membranes designed to operate in a particular environment, respond to specific controlled stimulus (pH, ionic strength, temperature, UV/Vis radiation or magnetic field) or interact specifically through molecular recognition with target solutes.

References

1 van Reis, R. and Zydney, A. (2007) *Journal of Membrane Science*, **297**, 16–50.

2 Hubbuch, J. and Kula, M.-R. (2007) *Journal of Non-Equilibrium Thermodynamics*, **32**, 99–127.

3 Charcosset, C. (2006) *Biotechnology Advances*, **24**, 482–492.

4 Giorno, L. and Drioli, E. (2000) *Trends in Biotechnology*, **18**, 339–348.

5 Lopez, J.L. and Matson, S. (1997) *Journal of Membrane Science*, **125**, 189–211.

6 Sousa, H.A., Afonso, C.A.M., Mota, J.P.B. and Crespo, J.G. (2005) *Chemical Engineering Research and Design*, **83** (A3), 285–294.

7 Vogelaar, L., Lammertink, R.G.H. and Wessling, M. (2006) *Langmuir*, **22**, 3125–3130.

8 Susanto, H. and Ulbricht, M. (2007) *Langmuir*, **23**, 7818–7830.

9 He, D.M. and Ulbricht, M. (2007) *Macromolecular Chemical Physics*, **208**, 1582–1591.

10 Gronda, A.M., Buechel, S. and Cussler, E.L. (2000) *Journal of Membrane Science*, **165** (2), 177–187.

11 Belfort, G., Pimbley, J.M., Greiner, A. and Chung, K.-Y. (1993) *Journal of Membrane Science*, **77**, 1–22.

12 Kaur, J. and Agarwal, G.P. (2002) *Journal of Membrane Science*, **196**, 1–11.

13 Kelder, J.D.H., Janssen, J.J.M. and Boom, R.M. (2007) *Journal of Membrane Science*, **304** (1–2), 50–59.

14 Santos, J.L.C., Geraldes, V., Velizarov, S. and Crespo, J.G. (2007) *Journal of Membrane Science*, **305** (1–2), 103–117.

15 Kane, R.S., Takayama, S., Ostuni, E., Ingber, D.E. and Whitesides, G.M. (1999) *Biomaterials*, **20**, 2363–2376.

16 Bacchin, P., Aimar, P. and Sanchez, V. (1995) *AIChE Journal*, **41**, 368–377.

17 Belfort, G., Davis, R.H. and Zydney, A.L. (1994) *Journal of Membrane Science*, **96**, 1–58.

18 Mcdonogh, R.M., Fane, A.G. and Fell, C.J.D. (1989) *Journal of Membrane Science*, **43**, 69–85.

19 Field, R.W., Wu, D., Howell, J.A. and Gupta, B.B. (1995) *Journal of Membrane Science*, **100**, 259–272.

20 Howell, J.A. (1995) *Journal of Membrane Science*, **107**, 165–171.

21 Bacchin, P., Aimar, P. and Field, R.W. (2006) *Journal of Membrane Science*, **281** (1–2), 42–69.

22 Sandblom, R.M. (1978) SW patent 74,16,257, 1974 and US patent 4,105,547.

23 Baruah, G.L., Nayak, A. and Belfort, G. (2006) *Journal of Membrane Science*, **274**, 56–63.

24 Lim, H.S., Han, J.T., Kwak, D., Jin, M. and Cho, K. (2006) *Journal American Chemical Society*, **128**, 14458–14459.

25 Nayak, A., Liu, H. and Belfort, G. (2006) *Angewandt Chemie International*, **45**, 4094–4098.

26 Pujar, N.S. and Zydney, A.L. (1998) *Journal of Chromatography A*, **796**, 229–238.

27 Saksena, S. and Zydney, A.L. (1994) *Biotechnology & Bioengineering*, **43**, 960–968.

28 Garde, A., Rype, J.-U. and Jonsson, G. (2002) WO patent 02/48044 A2.

29 Zelic, B., Gostovic, S., Vuorilehto, K., Vasic-Racki, D. and Takors F R. (2004) *Biotechnology & Bioengineering*, **85** (6), 638–646.

30 Poulin, J.F., Amiot, J. and Bazinet, L. (2006) *Journal of Biotechnology*, **123** (3), 314–328.

31 Schäfer, T., Bengtson, G., Pingel, H., Böddeker, K.W. and Crespo, J.G. (1999) *Biotechnology & Bioengineering*, **62** (4), 412–421.

32 Schäfer, T. and Crespo, J.G. (2007) *Journal of Membrane Science*, **301** (1–2), 46–56.

33 Stefer, B. and Kunz, B. (2002) *Chemie Ingenieur Technik*, **74** (7), 1029–1034.

34 Maume, K.A. and Cheetham, P.S.J. (1991) *Biocatalysis*, **5**, 79.

35 Rutherford, S.W., Kurtz, R.E., Smith, M.G., Honnell, K.G. and Coons, J.E. (2005) *Journal of Membrane Science*, **263** (1–2), 57–65.

36 Schäfer, T., Vital, J. and Crespo, J.G. (2004) *Journal of Membrane Science*, **241** (2), 197–205.

37 Marin, M., Hammami, C. and Beaumelle, D. (1996) *Journal of Food Engineering*, **28** (3–4), 225–238.

38 Brüschke, H.E.A., Schneider, W. and Tusel, G.F. (1989) EP patent 0 332 738.

39 Baudot, A. and Marin, M. (1997) *Food and Bioproducts Processing*, **75** (C2), 117–142.

40 Borjesson, J., Karlsson, H.O.E. and Tragardh, G. (1996) *Journal of Membrane Science*, **119** (2), 229–239.

41 Pereira, C.C., Rufino, J.R.M., Habert, A.C., Nobrega, R., Cabral, L.M.C. and Borges, C.P. (2005) *Journal of Food Engineering*, **66** (1), 77–87.

42 Alvarez, S., Riera, F.A., Alvarez, R., Coca, J., Cuperus, F.P., Bouwer, S.T., Boswinkel, C., van Gemert, R.W., Veldsink, J.W., Giorno, L., Donato, L., Todisco, S., Drioli, E., Olsson, J., Tragardh, G., Gaeta, S.N. and Panyor, L. (2000) *Journal of Food Engineering*, **46** (2), 109–125.

43 Souchon, I., Pierre, F.X., Athes-Dutour, V. and Marin, A. (2002) *Desalination*, **148** (1–3), 79–85.

44 Kattenberg, H.R. and Willemsen, J.H.A. (1999) WO patent 00/38540.

45 Sarney, D.B., Hale, C., Frankel, G. and Vulfson, E.N. (2000) *Biotechnology and Bioengineering*, **69** (4), 461–467.

46 Butylina, S., Luque, S. and Nystrom, M. (2006) *Journal of Membrane Science*, **280** (1–2), 418–426.

47 Visioli, F. and Galli, C. (1998) *Journal of Agricultural and Food Chemistry*, **46** (10), 4292–4296.

48 Petroni, A., Blasevich, M., Salami, M., Papini, N., Montedoro, G.F. and Galli, C. (1995) *Thrombosis Research*, **78** (2), 151–160.

49 Manna, C., Galleti, P., Maisto, G., Cucciola, V., D'Angelo, S. and Zappia, V. (2000) *FEBS Letters*, **470** (3), 341–344.

50 Da Ponte, M.N., Santos, J.L., Matias, A., Nunes, A., Duarte, C. and Crespo, J.G. (2007) patent WO2007013032.

51 van Reis, R., Gadam, S., Frautschy, L.N., Orlando, S., Goodrich, E.M., Saksena, S., Kuriyel, R., Simpson, C.M., Pearl, S. and Zydney, A.L. (1997) *Biotechnology and Bioengineering*, **56** (1), 71–82.

52 Ebersold, M.F. and Zydney, A.L. (2004) *Biotechnology Progress*, **20** (2), 543–549.

53 van Reis, R., Brake, J.M., Charkoudian, J., Burns, D.B. and Zydney, A.L. (1999) *Journal of Membrane Science*, **159** (1–2), 133–142.

54 van Reis, R. and Saksena, S. (1997) *Journal of Membrane Science*, **129** (1), 19–29.

55 Crespo, J.G., Trotin, M., Hough, D. and Howell, J.A. (1999) *Journal of Membrane Science*, **155** (2), 209–230.

56 Portugal, C.A.M., Crespo, J.G. and Lima, J.C. (2006) *Journal of Photochemistry and Photobiology B-Biology*, **82** (2), 117–126.

57 Portugal, C.A.M., Lima, J.C. and Crespo, J.G. (2006) *Journal of Membrane Science*, **284** (1–2), 180–192.

58 Portugal, C.A.M., Crespo, J.G. and Lima, J.C. (2007) *Journal of Membrane Science*, **300** (1–2), 211–223.

59 Portugal, C.A.M., Lima, J.C. and Crespo, J.G. (2008) *Journal of Membrane Science*, **321** (1), 69–80.

60 Gottschalk, U. (2005) *Biopharm International*, **18** (3), 42–58.

12
Integrated Membrane Processes

Enrico Drioli and Enrica Fontananova

12.1
Introduction

Separation processes are intensively used in practically all industrial processes. About 40–50% of the energy use in major commodity-producing industries is utilized in separations, but most of these are still carried out by traditional thermally driven processes [1].

The necessity to realize a *sustainable growth*, also by a more rational and efficient energy use, calls for additional and substantial developments in the separations field.

Sustainable growth focuses on making progress to satisfy global human needs without damaging the environment. A promising way for the realization of a sustainable growth, is the strategy of *process intensification* [2].

This consists of innovative equipments design and process development methods that are expected to bring relevant improvements in manufacturing and processing, decreasing production costs, energy consumption, waste generation, equipment size and improving remote control, scale-up, design and process flexibility resulting in cheaper, sustainable technical solutions [2].

Because of its intrinsic properties that well fit the requirements of process-intensification strategy (efficiency, modularity, reduced energy consumption, etc.), membrane-separation processes have well-established applications in various industrial fields and more progresses can be anticipated for the near future [3].

Various membrane operations are available today for a wide spectrum of industrial applications. Microfiltration (MF), ultrafiltration (UF), nanofiltration (NF), reverse osmosis (RO), gas and vapor separation (GS, VS), pervaporation (PV), dialysis (D), electrodialysis (ED) and membrane contactors (MCs) are only some of the best-known membrane unit operations.

However, the possibility to integrate various membrane operations in the same process or in combination with conventional separation units, allows, in many cases better performance in terms of product quality, plant compactness, environmental impact, and energy use to be obtained.

Membrane Operations. Innovative Separations and Transformations. Edited by Enrico Drioli and Lidietta Giorno

ISBN: 978-3-527-32038-7

In this chapter, some important examples of integrated membrane systems will be presented starting from the most well-known application of membrane technology, that is, water desalination.

12.2
Integrated Membrane Processes for Water Desalination

Membrane technology is already recognized as the best choice for water desalination since thermal options for desalting are about 10 times less energy efficient [1, 4].

The number of membrane desalination installations accounts for about the 80% of the total number of desalination plants [5] and for about the 50% of the total capacity [5] (Table 12.1).

Although RO water desalination is today considered as the most cost-effective solution [4], key factors for further improvements in membrane-based desalination systems are: enhancement of water-recovery factor, cost reduction, improvement of water quality, new brine-disposal strategies. All these issues can be addressed by an integrated approach.

Traditional pretreatments make an extensive use of chemicals (NaClO as disinfection, $FeCl_3$ as flocculants, H_2SO_4 as antiscaling agent) and mechanical filtration units (sand filtration, media filtration, cartridge filtration).

Another interesting possibility is the use of pressure-driven membrane processes, in particular MF and UF are becoming standard and very efficient pretreatment options for sea- and brackish-water desalination. Also, for wastewater treatment, MF/UF pretreatment technology can efficiently reduce the highly fouling nature of the feed.

UF is typically used to retain macromolecules, colloids, solutes with molecular weight higher than a few thousand Daltons. MF is a low-pressure membrane process for separating colloidal and suspended micrometer-size particles [6].

Table 12.1 Membrane technology vs. thermal technologies for water desalination (Data from Ref. 5.)

Desalination plant type		% over the total number of desalination plants worldwide (~14 000)		% over the total capacity of desalination plants worldwide (~7 000 000 MGD)	
Thermal desalination	Multistage flash (MSF)	20	45	50	85
	Vapor compression (VC)		30		10
	Multieffect distillation (MED)		25		5
Membrane desalination	Reverse osmosis (RO)	80	90	50	90
	Electrodialysis reversal (EDR)		10		10

Although the capital cost of membrane pretreatment is usually higher than that of a conventional pretreatment, the additional cost of MF/UF is paid and also exceeded by reducing RO replacement and chemical cost for both dosing and RO cleaning [7].

Other potential benefits arise from the 33% space saving of MF/UF and the opportunity to increase RO flux and water recovery. Moreover MF/UF provides a more reliable system, and is tolerant of feed-quality variations [7].

An UF system utilizing hollow-fiber (HF) membranes has been successfully used as pretreatment prior to seawater reverse osmosis (SWRO) desalination without any chemical treatments [8]. The quality of UF permeate was good and satisfied the need of SWRO feed water [8].

In this pilot plant the UF pretreatment system is arranged in 2 trains, each housing 3 modules (PAN HF membranes, nominal pore size of 0.02 μm, MWCO 50 000, total effective surface area of 30 m^2). Raw seawater (samples from Qingdao Jiaozhou Bay, the Yellow Sea of China) was first passed into a cartridge sand filter and successively feed to UF system, the UF permeate was then pumped to the RO system (spiral-wound composite polyamide) (Figure 12.1).

The advantages of UF pretreatment in comparison with traditional ones, has been also examined for an Eastern Mediterranean feed water [7]. In this study it has been shown that the UF/MF can be cheaper than conventional pretreatment (dual-media filters followed by cartridges) by 0.7 cents/m^3 for total water cost [7].

Although the integration of RO with other pressure-driven membrane processes has led to significant improvements in membrane-based desalination process economics, another fundamental problem is the environmental aspects of brine discharge from reverse-osmosis desalination plants.

The most frequent disposal practice is the direct discharge in the sea. However, the more and more stringent environmental regulations preclude, in many cases, this low-cost possibility in order to protect the aquatic environment.

Various process engineering strategies have been investigated in order to have a more environmentally friendly strategy for brine disposal in reverse-osmosis desalination.

A suitable solution is the possibility to redesign completely a desalination system by also introducing MC operations [9–11].

Membrane contactors are systems in which the membrane function is to facilitate diffusive mass transfer between two contacting phases (liquid–liquid, liquid–gas, etc.) without dispersion of one phase within another [12]. The membrane does not act as a selective barrier, but creates and sustains the interfaces immobilized at the

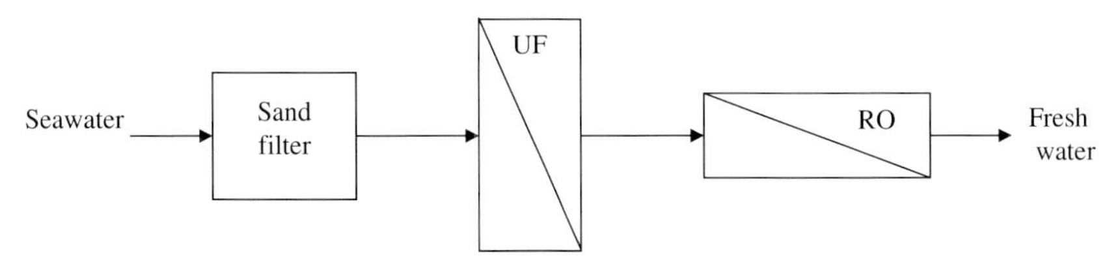

Figure 12.1 Scheme of an UF-RO membrane integrated desalination system [8].

mouth of the pores; the separation process is based on the principles of phase equilibria.

Membrane distillation (MD) is an example of membrane contactors applied to the concentration of aqueous solutions of nonvolatile solutes.

In a MD process, a microporous hydrophobic membrane is in contact with an aqueous heated solution on the feed or retentate side. The hydrophobic nature of the membrane prevents the mass transfer in liquid phase and creates a vapor/liquid interface at the entrance of each pore. Here, volatile compounds (typically water) evaporate, diffuse and/or convect across the membrane, and are condensed and/or removed on the permeate or distillate side.

With respect to RO process, MD does not suffer osmotic-pressure limitation and can be therefore employed when high permeate recovery factors or retentate concentrations are required.

Membrane crystallization (MCr) has been recently proposed as one of the most interesting and promising extensions of the MD concept [13].

Evaporative mass transfer of volatile solvents through microporous hydrophobic membranes is employed in order to concentrate feed solutions above their saturation limit, thus obtaining a supersaturated environment where crystals may nucleate and grow. In addition, the presence of a polymeric membrane increases the probability of nucleation with respect to other locations in the system (heterogeneous nucleation) [14].

An integrated membrane desalination system based on NF, RO, and MD has been proposed [11]. In this system a NF unit has been used as pretreatment, while the MD contributed to concentrate the two brine streams from both NF and RO (Figure 12.2).

The water production cost estimated for the integrated system was 0.92 $\$/m^3$ with a recovery factor of 76.2% [11].

Another integrated MF–NF–RO system having MD/MCr units operating on the NF and/or RO retentate, has also been proposed [10].

Five different integrated membrane system (IS) configurations have been considered: in the IS1 MF and NF are used as pretreatment to RO; in the IS2 a MCr unit operates on the NF retentate; in the IS3 a MCr works on the RO retentate; in the IS4 a MCr and a MD operates, respectively, on the NF retentate and on the RO retentate; finally in the IS5 two MCr operates both on the NF and RO retentate (Figure 12.3).

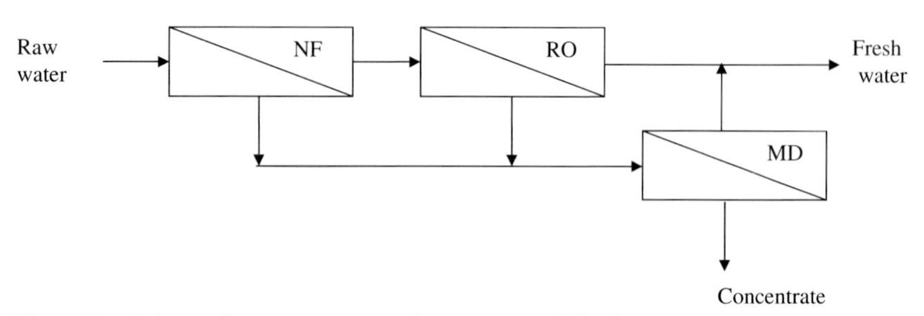

Figure 12.2 Scheme of a NF-RO-MD membrane integrated desalination system [11].

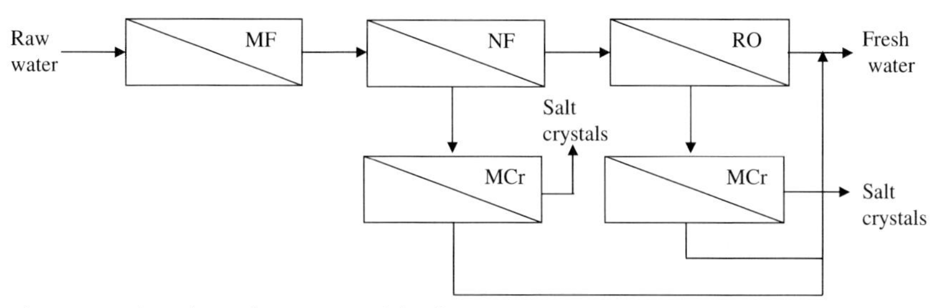

Figure 12.3 Flow sheet of an integrated desalination system utilizing membrane crystallization units (IS5, Ref. 10).

In addition to the MF pretreatment, used to remove particulates and microorganisms from raw water, reducing membrane fouling in the successive steps, the NF step is able to reduce water hardness decreasing osmotic pressure of the RO feed, allowing operation at higher water recovery level.

Moreover, when a MCr/MD is fed with the NF and/or RO retenates, salts crystals can grow in the high concentrated brines, or at least, a more concentrate solution is obtained, increasing the water recovery.

Although the MCr/MD units introduce a thermal energy requirement, the water-recovery factor is increased up to 92.8% for the IS5 without a relevant increase of the cost if waste thermal energy is already available for the process; at the same time brine flow rate is significantly reduced while the fresh water flow rate is increased (Table 12.2) [10].

Moreover, the sale of salt crystals grown in highly concentrated brines (in particular $MgSO_4{\cdot}7\ H_2O$), might potentially reduce the overall desalination cost, thus confirming the potential interest for the proposed approach.

In fact, production of solid materials of high quality and controlled properties (specific polymorphs) with important added value, could transform the traditional brine-disposal cost in a potentially new profitable market; reducing, moreover, the environmental problems of the brine disposal [10].

Table 12.2 Comparison of five integrated desalination systems (Data from Ref. [10].)

	IS1	IS2	IS3	IS4	IS5
Fresh-water recovery (%)	49.2	71.6	70.4	88.6	92.8
Water unit cost ($\$/m^3$)	0.46	0.68–0.55[a]	0.59–0.47[a]	0.74–0.55[a]	0.73–0.54[a]
Brine flow rate ($m^3\,h^{-1}$)	531.9	296.6	309.9	118.5	74.6
Brine concentration ($g\,L^{-1}$)	68.02	95.94	76.53	240.0	214.4
Fresh-water flow rate ($m^3\,h^{-1}$)	517.6	753.0	739.6	931.5	974.9
Fresh-water concentration ($g\,L^{-1}$)	0.270	0.186	0.189	0.150	0.143
$CaCO_3$ produced ($kg\,m^{-3}$)	—	1.19	0.12	0.96	1.00
NaCl produced ($kg\,m^{-3}$)	—	9.79	16.86	7.91	20.35
$MgSO_4{\cdot}7\ H_2O$ produced ($kg\,m^{-3}$)	—	1.25	0.00	1.01	0.96

[a] If thermal energy is already available in the plant.

An integrated water-treatment system designed to use different water sources and different treatment processes, including membrane processes, has been realized in Terneuzen (The Netherlands) [15]. Raw-water sources and treatments include: seawater and integrated membrane system to produce demineralized water; fresh water, and ion exchange to produce demineralized water; effluent industrial wastewater-treatment plant (WWTP) and media filtration to produce cooling tower supply water.

The plant, operating from 2000, produces an 750 m^3/h demineralized water, 650 m^3/h cooling tower supply water and 1.050 m^3/h ultrapure water [15].

In the integrated membrane system fed with seawater (Figure 12.4A) the pretreatment consists of two rotating microscreens of 150 μm used to remove the larger suspended particles from the water. Successively 8 MF units (polypropylene (PP) hollow membranes placed in a vertical position and operated according to the deadend principle) are used to remove the suspended solids completely, algae and to disinfect the water. The MF section is designed to filter 700–750 m^3/h [15].

The permeate of the MF units is fed to two SWRO units equipped with high-pressure pumps with an energy-recovery system (Pelton wheel). The SWRO units comprise 44 pressure vessels loaded with 6 Dow/FilmTec SW30 membranes each. The designed permeate capacity of each unit is 210 m^3/h per unit with a recovery of 50–55%. Antiscalant is dosed to the feed stream of the SWRO with a concentration of 3 to 4 ppm [15].

While the SWRO permeate flows into a reservoir and is successively fed to two BWRO units, the concentrate is directly discharged into the Westerschelde.

The first BWRO unit consists of 16 pressure vessels; the second BWRO unit consists of 6 pressure vessels. Each vessel contains 6 Dow/FilmTec BW30 membranes. The designed permeate capacity of each BWRO unit is 175 m^3/h with a recovery of 85% [15].

The permeate of the BWRO (conductivity of 10–15 μS/cm) is mixed with the demineralized water produced by the ion-exchange process and consecutively

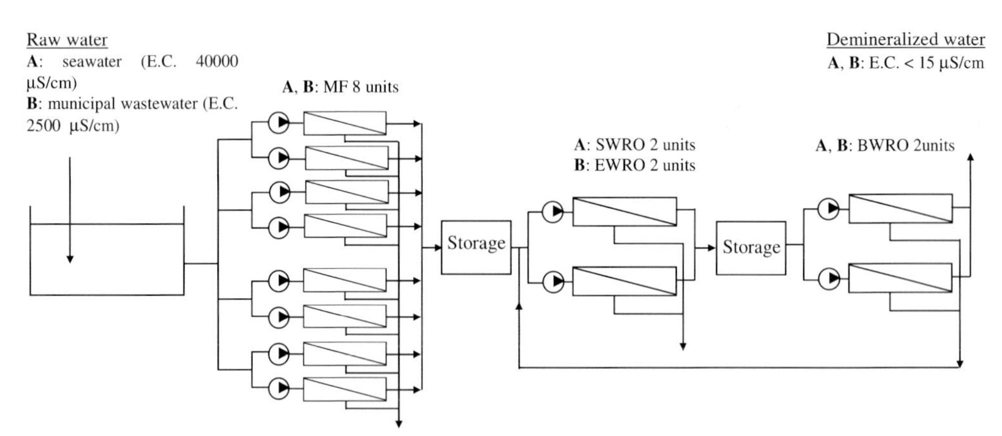

Figure 12.4 Process flow diagram of the water-treatment plant fed with seawater (A, period 2000–2006) and now operating on treated secondary communal wastewater (B, from 2007) [15].

supplied at the Dow plant. The concentrate of the BWRO is mixed with the pretreated seawater and fed to the SWRO.

In consideration of some operational difficulties, like biofouling and corrosion, directly related to the feed-water source, more reliable water sources have been evaluated by the Evides Industriewater that owns and operates this plant, like anaerobic groundwater and sweet tertiary wastewater.

From the 2007 treated secondary communal wastewater (effluent from the communal wastewater treatment plant in the city of Terneuzen) is used as the new feed-water source [16].

The integrated membrane system was reengineered (Figure 12.4B). The MF pretreatment section was not subject to hardware changes, but operated in a different regime [16].

The two SWRO units have been reengineered to low-pressure RO systems (EWRO array: 28–16 with membranes Dow Filmtec BW30-FR). The two BWRO units remained unchanged. The designed permeate capacity of the BWRO is 150–175 m^3/h per unit operating at a recovery of 85%. The permeate of the BWRO has a conductivity of 10 μS/cm and is mixed with demineralized water originating from the ion-exchange process and consecutively supplied to Dow [15]. The concentrate of the integrated membrane system is mixed and discharged into the Westerschelde.

12.3 Integrated Membrane Process for Wastewater Treatment

As reported in the previous example, integrated membrane processes represent a viable solution also for wastewater treatment. Comparing three different treatments system (Figure 12.5): traditional treatment using coagulation/flocculation, sand filtration, physicochemical softening, activated carbon adsorption, and disinfection (A); spiral-wound nanofiltration with ultrafiltration pretreatment followed by marble filtration and disinfection (B); and direct capillary nanofiltration with only a limited pretreatment and post-treatment by marble filtration and disinfection (C), the solution B and C showed better performance for ‘quality and public health’ and ‘operational aspects’ than the solution A [17]. Taking into account economical aspects and environment impact criteria the process C was more advantageous than B; however; the traditional treatment A resulted to be more advantageous for economical aspects [17].

More progresses can be anticipated in the near future by promoting the integration of different membrane operations, including MCs and membrane bioreactor (MBRs), also for wastewater treatment.

MBR are already considered by the European Union as one of the best available technologies for municipal and industrial wastewater treatment.

In a MBR, biological treatment is integrated with membrane filtration, providing an effective alternative to conventional processes such as activated sludge for municipal and industrial effluents [18].

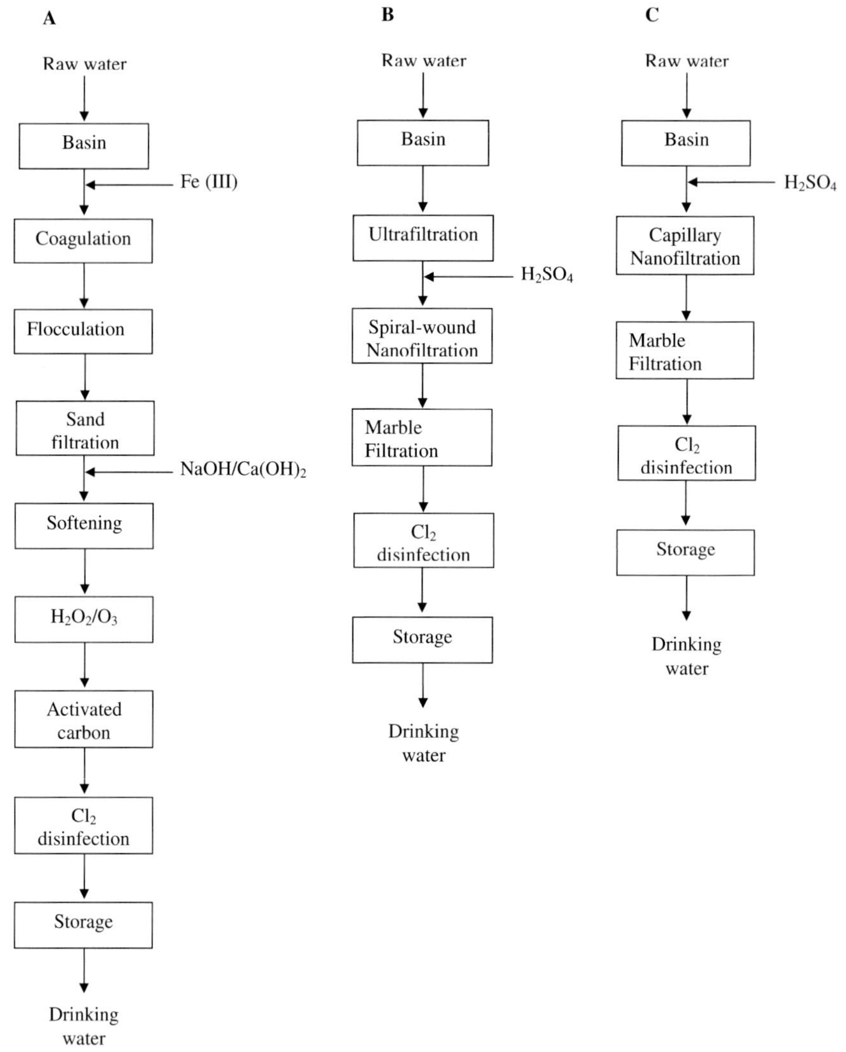

Figure 12.5 Scheme of three different drinking-water production plants: traditional (A); using ultrafiltration pretreatment and spiral-wound nanofiltration (B) and using capillary nanofiltration (C) [17].

Membrane bioreactors are composed of two fundamentals parts, the biological reactor in which the reaction occurs (active sludge containing purifying bacteria) and the membrane module for the separation of the different compounds.

Membrane bioreactors can be classified into two main groups according to their configuration. The first group is commonly known as recirculated or external MBR and involves the recirculation of the solution through a membrane module that is outside the bioreactor. Both inner-skin and outer-skin membranes can be used in this application. The second configuration is the integrated or submerged MBR that involves outer skin membranes internal to the bioreactor.

In the external MBR the driving force is due to a difference in transmembrane pressure obtained by high-pressure crossflow along the membrane surface.

In the submerged MBR the driving force is achieved by pressurizing the bioreactor or creating negative pressure on the permeate side. A diffuser is usually placed directly beneath the membrane module to facilitate scouring on the filtration surface. Aeration and mixing are also achieved by the same unit.

The first example of an MBR operated with tubular membranes placed in external recirculation loops. However, the use of recirculation loops leads to relative high energy costs. In addition, the high shear stresses in the tubes and recirculation pumps can contribute to the loss of biological activity of the system [19]. Submerged MBRs are alternative systems to overcome these limits. This operating mode limits the energy consumption associated with the recirculation cost [20]. Moreover, the use of submerged membranes allow operation at low transmembrane pressures. This makes MBRs well suited to relatively large-scale applications.

Submerged MBRs were first introduced for decentralized sanitation in the US and in building-water reuse in Japan, and are now widely applied in different sectors.

One of the largest membrane bioreator unit in the world was recently built in Porto Marghera (Venice, Italy) in order to extract remaining pollutants in tertiary water prior to disposal into the Venetian Lagoon [21].

The ultrafiltration unit, containing submerged polyvinylidenefluoride (PVDF) hollow-fiber membranes (ZeeWeed by Zenon), is designed to treat 1600 m^3/h of wastewater with a COD/h of 445 kg and the suspended matter of the treated water is <1 mg/L [21].

There are two interconnected UF lines, each line contains 4 unit composed by 9 ZeeWeed modules and the total membrane area is 100 000 m^2 [21].

The application of membrane processes in the treatment of aqueous effluents offers very interesting potentials also for the leather industry [22], traditionally considered one of the most polluting industries, being generally characterized by a low technological level of its operations. Wastes coming from the processing cycles of the leather are characterized by a high level of organic and inorganic pollutants, originating from natural skins or introduced during the treatment operations.

The application of membrane-separation processes in the treatment of wastewater of the leather industry can give a reduction of the environmental impact, a simplification of cleaning-up procedures of aqueous effluents, an easy re-use of sludge, a decrease of disposal costs, and a saving of chemicals, water, and energy [22].

The separation operations of the leather cycle can be combined with or modified by membrane processes such as MF, UF, NF, and RO, showing in many cases a significant improvement from an energetic point of view [22].

Experiments carried out in pilot-scale plants on liming, degreasing, and chromium tannage show the possibility to realize more efficient operations for recovering by-products (e.g., proteins and fats) and chemicals (e.g., chromium) [22].

Recovering chromium salts from spent tanning liquors has been carried out with an integrated process based on preliminary UF followed by NF [23].

The UF pretreatment (spiral-wound membrane module with a molecular weight cutoff of 50–100 kDa) allows a reduction of suspended solids (84%) and fat substances (71%) [23].

The NF step (spiral-wound membrane module with a molecular weight cutoff of 150–300 Da) gives a rejection 97% of the chromium and 98% of the sulfates in the feed stream, whereas the organic matter retention was 51% [23].

The concentrated chromium solutions obtained (final value of about 10 g/L) was used in chrome tannage experiments performed on pickled sheepskins. The physical properties of these samples were compared with those of control skins treated with the conventional method using basic chromium sulfate containing 26% Cr_2O_3, observing improved characteristics of the first ones (same hydrothermal stability grain crack and bursting strength values higher) [23].

A higher percentage of chromium recovery was obtained in the tanning operation performed with the NF retentate. In this process, the reuse of the NF permeate in the pickling step is also possible, adjusting the chloride concentration through the addition of NaCl, because the 58% of the chlorides is recovered in the permeate [23]. Alternatively, this solution can be used directly in the chrome tannage step according to a one-bath process in which skins are pickled at a pH of 3 or lower and then chrome tanning materials are introduced and the pH is raised (Figure 12.6).

The main advantages of this integrated processes are: reduction of the environmental impacts, improving the leather quality, reducing the wastewater discharge and pollution load, saving of chemicals and water, allowing the reuse of sludges and decreasing disposal costs.

12.4
Integrated Membrane System for Fruit-Juices Industry

Integrated membrane processes are today proposed also in the dairy, food, and fruit-juices industries.

The RO potentialities as a concentration technique to remove water from fruit juices for the production of high-quality fruit-juice concentrate are well known [24]. The most relevant advantages of the RO process over traditional evaporation are in the reduced thermal damage of the product, increase of aroma retention, and lower energy consumption, since the process is carried out at low temperature.

Disadvantages of RO come from its inability to reach the high concentration of juices typically obtained by evaporation, because of osmotic-pressure limitation. As a consequence concentrations larger than 25–30° Brix total soluble solid content (TSS) cannot be reached with a single-stage RO [25].

However, technological advances related to the development of new membranes operations and innovative strategies of process design, have partially overcome this limitation. Membrane distillation, for example, is not subject to osmotic-pressure limitation and can be therefore employed in integrated systems when high permeate recovery factors or retentate concentrations are requested.

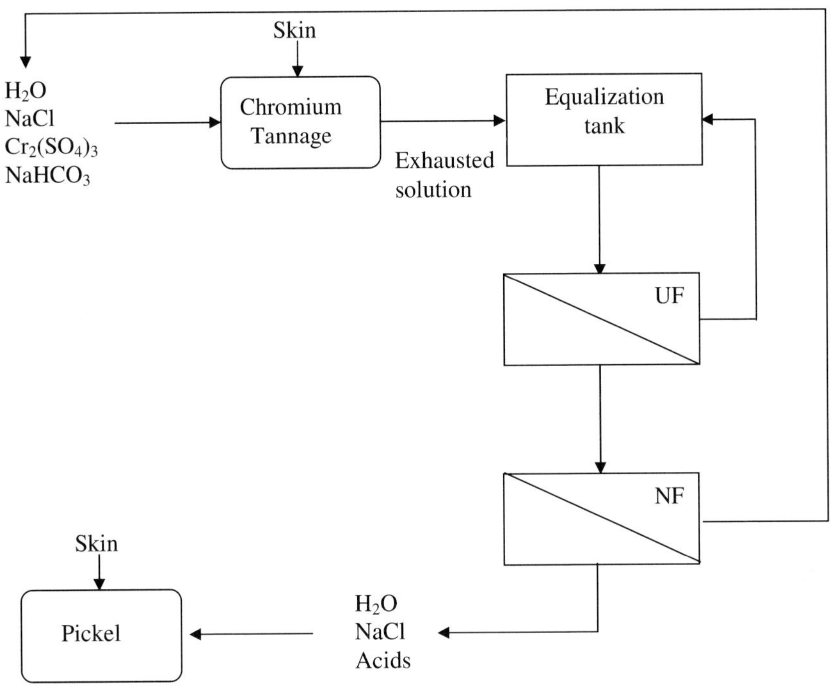

Figure 12.6 Scheme of an integrated UF/NF system for the recovery of chromium from spent tanning effluents [23].

Moreover, MD typically works at temperatures underneath the boiling point of the solution, for this reason it is appropriate for concentrating thermolabile solutes, as in the case of fruit juices.

The driving force, that is, a partial pressure gradient, is obtained by a temperature gradient between the feed and the distillate side. In the osmotic distillation (OD), a partial pressure gradient is activated by a difference in concentration. As a consequence, OD can proceed at ambient temperature and flavor and fragrance compounds can be more conveniently preserved, than in thermally activated concentration processes.

Because of the high content of suspended solids and pectins in fruit juices, the use of a MF or UF pretreatment before the RO unit is also able to reduce the viscosity of the feed stream, increasing the transmembrane flux.

In this logic, a new integrated membrane process for the production of concentrated blood orange juice has been proposed as an alternative to thermal evaporation (Figure 12.7) [26].

Freshly squeezed juice is initially clarified by UF, then it is concentrated by RO up to about 25° Brix and finally by an OD step to a final concentration of about 60° Brix [26].

The comparison in terms of preservation of the total antioxidant activity and of the bioactive antioxidant components of the juice (ascorbic acid, anthocyanins, hydroxycinnamic acids, flavanones) demonstrates that this integrated membrane

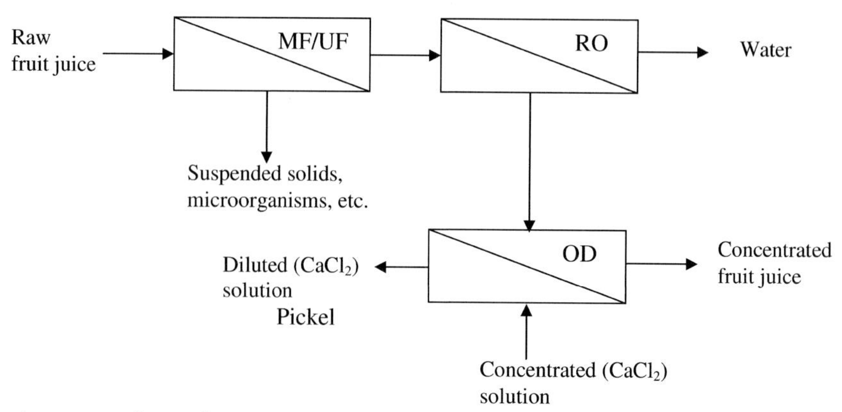

Figure 12.7 Scheme of an integrated membrane process for fruit-juice concentration [26].

process is a valid alternative to the conventional thermal process to obtain high-quality concentrated juice with a reduction in thermal damage and energy consumption.

In the thermally concentrated juice the total antioxidant activity (TAA) is reduced by about −26%; ascorbic acid −30%, anthocyanins −36% with respect to the fresh squeezed juice. In the membrane processes the TAA reduction was about −15, ascorbic acid −15% and anthocyanins −20% [26].

A similar integrated system has been employed to produce highly concentrated black currant juice. The main difference is the use of MF as pretreatment for the disinfection of the raw juice.

The concentrated juice (63–72° Brix) has more than three times higher anthocyanin content, than the raw juice (15–18° Brix) and a good taste [27].

An UF/OD integrated system has been used for the concentration of the cactus pear juice [28]. This juice is characterized by a high micro-organism load and it is typically clarified by thermal treatment (>110 °C) that significantly changes the original color, flavor, and taste.

UF operations are able to produce a juice with a good microbiological stability, preserving organoleptic properties. The clarified juice is then concentrated by OD up to about 60° Brix, maintaining its nutraceutical characteristics [28].

12.5
Integrated Membrane Processes in Chemical Production

Besides previously described examples of integrated membrane systems and much more reported in the literature, including applications in gas separation and the petrochemical industry [29], a special case of integrated or hybrid membrane systems, with a lot of interest in the logic of the *sustainable growth*, is represented by the catalytic membranes reactors (CMRs).

Catalysis is today one of chemistry's most important and powerful technologies. Currently, about 90% of chemical manufacturing processes and more than 20% of all industrial products involve catalytic steps [30]. Catalytic reactions are intensively used in chemical industry, energy conversion, wastewater treatment and in many other processes.

In the perspective to realize a *sustainable growth*, the application of membrane technology in integrated catalytic processes is very promising [31, 32].

Among the different heterogenization strategies, the entrapping of catalysts in membranes or, in general, the use of a catalyst confined by a membrane in the reactor, offers new possibility for the design of new catalytic processes.

Because there are many different ways to combine a catalyst with a membrane, there are numerous possible classifications of the CMRs. However, one of the most useful classifications is based on the role of the membrane in the catalytic process: we have a catalytically active membrane if the membrane has itself catalytic properties (the membrane is functionalized with a catalyst inside or on the surface, or the material used to prepare the membrane is intrinsically catalytic); otherwise if the only function of the membrane is a separation process (retention of the catalyst in reactor and/or removal of products and/or dosing of reagents) we have a catalytically passive membrane. The process carried out with the second type of membrane is also known as membrane-assisted catalysis (a complete description of the different CMRs configurations will be presented in a specific chapter).

CMRs can offer viable solutions to the main drawback of homogeneous catalysis: catalyst recycling. In addition, the membrane can actively take part in the reactive processes by controlling the concentration profiles thanks to the possibility to have membranes with well-defined properties by the modulation of the membrane material and structure.

Moreover, the selective transport properties of the membranes can be used to shift the equilibrium conversion by the removal of one product from the reaction mixture (e.g., hydrogen in dehydrogenation reactions), or to increase the reaction selectivity by the controlled supply of the reagents (e.g., oxygen for partial oxidation reactions).

The membrane can also define the reaction volume, for example, by providing a contacting zone for two immiscible phases, as in phase-transfer catalysis, excluding polluting solvents and reducing the environmental impact of the process.

In numerous cases, membrane-separation processes operate at much lower temperature, especially when compared with thermal processes such as reactive distillation. As a consequence they might provide a solution for the limited thermal stability of either catalyst or products. Furthermore, by membrane-separation processes is possible also to separate nonvolatile components.

The downstream processing of the products can be substantially facilitated when they are removed from the reaction mixture by means of a membrane [31].

In some cases, the heat dissipated in an exothermic reaction can be used in an endothermic reaction taking place at the opposite side of the membrane. Typical examples are hydrogenation/dehydrogenation reactions carried out by palladium or Pd-alloy membranes characterized by a 100% theoretical selectivity towards the hydrogen.

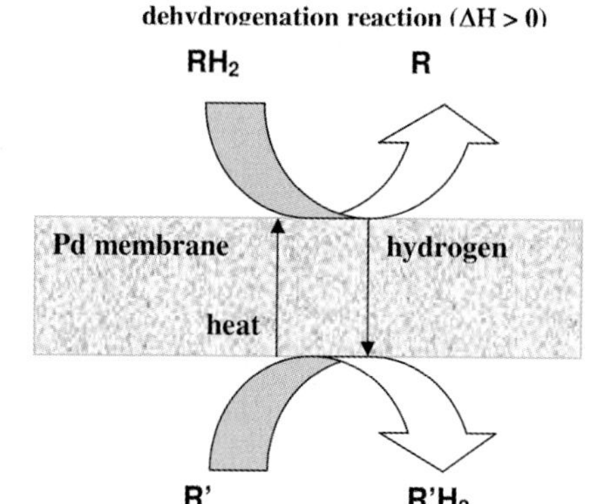

Figure 12.8 Scheme of the combination of a dehydrogenation reaction (endothermic) on one surface of a Pd-based catalytic membrane and a hydrogenation reaction (exothermic) by the diffused hydrogen on the other membrane surface.

Combination of the dehydrogenation on one surface of the Pd-based catalytic active membrane and hydrogenation by the diffused hydrogen on the other surface has been proposed [33]. Thanks to the excellent thermal conductivity of these membranes, the heat released by the hydrogenation can be utilized to drive the endothermic dehydrogenation (Figure 12.8).

The use of a mixed oxygen ion–electronic conductor membrane for oxygen separation with direct reforming of methane, followed by the use of a mixed protonic–electronic membrane conductor for hydrogen extraction has also been proposed in the literature [34]. The products are thus pure hydrogen and synthesis gas with reduced hydrogen content, the latter suitable, for example, in the Fisher–Tropsch synthesis of methanol [34].

Improved selectivity in the liquid-phase oligomerization of *i*-butene by extraction of a primary product (i-octene C_8) in a zeolite membrane reactor (acid resin catalyst bed located on the membrane tube side) with respect to a conventional fixed-bed reactor has been reported [35]. The MFI (silicalite) membrane selectively removes the C_8 product from the reaction environment, thus reducing the formation of other unwanted by-products. Another interesting example is the isobutane (iC4) dehydrogenation carried out in an extractor-type zeolite CMR (including a Pt-based fixed-bed catalyst) in which the removal of the hydrogen allows the equilibrium limitations to be overcome [36].

Catalytic reactions can be combined in membrane-assisted integrated catalytic processes with practically all the membrane unit operations available today. Many examples of integration of membrane contactors, pervaporation, gas separation, nanofiltration, microfiltration, and ultrafiltration operations together with catalytic reactions, have been proposed in the literature.

One of the most investigated fields is the pervaporation-assisted catalysis applied to equilibrium-limited reactions.

PV is today considered as an interesting alternative for the separation of liquid mixtures that are difficult or not possible to separate by conventional distillation methods.

PV-assisted catalysis in comparison with reactive distillation has many advantages: the separation efficiency is not limited by relative volatility as in distillation; in pervaporation only a fraction of the feed is forced to permeate through the membrane and undergoes the liquid- to vapor-phase change and, as a consequence, energy consumption is generally lower compared to distillation.

In the PV-assisted catalysis, pervaporation is usually used to extract continuously one of the formed products in order to improve conversion of the reactants or to increase reaction selectivity.

By far the most studied reactions combined with pervaporation is esterification. It is a typical example of an equilibrium-limited reaction with industrial relevance.

The esterification of acetic acid with ethanol has been investigated using zeolite membranes grown hydrothermally on the surface of a porous cylindrical alumina support (the catalyst used was a cation exchange resin) [37]. The conversion exceeded the equilibrium limit, by the selective removal through the membrane of water and reached to almost 100% within 8 h [37].

Nanofiltration-coupled catalysis was also widely used for catalyst compartmentalization in CMRs. A NF step has been used for the arylation of olefins using as catalyst $Pd(OAc)_2(PPh_3)_2$ with $P(o\text{-tolyl})_3$ as stabilizing agent, obtaining catalyst recycling, preventing metal contamination of the products and increasing reactor productivity [38].

The coupling of photocatalysis and polymeric membranes has been carried out using TiO_2 as photocatalyst compartmentalized in the reactor by a membrane [39]. Various types of commercial membranes (ranging from UF to NF) and reactor configurations have been investigated [39].

The configurations with irradiation of the recirculation tank and catalyst in suspension confined by means of the membrane, has been reported as the more promising for the 4-nitrophenol mineralization [39] The membrane function in this case is the confining of the photocatalyst and maintaining the pollutants in the reaction environment until their complete mineralization.

Photocatalysts have been also successfully heterogenized in polymeric (catalytically active) membranes.

Novel photocatalytic membranes have been prepared by the heterogenization in PVDF membrane of the decatungstate ($W_{10}O_{32}^{4-}$), a polyanionic metal-oxide cluster used as photocatalysts for oxidation reactions.

Decatungstate exhibits especially interesting properties for the photocatalytic detoxification of wastewater since its absorption spectrum ($\lambda_{max} = 324$ nm) partially overlaps the UV solar emission spectrum opening the potential route for an environmentally benign solar-photoassisted application [40].

However, decatungstate has also some relevant limitations: it is characterized by low quantum yields, small surface area, poor selectivity and limited stability at pH higher than 2.5 [41].

Membrane technology could offer interesting possibilities in order to overcome these limitations and to improve the advantages of catalysis mediated by the decatungstate by: the multiturnover recycling associated to heterogeneous supports, the selectivity tuning as a function of the substrate affinity towards the membrane, the effect of the polymeric microenvironment on catalyst stability and activity.

Decatungstate, in the form of a lipophilic tetrabutilamonium salt ($(n\text{-}C_4H_9N)_4W_{10}O_{32}$), has been homogeneously dispersed in porous membranes made of PVDF (PVDF-W10). Solid-state characterization techniques confirmed that catalyst structure and spectroscopic properties of decatungstate have been preserved once immobilized within the membranes [42–44].

These catalytic membranes have been successfully used in a photocatalytic membrane reactor for the mineralization of the phenol, one of the main organic pollutants in wastewater, demonstrating that these catalytic membranes are stable and recyclable [42].

The rate of phenol degradation as a function of catalyst loading in PVDF membranes and the effect of transmembrane pressure that influences the contact time substrate/catalyst, have been investigated.

The rate of phenol degradation catalyzed by decatungstate in homogeneous phase and in heterogeneous phase (PVDF-W10 membrane) was similar; however, when the catalyst is immobilized in the polymeric membranes a higher mineralization degree of the phenol was observed [42].

The high mineralization activity of the PVDF-W10 membrane in comparison to the homogeneous catalyst can be ascribed to the selective absorption of the organic substrate from water on the hydrophobic PVDF polymer membrane that increases the effective phenol concentration around the catalytic sites. Moreover, the polymeric hydrophobic environment protects the decatungstate from the conversion over longer time to a less-reactive isomer that has a maximum absorption at a wavelength of 280 nm.

The possibility of linking a catalyst only on the external surface of PVDF membranes modified using plasma treatments has also been investigated [45, 46] Photocatalytic membranes were prepared from the self-assembly of phosphotungstic acid ($H_3PW_{12}O_{40}$, a polyoxometallate able to promote photo-oxidation reactions or acid-catalyzed reactions) on the surface of PVDF membranes modified by an Ar/NH_3 plasma discharge [45, 46]. This new method is very versatile, and can be easily extended to the heterogenization of other catalysts on membrane surfaces.

Decatungstate has also been heterogenized by phase-inversion techniques in membranes made of Hyflon, an amorphous perfluoropolymer [47]. Because of the low affinity of the tetrabutilamonium salt of decatungstate with this inert polymeric matrix, the formation of irregular catalyst aggregates were observed. However, it was possible to improve the polymer–catalyst affinity by functionalizing the catalyst with a fluorous-tagged decatungstate ($[CF_3(CF_2)_7(CH_2)_3]_3CH_3N)_4W_{10}O_{32}$ (indicated as R_fN_4W10) [47]. In this form, the catalyst was successfully and homogeneously dispersed as spherical clusters of uniform size in Hyflon membranes.

The cationic amphiphile R_fN^+ groups induced the self-assembly of the surfactant-encapsulated clusters (RfN^+ groups capped on $W_{10}O_{32}^{4-}$) that, during membrane formation, caused the formation of supramolecular catalyst assemblies, which were then stabilized within the polymeric matrix. It was therefore possible to tune the decatungstate self-assembling process through the proper choice of the conditions used during membrane preparation.

These Hyflon-based catalytic membranes have been used to catalyze the photo-oxidation of ethylbenzene (neat), showing superior catalytic performance with higher turnover number and better selectivity when compared to homogeneous catalysts [47]. When dispersed in the Hyflon matrix, the efficacy of fluoro-containing decatungstate depends on the specific electrochemical environment of the catalytic sites, the high solubility of O_2 in the Hyflon matrix, and the selectivity of the perfluorinated polymeric material towards the ethylbenzene reagent and the products 2-phenylethanol and acetophenone.

12.6 Conclusions

Membrane unit operations are today largely used in many different applications for their higher efficiency in comparison with traditional separation systems. Moreover the integration of different membrane operations in the same unit, or in combination with conventional ones, offers important benefits in terms of product quality, plant compactness, environmental impact, and energetic aspects.

Microfiltration, ultrafiltration, and nanofiltration are becoming standard in feed pretreatment for water desalination, wastewater treatment and fruit-juice concentration.

The availability of new membrane processes such as membrane contactors and catalytic membrane reactors, the progresses in membrane-fouling control and the development of new membranes with well-controlled structures and properties, are recognized as key factors for the design of alternative production systems.

In any case, a more systematic analysis of all the possible advantages and limitations caused by the introduction of one or more membrane operations in a process, is necessary. This can be realized by considering specific indicators that allow the progress of industrial processes towards sustainability to be quantified and their impact on the environment, economy and society to be measured.

References

1 Koros, W.J. (2007) *Journal of Membrane Science*, **300**, 1.

2 Charpentier, J.-C. (2007) *Industrial & Engineering Chemistry Research*, **46**, 3465–3485.

3 Drioli, E. and Fontananova, E. (2004) *Chemical Engineering Research & Design*, **82** (A12), 1557–1562.

4 Raluy, G., Serra, L. and Uche, J. (2006) *Energy*, **31**, 2025–2036.

5 Frenkel, V. (2008) *International Desalination and Water Reuse Quarterly*, **17**, 47–50.
6 Drioli, E., Curcio, E. and Fontananova, E. (2006) Mass Transfer Operation–Membrane Separations, in Chemical Engineering, in *Encyclopedia of Life Support Systems (EOLSS)*, Developed under the Auspices of the UNESCO (eds J. Bridgwater, M. Molzahn and R. Pohorecki), Eolss Publishers, Oxford, UK, http://www.eolss.net.
7 Pearce, G.K. (2007) *Desalination*, **203**, 286–295.
8 Xu, J., Ruan, G., Chu, X., Yao, Y., Su, B. and Gao, C. (2007) *Desalination*, **207**, 216–226.
9 Drioli, E., Curcio, E., Criscuoli, A. and Di Profio, G. (2004) *Journal of Membrane Science*, **239**, 27–38.
10 Drioli, E., Curcio, E., Di Profio, G., Macedonio, F. and Criscuoli, A. (2006) *Chemical Engineering Research & Design*, **84** (A3), 209–220.
11 El-Zanati, E. and El-Khatib, K.M. (2007) *Desalination*, **205**, 15–25.
12 Drioli, E., Criscuoli, A. and Curcio, E. (2006) *Membrane Contactors: Fundamentals, Applications and Potentialities, Membrane Science and Technology Series*, vol. **11**, Elsevier, Amsterdam, Boston.
13 Curcio, E., Di Profio, G. and Drioli, E. (2002) *Desalination*, **145**, 173–177.
14 Curcio, E., Fontananova, E., Di Profio, G. and Drioli, E. (2006) *The Journal of Physical Chemistry. B*, **110** (25), 12438–12445.
15 van Agtmaal, J., Huiting, H., de Boks, P.A. and Paping, L.L.M.J. (2007) *Desalination*, **205**, 26–37.
16 http://www.evides.nl/industriewater/media/m260.pdf.
17 van der Bruggen, B., Verberk, J.Q.J.C. and Verhack, J. (2004) *Water SA*, **30**, 413–419.
18 Drioli, E. and Giorno, L. (1999) *Biocatalytic Membrane Reactors: Application in Biotechnology and the Pharmaceutical Industry*, Taylor & Francis Publisher, Padstow, UK.
19 Brockmann, M. and Seyfried, C.F. (1997) *Water Science and Technology*, **35** (10), 173–181.
20 him, J.K., Yoo, I.-K. and Lee, Y.M. (2002) *Process Biochemistry*, **38**, 279–285.
21 Vigiano, C. (2007) *La Chimica e l'Industria*, **5**, 90–94.
22 Cassano, A., Molinari, R., Romano, M. and Drioli, E. (2001) *Journal of Membrane Science*, **181**, 111–126.
23 Cassano, A., Della Pietra, L. and Drioli, E. (2007) *Industrial & Engineering Chemistry Research*, **46**, 6825–6830.
24 Rao, M.A., Acree, T.E., Cooley, H.J. and Ennis, R.W. (1987) *Journal of Food Science*, **52**, 375–378.
25 Jiao, B., Cassano, A. and Drioli, E. (2004) *Journal of Food Engineering*, **63**, 303–324.
26 Galaverna, G., Di Silvestro, G., Cassano, A., Sforza, S., Dossena, A., Drioli, E. and Marchelli, R. (2008) *Food Chemistry*, **106**, 1021–1030.
27 Kozàk, A., Bànvölgyi, S., Vincze, I., Kiss, I., Bèkàssy-Molnàr, E. and Vatai, G. (2008) *Chemical Engineering and Processing*, **47**, 1171–1177.
28 Cassano, A., Conidi, C., Timpone, R., D'Avella, M. and Drioli, E. (2007) *Journal of Food Engineering*, **80**, 914–921.
29 Drioli, E. and Curcio, E. (2007) *Journal of Chemical Technology and Biotechnology (Oxford, UK, 1986)*, **82**, 223–227.
30 Zhao, X.S., Bao, X.Y., Guo, W. and Lee, F.Y. (2006) *Materials Today*, **9**, 32–39.
31 Vankelecom, I.F.J. (2002) *Chemical Reviews*, **102**, 3779–3810.
32 Miachon, S. and Dalmon, J.-A. (2004) *Topics in Catalysis*, **29**, 59–65.
33 Gryaznov, V. (1999) *Catalysis Today*, **51**, 391–395.
34 Norby, T. (1999) *Solid State Ionics*, **125**, 1–11.
35 Piera, E., Téllez, C., Coronas, J., Menéndez, M. and Santamaria, J. (2001) *Catalysis Today*, **67**, 127–138.
36 van Dyk, L., Miachon, S., Lorenzen, L., Torres, M., Fiaty, K. and Dalmon, J.-A. (2003) *Catalysis Today*, **82**, 167–177.

37 Tanaka, K., Yoshikawa, R., Ying, C., Kita, H. and Okamoto, K. (2001) *Catalysis Today*, **67**, 121–125.

38 Nair, D., Scarpello, J.T., White, L.S., Freitas dos Santos, L.M., Vankelecom, I.F.J. and Livingston, A.G. (2001) *Tetrahedron Letters*, **42**, 8219–8222.

39 Molinari, R., Palmisano, L., Drioli, E. and Schiavello, M. (2002) *Journal of Membrane Science*, **206**, 399–415.

40 Texier, J., Giannotti, C., Malato, S., Richter, C. and Delaire, J. (1999) *Catalysis Today*, **54**, 297–307.

41 Mylonas, A. and Papaconstantinou, E. (1996) *Polyhedron*, **15**, 3211–3217.

42 Fontananova, E., Drioli, E., Donato, L., Bonchio, M., Carraro, M. and Scorrano, G. (2006) *Chinese Journal of Process Engineering*, **6**, 645–650.

43 Bonchio, M., Carraro, M., Scorrano, G., Fontananova, E. and Drioli, E. (2003) *Advanced Synthesis & Catalysis*, **345**, 1119–1126.

44 Bonchio, M., Carraro, M., Gardan, M., Scorrano, G., Drioli, E. and Fontananova, E. (2006) *Topics in Catalysis*, **40**, 133–140.

45 Fontananova, E., Donato, L., Drioli, E., Lopez, L., Favia, P. and d'Agostino, R. (2006) *Chemistry of Materials*, **18**, 1561–1568.

46 Lopez, L.C., Buonomenna, M.G., Fontananova, E., Iacoviello, G., Drioli, E., d'Agostino, R. and Favia, P. (2006) *Advanced Functional Materials*, **16**, 1417–1424.

47 Carraro, M., Gardan, M., Scorrano, G., Drioli, E., Fontananova, E. and Bonchio, M. (2006) *Chemical Communications*, **43**, 4533–4535.

Part Two
Transformation

This Part will be focused on the fundamentals and applications of membrane-assisted transformation processes, i.e. membrane reactors. Two separate chapters are dedicated to the fundamentals of membrane reactors using traditional chemical catalysts at high temperature (>200 °C) and catalysts of biological origin or biomimetic at low temperature (<100 °C).

Application of membrane reactors for degradation of organic compounds via photocatalysis, treatment of wastewater, production of bioactive high added value compounds, and biomedical treatments are discussed.

The possibility of having membrane systems also as tools for a better design of chemical transformation is today becoming attractive and realistic. Catalytic membranes and membrane reactors are the subject of significant research efforts at both academic and industrial levels. For biological applications, synthetic membranes provide an ideal support to catalyst immobilization due to their biomimic capacity; enzymes are retained in the reaction side, do not pollute the products and can be continuously reused. The catalytic action of enzymes is extremely efficient, selective and highly stereospecific if compared with chemical catalysts; moreover, immobilization procedures have been proven to enhance the enzyme stability. In addition, membrane bioreactors are particularly attractive in terms of eco-compatibility, because they do not require additives, are able to operate at moderate temperature and pressure, and reduce the formation of by-products.

Potential applications have been at the origin of important developments in various technology sectors, including genetic engineering of micro-organisms to produce specific enzymes, techniques for enzyme purification, bioengineering techniques for enzyme immobilization, design of efficient productive cycles.

The development of catalytic membrane reactors for high-temperature applications became realistic only in the last few years with the development of high-temperature-resistant membranes. Due to the generally severe conditions of heterogeneous catalysis, most catalytic membrane reactors applications use inorganic membranes that can be dense or porous, inert or catalytically active. The scientific literature on catalytic membrane reactors is significant today; however, practically no large-scale industrial applications have been reported so far because of the relatively high price of membrane units. However, current and future advancements in material engineering might significantly reverse this trend.

Membrane Operations. Innovative Separations and Transformations. Edited by Enrico Drioli and Lidietta Giorno

ISBN: 978-3-527-32038-7

The good H_2 selectivity and permeability of the last-generation dense (Pd-based) and almost dense SiO_2 membranes have been successfully exploited for a number of H_2-consuming or -generating reactions; for some applications, the thermochemical instability of Pd membranes and the hydrothermal instability of silica remain the main problems to solve. Concerning O_2-generating or -consuming reactions, the development of O_2 permselective membranes with good fluxes in the range of 400–700 °C is still a challenge.

Direct catalytic conversion of natural gas and light alkanes into oxygenates, fuels and higher hydrocarbons is currently one of the most strategic research topics for fundamental and industrial catalysis. The challenge to develop viable processes for the valorization of large reserves of natural gas and light alkanes along with the stringent need to innovate the conventional processes for the catalytic oxidation of hydrocarbons have resulted in a great research effort in this area since the 1970s. In this context, syngas production is a key step for the production of chemicals and fuels from natural gas. The current technology is based on steam or autothermal reforming, the partial oxidation with O_2 being costly owing to the associated air-separation plant; for this application, oxygen-selective membranes integrated with the reactor represent an interesting alternative. Other innovative catalytic membrane reactors have been proposed for the direct conversion of natural gas into higher added value products, such as that based on the adoption of a proton conductor ceramic membrane that ensures H_2 removal in the aromatization of CH_4.

13
Fundamental of Chemical Membrane Reactors

Giuseppe Barbieri and Francesco Scura

13.1
Introduction

Membrane reactors (MRs) are an interesting alternative to traditional reactors (TRs) owing to their characteristic of product separation during the reaction progress. The simultaneous separation shows some advantages related to the process of both permeate and retentate downstreams and on the reaction (rate) itself. In fact, the load of the downstream separation is significantly lower because both (permeate and retentate) streams leaving the MR are concentrated in more and fewer permeable species, respectively. In addition, separation/purification is not required in the special case of pure permeate.

From the reaction point of view, the product removal (1) reduces the flow rate of the reactant stream, in the meantime increasing the residence time; (2) increases the reactant concentration and hence the forward reaction rate; (3) reduces product concentration, reducing the reverse reaction rate. The rate-determining steps of the reaction could change because, even though the species present are the same, they have a very different concentration with respect to a TR. The overall effect is a higher net reaction rate and residence time, both improve the performance (conversion, selectivity, yield, etc.) of the MRs itself. This implies a large reduction of the reaction volume [1, 2] and catalyst amount in the case of catalytic reaction. The improved performance indicates also the possibility of operating in conditions (e.g., of temperature and pressure) better than those used in the present industrial processes. The advantages of MR (volume reduction, improved conversion, two more valuable streams instead of one, etc.) pursue the logic of process intensification [3, 4] (Figure 13.1) today the better strategy for sustainable growth compatible with a high-quality lifestyle, mitigating greenhouse-gas emissions, reducing investment and operating costs of chemical plants. It is a new design philosophy aimed at significant reduction (by a factor of 10, 100 or more) in plant volume also because process equipment, piping, and so on are 20% of the plant costs.

Membrane Operations. Innovative Separations and Transformations. Edited by Enrico Drioli and Lidietta Giorno

ISBN: 978-3-527-32038-7

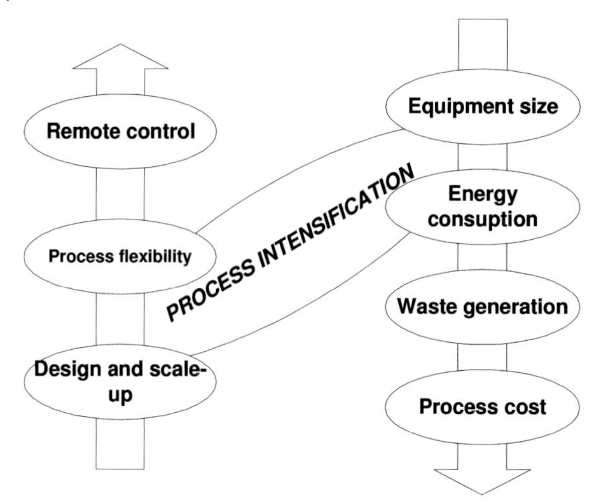

Figure 13.1 Process intensification strategy: the trend required to variables.

Gas-phase reactions and in particular hydrogenation/dehydrogenations (e.g., hydrogen production, upgrade and clean-up) will be considered as an example in this chapter since they have been widely studied [5–34] and some investigations are also in progress both experimentally and by modeling [35] as well as the definition of the upper limit of conversion from a thermodynamic point of view is of specific interest [36, 37]. Porous and dense, ceramic or metallic membranes have only a separation function; they do not have catalytic properties. The scheme of a catalytic MR as shown in Figure 13.2 presents a tube-in-tube configuration, where the inner tube is the selective membrane. The permeating species are collected inside the core of the inner tube. The permeation can be driven by a sweep gas; the use of feed pressure for permeation promoting is better because the permeate stream is concentrated in the permeating species and an extra separation of the sweep component is not required. Both (annular space or tube core) volumes can be used for reaction but the choice depends also on the energy transport [38]. If the reaction is energy intensive for example, methane steam reforming, the annular space gives an overall heat-exchange coefficient higher than that shown by the same MR geometry with the catalyst packed inside the membrane.

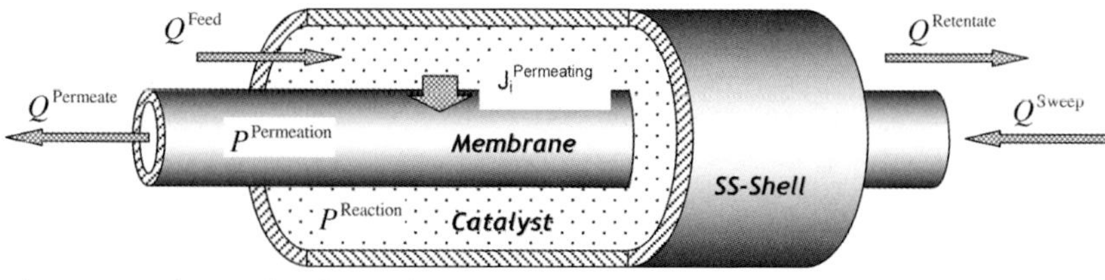

Figure 13.2 Scheme of a tube-in-tube MR.

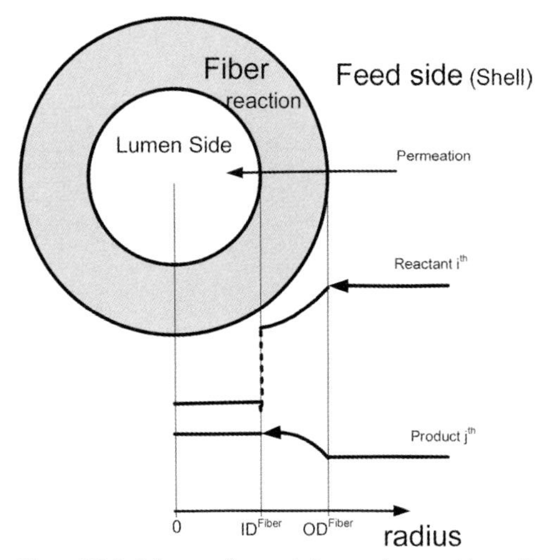

Figure 13.3 Scheme of a catalytic membrane with a cylindrical geometry (tubular or hollow fiber).

Catalytic membranes are also very interesting. The reactants flowing through the membrane pores pass at a short distance from the catalytic sites and the probability of an interaction is much higher (see Figure 13.3). A catalytic membrane leaves very little room to by-pass: the reactant stream has to flow inside the membrane. In addition, very high conversion might also be achieved in the purification process [39] operated by reaction when the component to be removed has a lower concentration with respect to the other desired species.

Table 13.1 reports, as an example, a list of some hydrogenation and dehydrogenation reactions investigated in MRs. It also reports reactions of hydrogen production and its upgrade (MSR, WGS, etc.), because they are widely studied owing to their deep interest in the energy field, hydrogen being one of the most used energy carriers. All these reactions are typically operated at a temperature higher than 200 °C. In addition, reactions carried out using polymeric membranes (at a temperature lower than 100 °C) are cited just as an example of the use of polymeric membranes in fine-chemicals production.

13.2 Membranes

A fundamental innovative aspect of chemical MRs is the separation operated by means of the membrane. In fact, the mass-balance equations have to take into account the mass transfer due to the permeation through the membrane; the other terms being the same present in mass-balance equations of TRs. Figure 13.4 introducing reaction paths for MSR at different values of hydrogen permeance shows a very large difference with the permeance and also with the path of a TR.

Table 13.1 Examples of reactions investigated in catalytic MRs and catalytic membranes.

Reaction	Membrane	
Hydrogenation of ...		
butanes	dense	Pd-Sb
butadiene	dense	Pd
ethylene to ethane	dense; porous	Pd; Al_2O_3
Dehydrogenation of ...		
ethane to ethylene	dense; porous	Pd-Ag; Al_2O_3
ethylbenzene to styrene	dense	Pd-Ag
1-butene to butadiene	dense	Pd
iso-butene	dense	Silica
n-butane	porous	γ-Al_2O_3
methanol	porous	γ-Al_2O_3
Other		
Methane steam reforming, *MSR*	dense	Pd-alloy
Methane dry reforming	dense	Pd- alloy
Partial oxidation of methane, POM	dense	Pd-based, perovskite
Water gas shift, *WGS*	dense	Pd- alloy
CO clean-up	porous	zeolite
Decomposition of H_2S	porous	γ-Al_2O_3
Oxidation of secondary ammine (liquid phase)	porous, dense	polymeric
Photo-oxidation of n-pentanol (liquid phase)	porous, dense	Polymeric

Therefore, the peculiarity of the permeation mechanisms will be discussed because it is introduced later in the balance equations.

Dense and porous, metallic, polymeric, zeolitic, of zirconia, alumina, and so on membranes were used and are in use for MR investigations (Table 13.1). The choice depends on the reaction and operating conditions as well as on the separation desired. The peculiarities of some membranes that are important for the subsequent discussion are recalled here. Pd-based membranes are the most used dense membranes because they have infinite hydrogen selectivity. The permeation follows Sieverts' law (13.1) that identifies the hydrogen diffusion in the metal bulk as the rate-determining step and the difference of the partial pressure of the hydrogen on both membrane sides as the driving force. The permeation is due to several elementary steps [40, 41]. Hydrogen permeance (Figure 13.5) can be reduced by interaction between the membrane surface and some chemical species such as CO [42]. Perovskite membranes transport (13.2) the oxygen [43] through a dense layer of La, Fe, Co, Sr, and so on oxides; the operating temperature is very high (800–900 °C). These membranes were used in partial oxidation reactions [44] where diffused oxygen feed is a significant improvement for driving the reaction selectively towards the desired products. Polymeric dense membranes are less selective than Pd-alloy or perovskitic membranes, the permeation is due to the diffusion (13.3) of the solubilized species in

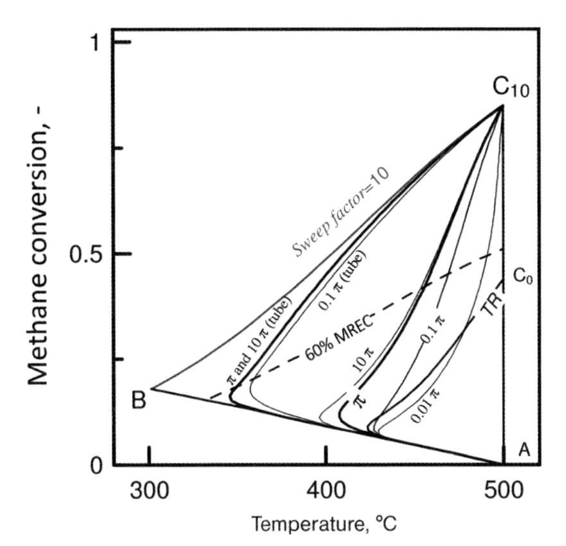

Figure 13.4 Effect of the permeance (ranging from 0.01 to 10 times π, used as a reference value of the permeance) on the MR performance. Two configurations (reaction in the annular space or in the core of the tube) are considered. The curves indicated with π and 10 π (tube) and 0.1 π (tube) are the reaction paths when the catalyst is packed inside the tube and thus the reaction takes place inside the membrane. *TR* is the TR path. The dashed curve represents the conditions of 60% of the MREC; MREC being the curve B-$C_{10}$$T^{Feed} = T^{Sweep} = 500\,°C$, $P^{Reaction} = P^{Permeation} = 100$ kPa, H_2O/CH_4 feed molar ratio = 3, $Q^{Feed}_{CH4} = 400\,cm^3(STP)\,min^{-1}$; $Q^{Sweep} = 10\,Q^{Feed}_{CH4}$, $U^{Shell} = 227\,W\,m^{-2}\,K^{-1}$, $U^{Membrane} = 2.4\,W\,m^{-2}\,K^{-1}$.

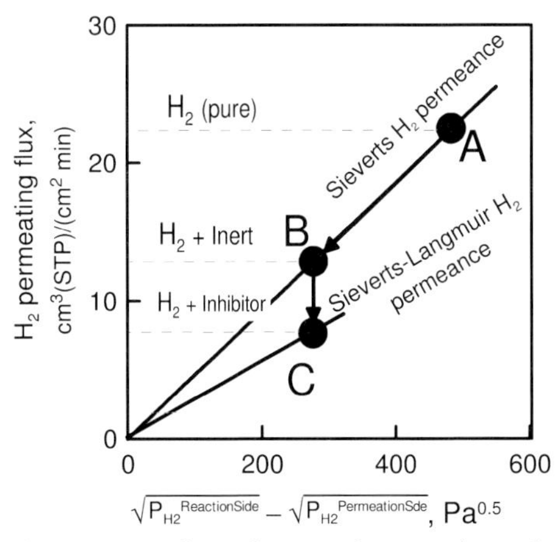

Figure 13.5 H_2 flux as function of *Sieverts*' driving force. Scheme summarizing dilution and inhibition effects owing to inert and inhibitor species present in mixture with hydrogen.

the polymeric bulk. These membranes were also used for catalyst heterogenization for, for example, photo-oxidation and secondary ammine oxidations [45, 46]. Knudsen is a transport mechanism (13.4) of species in porous structures. Zeolitic membranes were used in hydrogen purification of stream containing CO [39].

Sieverts' law

$$J_{H_2}^{\text{Permeating}} = \frac{Q_{H_2} e^{(-Ep/RT)}}{\text{Thickness}} \left(\sqrt{P_{H_2}^{\text{Reaction side}}} - \sqrt{P_{H_2}^{\text{Permeation side}}} \right) \quad (13.1)$$

with CO inhibition effect

$$J_{H_2}^{\text{Permeating}} = \left(1 - \alpha(T) \frac{k_{CO} P_{CO}}{1 + k_{CO} P_{CO}} \right) \frac{Q_{H_2} e^{(-Ep/RT)}}{\text{Thickness}} \times \left(\sqrt{P_{H_2}^{\text{Reaction side}}} - \sqrt{P_{H_2}^{\text{Permeation side}}} \right) \quad (13.2)$$

Perovskite membranes

$$J_{O_2}^{\text{Permeating}} = \frac{k_r}{k_p} \left(\frac{1}{k_{\text{external}}^{\text{Reaction side}}} + \frac{2\,\text{Thickness}}{D_V} + \frac{1}{k_{\text{external}}^{\text{Permeation side}}} \right)^{-1} \times \left(\sqrt{P_{O_2}^{\text{Reaction side}}} - \sqrt{P_{O_2}^{\text{Permeation side}}} \right)$$

Solution-diffusion

$$J_i^{\text{Permeating}} = \frac{\text{Solubility}_i \, \text{Diffusivity}_i}{\text{Thickness}} \left(P_i^{\text{Reaction side}} - P_i^{\text{Permeation side}} \right) \quad (13.3)$$

Knudsen flux

$$J_i^{\text{Permeating}} = d_{\text{pore}} \frac{\varepsilon}{\tau} \sqrt{\frac{1}{3\,RTM_i}} \frac{\left(P_i^{\text{Reaction side}} - P_i^{\text{Permeation side}} \right)}{\text{Thickness}} \quad (13.4)$$

Some of the variables that are important for the subsequent discussion are recalled here. The membrane properties are related to the mass transport of the different chemical species through the membrane itself or its separating layer (for an asymmetric or multilayer membrane). Permeability and selectivity were defined for the mass transport by permeation; both depend on the membrane nature and morphology that impose the specific transport mechanism driving the permeation of which it is characteristic. Table 13.2 reports the permeability coefficient, selectivity and permeating driving force of some permeation mechanisms.

Table 13.2 Permeability, selectivity and permeating driving force of some transport mechanisms.

Transport mechanism		Permeability	Permeation driving force	Selectivity α_{ij}, -
Sievert's	(1)	$Q_0 e^{(-Ep/RT)}$ (mol m^{-1}s^{-1} Pa$^{-0.5}$)	$\left(\sqrt{P_{H_2}^{Reaction}} - \sqrt{P_{H_2}^{Permeate}}\right)$, Pa$^{0.5}$	Infinite to hydrogen
Ion transport	(2)	$\dfrac{\frac{k_r}{k_p}\left(\frac{1}{k_{external}^{Reaction\ side}} + \frac{2\,Thickness}{D_V} + \frac{1}{k_{external}^{Permeation\ side}}\right)^{-1}}{Thickness}$ (mol m^{-1} s^{-1} Pa$^{-0.5}$)	$\sqrt{P_{O_2}^{Reaction\ side}} - \sqrt{P_{O_2}^{Permeation\ side}}$, Pa$^{0.5}$	Infinite to oxygen
Solution-diffusion	(3)	$Solubility_i\ Diffusivity_i$ (mol m^{-1} s^{-1} Pa^{-1})	$(P_i^{Reaction} - P_i^{Permeate})$, Pa	$\dfrac{Solubility_i\ Diffusivity_i}{Solubility_j\ Diffusivity_j}$
Knudsen	(4)	$d_{pore}\dfrac{\varepsilon}{\tau}\sqrt{\dfrac{1}{3RTM_i}}$ (mol m^{-1} s^{-1} Pa^{-1})	$(P_i^{Reaction} - P_i^{Permeate})$, Pa	$\sqrt{\dfrac{M_j}{M_i}}$

The permeability value of the specific species (ith) can be evaluated as the ratio of the permeating flux (experimentally measurable) and the gradient of the permeating driving force:

$$\text{Permeability}_i = \frac{\text{Permeating flux}_i}{\mathrm{d}P_i/\mathrm{d(Thickness)}}, \quad \mathrm{mol\,m^{-1}\,s^{-1}\,Pa^{-1}} \tag{13.5}$$

When membrane thickness is unknown, the permeance, which is the ratio between the permeability and membrane thickness can be used.

$$\text{Permeance}_i = \frac{\text{Permeability}_i}{\text{Thickness}}, \quad \mathrm{mol\,m^{-2}\,s^{-1}\,Pa^{-1}} \tag{13.6}$$

The performance of a specific already produced membrane characterized by a defined (known or unknown) thickness depends on permeance, which can be evaluated by means of a permeation measurement.

The permeability is a property of the membrane material of the separating layer, whereas the permeance is also a property of the 'product' membrane in its entirety considering thickness, the eventual support, defects, and so on. For this reason, the permeance will be used instead of permeability in subsequent mass-balance equations.

The ratio of the permeability (or permeance) of two species (ith and jth) defines the selectivity:

$$\text{Selectivity}_{i,j} = \frac{\text{Permeability}_i}{\text{Permeability}_j} = \frac{\text{Permeance}_i}{\text{Permeance}_j} \tag{13.7}$$

The permeation (mass transport through the membrane) generates a concentration gradient in the orthogonal direction to the membrane surface. Thus, the concentration of the faster permeating species (generally, it is desired in the permeate) reduces, and in the meantime the concentration of the less-permeating compound increases. Thus, the permeation of the desired species is reduced too. This phenomenon, already known as concentration polarization, affects the permeation as well as the system performance. It can be taken into account in the permeance considering the concentration-polarization coefficient [47]. An onerous alternative solution is 2D mathematical models including radial diffusion [35].

13.3 Membrane Reactors

13.3.1 Mass Balance

The mass balances for all the species involved, for both the tube and shell side of the system shown in Figure 13.2 (shell-side feed configuration, system with cylindrical symmetry) can be written as:

Shell side with reaction, species *i*th

$$\frac{\partial c_i^{\text{reaction}}}{\partial t} + \left(\frac{1}{r}\frac{\partial}{\partial r}\left(rN_{i,r}^{\text{reaction}}\right) + \frac{\partial N_{i,z}^{\text{reaction}}}{\partial z}\right) = \sum_{j=1}^{N_{\text{Reactions}}} \upsilon_{i,j} r_j \tag{13.8}$$

Tube-side only permeation, species *i*th

$$\frac{\partial c_i^{\text{permeation}}}{\partial t} + \left(\frac{1}{r}\frac{\partial}{\partial r}\left(rN_{i,r}^{\text{permeation}}\right) + \frac{\partial N_{i,z}^{\text{permeation}}}{\partial z}\right) = 0 \tag{13.9}$$

In particular, for the species A in the case of *Fick* diffusion of binary mixtures [48] the flux is given:

$$\bar{N}_{\text{A}} = x_{\text{A}} c_{\text{total}} \bar{\nu} - c_{\text{total}} D_{\text{AB}} \nabla x_{\text{A}} \tag{13.10}$$

$x_{\text{A}} c_{\text{total}} \bar{\nu}$ being the convective contribute and $(-c_{\text{total}} D_{\text{AB}} \nabla x_{\text{A}})$ the diffusive contribute. In a more general case, such as a multicomponent system, the flux$\bar{N}_i$ could be expressed by means of Maxwell–Stefan equations.

However, also momentum balance (Equation (13.11), [48]) has to be solved for both the shell and tube sides in addition to the mass balances (the molar flux,$\bar{N}_i$, strictly depends on the velocity field, $\bar{\nu}$).

$$\frac{\partial \rho \bar{\nu}}{\partial t} = -[\nabla \cdot \{\rho \bar{\nu} \bar{\nu} + \bar{\bar{\pi}}\}] \tag{13.11}$$

A set of boundary and initial conditions (BCs and ICs) is necessary to solve the system of Equations 13.8, 13.9 and 13.11. The specific contribution of the permeation is expressed by means of a BC related to the membrane surface for both reaction (13.8) and permeation side (13.9). It is equal to the permeating flux.

Reaction side

$$N_{i,r}^{\text{reaction}}\Big|_{r=\frac{OD^{\text{Membrane}}}{2}} = J_i = -\text{Permeance}_i \cdot \text{Driving force}_i \tag{13.12}$$

Permeation side

$$N_{i,r}^{\text{reaction}}\Big|_{r=\frac{ID^{\text{Membrane}}}{2}} = \text{Permeance}_i \cdot \text{Driving force}_i \cdot \left(\frac{\text{OD}^{\text{Membrane}}}{\text{ID}^{\text{Membrane}}}\right) \tag{13.13}$$

A 2D model requires a significant computational effort. If the radial diffusion ($N_{i,r} = 0$) can be considered negligible the balance equations are much simpler: the set of equations is of PDEs, 1D of the second order. Thus, the computational requirement is also reduced.

Shell (reaction zone)

$$\frac{1}{RT^{\text{Reaction}}}\frac{\partial P_i^{\text{Reaction}}}{\partial t} = \frac{1}{RT^{\text{Reaction}}} D_i \frac{\partial^2 P_i^{\text{Reaction}}}{\partial z^2} - \frac{\partial N_i^{\text{Reaction}}}{\partial z} + \sum_{j=1}^{N_{\text{Reactions}}} \upsilon_{i,j} r_j - \frac{A^{\text{Membrane}}}{V^{\text{Reaction}}} J_i^{\text{Permeating}}$$

$$\text{I.C.} \quad P_i^{\text{Reaction}}(z) = P_i^{\text{Reaction, Initial}}$$

$$\text{B.C.} \quad 1;2 \quad P_i^{\text{Reaction}}\Big|_{z=0} = P_i^{\text{Feed}}; \quad \frac{\partial P_i^{\text{Reaction}}}{\partial z}\Big|_{z=L} = 0 \tag{13.14}$$

Tube (permeation zone)

$$\frac{1}{RT^{\text{Permeation}}}\frac{\partial P_i^{\text{Permeation}}}{\partial t} = \frac{1}{RT^{\text{Permeation}}} D_i \frac{\partial^2 P_i^{\text{Permeation}}}{\partial z^2} - \frac{\partial N_i^{\text{Permeation}}}{\partial z} + \frac{A^{\text{Membrane}}}{V^{\text{Permeation}}} J_i^{\text{Permeating}}$$

$$\text{I.C.} \quad P_i^{\text{Permeation}}(z) = P_i^{\text{Permeation, Initial}}$$

$$\text{B.C.}1;2 \quad P_i^{\text{Permeation}}\Big|_{z=0} = P_i^{\text{Sweep}}; \quad \frac{\partial P_i^{\text{Permeation}}}{\partial z}\Big|_{z=L} = 0 \tag{13.15}$$

A 1D model does not require any BC in the radial direction and the term related to the permeation ($J_i^{\text{Permeating}}$) is now directly present in the mass-balance equations. Any 1D model for a tubular MR, not considering any gradient along the permeation direction, could lead to too strong an approximation. In this case, the concentration-polarization coefficient can be introduced as a reducing factor of the permeance and hence the permeating flux is consequently reduced. This solution allows the introduction of the concentration-polarization coefficient for the actual permeance value and by integrating a 1D model good model results to be obtained. The determination/estimation of the concentration-polarization coefficient remains an important task to be solved, in this case. The polarization in the gas-phase reaction is not so important; thus, it can be neglected in most cases. It has to be considered only if the permeation is faster than product formation and its incoming flow [47].

13.3.2 Energy Balance

Usually, heat is developed during chemical reactions, changing the temperature along the reactor. Any temperature variation means variations of kinetics (reactions rate) and, specifically for MRs, permeation. Therefore, the energy balance has to be considered as part of the equation set in addition to the mass balances. Below, the energy balance is written down for a 1D system: the same as used for writing the last mass-balance equations. The following equations contain the heat exchange between the two membrane sides and that transported by the permeated species, in addition to the typical terms of the TRs. The contribution owing to the permeated species is different from zero on the permeate side (it contributes to the increase in the temperature), but is null on the other side, because it leaves the reaction side at the same temperature. Figure 13.6 shows a very different behavior of MRs depending on heat-transfer coefficients owing to different temperature profiles developed inside the MR.

$$\sum_{i}^{Nspecies} C_i Cp_i \frac{\partial T^{\text{Annulus}}}{\partial t} = -\sum_{i=1}^{Nspecies} N_i Cp_i \frac{\partial T^{\text{Annulus}}}{\partial z} + k_z \frac{\partial^2 T^{\text{Annulus}}}{\partial z^2}$$

$$\frac{U^{\text{Shell}} A^{\text{Shell}}}{V^{\text{Annulus}}} (T^{\text{Furnace}} - T^{\text{Annulus}}) - \frac{U^{\text{Membrane}} A^{\text{Membrane}}}{V^{\text{Annulus}}}$$

$$(T^{\text{Annulus}} - T^{\text{Lumen}}) + \Psi + \Phi \frac{A^{\text{Membrane}}}{V^{\text{Annulus}}}$$

$$\text{I.C.} \quad T^{\text{Annulus}}\Big|_{t=0} = T^{\text{Annulus, Initial}}$$

$$\text{B.C.1; 2} \quad T^{\text{Annulus}}\Big|_{z=0} = T^{\text{Feed}} \text{ or } T^{\text{Sweep}}; \quad \left.\frac{\partial T^{\text{Annulus}}}{\partial z}\right|_{z=L} = 0 \tag{13.16}$$

$$\sum_{i}^{Nspecies} C_i Cp_i \frac{\partial T^{\text{Lumen}}}{\partial t} = -\sum_{i=1}^{Nspecies} N_i Cp_i \frac{\partial T^{\text{Lumen}}}{\partial z} + k_z \frac{\partial^2 T^{\text{Lumen}}}{\partial z^2}$$

$$+ \frac{U^{\text{Membrane}} A^{\text{Membrane}}}{V^{\text{Lumen}}} (T^{\text{Annulus}} - T^{\text{Lumen}}) + \Psi + \Phi \frac{A^{\text{Membrane}}}{V^{\text{Lumen}}} \tag{13.17}$$

$$\text{I.C.} \quad T^{\text{Lumen}}\Big|_{t=0} = T^{\text{Lumen, Initial}}$$

$$\text{B.C.1; 2} \quad T^{\text{Lumen}}\Big|_{z=0} = T^{\text{Feed}} \text{ or } T^{\text{Sweep}}; \quad \left.\frac{\partial T^{\text{Lumen}}}{\partial z}\right|_{z=L} = 0$$

Temperature variation owing to enthalpy flux associated with permeation

$$\Phi = \begin{cases} 0 \text{ on reaction side} \\ J_i^{\text{Permeating}} \left(h_i^{T^{\text{Reaction side}}} - h_i^{T^{\text{Permeation side}}}\right) \text{ on permeation side} \end{cases} \tag{13.18}$$

Heat produced by chemical reactions

$$\Psi = \begin{cases} \sum_{j=1}^{N^{\text{Reactions}}} r_j(-\Delta H_j) \text{ on reaction side} \\ 0 \text{ on permeation side} \end{cases} \tag{13.19}$$

Analysis of the effect of permeation, temperature profile and sweep gas will be proposed hereafter considering a steady-state MR modeled by a 1D, first-order model. The model can be extracted from the mass and energy balance, Equations 13.14

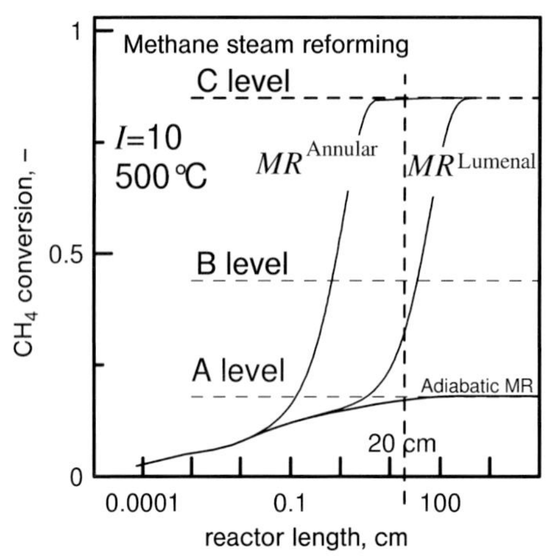

Figure 13.6 Conversion versus reactor length for annular, luminal and adiabatic MRs. Inlet conditions: $p^{\text{Reaction}} = p^{\text{Permeation}} = 100$ kPa, Sweep factor = 10, H_2O/CH_4 feed molar ratio = 3 and $Q_{\text{CH4}}^{\text{Feed}} = 200\ \text{cm}^3(\text{STP})\ \text{min}^{-1}$.

and 13.16, respectively, deleting the transient and second-order terms and the related ICs and BCs.

Figure 13.6 shows the different (higher) conversion reached by an MR with respect to a TR. The MR length for a set conversion value depends strongly on the overall heat-exchange coefficient (annular or luminal MR) for the MSR reaction (highly endothermic). When the catalyst is packed in the annular volume the energy is supplied faster for two reasons: (1) the energy transfer has to pass through the stainless steel shell and both gaseous films adjacent to the shell itself; (2) the thermal resistance is not so high: it depends mainly on gaseous films. On the contrary in the other case (catalyst packed in the core of the tube), the energy required by the reactions also has to cross the membrane. If the membrane is supported on porous alumina, as in this specific case, the overall heat-transfer coefficient covering the supported membrane is very low. The reaction, requiring a lot of energy, cannot proceed as fast as the other (annular) case. In any case, both MR configurations achieved MREC (for a sufficiently long reactor length) and exceed the TR equilibrium conversion (TREC), the maximum values achievable, which is also reported. Only in the case of adiabatic condition does the MR show not a good performance because the temperature profile goes down along the reactor length and the final conversion is that of equilibrium at the corresponding temperature.

The same behavior can be observed in another diagram type (Figure 13.4) in which the methane conversion is plotted against temperature. The annular MR shows, at any point, a temperature always higher than that showed by a tubular MR. This means a higher distance from the MREC (curve B-C_{10}), lower reaction rate and permeance (owing to the temperature), and so on. The same figure also shows the effect owing to the permeation. A significant reduction (100-fold) of permeance (for the annular MR) gives MR behavior close to that of a TR: low permeance leads the MR to approaches a TR. For a tubular MR this effect is not significant because the

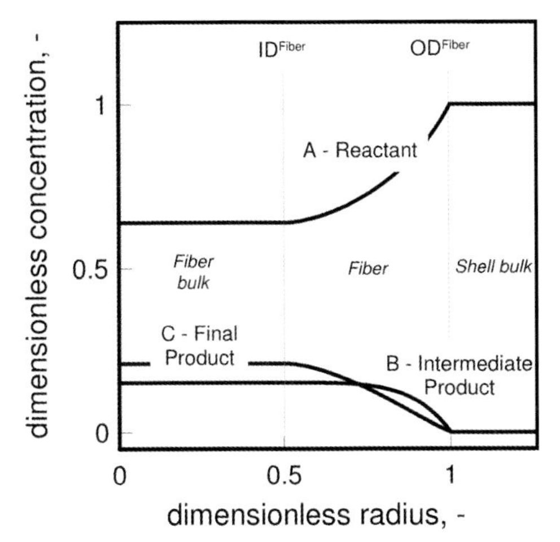

Figure 13.7 Radial concentration profiles for an A → B → C reaction in the case of convection-diffusion-reaction in a catalytic hollow-fiber membrane.

rate-determining step is the energy supply. All the reaction paths reach the MREC value (point C_{10}) at the furnace temperature. The permeance reduction could be due to a different chemical nature or thickness of the separating layer, or surface phenomena such as that discussed with regard to Figure 13.7.

Another fundamental aspect of an MR is related to the permeation driving force. Any system with a permeance value different from zero gives permeation in the desired direction under a suitable driving force. It can be generated by means of an appropriate value of feed pressure or using a sweep gas. Figure 13.8 shows the MREC

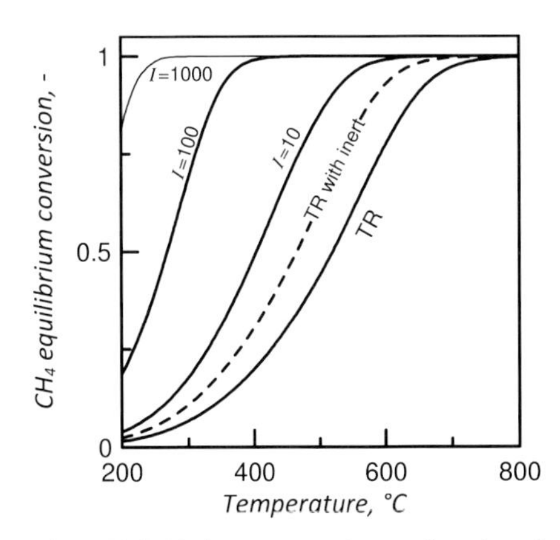

Figure 13.8 Methane conversion as a function of the temperature at several sweep factors (solid lines). H_2O/CH_4 feed molar ratio = 3, $p^{Reaction} = p^{Permeation} = 100$ kPa. TR with inert: equilibrium of a TR when an inert stream with a flow rate equal to 10 times that of methane was added to the feed. The points *BI* and *CI* are the final points of the TR and MRs obtained by simulation in adiabatic, isothermal, and nonisothermal reactors, respectively, with the following conditions: $T^{Reaction} = T^{Permeation} = 500\,^{\circ}C$.

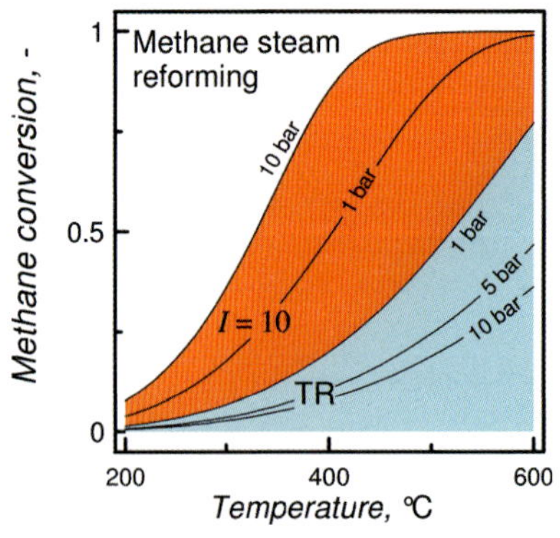

Figure 13.9 MSR reaction. CH_4 equilibrium conversion for both traditional and membrane reactors. I is the ratio of the sweep flow rate to the CH_4 feed flow rate. H_2O/CH_4 feed molar ratio = 3, permeate pressure = 100 kPa.

as a function of the temperature at different values of the sweep factor. This new parameter is the ratio between the sweep-gas flow rate and the reference reactant flow rate; it is defined in an analogous way to the feed molar ratio of a TR. The MREC trend is the same as that of a TR with the temperature: the temperature does not change any functionality of reaction.

On increasing the sweep factor the permeate side has a higher removal capacity and hence the conversion is increased too. If the same sweep flow is fed into a TR together with the reactants the conversion, represented by the dashed lines, even though higher than that of a TR, is significantly lower than that shown by an MR. The effect of the feed pressure is shown in Figures 13.9 and 13.10 for equilibrium conversion. Any pressure increase produces an increase of the MREC. The non-equilibrium conversion goes in the same direction.

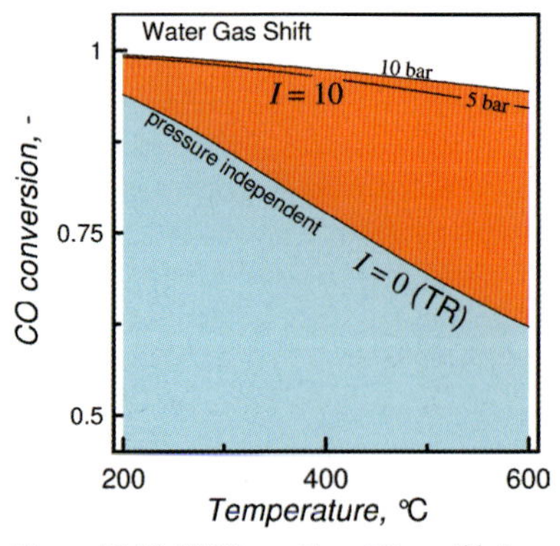

Figure 13.10 WGS reaction. CO equilibrium conversion for both traditional and membrane reactors. I is the ratio of the sweep flow rate to the CO feed flow rate. H_2O/CO feed molar ratio = 1, permeate pressure = 100 kPa.

13.4 Catalytic Membranes

Catalytic membranes are characterized by reaction and permeation at the same point inside the membrane itself. This very interesting case is depicted in Figure 13.3 for a cylindrical (hollow-fiber or tubular) membrane. Any cross-section for the mass transport has a cylindrical shape (circumference in Figure 13.3). Figure 13.3 shows the membrane section along the permeation and reaction direction, the figure also shows an indicative profile of a reactant and product. The mass-transport equations for this system, focusing attention inside the membrane and neglecting the axial profile along the fiber, are:

$$\begin{aligned}
&\varepsilon\frac{\partial c_i}{\partial t} + v_r(r)\frac{\partial c_i}{\partial r} = D_{\text{effective}}\left(\frac{1}{r}\frac{\partial c_i}{\partial r} + \frac{\partial c_i}{\partial r^2}\right) - \varepsilon\sum_{j=1}^{N_{\text{Reactions}}} \upsilon_{i,j} r_j \\
&\text{I.C.} \quad t = 0 \quad c_i = 0; \\
&\text{B.C.1} \quad r = \text{OD}^{\text{Membrane}} \quad c_i = c_i^{\text{Feed}}; \\
&\text{B.C.2} \quad r = \text{ID}^{\text{Membrane}} \quad \frac{\partial c_i}{\partial r} = 0
\end{aligned} \tag{13.20}$$

$$v_r(r) = \frac{Vr}{r}\left(\frac{\text{OD}^{\text{Membrane}}}{2}\right) \tag{13.21}$$

An overall mass balance of the whole system has to be coupled to previous equations for comparison with measurements.

Species-concentration profiles, obtained by integrating Equation (13.20), are plotted in Figure 13.7 for the case of two reactions in series (A → B → C) occurring in a catalytic hollow-fiber membrane. The reactant A contained in the bulk phase on shell-side flow through the membrane where reacting produces the intermediate product B; then, B is converted in the final product C. Variations on concentration profiles are present only inside the fiber, outside the fiber there is no variation due to the reaction. No diffusion limitation in the films were considered in the present model in order to focus on transformation inside the membrane.

13.5 Thermodynamic Equilibrium in Pd-Alloy Membrane Reactor

The product removal from the reaction volume drives the conversion that can exceed that imposed by thermodynamics to the TR (TREC, TR equilibrium conversion). An MR has to respect the thermodynamic law, even if it exceeds the TREC. Therefore, an upper limit to conversion of an MR has to be identified (*MREC, MR equilibrium conversion*). The permeation equilibrium has to be reached in an MR in addition to the reaction equilibrium typical of a TR. This means no further permeation – no net permeating flux through the membrane – and it can be expressed:

$$J_i^{\text{Permeating}} = 0 \Leftrightarrow f\left(P_i^{\text{Reaction}}, P_i^{\text{Permeation}}\right) = 0 \Leftrightarrow P_i^{\text{Reaction}} = P_i^{\text{Permeation}}$$
$$\forall \text{ permeable specie } i \tag{13.22}$$

This relation has to be coupled to the chemical equilibrium law

$$Kp_j(T) = \prod_i^{N_{\text{species}}} P_i^{\nu_{i,j}} \quad \forall \text{ reaction } j \tag{13.23}$$

Since the permeance and permeability are always different from zero, no permeation is equivalent to zero permeation driving force, which occurs when the species partial pressures on both membrane sides are equal to each other. It must be noted that the equilibrium conversion of an MR is independent of the permeation law that expresses the penetrant velocity through the membrane materials.

The *MREC* is a function of the thermodynamic variables (i.e., temperature and pressure) and initial compositions on both sides of the Pd-alloy membranes.

$$\begin{aligned}\text{MREC} &= \text{MR equilibrium conversion}\\ &= f\left(Kp, T^{\text{Reaction}}, P^{\text{Reaction}}, Y_i^{\text{Feed}}, T^{\text{Permeation}}, P^{\text{Permeation}}, F^{\text{Feed}}/F^{\text{Sweep}}, Y_i^{\text{Sweep}}\right)\end{aligned} \tag{13.24}$$

MREC, as the TREC, does not depend on the reaction path. In addition, there is no dependence on the membrane-permeation properties (related to the time required for species permeation).[1] In any case, the final value reached depends on the extractive capacity of the system, for example, the pressure and composition on the permeate side. The composition on the permeate side, similarly to the feed molar ratio, can be expressed by considering the ratio (named sweep factor) between the initial molar number of nonpermeating species (present on the permeate side) and the initial molar number of the reference reactant, for example, methane for methane steam reforming, or carbon monoxide for water gas shift). The sweep factor was defined for a closed MR as:

$$I = \text{Sweep factor} = \left.\frac{n^{\text{Sweep}}}{n^{\text{Feed}}_{\text{Reference component}}}\right|_{\text{Initial time}} \tag{13.25}$$

The sweep factor can be defined in analogous way for an open MR such as a plug-flow, in this case flow rates of feed and sweep streams are used instead of the number of moles:

$$I = \text{Sweep factor} = \frac{F^{\text{Sweep}}}{F^{\text{Feed}}_{\text{Reference component}}} \tag{13.26}$$

The MREC was evaluated for some dehydrogenation reactions (e.g., methane steam reforming and water gas shift) in a Pd-based MR where the membranes are characterized by infinite selectivity towards hydrogen [36, 37]. The significant

1) A very long time, theoretically infinite, is required by a closed TR and MR to reach equilibrium conversion.

importance of the permeation on conversion is shown in Figure 13.9 reporting the MREC in Pd-based MRs for methane steam reforming and water gas shift reactions. MREC functionality with the temperature is the same as that of a TR, but the pressure always has a positive effect, since it drives the permeation. In particular, for methane steam reforming the feed pressure has an opposite effect to that shown by a TR.

13.6 Conclusions

Fundamental aspects of chemical membrane reactors (MRs) were introduced and discussed focusing on the peculiarity of MRs. Removal by membrane permeation is the novel term in the mass balance of these reactors. The permeation through the membrane is responsible for the improved performance of an MR; in fact, higher (net) reaction rates, residence times, and hence improved conversions and selectivity versus the desired product are realized in these advanced systems. The permeation depends on the membranes and the related separation mechanism; thus, some transport mechanisms were recalled in their principal aspects and no deep analysis of these mechanisms was proposed.

Owing to the permeation, the energy transport in MRs requires further consideration on the configuration of an MR to be used, specifically, in energy-intensive reactions for example, methane steam reforming. The energy transport drives to the right MR configuration to be used: the catalyst inside the membrane core or in the annular space. The two volumes have very different heat transfer from the energy source and reaction volume. In addition, a higher conversion of an MR also means a higher energy demand. Therefore, the energy management in an MR is also more important than that in a TR; the temperature being a fundamental variable for the reaction and permeation rate.

The permeation effect is also shown by the suited reduction of the reaction volume (or catalyst amount) (Figure 13.11); in fact, the MR reaction volume is significantly lower than that of a TR.

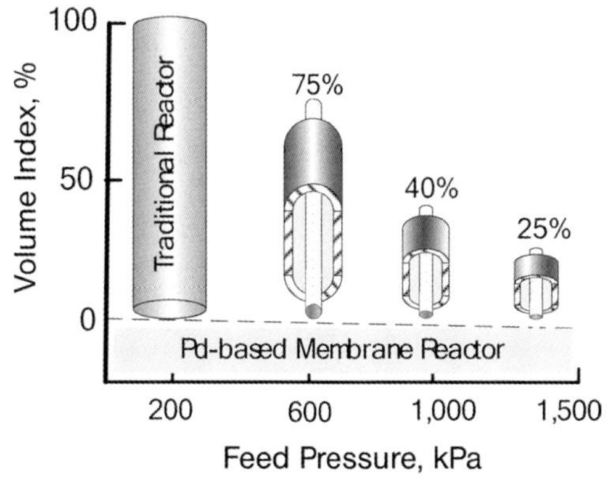

Figure 13.11 Reaction volume (catalyst amount) of an MR with respect to a TR for water gas shift reaction.

List of Symbols

A	Surface area (m^2)
C	Concentration ($mol\,m^{-3}$)
C_p	Specific heat ($J\,mol^{-1}\,K^{-1}$)
d	Diameter (m)
D	Diffusivity ($m^2\,s^{-1}$)
Driving force	Pa ($Pa^{0.5}$)
$\Delta P_{H2}^{Sievert}$	H_2 permeation Sievert's driving force ($Pa^{0.5}$)
E	Activation energy ($J\,mol^{-1}$)
F	Molar flow rate, ($mol\,s^{-1}$)
h	Enthalpy ($J\,mol^{-1}$)
I	Sweep factor (–)
ID	Inner diameter (m)
J	Permeating flux ($mol\,m^{-2}\,s^{-1}$)
K_{eq}	Equilibrium constant (–)
K_p	Equilibrium constant in terms of partial pressures (–)
k_z	Axial thermal conductivity ($J\,m^{-1}\,s^{-1}\,K^{-1}$)
L	Length (m)
n	Number of mole (–)
N	Molar flux ($mol\,m^{-2}\,s^{-1}$)
OD	Outer diameter (m)
P	Pressure (Pa)
Pe_0	Permeability pre-exponential factor ($mol\,m^{-1}\,s^{-1}\,Pa^{-0.5}$)
Permeability	$mol\,m^{-1}\,s^{-1}\,Pa^{-0.5}$
Permeance	$mol\,m^{-2}\,s^{-1}\,Pa^{-0.5}$
Permeating flux	$mol\,m^{-2}\,s^{-1}$
R	Gas law constant ($82.05\,cm^3\,atm\,g\text{-}mol^{-1}\,K^{-1}$)
r	Radial coordinate (m)
r_{ij}	jth reaction rate for ith species ($mol\,m^{-3}\,s^{-1}$)
T	Temperature (°C or K)
t	Time (s)
U	Overall heat-transfer coefficient ($W\,m^{-2}\,K^{-1}$)
V	Volume (m^3)
VI	Volume index (–)
X	Conversion (–)
Y_i	Molar fraction of the species ith (–)
$\bar{Y}^j$	Molar fraction vector of the jth stream (–)
z	Axial coordinate (m)

Greek Letters

Φ	Enthalpy flux associated to hydrogen permeation ($J\,m^{-2}\,s^{-1}$)
Π	Permeance ($mol\,m^{-2}\,s^{-1}\,Pa^{-0.5}$)

$\bar{\bar{\pi}}$	Pressure matrix
Ψ	Heat generated by chemical reactions, $\mathrm{J\,m^{-3}\,s^{-1}}$
δ	Membrane thickness (m)
ε	Porosity (–)
$\nu_{I,\varphi}$	Stoichiometric coefficient with respect to the reference component of *i*th species in *j*th reaction (–)
ρ	Density ($\mathrm{g\,m^{-3}}$)
τ	Space time, s and tortuosity (–)

Superscripts

Annulus	Annulus side in a luminal (tubular) MR
Feed	Membrane module inlet stream referred to
Lumen	Lumen side in a tubular MR
Membrane	Membrane phase referred to
Permeate	Membrane module permeate stream referred to
Permeating	Membrane module permeating stream referred to
Permeation, permeation side	Membrane module stream on the permeation volume referred to
Reaction, Reaction side	Membrane module stream on the reaction volume referred to
Shell	Membrane module shell side referred to
Sweep	Membrane module inlet stream on permeate side referred to

Acronyms

B.C.	Boundary condition
I.C.	Initial condition
MR	Membrane reactor
MREC	Membrane reactor equilibrium conversion
MSR	Methane steam reforming
PDE	Partial differential equation
TR	Traditional reactor
TREC	Traditional reactor equilibrium conversion
WGS	Water gas shift

Acknowledgments

The *Italian Ministry of Education, University and Research,* Progetto ‘FIRB–CAMERE RBNE03JCR5 – Nuove membrane catalitiche e reattori catalitici a membrana per reazioni selettive come sistemi avanzati per uno sviluppo sostenibile' is gratefully acknowledged for cofunding this work.

References

1 Brunetti, A., Caravella, C., Barbieri, G. and Drioli, E. (2007) Simulation study of water gas shift in a membrane reactor. *Journal of Membrane Science*, **306** (1–2), 329–340.

2 Barbieri, G., Brunetti, A., Tricoli, G. and Drioli, E. (2008) An innovative configuration of a Pd-based membrane reactor for the production of pure hydrogen. Experimental analysis of water gas shift. *Journal of Power Sources*, **182** (1), 160–167, http://dx.doi.org/10.1016/j.jpowsour.2008.03.086.

3 Stankiewicz, A. and Moulijn, J.A. (2002) Process intensification. *Industrial & Engineering Chemistry Research*, **41**, 1920.

4 Drioli, E. and Romano, M. (2001) Progress and new perspectives on integrated membrane operations for sustainable industrial growth. *Industrial & Engineering Chemistry Research*, **40**, 1277–1300.

5 Brunetti, A., Barbieri, G. and Drioli, E. (2008) A PEM-FC and H_2 membrane purification integrated plant. *Chemical Engineering and Processing: Process Intensification*, **47** (7), 1081–1089, http://ds.doi.org/10.1016/j.cep.2007.03.015, special issue 'Euromembrane 2006'.

6 Kikuchi, E., Uemiya, S., Sato, N., Inoue, H., Ando, H. and Matsuda, T. (1989) Membrane reactor using microporous glass-supported thin film of palladium. Application to the water–gas shift reaction. *Chemistry Letters*, 489.

7 Uemiya, S., Sato, N., Inoue, H., Ando, H. and Kikuchi, E. (1991) The water–gas shift reaction assisted by palladium membrane reactor. *Industrial & Engineering Chemistry Research*, **30**, 585.

8 Criscuoli, A., Basile, A. and Drioli, E. (2000) An analysis of the performance of membrane reactors for the water–gas shift reaction using gas feed mixtures. *Catalysis Today*, **56**, 53.

9 Barbieri, G. and Bernardo, P. (2004) Experimental evaluation of hydrogen production by membrane reaction, in *Carbon Dioxide Capture for Storage in Deep Geologic Formations – Results from the CO_2 Capture Project*, vol. **1**, Elsevier, pp. 385–408, Chapter 22.

10 Onstot, W.J., Minet, R.G. and Tsotsis, T.T. (2001) Design aspects of membrane reactors for dry reforming of methane for the production of hydrogen. *Industrial & Engineering Chemistry Research*, **40**, 242–251.

11 Sjardin, M., Damen, K.J. and Faaij, A.P.C. (2006) Techno-economic prospects of small-scale membrane reactors in a future hydrogen-fuelled transportation sector. *Energy*, **31** (14), 2187–2219.

12 Bracht, M., Alderliesten, P.T., Kloster, R., Pruschek, R., Haupt, G., Xue, E., Ross, J.R.H., Koukou, M.K. and Papayannakos, N. (1997) Water gas shift membrane reactor for CO_2 control in IGCC systems: Techno-economic feasibility study. *Energy Conversion and Management*, **38** (9999), S159–S164.

13 Yeung, K.L., Zhang, X., Lau, W.N. and Martin-Aranda, R. (2005) Experiments and modeling of membrane microreactors. *Catalysis Today*, **110** (1–2), 26–37.

14 Dudukovic, M.P. (1999) Trends in catalytic reaction engineering. *Catalysis Today*, **48** (1–4), 5–15.

15 Sun, J., Desjardins, J., Buglass, J. and Liu, K. (2005) Noble metal water gas shift catalysis: Kinetics study and reactor design. *International Journal of Hydrogen Energy*, **30** (11), 1259–1264.

16 Majumder, D. and Broadbelt, L.J. (2006) A multiscale scheme for modeling catalytic

flow reactors. *AIChE Journal*, **52** (12), 4214–4228.

17 Collins, J.P., Way, J.D. and Kraisuwansarn, N. (1993) A mathematical model of a catalytic membrane reactor for the decomposition of NH_3. *Journal of Membrane Science*, **77** (2–3), 265–282.

18 Chmielewski, D., Ziaka, Z. and Manousiouthakis, V. (1999) Conversion targets for plug flow membrane reactors. *Chemical Engineering Science*, **54**, 2979–2984.

19 Belgued, M., Pareja, P., Amariglio, A. and Amariglio, H. (1991) Conversion of methane into higher hydrocarbons on platinum. *Nature*, **352**, 789–790.

20 Koerst, T., Deelen, M.J.G. and Van Santen, R.A. (1992) Hydrocarbon formation from methane by a low-temperature two-step reaction sequence. *Journal of Catalysis*, **138**, 101.

21 Gryaznov, V. (1999) Membrane catalysis. *Catalysis Today*, **51**, 391–395.

22 Armor, J.N. (1998) Applications of catalytic inorganic membrane reactors to refinery products. *Journal of Membrane Science*, **147**, 217–233.

23 Coronas, J. and Santamaria, J. (1999) Catalytic reactors based on porous ceramic membranes. *Catalysis Today*, **51**, 377–389.

24 Sirkar, K.K., Shanbhag, P.V. and Kovvali, A.S. (1999) Membrane in a reactor: a functional perspective. *Industrial & Engineering Chemistry Research*, **38**, 3715–3727.

25 Drioli, E., Basile, A. and Criscuoli, A. (2000) High temperature membrane reactors for clean productions. *Clean Products and Processes*, **2**, 179–186.

26 Julbe, A., Farrusseng, D. and Guizard, C. (2001) Porous ceramic membranes for catalytic reactors - overview and new ideas. *Journal of Membrane Science*, **181**, 3–20.

27 Tosti, S., Bettinali, L. and Violante, V. (2000) Rolled thin Pd and Pd–Ag membranes for hydrogen separation and production. *International Journal of Hydrogen Energy*, **25**, 319–325.

28 Belgued, M., Amariglio, A., Pareja, P. and Amariglio, H. (1996) Oxygen-free conversion of methane to higher alkanes through an isothermal two-step reaction on ruthenium. *Journal of Catalysis*, **161**, 282–291.

29 She, Y., Dardas, Z., Gummalla, M., Vanderspurt, T. and Emerson, S. (2005) Integrated water gas shift (WGS) Pd membrane reactors for compact hydrogen production systems from reforming of fossil fuels. *ACS Division of Fuel Chemistry, Preprints*, **50** (2), 561.

30 Xu, Z.-Q., Chen, Q.-L. and Lu, G.-Z. (1999) The use of membrane reactor in the reaction of ethylbenzene dehydrogenation to styrene. *Petrochemical Technology*, **28** (6), 362.

31 Xu, Z., Chen, Q. and Lu, G. (1999) Use of membrane reactor in reaction of ethylbenzene dehydrogenation to styrene. *Shiyou Huagong/Petrochemical Technology*, **28** (6), 358–362.

32 Assaf, E.M., Jesus, C.D.F. and Assaf, J.M. (1998) Mathematical modelling of methane steam reforming in a membrane reactor: An isothermic model. *Brazilian Journal of Chemical Engineering*, **15** (2), 160–166.

33 Lin, Y.-M., Liu, S.-L., Chuang, C.-H. and Chu, Y.-T. (2003) Effect of incipient removal of hydrogen through palladium membrane on the conversion of methane steam reforming: Experimental and modelling. *Catalysis Today*, **82** (1–4), 127–139.

34 Capobianco, L., Del Prete, Z., Schiavetti, P. and Violante, V. (2006) Theoretical analysis of a pure hydrogen production separation plant for fuel cells dynamical applications. *International Journal of Hydrogen Energy*, **31**, 1079–1090.

35 Barbieri, G., Scura, F. and Brunetti, A. (2008) Mathematical modelling of Pd-alloy membrane reactors, in *Inorganic Membranes: Synthesis, Characterization and Applications*, vol. **13** (eds R. Mallada and M. Menendez), Membrane Science and Technology, Elsevier BV, Chapter 9 (ISSN

0927-5193, 10.1016/S0927-5193(07)13009-6).

36 Marigliano, G., Barbieri, G. and Drioli, E. (2003) Equilibrium conversion for a Pd-alloy membrane reactor. Dependence on the temperature and pressure. *Chemical Engineering and Processing*, **42** (3), 231–236.

37 Barbieri, G., Scura, F. and Drioli, E. (2006) Equilibrium of a Pd-alloy membrane reactor. *Desalination*, **200** (1–3), 679–680.

38 Marigliano, G., Barbieri, G. and Drioli, E. (2001) Effect of energy transport in a palladium based membrane reactor for methane steam reforming process. *Catalysis Today*, **67** (1–3), 85–99.

39 Bernardo, P., Algieri, C., Barbieri, G. and Drioli, E. (2008) Hydrogen purification from carbon monoxide by means of selective oxidation using zeolite membranes. *Separation and Purification Technology*, **62**, 631–637.

40 Ward, T.L. and Dao, T. (1999) Model of hydrogen permeation behavior in palladium membranes. *Journal of Membrane Science*, **153**, 211–231.

41 Caravella, A., Barbieri, G. and Drioli, E. (2008) Modelling and simulation of hydrogen permeation through supported Pd-alloy membranes with a multicomponent approach. *Chemical Engineering Science*, **63** (8), 2149–2160, 10.1016/j.ces.2008.01.009.

42 Barbieri, G., Scura, F., Lentini, F., De Luca, G. and Drioli, E. (2008) A novel model equation for the permeation of hydrogen in mixture with carbon monoxide through Pd-Ag membranes. *Separation and Purification Technology*, **61** (2), 217–224, 10.1016/j.seppur.2007.10.010.

43 Xu, S.J. and Thomson, W.J. (1999) Oxygen permeation rates through ion-conducting perovskite membranes. *Chemical Engineering Science*, **54**, 3839–3850.

44 Wang, h., Tablet, C., Schiestel, T., Werth, S. and Caro, J. (2006) Partial oxidation of methane to syngas in perovskite hollow fiber membrane reactor. *Catalysis Communications*, **7**, 907–912.

45 Buonomennna, M.G., Lopea, L.C., Barbieri, G., Favia, P., d'Agostino, R. and Drioli, E. (2007) Sodium tungstate immobilized on plasma-treated PVDF membranes: new efficient heterogeneous catalyst for oxidation of secondary amines to nitrones. *Journal of Molecular Catalysis. A, Chemical*, **273**, 32–38, 10.1016/j.molcata.2007.03.065.

46 Bonchio, M., Carraro, M., Gardan, M., Scorrano, G., Drioli, E. and Fontananova, E. (2006) Hybrid photocatalytic membranes embedding decatungstate for heterogeneous photooxygenation. *Topics in Catalysis*, **40** (1–4), 133–140.

47 Caravella, A., Barbieri, G. and Drioli, E., (2009) Concentration polarization analysis in self-supported Pd-based membranes. *Separation and Purification Technology*, in press, http://dx.doi.org/10.1016/j.seppur.2009.01.008.

48 Bird, R.B., Stewart, W.E. and Lightfoot, E.N. (1979) *Fenomeni di Trasporto*, Casa Editrice Ambrosiana, Milano.

14
Mathematical Modeling of Biochemical Membrane Reactors

Endre Nagy

14.1
Introduction

Membrane bioreactor (MBR) technology is advancing rapidly around the world both in research and commercial applications [1–4]. Integrating the properties of membranes with biological catalyst such as cells or enzymes forms the basis of an important new technology called membrane bioreactors. The membrane layer is especially useful for immobilizing whole cells (bacteria, yeast, mammalian, and plant cells) [5, 6], bioactive molecules such as enzymes [7–9] to produce a wide variety of chemicals and substances. The MBR were introduced over 30 years ago and until now they are recommended or applied for production of foods, biofuels, plant metabolites, amino acids, antibiotics, anti-inflammatories, anticancer drugs, vitamins, proteins, optically pure enantiomers, isomers, fine chemicals, as well as for treatment of wastewater (e.g., industrial, domestic, and municipal [2, 10]). Our work is focused primarily on the hollow-fiber bioreactor with biocatalyst, either live cells or enzymes, inoculated into the shell or immobilized within the membrane matrix or in a thin layer at the membrane matrix-shell interface. Membrane bioreactors for immobilized whole cells can provide a suitable environment for high cell densities [8, 11, 12]. Cells are either grown in the extracapillary space with medium flow through the fibers and supplied with oxygen and nutrients, or grown within the fibers with medium flow outside or across the fibers, while wastes and desired products are removed. The main advantages of the hollow-fiber bioreactor are the large specific surface area (internal and external surface of the membrane) for cell adhesion or enzyme immobilization; the ability to grow cells to high density; the possibility for simultaneous reaction and separation; relatively short diffusion path in the membrane layer; the presence of convective velocity through the membrane if necessary in order to avoid the nutrient limitation [13, 14].

The performance of a hollow-fiber or sheet bioreactor is primarily determined by the momentum and mass-transport rate [15, 16] of the key nutrients through the biocatalytic membrane layer. Thus, the operating conditions (transmembrane pressure, feed velocity), the physical properties of membrane (porosity, wall thickness, lumen radius, matrix structure, etc.) can considerably influence the performance of a bioreactor, the

Membrane Operations. Innovative Separations and Transformations. Edited by Enrico Drioli and Lidietta Giorno

ISBN: 978-3-527-32038-7

effectiveness of the reaction. The main technological difficulties in using membrane bioreactors on an industrial level are related to rate-limiting aspects and scale-up difficulties of this technology. The limited transport of nutrients can cause serious damage in production [12, 14]. The introduction of convective transport is crucial in overcoming diffusive mass-transport limitation of nutrients [17] especially of the sparingly soluble oxygen. The mathematical description of the transport processes enables us to predict the concentration distributions of nutrients in the catalyst membrane layer, and thus, it makes it possible to choose the correct operating conditions that provide a sufficient level of nutrient concentration in the membrane layer. The main aim of this study to give closed, as simple as possible, mathematical equations in order to predict the concentration distribution and the mass-transfer rate through a biocatalyst membrane layer as well as the concentration variation in the lumen (or shell) side of a capillary membrane with particular regard to the variable transport parameters (diffusion coefficient, convective velocity, reaction rate constant) due to the anisotropy of the membrane and/or cell colony in or around the membrane [13, 14].

14.2
Membrane Bioreactors with Membrane as Bioreactor

Membrane bioreactors have been reviewed previously in every detail [3, 4, 7, 8, 18]. There are two main types of membrane bioreactors: (i) the system consists of a traditional stirred-tank reactor combined with a membrane separation unit (Figure 14.1); (ii) the membrane contains the immobilized biocatalysts such as enzymes, micro-organisms and antibodies and thus, acts as a support and a separation unit (Figure 14.2). The biocatalyst can be immobilized in or on the membrane by entrapment, gelification, physical adsorption, ionic binding, covalent binding or crosslinking [3, 7, 18]. Our attention will be primarily focused on the second case where the membrane acts as a support for biocatalyst and as a separation unit, in this study. The momentum and mass-transport process, in principle, are the same in both cases, namely when there is

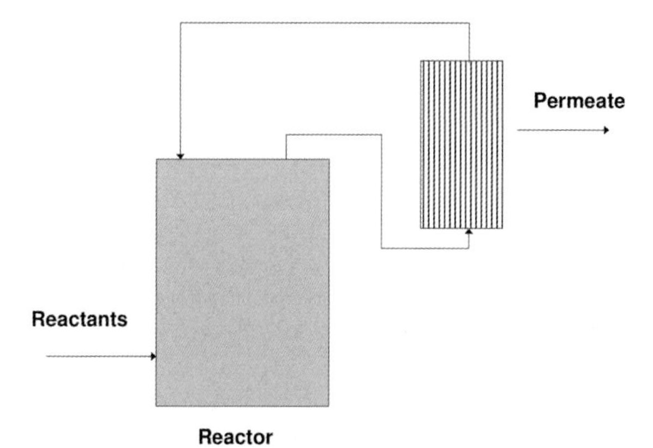

Figure 14.1 Schematic illustration of the external membrane bioreactor configuration.

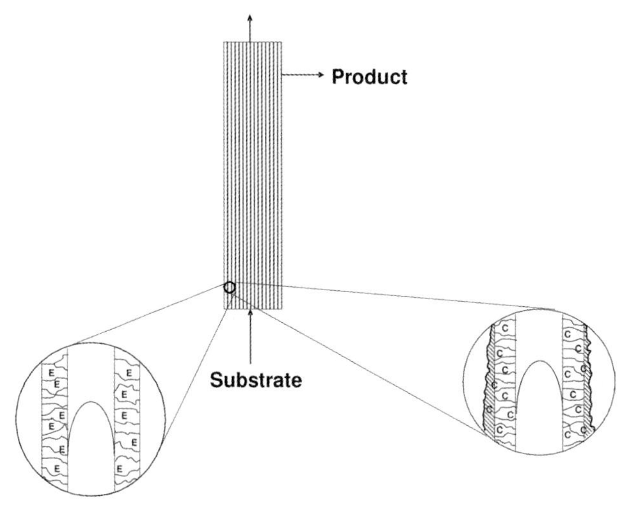

Figure 14.2 Membrane bioreactor with immobilized biocatalysts (enzyme or micro-organism).

no biochemical reaction in the membrane (Figure 14.1) or when a biochemical reaction occurs in the membrane due to its catalytic activity (Figure 14.2).

14.2.1 Enzyme Membrane Reactor

The enzymatic bioconversion processes are of increasing use in the production, transformation and valorization of raw materials [7]. Important applications have been developed in the field of food industries, fine chemicals or even for environmental purposes. Several important applications of the enzyme membrane reactor are given in Charosset's [8] and Rios's papers [7] as well as an excellent summary given by Giorno and Drioli [3] and Miguel *et al.* [9]. They discussed the different enzyme–membrane configurations (recycle-, dialysis-, diffusion-, multiphase membrane reactors) and the different types of enzyme retention (trapping, chemical coupling, adsorption electrostatic interactions, etc.). The enzyme membrane reactor has several advantages (continuous mode, retention and reuse of catalyst, reduction in substrate/product inhibition, enzyme-free product, integrated process, etc.) and also some disadvantages (decreasing enzyme activity as a function of time, membrane fouling, low substrate concentration, etc.). The enzyme can be immobilized into the membrane matrix or on the membrane interface. In this latter case, the membrane surface is covered by a gel layer and the enzyme binds to this layer [19, 20]. In this case the substrate solution has direct contact with the catalyst, while in the other case, namely when the enzyme is placed in the membrane matrix, the substrate has to flow through the membrane layer to come into contact with the biocatalyst. Concerning the mathematical description of these enzymatic processes, basically two different cases can be distinguished, namely the enzyme or live cells are immobilized onto or in the membrane layer or the biocatalyst is dissolved homogeneously or mixed heterogeneously (it is immobilized in porous particles) in the feed phase. The description of these two biochemical processes is different (discussed later). The structure of the membrane matrix does not practically change during the biocatalytic process, thus, the transport parameters remain constant. Obviously, due to the fouling or cake forming, if it is the case, the increase of the external mass-transfer resistance can alter parameters such as convective velocity, external mass-transfer coefficient. This change has to be taken into account to describe the transport process.

14.2.2 Whole-Cell Membrane Bioreactor

Membrane bioreactors for immobilized whole cells [6, 12, 21] provide an advantages environment for increased cell densities. The cells are perfused via a membrane with a steady continuous flow of medium containing the oxygen and other nutrients. The cells are either grown in the extracapillary space (to form a biofilm), or grown within the fibers (Figure 14.2). It was shown that a mass-transfer limitation for oxygen or other nutrient could occur, especially at higher cell density [5, 14, 21]. Due to the change of the nutrient concentration in the axial direction, the density of the cell culture, the thickness of the biofilm on the membrane interface can also change. This fact can alter the values of transport parameters (diffusion coefficient, convective velocity, it can even alter the biochemical reaction rate (the consumption rate of nutrient). Theoretical studies also confirm that nutrient limitation can often occur in hollow-fiber biocatalytic membranes [10, 13, 16]. To avoid this limitation the mass-transfer rate through the cell culture should be increased. This can be achieved by a construction change, by a change of the membrane structure, thickness or for example, by the increase of the transmembrane pressure. Due to this, the radial convective velocity also increases.

Applications of whole-cell biocatalytic membrane reactors, in the agro-food industry and in pharmaceutical and biomedical treatments are listed by Giorno and Drioli [3]. Frazeres and Cabral [9] have reviewed the most important applications of enzyme membrane reactors such as hydrolysis of macromolecules, biotransformation of lipids, reactions with cofactors, synthesis of peptides, optical resolution of amino acids. Another widespread application of the membrane bioreactor is the wastewater treatment will be discussed in a separate section.

14.3 Membrane Bioreactors with Membrane as Separation Unit

14.3.1 Moving-Bed Biofilm Membrane reactor

The basic advantage of the suspended biofilm membrane reactors over the suspended biomass system (either with dispersed cells or with flocs) is that the former are able to retain much more biomass [22]. It substantially reduces the biomass wash out and allows a more stable operation with higher biomass concentration. A moving-bed-biofilm membrane reactor is illustrated in Figure 14.5. The biomass is immobilized inside and outside of the fluidized particles. The structure of the biofilm continuously changes during the growing period of the micro-organism. In this study we do not analyze in detail the mass transfer into the suspended particles. The general equations are known and they are excellently summarized by Melo and Oliviera [22].

14.3.2
Wastewater Treatment by Whole-Cell Membrane Reactor

In this case, a biological step and a membrane module are integrated where both of them have specific functions [22]: (i) biological degradation of organic pollutant is carried out in a traditional bioreactor by micro-organisms; (ii) separation of micro-organisms from the treated wastewater is performed by the membrane module. The membrane enables recycling of the activated sludge to the bioreactor as well as the production of cleaned water. There are here also two types of configurations for the membrane array in the wastewater treatment: the membrane can be placed outside (Figure 14.1) or inside of the bioreactor (Figure 14.3). For the external configuration, the treated liquid is filtered under pressure in a specific membrane module, whereas for the submerged configuration, the filtration is carried out under vacuum. The latter configuration seems to be more economical based on energy consumption: a recycle pump is not needed since the aeration generates a tangential flow around the submerged membrane fibers [22]. The cells here are not immobilized in/or on the membrane layer, but biofilm can be formed on the membrane interface causing mass-transport difficulties through the membrane. The transport through the membrane is simple filtration that can be altered by fouling and/or the biofilm formed during the treatment.

The wastewater treatment is widely applied for industrial (e.g., food [23], beverage [24], dairy industry [25], municipal [26, 27]) as well as domestic wastewater.

14.3.3
Membrane Fouling

As the liquid passes through the membrane in crossflow filtration, the particles, macromolecules, colloids, and so on, rejected by the membrane will accumulate in

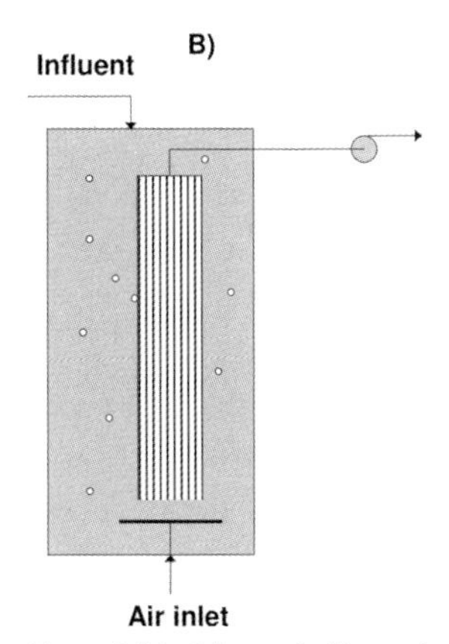

Figure 14.3 Schematic illustration of immersed membrane bioreactor configurations.

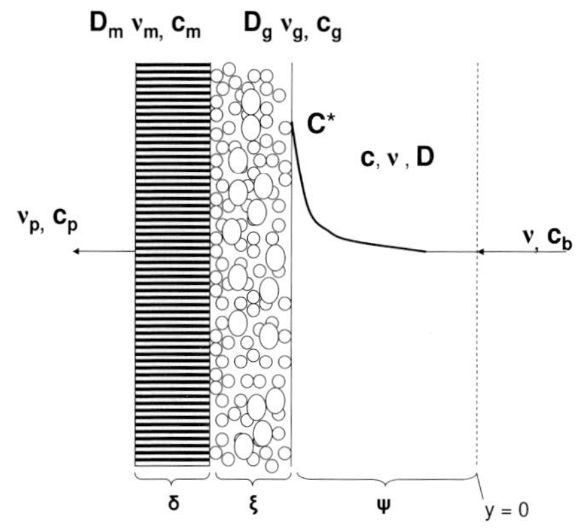

Figure 14.4 Schematic of the membrane, cake (gel) and concentration-polarization layer with their parameters.

the immediate vicinity of the membrane surface to form a dense layer (cake layer) of retained particles (Figure 14.4). The fouling components of pore clogging, deposition of large particles, sludge cake growth during wastewater treatment are dynamic processes, it induces transmembrane flux reduction. When the permeation rate reaches a critical value, membrane washing becomes necessary. Many studies have focused on the above problems [1, 28–31]. It is shown that the forming of a cake layer, gel layer or biofilm depends strongly on the permeate flux as well as on the membrane structure. Conventional techniques for limiting membrane fouling: its reduction by aeration in the vicinity of the membrane, by filtration below the critical flux, by the addition of coagulants, by high-frequency backpulsing, or by utilizing a high recycle velocity and/or removal of the fouling material by chemical washing (backwashing or backpulsing).

Figure 14.3 also shows the concentration-polarization layer that also forms during the filtration. Due to it, there is a backdiffusion of the retained compound that has higher concentration on the membrane surface. These phenomena can also decrease the efficiency of the filtration. This effect should also be taken into account during the mathematical modeling of the transport processes of the membrane bioreactor, as will be discussed later.

14.4 Mathematical Modeling of Membrane Bioreactor

14.4.1 Modeling of Enzyme Membrane Layer/Biofilm Reactor

The principle of the mass transport of substrates/nutrients into the immobilized enzyme/cells, through a solid, porous layer (membrane, biofilm) or through a gel layer of enzyme/cells is the same. The structure, the thickness of this mass-transport layer can be very different, thus, the mass-transport parameters, namely diffusion coefficient, convective velocity, the bioreaction rate constant, their dependency on the concentration and/or space coordinate is characteristic of the porous layer and the nature of the biocatalysts. Several investigators modeled the mass transport through this biocatalyst layer, through enzyme membrane layer [32–36] or cell-culture membrane layer [5, 12–14, 22, 37–39]. Some assumptions made for expression of the differential mass-balance equation to the biocatalytic membrane layer:

- Reaction occurs at every position within the biocatalyst layer;
- Reaction has one rate-limiting substrate/nutrient;
- Mass transport through the biocatalyst layer occurs by diffusion and convection;
- The partitioning of the components (substrate, product) is negligible;
- The mass-transport parameters (diffusion coefficient, convective velocity, bioreaction rate constant) can vary as a function of the space coordinate;
- The external mass-transfer resistance is to be taken into account.

Thus, the mass-balance equation obtained for various geometries, perpendicular to the membrane interface, can be given as follows:

$$\frac{1}{A}\left\{\frac{\partial}{\partial r}\left(DA\frac{\partial c}{\partial r}\right)+\frac{(p+1)}{r}\frac{\partial(ADc)}{\partial r}-\frac{\partial(Avc)}{dr}\right\}-Q=\frac{\partial c}{\partial t} \tag{14.1}$$

where A is the real area for mass transport, c is the substrate concentration, D is the diffusion coefficient, r is the radial space coordinate, t is the time, Q is the reaction rate, p is a geometrical factor with values of 1 for spherical pellets, 0 for cylindrical coordinates and -1 for rectangular membranes [32]. The value of A can be changed with the porosity of the biocatalyst membrane layer. The variation of the mass-transport parameters can especially occur during the growth of cells around/in hollow fibers [12, 14] because of their inhomogeneous growth due to the variation of the nutrient concentration. Close to the entrance of the bioreactor the density of nutrients (and the thickness of the biofilm formed on the membranc surface) could be much higher due to the higher nutrient concentration than that at the end of the reactor where the nutrient concentration can be much lower. The density of the cell can change not only the values of transport parameters but also the value of the reaction rate constant [12]. The variation of cell density is also true in the biocatalyst membrane layer perpendicular to the inlet surface. Increasing distance from the surface can mean decreasing nutrient concentration. This is why the variability of the transport parameters should also be taken into account. The source term can be different in biocatalytic reactions, the most often applied equations are listed in Table 14.1. The inhibited reaction can take place in both the enzymatic and microbial reactions.

For the sake of simplification, let us regard a steady-state reaction with constant area for mass transport ($A=$ constant) and let us use the Cartesian coordinate ($p=-1$), in the following. Thus, Equation 14.2 can be obtained by rewriting of Equation 14.1, as follows (y is here the transverse space coordinate, perpendicular to the membrane interface):

$$\left\{\frac{\mathrm{d}}{\mathrm{d}y}\left(D\frac{\mathrm{d}c}{\mathrm{d}y}\right)-\frac{\mathrm{d}(\upsilon c)}{\mathrm{d}y}\right\}-Q=0 \tag{14.2}$$

Table 14.1 Expressions of the important biocatalytic reactions.

Michaelis–Menten kinetics: $Q = \dfrac{r_{\max} c}{K_{Mi} + c}$
Substrate inhibition: $Q = \dfrac{r_{\max} c}{K_{Mi} + c + c^2/K_c}$
Substrate inhibition and competitive product inhibition: $Q = \dfrac{r_{\max} c}{K_{Mi}(1 + p/K_p) + c + c^2/K_c}$
Competitive product inhibition: $Q = \dfrac{r_{\max} c}{K_{Mi}(1 + p/K_p) + c}$
Noncompetitive product inhibition: $Q = \dfrac{r_{\max} c}{(K_{Mi} + c)(1 + p/K_p)}$

In general case, as was mentioned, the diffusion coefficient and/or convective velocity can depend on the space coordinate, thus $D = D(y)$, $\upsilon(y)$, [or on the concentration, $D = D(c)$ or both of them, $D = D(c, y)$]. In the boundary conditions the external mass-transfer resistance is also taken into account.

Thus, the boundary conditions will be as follows:

$$\text{if} \quad x = 0 \quad \text{then} \quad \upsilon_{\text{in}} c_{1L}^* + \beta_0 (c_1^{\text{o}} - c_{1L}^*) = -D_1 \left.\frac{\mathrm{d}c_1}{\mathrm{d}x}\right|_{y=0^+} + \upsilon_1 c_1 |_{y=0^+} \tag{14.3a}$$

$$\text{if} \quad x = \delta \quad \text{then} \quad \upsilon_{\text{out}} c_{ML}^* + \beta_\delta (c_{ML}^* - c_2^{\text{o}}) = -D_M \left.\frac{\mathrm{d}c_M}{\mathrm{d}x}\right|_{y=\delta^-} + \upsilon_M c_M |_{y=\delta^-} \tag{14.3b}$$

c^* denotes the concentration of liquid at the membrane interface, c_1^{o}, c_2^{o} denote the bulk concentration of feed and downstream, respectively, δ is the membrane thickness. The membrane concentration, c is given here in units of gmol/m^3. This can be easily obtained by means of the usually applied in, for example, g/g unit of measure with the equation of $c = w\rho/M$, where w concentration in kg/kg, ρ – membrane density, kg/m^3, M – molar weight, kg/mol. Its dimensionless form can be given by $C = c/(c_1^{\text{o}} H)$ [the H is the partition coefficient, its value is mostly unity during biochemical reactions].

A solution methodology of the above, a nonlinear differential equation, will be shown. In essence, this solution method serves the mass-transfer rate and the concentration distribution in closed, explicit mathematical expression. The method can be applied for Cartesian coordinates and cylindrical coordinates, as will be shown. For the solution of Equation 14.2, the biocatalytic membrane should be divided into M sublayers, in the direction of the mass transport, that is perpendicular to the membrane interface (for details see e.g., Nagy's paper [40]), with thickness of $\Delta\delta$ ($\Delta\delta = \delta/M$) and with constant transport parameters in every sublayer. Thus, for the mth sublayer of the membrane layer, using dimensionless quantities, it can be obtained:

$$D_m \frac{\mathrm{d}^2 c_m}{\mathrm{d}y^2} - \nu_m \frac{\mathrm{d}c_m}{\mathrm{d}y} - k_m c_m = 0 \qquad x_{m-1} < x < x_m. \tag{14.4}$$

The value of k_m can be obtained according, for example, to the Michaelis–Menten kinetics as follows:

$$k_m = \frac{r_{\max}}{K_{Mi} + c_m} \tag{14.5}$$

In dimensionless form one can get the following equation:

$$\frac{d^2 C_m}{dY^2} - \mathrm{Pe}_m \frac{dC_m}{dY} - \mathrm{Ha}_m^2 C_m = 0 \tag{14.6}$$

where $Y = y/\delta$; $\mathrm{Pe}_m = \upsilon_m \delta / D_m$; $\mathrm{Ha}_m = \sqrt{\delta^2 k_m / D_m}$.

The Ha number (Ha_m) defined here for the mth sublayer of the membrane layer corresponds to the well-known Thiele modulus defined for catalyst particles, while the Peclet number corresponds to the often used Bodenstein number.

Introducing the following equation:

$$\tilde{C} = C\exp\left(-\frac{\mathrm{Pe}Y}{2}\right) \tag{14.7}$$

One can get:

$$\frac{d^2 \tilde{C}_m}{dY^2} - \Theta_m^2 \tilde{C}_m = 0 \tag{14.8}$$

with

$$\Theta_m = \sqrt{\frac{\mathrm{Pe}_m^2}{4} + \mathrm{Ha}_m^2} \tag{14.9}$$

The solution of Equation 14.6 can be easily obtained by well-known mathematical methods as follows:

$$C_m = T_m e^{(\tilde{\lambda}_m Y)} + P_m e^{(\lambda_m Y)} \qquad Y_{m-1} < Y < Y_m \tag{14.10}$$

with

$$\lambda_m = \frac{\mathrm{Pe}_m}{2} - \Theta_m \qquad \tilde{\lambda}_m = \frac{\mathrm{Pe}_m}{2} + \Theta_m$$

The T_m and P_m parameters of Equation 14.10 can be determined by means of the boundary conditions for the mth sublayer (with $1 \le m \le M$). It will be shown in the following how the values of T_m and P_m can be obtained. The boundary conditions at the internal interfaces of the sublayers ($1 \le m \le M-1$; $Y_m = m\Delta Y$; $\Delta Y = 1/M$) can be obtained from the following two equations:

$$-\frac{dC_m}{dY} + \mathrm{Pe}_m C_m = \frac{D_{m+1}}{D_m}\left(-\frac{dC_{m+1}}{dY} + \mathrm{Pe}_{m+1} C_{m+1}\right) \qquad \text{at} \quad Y = Y_m \tag{14.11a}$$

$$H_m C_m = H_{m+1} C_{m+1} \qquad \text{at} \quad Y = Y_m \tag{14.11b}$$

The general solution of the algebraic equation system obtained by means of the internal [(14.11a) and (14.11b)] conditions with $m = 1,2,\ldots, M$ and external boundary conditions [(14.3a) and (14.3b)] are given in Appendix B. The mass-transfer rate on the upstream side of the membrane can be given, for that case, as follows:

$$J_{\mathrm{in}} = \frac{D_1 c_1^{\mathrm{o}}}{\delta}\left(-\frac{\mathrm{d}C}{\mathrm{d}Y} + \mathrm{Pe}_1 C\right)\bigg|_{Y=0} = \frac{D_1 c_1^{\mathrm{o}}}{\delta}(\lambda_1 T_1 + \tilde{\lambda}_1 P_1) \tag{14.12}$$

The outlet mass-transfer rate can be similarly given. This value should be as low as possible to avoid the loss of the substrate during the process.

14.4.2
Concentration Distribution and Mass-Transfer Rates for Real Systems

In this section the concentration distribution and the mass-transfer rate of a substrate is briefly discussed and shown under real operating conditions. The axial and radial depletion of substrate, for example, oxygen, nutrient, can often be critical scale-limiting factor in a cell-culture hollow-fiber reactor [12–14, 33]. In order to increase the substrate concentration in the membrane bioreactor, a sufficient diffusion rate and/or convective flow has to be provided through the lumen, in the axial direction, and through the membrane layer, in the radial direction, of the hollow fiber. Typical operating conditions of a hollow-fiber bioreactor were applied (Table 14.2) to calculate the inlet and outlet mass-transfer rates of a substrate. From this, the effectiveness of the biocatalytic reaction as well as the sufficiency of the nutrient supply can be estimated. The biochemical reaction rate depends on the amount of catalyst immobilized in the membrane or on the density of cells in the membrane structure. The oxygen consumption rate of cells can be estimated to be 0.1×10^{-12} mol/(cell h) [12–14], while the cell density may be about 2×10^{8} cell/cm^3. According to this, the consumption rate is equal to 6×10^{-9} mol/cm^3 s. From that we can get for the Ha number assuming the c_{m} value in Equation 14.5 is equal to zero: Ha = 0.1 ($\delta = 100\,\mu$m, $D = 1 \times 10^{-10}$ m^2/s, $k = 1.44 \times 10^{-6}$ 1/s) or Ha = 0.6 ($\delta = 500\,\mu$m, $D = 1 \times 10^{-10}$ m^2/s, $k = 1.44 \times 10^{-6}$ 1/s). The convective velocity through the membrane is an important means to avoid the substrate limitation (Figure 14.5). With increasing Pe number, that is, with the increase of the transmembrane pressure, the

Table 14.2 Membrane module characteristics and physical parameters applied for calculation of the mass-transfer rates into and out of a sheet membrane [12–14, 33].

Pressure difference: 20–30 kPa
Pe number: 0.1–10
Ha number: 0.1–5
Diffusion coefficient: 10^{-9}–10^{-10} m^2/s
Membrane thickness: $(100–1000) \times 10^{-6}$ m
Permeation velocity: 10^{-4}–10^{-6} m/s

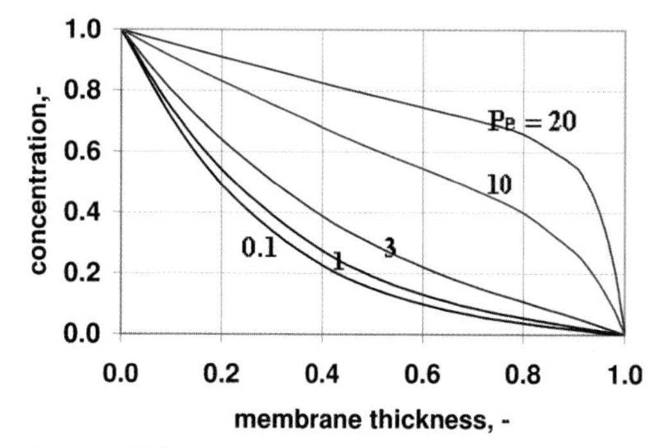

Figure 14.5 Typical concentration distribution in the membrane at different Pe numbers (Ha = 1, δ = 100 μm, $D = 4 \times 10^{-10}\,m^2/s$, $\beta_0 = \beta_\delta \rightarrow \infty$).

concentration can rapidly be increased in the biocatalytic membrane layer. The model presented is suitable to predict the substrate transport in the case of a nonlinear source term, that is, in the case of Michaelis–Menten (MM) kinetics or inhibition kinetics (Table 14.2). In Figures 14.6(a) and (b), the concentration distribution and the ratio of the outlet and the inlet mass-transfer rates are plotted, respectively. The calculation was made by the classical MM theory (continuous lines), and the two limiting cases, namely first-order ($K_{Mi} \gg c$, dotted lines) and zero-order ($K_{Mi} \ll c$, broken lines) kinetics, were applied. The outlet concentration of the substrate was chosen to be 0.2, which was considered as the critical concentration. Below this value the micro-organism was supposed to work insufficiently. Let us look at the values obtained at Ha = 1 and at Pe = 1 for the concentration distributions (Figure 14.6a). The three models show only slight differences, which will be higher at Ha = 1.9. There is a curve that has a minimum value in the figure. In order to maintain this concentration distribution, the substrate should be fed at both sides of the membrane. The substrate then might be transported form the downstream side by diffusion. Nagy [41] has investigated the diffusional mass transport through the enzyme membrane layer when the substrate can enter the membrane at both sides. The effect of an asymmetric membrane on the mass-transfer rate in the case of a first-order biochemical reaction has been shown. The ratio of the outlet mass-transfer rate and the inlet one is an important parameter in biochemical reactions, because an essential aim of these processes should be to reduce it as low as possible. This ratio is plotted in Figure 14.7(b) as a function of the Pe number at two values of the Ha number. This figure clearly shows the difference between the three models. The other important effect on the J_{out}/J_{in} is caused by the convective velocity. The solution method presented enables prediction of the value of the Pe number that one needs to avoid the limiting substrate/nutrient concentration.

The values obtained in the previous figures were predicted by constant parameters. In reality, the specific biomass concentration (and enzyme concentration) can vary as a function of space coordinate. As was mentioned, the activity of cells can be much higher at higher substrate concentration, that is, close to the feed membrane interface that can cause a higher cell density close to the interface. This can be true in the longitudinal

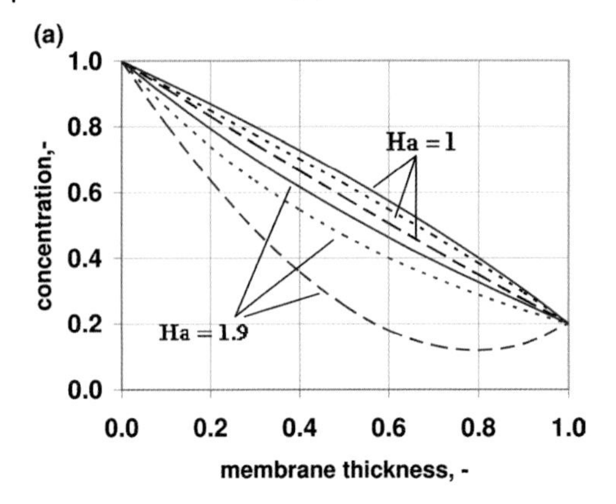

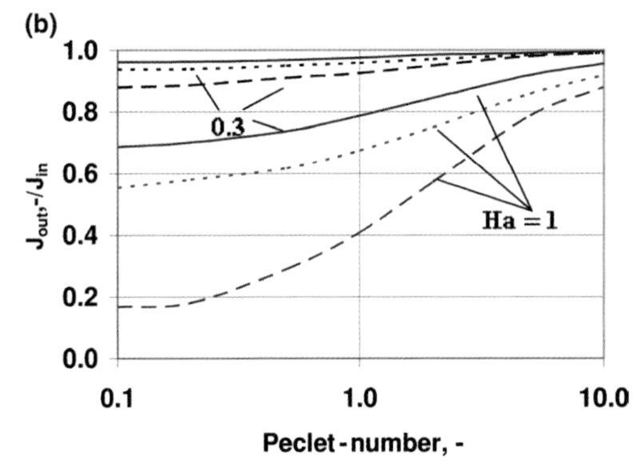

Figure 14.6 (a) Concentration vs. membrane thickness applying the Michaelis–Menten kinetics (continuous lines) and its limiting kinetics, namely first-order (. . . . . .) and zero-order (- - - -) ones. (Pe = 1, δ = 100 μm, $D = 4 \times 10^{-10}$ m^2/s, $\beta_O = \beta_\delta \to \infty$). (b) The relative values of the outlet mass-transfer rates as a function of Pe number at two different values of Ha number (the Michaelis–Menten kinetics (continuous lines) and its limiting kinetics, dotted lines, — namely first-order (. . . . . .) and zero-order (- -) ones; Pe = 1, δ = 100 μm, $D = 4 \times 10^{-10}$ m^2/s, $\beta_O = \beta_\delta \to \infty$).

direction of the hollow fiber, as well. This inhomogeneity of biomass properties can alter the value of diffusion coefficient and also that of convective velocity. These facts can be easily taken into account by means of the presented model. Figure 14.7 presents the predicted concentration in three different cases, namely for constant Pe number (Pe = 5, continuous line), for linearly increasing Pe number as a function of the y coordinate [Pe $(Y) = 10\,(1 + 0.9i/M)$, broken line] and for linearly decreasing Pe number [Pe$(Y) = 10\,(1 - 0.9i/M)$, dotted lines]. The average value of Pe number was the same in all the three cases, namely $Pe_{ave} = 5$. As can be seen, the change in Peclet number in the direction of the diffusion path of the membrane layer can essentially alter the concentration distribution in the membrane. In the case of decreasing value of Pe number, the higher starting value of Pe can substantially improve the nutrient supply across the biocatalyst layer, even 'far' from the inlet surface. Depending on the reaction rate, the concentration can even be higher than unity, in a wide range of the membrane layer. The increasing

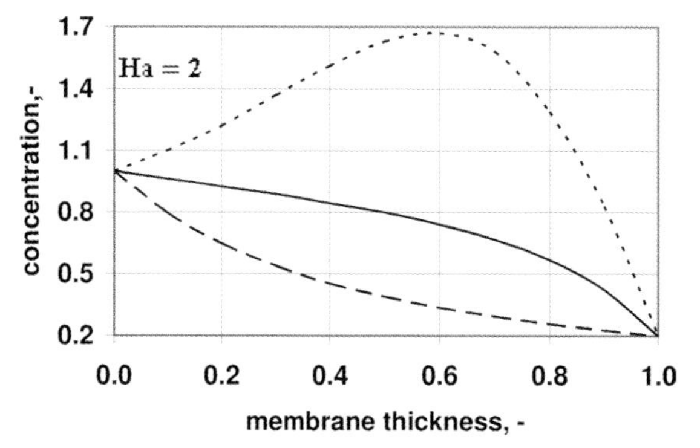

Figure 14.7 Concentration distributions with constant (Pe = 5, continuous line) and with linearly – with the space coordinate – variable Peclet number: . . . . . . decreasing Pe number $Pe = 10 - 9i/M$; $i = 1$–100, $M = 100$; – – – increasing Pe number, $Pe = 1 + 9i/M$. ($Ha = 2$, $\delta = 100\,\mu m$, $D = 4 \times 10^{-10}\,m^2/s$, $\beta_0 = \beta_\delta \rightarrow \infty$).

value of Pe, that is, the increasing value of convective velocity (or decreasing value of diffusion coefficient) lowers the concentration in the membrane layer.

14.4.3
Prediction of the Convective Velocity through Membrane with Cake and Polarization Layers

Generally, the pure-water flux through a membrane layer, υ_w is directly proportional to the applied hydrostatic pressure difference (transmembrane pressure, ΔP) according to Darcy's law as follows:

$$\upsilon_w = \frac{\Delta P}{\mu R_m} = L_p \Delta P \tag{14.13}$$

where μ is the viscosity, R_m is the hydrodynamic resistance of the membrane, υ_w is the hydrodynamic permeability [42]. The value of R_m is constant. However, when solutes are added to the water the behavior observed is completely different. The flux does not change linearly with the pressure difference, it tends to a limiting value as a function of ΔP. This maximum flux is called the limiting flux [42–45]. Change of permeate flux as a function of the transmembrane pressure difference, measured for example, by Ognier *et al.* [45] can be divided into three regions, a linearly increasing range (a permeate flux up to about $30\,dm^3/m^2\,h$), intermediate range (permeate flux > $30\,dm^3/m^2\,h$) and limiting flux range (here, more than about $45\,dm^3/m^2\,h$). In this last regime the permeate flux does not increase with increasing transmembrane pressure [43]. The pore diameter of the alumina, tubular, ultrafiltration membrane applied was $0.05\,\mu m$, its mass-transfer resistance, R_m, was $0.4 \times 10^{12}\,m^{-1}$. This anomaly of the flux curve is caused by the concentration-polarization layer and by forming a cake (or gel) layer on the membrane surface. In the general case, three mass-transfer resistance layers determine the permeation rate through the membrane (Figure 14.4). For membrane bioreactors two important cases can be distinguished, namely:

- The formation of a cake layer can be neglected (e.g., biocatalytic reaction without macromolecules); in this case the concentration-polarization layer can also exist;
- Both the concentration-polarization layer and membrane fouling are present (filtration in presence of macromolecules, proteins, cells in the liquid phase).

Generally, applying the resistance-in-series model, the following expression can be given [45]:

$$\upsilon = \frac{\Delta P}{\mu(R_m + R_g + R)} \tag{14.14}$$

where R_g is the mass-transfer resistance of the gel (cake) layer, R is the resistance of the concentration-polarization layer. For the prediction of the convective velocity through the membrane, the values of R_m, R_g, R should be determined. The filtration velocity through a porous medium can be obtained for example, by the well-known Carman–Kozeny equation [46]. Thus, this equation can often be applied for both the membrane layer and the cake layer. Thus, it can be given [47]:

$$R_g = \frac{\xi k A^2 (1-\varepsilon)^2}{\varepsilon^3} \tag{14.15}$$

where ξ is the layer thickness, k is the Kozeny constant, A is specific surface of the cake layer, ε is the porosity. Regarding only the mass-transfer rate through the membrane, the convection-diffusion model is widely used to describe the stationary permeate flux during filtration of small-size solutes at low concentration [43, 48–51]:

$$\frac{d}{dy}\left(D\frac{dc}{dy}\right) - \frac{d(\upsilon c)}{dy} = 0 \tag{14.16}$$

The solution of Equation 14.16 with variable parameters can be obtained by application of the method used for biochemical reaction in the previous subsection. The general solution of Equation 14.16 is as follows, in the case of constant parameters:

$$c = T\exp(\mathrm{Pe}Y) + S \tag{14.17}$$

where Y is a dimensionless coordinate, T and S are integration constants, the Peclet number, for example, for the membrane layer: $\mathrm{Pe} = \upsilon\delta/D$. The integration constants can be determined by proper boundary conditions. Let us look at the solution for the concentration-polarization layer (Figure 14.4). The boundary conditions can be given as: $y = 0$ then $c = c_b$ and $y = \psi$ the $c = c^*$. Thus, we get ($\mathrm{Pe} = \upsilon\psi/D$):

$$\frac{c - c_b}{c^* - c_b} = \frac{\exp(\mathrm{Pe}Y) - 1}{\exp(\mathrm{Pe}) - 1} \tag{14.18}$$

Introduction of the intrinsic retention coefficient, $\tilde{r} = 1 - c_p/c_b^*$, the value of the concentration modulus, c^*/c_b can be given as:

$$\frac{c^*}{c_b} = \frac{\exp(\mathrm{Pe})}{\tilde{r} + (1-\tilde{r})\exp(\mathrm{Pe})} \tag{14.19}$$

The value of the concentration modulus depends on the convective velocity and the mass-transfer coefficient of the concentration boundary layer (D/ψ) that means that on the membrane structure and the hydrodynamic conditions. If the retention coefficient is equal to 1, then $c^*/c_b = \exp(\mathrm{Pe})$. The larger convective velocity (or smaller diffusion coefficient) causes higher concentration polarization on the membrane interface.

If one wants to take into account all three mass-transfer resistances, the mass-balance equation [Equation 14.16] should be given for every layer. From that one can obtain the solution for every layer ($i = \mathrm{m}$ for membrane layer, $i = \mathrm{g}$ for gel layer, the concentration-polarization layer has no index according to Figure 14.4):

$$c = T_i \exp(\mathrm{Pe}_i Y) + S_i \tag{14.20}$$

The boundary conditions can be given as follows [40, 49]:

$$\upsilon_i C_i - D_i \frac{\mathrm{d}C_i}{\mathrm{d}y} = \upsilon_{i+1} C_{i+1} - D_{i+1} \frac{\mathrm{d}C_{i+1}}{\mathrm{d}y} \tag{14.21}$$

$$H_i C_i = H_{i+1} C_{i+1} \tag{14.22}$$

Equation 14.22 takes into account that the membrane and cake layers could have different partition coefficient. Thus, you can get an algebraic equation system with 6 equations, which can be solved relatively easily (as will be discussed elsewhere). Obviously, for prediction of the filtration efficiency, the values of transport parameters have to be known.

14.4.4
Convective Flow Profile in a Hollow-Fiber Membrane

14.4.4.1 Without Cake and Polarization Layers

The basic hydrodynamic equations are the Navier–Stokes equations [51]. These equations are listed in their general form in Appendix C. The combination of these equations, for example, with Darcy's law, the fluid flow in crossflow filtration in tubular or capillary membranes can be described [52]. In most cases of enzyme or microbial membrane reactors where enzymes are immobilized within the membrane matrix or in a thin layer at the matrix/shell interface or the live cells are inoculated into the shell, a cake layer is not formed on the membrane surface. The concentration-polarization layer can exist but this layer does not alter the value of the convective velocity. Several studies have modeled the convective-flow profiles in a hollow-fiber and/or flat-sheet membranes [11, 35, 44, 53–56]. Bruining [44] gives a general description of flows and pressures for enzyme membrane reactor. Three main modes of operations can be classified in this case: continuous open-shell mode; closed-shell mode; suction of permeate. Considering the velocity according to Equation 14.13, and applying the Hagen–Poiseuille's law for the longitudinal pressure drop, he defines a closed mathematical expression for the pressure drop in the longitudinal direction of the fiber. Knowing this pressure drop in the axial direction, the pressure drop through the membrane and the convective velocity can be predicted. Song [30, 43] used the filtration number to predict the pressure drop through the membrane. The simplified continuity and momentum equations for steady-state flow in the fiber lumen and shell can be given in dimensionless form as follows [11]:

$$\frac{\partial u}{\partial X} + \frac{1}{R}\frac{\partial (R\upsilon)}{\partial R} = 0 \qquad (14.23)$$

$$\frac{1}{R}\frac{\partial}{\partial R}\left(R\frac{\partial u}{\partial R}\right) = \frac{\partial P}{\partial X} \qquad (14.24)$$

with

$$R = r/R_o \qquad X = x/L$$

Where R_o is the lumen radius, L if the fiber length, u is the longitudinal convective velocity, P is pressure, X is the dimensionless axial space coordinate, R is the dimensionless radial coordinate. In writing Equation 14.24 it was assumed that flow is laminar and that entrance effects can be ignored. In addition, the axial stress terms have been neglected since the aspect ratio of the hollow fiber (R_o/L) is typically less than 0.01. The inertial terms have been neglected also, which is valid if the radial Reynolds number ($Re_R = \rho\upsilon R_o/\mu$) is much less than 1 [11]. The boundary conditions for the solution:

$$\frac{\partial u}{\partial R} = 0; \quad \upsilon = 0 \text{ at } R = 0 \qquad (14.25a)$$

$$u = 0; \quad \upsilon_o = L_p \Delta P \text{ at } R = 1 \qquad (14.25b)$$

The value ΔP can change in the axial direction in the hollow fiber (ΔP is the pressure drop in the membrane matrix due to the momentum transfer, the velocity through the membrane is $\upsilon_o\varepsilon$, where ε is the membrane porosity). Kelsey *et al.* [11] have solved the equation system in all three cases, namely for closed-shell operation, partial ultrafiltration and complete ultrafiltration and have plotted the dimensionless axial and radial velocities as well as the flow streamlines. Typical axial and radial velocity profiles are shown in the hollow-fiber membrane bioreactor at several axial positions in Figure 14.8 plotted by Kelsey *et al.* [11]. This figure illustrates clearly the change of the relative values of both the axial and the radial velocity [$V = \upsilon L/(u_o R_o)$, $U = u/u_o$ where u_o is the inlet centerline axial velocity].

14.4.4.2 With Cake and Polarization Layer

The principle of the calculation of the flow profiles is the same as in the previous section. Here, the increased mass-transfer resistance should be taken into account in

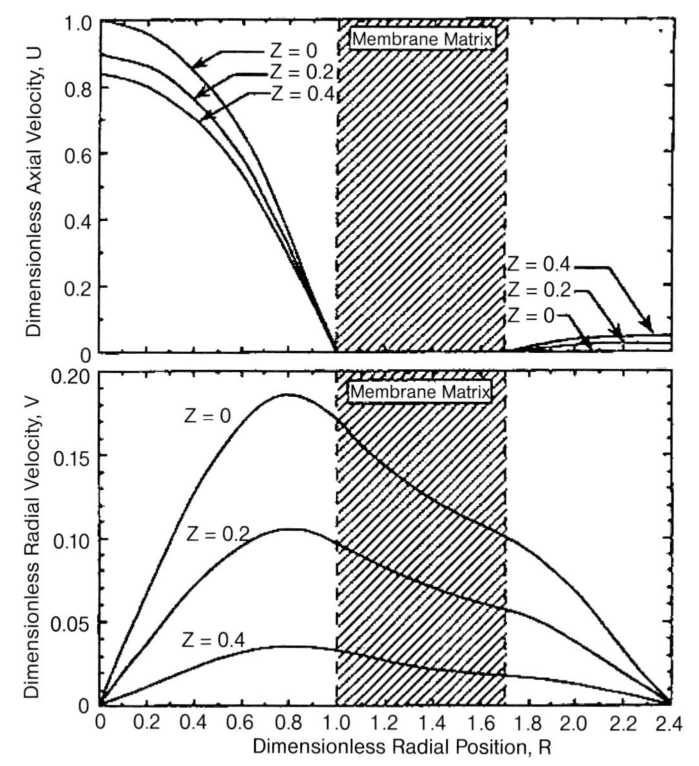

Figure 14.8 Axial (top panel and radial (bottom panel) velocity profiles in the hollow-fiber membrane bioreactor at several axial position for $\alpha \equiv 1 + \delta/R_o = 1.7$; $\beta/\alpha = 1.4$ (where $\beta = R_S/(R_o + \delta)$, R_S is the shell radius); $\kappa \equiv L_p L^2/R_o^3 = 0.1$ (κ is the dimensionless permeability, L_p is the membrane permeability), $f = 0$ (f is the retention factor). $U = u/u_o$ where u_o is the inlet centerline axial velocity; $V = \upsilon L/(u_o R_o)$.

order to calculate the pressure and the convective velocity distribution. In this case, due to the filtration effect, there exists here the concentration polarization and the change of the substrate, macromolecules, microparticle concentration. This concentration change can also alter the velocity profiles. In this case, additionally the differential mass-balance equation should also be solved together with the momentum equations. If this is not the case then the equation system of Equations 14.23 and 14.24 can be used in this mass-transport process, as well.

14.4.5
Mass Transport in the Feed Side of the Hollow-Fiber Membrane Bioreactor

In Section 4.1, the mass transport through a membrane layer as a bioreactor was discussed. Now, in this section, the concentration distribution in the feed side of the membrane layer (in most cases in its lumen side) will be discussed. When the momentum equation can be regarded as independent of the mass-transport equation, these equations can be treated separately. First, we predict the convective velocities by means of the momentum equations and applying them we can predict the mass transport in the feed side of the capillary or tubular or flat-plate membrane. The simplified form of the differential mass-balance equation, for example, for cylindrical coordinates, is as follows [13, 16, 57, 58]:

$$D_L\left(\frac{\partial^2 c}{\partial r^2}+\frac{1}{r}\frac{\partial c}{\partial r}\right)+\upsilon\frac{\partial c}{\partial r}=u\frac{\partial c}{\partial x} \tag{14.26}$$

where r is the radial space coordinate, x is the axial space coordinate, u is the axial convective velocity, υ is the radial convective velocity. When the laminar flow is perfectly developed in the lumen, the value of the axial convective velocity can be given as follows:

$$u = 2u_{max}(1-[r/R_o]^2) \tag{14.27}$$

The axial diffusion term can often be neglected, because the convective velocity can be much higher than the axial diffusion flow. The boundary conditions are as follows:

$$c = c_{in} \text{ at } x = 0 \text{ for all } r \tag{14.28a}$$

$$\frac{\partial c}{\partial r} = 0 \text{ at } r = 0 \text{ for all } x \tag{14.28b}$$

$$\upsilon_L c - D_L\frac{\partial c}{\partial r}\bigg|_{r=R_o^-} = \upsilon_m c - D_m\frac{dc}{dr}\bigg|_{r=R_o^+} \tag{14.28c}$$

where D, D_m as well as υ_L, υ_m are the diffusion coefficients in the feed fluid as well as the radial convective velocities in the fluid and in the membrane, respectively. R_o is the lumen radius.

The overall mass-transfer rates on both sides of the membrane can only be calculated when we know the convective velocity through the membrane layer. For this, Equation 14.2 should be solved. Its solution for constant parameters and for first-order and zero-order reaction have been given by Nagy [68]. The differential equation 14.26 with the boundary conditions (14.28a) to (14.28c) can only be solved numerically. The boundary condition (14.28c) can cause strong nonlinearity because of the space coordinate and/or concentration-dependent diffusion coefficient [40, 57, 58] and transverse convective velocity [11]. In the case of an enzyme membrane reactor, the radial convective velocity can often be neglected. Qin and Cabral [58] and Nagy and Hadik [57] discussed the concentration distribution in the lumen at different mass-transport parameters and at different $D_m(c)$ functions in the case of $\upsilon_L = 0$, that is, without transverse convective velocity (not discussed here in detail).

14.5 Modeling of the MBR with Membrane Separation Unit

In this group of MBR processes, the bioreaction takes place in a stirred-tank reactor and the purified liquid will be separated from the activated sludge or fermentation broth by a microfiltration or ultrafiltration module (Figures 14.1, 14.3 and 14.4). For modeling of this system, the mass transport and/or momentum transport may be separately described for both the liquid phase and the membrane module. After integration of these two models, the bioprocess can be calculated. Several research groups have investigated the performance of both the external [45] and the submerged MBR [27, 60–63] as well as moving-bed-biofilm [64–66], where the support particles are suspended by aeration (Figure 14.9). These studies primarily investigated the fouling, the cake-layer formation on the filtration membrane and its effect on the permeate velocity. In principle, the basic equations of the bioreactions of micro-organisms are known (see e.g., Moser's book [67]), these equations can be adapted to the moving-bed biofilm membrane reactor or to the aerated bioreactor with immersed membrane or with external membrane module.

14.5.1 Moving-Bed-Biofilm Membrane Reactor

The use of immobilized biocatalyst in large-scale industrial processes is presently a widespread technique. Thus, the description of the mass transport in spherical particles as support material is well known. Ferreira *et al.* [32] summarizes the most important variables and equations for the mass transport, the effect of the reaction rate on the concentration distribution in the spherical particles.

14.5.2 Submerged or External MBR Process

In this case the fluid phase is aerated (in the case of aerobic bioreactor) that maintains the turbulent hydrodynamic conditions on the one hand, and prevents the forming of the cake layer on the immersed membrane module, on the other hand. The reactor description is also well known [67], and is not discussed here.

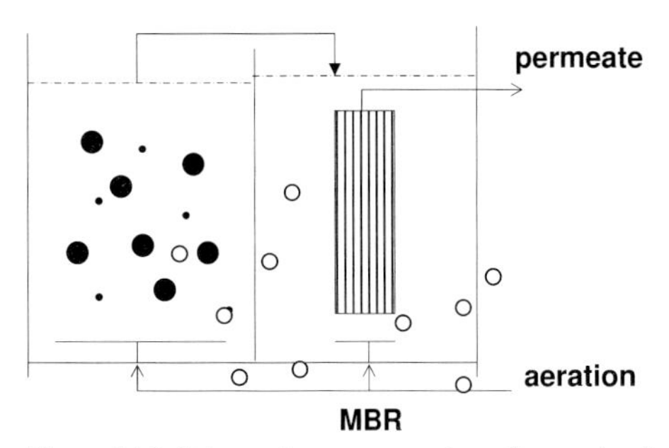

Figure 14.9 Schematic representation of a moving-bed biofilm membrane reactor.

14.5.3
Fouling in Submerged Membrane Module

The submerged membrane bioreactor is applied mostly in wastewater treatment. The activated sludge can often cause deposition of its soluble and particulate materials onto and into the membrane because of their relatively high concentration, during the wastewater treatment. Le-Clech *et al.* [59] discussed in detail, the fouling mechanism in their review and the different properties, namely that of membrane characteristic, biomass characteristic, operation conditions that could lead to forming a cake layer on the membrane surface. The analysis of this fouling process is not a task of this paper. The effect of the fouling on the permeate velocity can be characterized by the equations given in Section 4.3.

14.6
Conclusions and Future Prospects

In recent years, membrane bioreactors, bioreactors combined with membrane separation unit have established themselves as an alternative configuration for traditional bioreactors. The important advantages offered by membrane bioreactors are the several different types of membrane modules, membrane structures, materials commercially available. Membrane bioreactors seem particularly suited to carry out complex enzymatic/microbial reactions and/or to separate, *in situ*, the product in order to increase the reaction efficiency. The membrane bioreactor is a new generation of the biochemical/chemical reactors that offer a wide variety of applications for producing new chemical compounds, for treatment of wastewater, and so on.

Some topics regarding the membrane bioreactor technologies are listed here to where research efforts may be concentrated:

- Developing new, more effective membrane modules, and membrane material with the desired membrane structure that have narrow pore-size distribution and thus, better selectivity;
- Production of membrane layer with thinner skin layer;
- Producing charged membrane reducing significantly the fouling during the filtration of macromolecules and/or particles;
- Developing new membrane processes and hybrid processes in order to increase the efficiency of the biochemical reactions and the filtration;
- Producing more effective, and more stabile, heat-resisting enzyme biocatalyst;
- Improving the microbial activities, for example, by gene manipulation.

The modeling of membrane bioreactors is in the initial stage. There are not available more or less sophisticated mathematical tools to describe the complex biochemical processes. It is not known how the mass-transport parameters, diffusion coefficients, convective velocity, biological kinetic parameters might vary in function of the operating conditions, of the biolayer (enzyme/micro-organism membrane layer) properties. Widespread research activities are needed to establish correlations able to produce the values of mass-transport parameters and reaction kinetic parameters.

Acknowledgement

This work was supported by the Hungarian Research Foundation under Grants OTKA 63615/2006.

Appendix A

The differential mass-balance equation for state-state conditions, for cylindrical coordinate and for the mth sublayer is

$$D_m\left(\frac{d^2c}{dr^2}+\frac{1}{r_m}\frac{dc}{dr}\right)-\upsilon_m\frac{dc}{dr}-r_{max}\frac{c}{K_{Mi}+c_m}=0 \tag{14.A1}$$

Rearranging Equation 14.A1 the following equation is obtained to be solved:

$$\frac{d^2c}{dR^2}-\left(\frac{1}{R_m}-\mathrm{Pe}_m\right)\frac{dc}{dR}-k_mR_o^2c=0 \tag{14.A2}$$

where

$$k_m=\frac{\nu_{max}}{K_{Mi}+c_m}\qquad \mathrm{Pe}_m=\frac{\upsilon_m R_o}{D_m}$$

The solution of Equation 14.A2 is the same as Equation 14.6:

$$C_m=T_m e^{(\tilde{\lambda}_m R)}+P_m e^{(\lambda_m R)}\qquad R_{m-1}<R<R_m \tag{14.A3}$$

with

$$\tilde{\lambda}_m=\frac{-\left(\frac{1}{R_m}-\mathrm{Pe}_m\right)+\sqrt{\left(\frac{1}{R_m}-\mathrm{Pe}_m\right)^2+4\mathrm{Ha}^2}}{2}$$

$$\lambda_m=\frac{-\left(\frac{1}{R_m}-\mathrm{Pe}_m\right)-\sqrt{\left(\frac{1}{R_m}-\mathrm{Pe}_m\right)^2+4\mathrm{Ha}^2}}{2}$$

as well as

$$\mathrm{Ha}=\sqrt{\frac{k_mR_o^2}{D_m}}$$

where R_o is the radius of the cylindrical membrane, $R=1+m(R_o+\delta)/R_o$, δ is the membrane thickness, $R=r/R_o$, $C_m=c_m/c_{in}$, c_{in} is the inlet concentration.

The determination of T_m, P_m parameters is the same as in the case of Cartesian coordinates given by in Appendix B.

Appendix B

Applying the boundary conditions, defined by Equations 14.3a, 14.3b, 14.11a, 14.11b, an algebraic equation system can be obtained. The form of these equations is the same as are given here by Equation 14.B1 to Equation 14.10. The value of the parameters, a_m, $\tilde{a}_m$, b_m, $\tilde{b}_m$, c_m, $\tilde{c}_m$, d_m, $\tilde{d}_m$, S_m, $\tilde{S}_m$ (with $m = 1,2, \ldots, M$) is defined by the boundary conditions. The expressions of these parameters are easy to get from the boundary conditions (they are not shown here). The general form of the algebraic equation system to be solved can be given as follows:

$$S_0 = a_1 T_1 + \tilde{a}_1 P_1 \quad (14.B1)$$

$$S_1 = b_1 T_1 + \tilde{b}_1 P_1 + a_2 T_2 + \tilde{a}_2 P_2 \quad (14.B2)$$

$$\tilde{S}_1 = c_1 T_1 + \tilde{c}_1 P_1 + b_2 T_2 + \tilde{b}_2 P_2 \quad (14.B3)$$

$$S_2 = c_2 T_2 + \tilde{c}_2 P_2 + a_3 T_3 + \tilde{a}_3 P_3 \quad (14.B4)$$

$$\tilde{S}_2 = d_2 T_2 + \tilde{d}_2 P_2 + b_3 T_3 + \tilde{b}_3 P_3 \quad (14.B5)$$

$$\vdots$$

$$S_m = c_m T_m + \tilde{c}_m P_m + a_{m+1} T_{m+1} + \tilde{a}_{m+1} P_{m+1} \quad (14.B6)$$

$$\tilde{S}_m = d_m T_m + \tilde{d}_m P_m + b_{m+1} T_{m+1} + \tilde{b}_{m+1} P_{m+1} \quad (14.B7)$$

$$\vdots$$

$$S_{M-1} = c_{M-1} T_{M-1} + \tilde{c}_{M-1} P_{M-1} + a_M T_M + \tilde{a}_M P_M \quad (14.B8)$$

$$\tilde{S}_{M-1} = d_{M-1} T_{M-1} + \tilde{d}_{M-1} P_{M-1} + b_M T_M + \tilde{b}_M P_M \quad (14.B9)$$

$$S_M = c_M T_M + \tilde{c}_M P_M \quad (14.B10)$$

The value of T_1 can be obtained by solving Equation 14.B11, where Chart 1 means the solution of the determinant obtained with the parameters of the of the right hand side of Equations 14.B1–14.B10 and Chart 2 is the solution of the determinant obtained from Chart 1 replacing its first column with the left-hand side of Equations 14.B1–14.B10, according to the well-known Cramer role [69].

$$T_1 = \frac{\text{Chart 2}}{\text{Chart 1}} \quad (14.B11)$$

Applying the well-known Cramer roles for the solution of a determinant, the value of T_1 can be obtained as follows:

$$T_1 = \frac{R_M}{c_1 E_{M-1} \alpha_M} \quad (14.B12)$$

with

$$\alpha_1 = \frac{a_M}{c_M} - \frac{\tilde{a}_M}{\tilde{c}_M} \qquad \beta_1 = \frac{b_M}{c_M} - \frac{\tilde{b}_M}{\tilde{c}_M}$$

$$R_1 = S_{M-1} - S_M \frac{\tilde{a}_M}{\tilde{c}_M} \qquad Q_1 = \tilde{S}_{M-1} - S_M \frac{\tilde{b}_M}{\tilde{c}_M}$$

$$\alpha_m = \frac{a_{M+1-m}}{d_{M+1-m} E_{m-1}} - \frac{\tilde{a}_{M+1-m}}{\tilde{d}_{M+1-m} B_{m-1}} \qquad m = 2, 3, \ldots, M$$

$$\beta_m = \frac{b_{M+1-m}}{d_{M+1-m} E_{m-1}} - \frac{\tilde{b}_{M+1-m}}{\tilde{d}_{M+1-m} B_{m-1}} \qquad m = 2, 3, \ldots, M-1$$

$$E_m = \frac{c_{M-m}}{d_{M-m}} - \frac{\alpha_m}{\beta_m} \qquad B_m = \frac{\tilde{c}_{M-m}}{\tilde{d}_{M-m}} - \frac{\alpha_m}{\beta_m} \qquad m = 1, 2, \ldots, M-1$$

$$R_m = S_{M-m} - \frac{\tilde{a}_{M+1-m}}{\tilde{d}_{M+1-m} B_{m-1}} \left(R_{m-1} - Q_{m-1} \frac{\alpha_{m-1}}{\beta_{m-1}} \right) \qquad m = 2, 3, \ldots, M-1$$

$$Q_m = \tilde{S}_{M-m} - \frac{\tilde{b}_{M+1-m}}{\tilde{d}_{M+1-m} B_{m-1}} \left(R_{m-1} - Q_{m-1} \frac{\alpha_{m-1}}{\beta_{m-1}} \right) \qquad m = 2, 3, \ldots, M-1$$

as well as

$$R_M = S_0 - \frac{\tilde{a}_1}{\tilde{c}_1 B_{M-1}} \left(R_{M-1} - Q_{M-1} \frac{\alpha_{M-1}}{\beta_{M-1}} \right)$$

$$\alpha_M = \frac{a_1}{c_1 E_{M-1}} - \frac{\tilde{a}_1}{\tilde{c}_1 B_{M-1}}$$

The solution methodology of the determinants is similar to that of the well-known Thomas algorithm used for the numerical solution of a differential equation with the finite-difference method [50]. An essential difference from the Thomas algorithm is that the first step of the algorithm here is a so-called backward process. This means that the calculation of T_1 starts from the last sublayer, that is, from the Mth sublayer of the determinant and it is continued down to the 1st sublayer. Thus, the value of T_1 is obtained directly, in the fist calculation step. Then, applying the known value of T_1, the value of P_1 can be obtained by means of the fist boundary condition at $X = 0$, namely:

$$P_1 = \frac{S_0 - a_1 T_1}{\tilde{a}_1} \tag{14.B13}$$

The values of T_m and P_m can then be obtained by means of the internal boundary conditions ($m = 2, \ldots, M$) with a forward sweep. Thus, the values of T_{m+1} and P_{m+1} should be determined by Equations (14.B6) and (14.B7). Thus, the values of T_{m+1} and P_{m+1} can be easily obtained as follows, with values of $m = 2, 3, \ldots, M-1$:

$$T_{m+1} = \frac{\eta_m \tilde{b}_{m+1} - \tilde{\eta}_m \tilde{a}_{m+1}}{Y_{m+1}} \tag{14.B14}$$

$$P_{m+1} = \frac{\eta_m a_{m+1} - \tilde{\eta}_m b_{m+1}}{Y_{m+1}} \tag{14.B15}$$

with

$$\eta_m = S_m - c_m T_m - \tilde{c}_m P_m \tag{14.B16}$$

$$\tilde{\eta}_m = \tilde{S}_m - d_m T_m - \tilde{d}_m P_m \tag{14.B17}$$

as well as

$$Y_{m+1} = a_{m+1} \tilde{b}_{m+1} - b_{m+1} \tilde{a}_{m+1} \tag{14.B18}$$

Appendix C

The mass and momentum equations, that is, the Navier–Stokes approximation expressed in cylindrical coordinates with axisymmetry assumption, are [50, 51]:

$$\frac{1}{r}\frac{\partial}{\partial r}(r\upsilon) + \frac{\partial u}{\partial x} = 0 \tag{14.C1}$$

$$\rho\left(\upsilon\frac{\partial u}{\partial r} + u\frac{\partial u}{\partial x}\right) = -\frac{\partial P}{\partial x} + \mu\left\{\frac{\partial}{\partial r}\left[\frac{1}{r}\frac{\partial (ru)}{\partial r}\right] + \frac{\partial^2 u}{\partial x^2}\right\} \tag{14.C2}$$

$$\rho\left(\upsilon\frac{\partial \upsilon}{\partial r} + u\frac{\partial \upsilon}{\partial x}\right) = -\frac{\partial P}{\partial x} + \mu\left\{\frac{\partial}{\partial r}\left[\frac{1}{r}\frac{\partial (r\upsilon)}{\partial r}\right] + \frac{\partial^2 \upsilon}{\partial x^2} - \frac{\upsilon}{r^2}\right\} \tag{14.C3}$$

where u is the longitudinal convective velocity, υ is the radial convective velocity, ρ is the fluid density, μ is the fluid viscosity, P is the pressure, x is the axial space coordinate.

References

1. Strathmann, H., Giorno, L. and Drioli, E. (2006) *An Introduction to Membrane Science Technology*, Institute of Membrane Technology, Italy, University of Calabria, Rende (CS).
2. Yang, W., Cicek, N. and Ilg, J. (2006) *Journal of Membrane Science*, **270**, 201–211.
3. Giorno, L. and Drioli, E. (2000) *Trends in Biotechnology*, **18** (8), 339–349.
4. Marcano, J.G.S. and Tsotsis, T.T. (2002) *Catalytic Membranes and Membrane Reactors*, Wiley-VCH, Weinheim.
5. Brotherton, J.D. and Chau, P.C. (1990) *Biotechnology and Bioengineering*, **35**, 375–394.
6. Sheldon, M.S. and Small, H.J. (2005) *Journal of Membrane Science*, **263**, 30–37.

7 Rios, G.M., Belleville, M.-P. and Paolucci-Jeanjean, D. (2007) *Trends in Biotechnology*, **25**, 242–246.

8 Charcosset, C. (2006) *Biotechnology Advances*, **24**, 482–492.

9 Frazeres, D.M.F. and Cabral, J.M.S. (2001) Enzymatic membrane reactors, in *Multiphase Bioreactor Design* (eds J.M.S. Cabral, M. Mota and J. Tramper), Taylor and Francis, pp. 135–184, London.

10 Belfort, G. (1989) *Biotechnology and Bioengineering*, **33**, 1047–1066.

11 Kelsey, L.J., Pillarella, M.R. and Zydney, A.L. (1990) *Chemical Engineering Science*, **45** (11), 3211–3220.

12 Schonberg, J.A. and Belfort, G. (1987) *Biotechnology Progress*, **3** (2), 81–89.

13 Piret, J.M. and Cooney, C.L. (1991) *Biotechnology and Bioengineering*, **37**, 80–92.

14 Sardonini, C.A. and DiBiasio, D. (1992) *Biotechnology and Bioengineering*, **40**, 1233–1242.

15 Godongwana, B., Sheldon, M.S. and Solomons, D.M. (2007) *Journal of Membrane Science*, **303**, 86–99.

16 Calabro, V., Curcio, S. and Iorio, G. (2002) *Journal of Membrane Science*, **206**, 217–241.

17 Nakajima, M. and Cardoso, J.P. (1989) *Biotechnology and Bioengineering*, **33**, 856.

18 Giorno, L., De Bartolo, L. and Drioli, E. (2003) Membrane bioreactors for biotechnology and medical applications, in *New Insight into Membrane Science and Technology: Polymeric and Bifunctional Membranes* (eds D. Bhattacharyya and D.A. Butterfield), Elsevier, Chapter 9, Oxford.

19 Habulin, M. and Knez, Z. (1991) *Journal of Membrane Science*, **61**, 315–324.

20 Lozano, P., Perez-Marin, A.B., De Diego, T., Gomez, D., Paolucci-Jeajean, D., Belleville, M.P., Rios, G.M. and Iborra, J.L. (2002) *Journal of Membrane Science*, **201**, 55–64.

21 Chung, T.P., Wu, P.C. and Jung, R.S. (2005) *Journal of Membrane Science*, **258**, 55–63.

22 Melo, L.F. and Oliveira, R. (2001) Biofilm reactors, in *Multiphase Bioreactor Design* (eds J.M.S. Cabral, M. Mota and J. Tramper), Taylor and Francis, pp. 271–308, London.

23 Marrot, B., Barnos-Martinez, A., Moulin, P. and Roche, N. (2004) *Environmental Progress*, **23**, 59–68.

24 Mavrov, V. and Bélirés, E. (2000) *Desalination*, **131**, 75–86.

25 Chmiel, H., Kaschek, M., Blöcher, C., Noronha, M. and Mavrov, V. (2003) *Densalination*, **152**, 307–314.

26 Sarkar, B., Chakrabarti, P.P., Vijaykumar, A. and Kale, V. (2006) *Desalination*, **195**, 141–152.

27 Rosenberger, S., Krüger, U., Witzig, R., Manz, W., Szewzyk, U. and Kraume, M. (2002) *Water Research*, **36**, 413–420.

28 Li, X.-y. and Wang, X.-m. (2006) *Journal of Membrane Science*, **278**, 151–161.

29 Bacchin, P., Si-Hassen, D., Starov, V., Clifton, M.J. and Aimar, P. (2002) *Chemical Engineering Science*, **57**, 77–91.

30 Song, L. (1998) *Journal of Membrane Science*, **144**, 173–185.

31 Defrance, L. and Jaffrin, M.Y. (1999) *Journal of Membrane Science*, **152**, 203–210.

32 Ferreira, B.S., Fernandes, P. and Cabral, J.M.S. (2001) Design and modeling of immobilized biocatalytic reactors, in *Multiphase Bioreactor Design* (eds J.M.S. Cabral, M. Mota and J. Tramper), Taylor and Francis, London, UK, pp. 85–180.

33 Long, S.W., Bhatia, S. and Kamaruddin, A. (2003) *Journal of Membrane Science*, **219**, 69–88.

34 Hossain, M.M. and Do, D.D. (1989) *Biotechnology and Bioengineering*, **33**, 963–975.

35 Salzman, G., Tadmor, R., Guzy, S., Sideman, S. and Lotan, N. (1999) *Chemical Engineering Progress*, **38**, 289–299.

36 Waterland, L.R., Robertson, C.R. and Michaelis, A.S. (1975) *Chemical Engineering Communications*, **2**, 37–47.

37 Cabral, J.M.S. and Tramper, J. (1994) Bioreactor Design, in *Applied Biocatalysis* (eds J.M.S. Cabral, D. Best, L. Boross and J. Tramper), Harwood Academic Publishers, Switzerland, pp. 330–370.

38 Brotherton, J.D. and Chau, P.C. (1996) *Biotechnology Progress*, **12** (5), 575–590.

39 Lu, S.G., Imai, T., Ukita, M., Sekine, M., Higouchi, T. and Fukagawa, M. (2001) *Water Research*, **35** (8), 2038–2048.

40 Nagy, E. (2006) *Journal of Membrane Science*, **274**, 159–168.

41 Nagy, E. (1999) Diffusion mass transfer in enzyme membrane reactor, in *Integration of Membrane Processes into Bioconversions* (eds K. Bélafi-Bakó, L. Gubicza and M. Mulder), Kluwer Academic/Plenum Publishers, New York, pp. 211–221.

42 Mulder, M.H.V. (1995) Polarization phenomena and membrane fouling, in *Membrane Separation Technology, Principles and Applications* (eds R.D. Noble and I. Stern), Elsevier, Oxford.

43 Song, L. (1999) *Journal of Colloid and Interface Science*, **214**, 251–263.

44 Bruining, W.J. (1989) *Chemical Engineering Science*, **44**, 1441–1447.

45 Ognier, S., Wisniewski, C. and Grasmick, A. (2004) *Journal of Membrane Science*, **229**, 171–177.

46 Yeh, H.-M. (2002) *Desalination*, **145**, 153–157.

47 Posh, C. and Schiewer, S. (2006) *Journal of Membrane Science*, **280**, 284–297.

48 Bacchin, P., Hassen, D.S., Stratov, V., Clifton, M.J. and Aimar, P. (2002) *Chemical Engineering Science*, **57**, 77–91.

49 Davis, R.H. and Scherwood, J.D. (1990) *Chemical Engineering Science*, **45**, 3203–3209.

50 De, S. and Bhattacharya, P.K. (1997) *Journal of Membrane Science*, **136**, 57–69.

51 Cebeci, T., Shao, J.P., Kafyeke, F. and Laurendeau, E. (2005) *Computational Fluid Dynamics for Engineers*, Horizons Publishing Inc. and Springer, Heidelberg.

52 Damak, K., Ayadi, A., Zeghamati, B. and Schmitz, P. (2004) *Desalination*, **161**, 67–77.

53 Song, L. and Elimelech, M. (1995) *Journal of the Chemical Society-Faraday Transactions*, **91** (19), 3389–3398.

54 Kim, J.-I. and Stroeve, P. (1988) *Chemical Engineering Science*, **43**, 247–257.

55 Zhang, M. and Song, L. (2000) *Journal of Environmental Engineering*, **126**, 667–674.

56 Park, J.K. and Chang, H.N. (1986) *AIChE Journal*, **32**, 1937–1974.

57 Nagy, E. and Hadik, P. (2002) *Desalination*, **145**, 147–152.

58 Qin, Y. and Cabral, J.M.S. (1998) *AIChE Journal*, **41**, 836–848.

59 Le-Clech, P., Chen, V. and Fane, T.A.G. (2006) *Journal of Membrane Science*, **284**, 17–53.

60 Al-Malack, M.A. (2007) *Desalination*, **214**, 112–127.

61 Wintgens, T., Rosen, J., Melin, T., Brepols, C., Drensla, K. and Engelhardt, N. (2003) *Journal of Membrane Science*, **216**, 55–65.

62 Busch, J., Cruse, A. and Marquardt, W. (2007) *Journal of Membrane Science*, **288**, 94–111.

63 Ng, A.N.L. and Kim, A.S. (2007) *Desalination*, **212**, 261–328.

64 Leiknes, T. and Odeggard, H. (2007) *Desalination*, **202**, 135–143.

65 Artiga, P., Oyanedel, V., Garrido, J.M. and Méndez, R. (2005) *Desalination*, **179**, 171–179.

66 Sheldon, M.S. and Small, H.J. (2005) *Journal of Membrane Science*, **263**, 30–37.

67 Moser, A. (1988) *Bioprocess Technology*, Springer, Wien.

68 Nagy, E. (2008) *Asia-Pac. J. Chem. Eng.*, doi:10.1002/apj.242.

69 Nagy, E. (2008) *Chem. Eng. Res. Des.*, **86**, 723–730.

15
Photocatalytic Membrane Reactors in the Conversion or Degradation of Organic Compounds

Raffaele Molinari, Angela Caruso, and Leonardo Palmisano

15.1
Introduction

The principles of green chemistry are based on the necessity to develop chemical products and industrial processes that reduce or eliminate the use and the generation of toxic substances along with the risk for the human health and for the environment.

In this context photocatalytic processes in membrane reactors represent a technology of great scientific interest because they allow chemical reactions and separation process to be obtained in one step, minimizing environmental and economic impacts.

Heterogeneous photocatalysis is a technology that has been extensively studied for about three decades, since Fujishima and Honda discovered the photocatalytic splitting of water on TiO_2 electrodes in 1972.

Photocatalysis includes a large variety of reactions such us partial or total oxidations, hydrogen transfer, functionalization, rearrangements, dehydrogenation, mineralization, and so on [1].

These processes, included in a special class of oxidation techniques defined as advanced oxidation processes (AOPs), are based on the irradiation of a semiconductor photocatalyst with UV light that leads to the formation of highly reactive hydroxyl radicals.

In particular, water treatment in principle constitutes one of the most important fields of application of the photocatalytic processes that involve both degradation reactions for the removal of environmental pollutants and selective reactions for the synthesis of organic compounds.

Coupling a membrane process to this technology it is possible to obtain the separation of the clarified solution or the reaction product and also the recovery and the reuse of the catalyst. In fact, the choice of an appropriate membrane allows a selective separation of the product to be performed and to maintain the catalyst in the reaction environment in a continuous process that increases the efficiency of the whole system.

Membrane Operations. Innovative Separations and Transformations. Edited by Enrico Drioli and Lidietta Giorno

ISBN: 978-3-527-32038-7

Moreover, the use of solar light as a source of activating radiation represents an interesting future perspective on which many studies are focused.

15.2
Fundamentals on Heterogeneous Photocatalysis

Due to its high efficiency and the generation of harmless products, The heterogeneous photocatalytic process has caught the attention of many researchers all over the world.

In recent years many studies have been realized with the purpose to understand the fundamentals of photocatalytic mechanisms and to increase the application fields [2].

15.2.1
Mechanism

The electronic structure of a semiconductor is characterized by a filled conduction band and an empty valence band separated by a bandgap of energy (E_G). When the catalyst is illuminated with photons whose energy is equal to or greater than this bandgap, the promotion of an electron from the valence band to the conduction band occurs with the creation of electron–hole pairs ($e^-_{cb}-h^+_{vb}$). The valence-band hole can oxidize electron-donor molecules (water or hydroxyl ions) to produce oxidizing hydroxyl radicals, whereas the conduction-band electron can reduce acceptor molecules such as O_2 (to yield a superoxide ion) or a metal ion (reduced to its lower valence states) (Figure 15.1).

The lowest energy level of the conduction band defines the reduction potential of the photoelectrons, while the highest one of the valence band determines the oxidizing power of the photoholes, respectively. When the reagents spread on the catalyst surface they are adsorbed on the active site and they can participate in redox reactions.

The adsorbed molecule can be reduced if its reduction potential is higher than that of the photoelectrons or it can be oxidized if its potential is lower than that of the photoholes.

The photonic excitation of the catalyst represents the initial step of the activation of the photocatalytic process and the hydroxyl radicals are the primary oxidant in these systems, although the recombination of the $e^-_{cb}-h^+_{vb}$, that produces thermal energy, can occur, with a reduction of photocatalytic activity.

15.2.2
Photocatalysts: Properties and New Semiconductor Materials Used for Photocatalytic Processes

A semiconductor must exhibit some characteristics such as suitable bandgap energies, chemical and physical stability, nontoxic nature, availability and low cost, and physical characteristics that allow it to operate as photocatalyst.

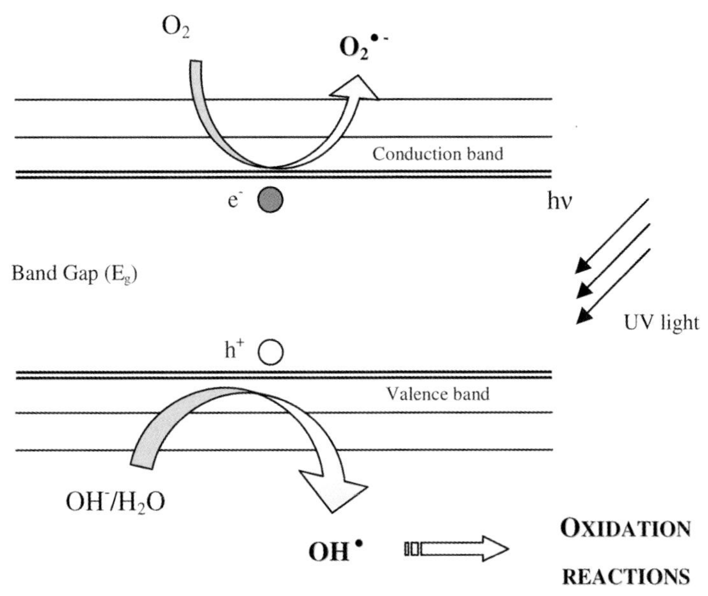

Figure 15.1 Energy bandgap of a semiconductor.

In the literature many semiconductor materials acting as catalyst are used in photocatalytic reactions.

The redox potentials of the valence and conduction bands for different semiconductors varies between +4.0 and −1.5 volts vs. the normal hydrogen electrode (NHE), respectively. Therefore, by careful selection of the photocatalyst a wide range of molecules can be converted via these processes [3].

The classic photocatalysts are generally oxides (TiO_2, ZnO, ZrO_2, CeO_2, WO_3, etc.) or sulfides (CdS, ZnS, WS_2, etc.). In Table 15.1 are reported a list of semiconductor

Table 15.1 Band positions of some common semiconductors used for photocatalytic processes [2, 3, 6, 7, 95].

Semiconductor	Valence band (V vs. NHE)	Conduction band (V vs. NHE)	Bandgap (eV)	Bandgap wavelength (nm)
TiO_2 anatase	+3.1	−0.1	3.2	387
TiO_2 rutile	+3.1	+0.1	3.0	380
SnO_2	+4.1	+0.3	3.8	318
ZnO	+3.0	−0.2	3.2	387–390
ZnS	+1.4	−2.3	3.7	335–336
Fe_2O_3	+2.6	−0.4	2.2	560
ZrO_2	+4.2	−0.8	5.0	460
WO_3	+3.0	+0.2	2.8	443
CdS	+2.1	−0.4	2.5	496–497
CdSe	+1.6	−0.1	1.7	729–730
GaAs	+1.0	−0.4	1.4	886–887
GaP	+1.3	−1.0	2.3	539–540

materials with their valence band and conduction-band positions, the bandgaps and the wavelength of radiation required to activate the catalyst. Some values can be different from other bibliographic sources.

15.2.2.1 Titanium Dioxide

Among the various catalysts employed in the photocatalytic processes, the most popular and used is titanium dioxide, TiO_2, thanks to its strong catalytic activity, high chemical stability in aqueous media and in a large range of pH (0–14), low cost due to the abundance of Ti (0.44% of the Earth's crust) and long lifetime of electron–hole pairs.

It occurs in nature in three forms: rutile (a tetragonal mineral usually of prismatic habit), anatase (a tetragonal mineral of dipyramidal habit that exhibits higher photocatalytic activity) and brookite (a rare orthorhombic mineral).

The different photoactivity between anatase and rutile type is due to a difference in the energy structure. In both forms, the position of the valence band is deep, and the resulting positive holes show sufficient oxidative power. Nevertheless, the conduction band in the anatase type is closer to the negative position than in the rutile type; therefore, the reducing power of anatase type is stronger than that of rutile type.

Thus, in order to avoid photodamage, the rutile is the form employed in industrial applications as a pigment to provide whiteness and opacity to products such as plastics, papers, coatings, paints, inks, foods, or in cosmetic and skin-care products as a thickener.

Instead, thanks to its high photoactivity, the anatase type is used in a wide range of applications as photocatalytic coatings on various substrates such as glass and ceramic tiles that can photodegrade various noxious or malodorous chemicals, smoke and cooking oil residues under low-intensity near-UV light.

15.2.2.2 Modified Photocatalysts

Despite the wide range of application of the photocatalytic processes, their use at industrial level is still limited due to different reasons:

- recombination of photogenerated electron–hole pairs: holes and excited electrons are unstable species that can quickly recombine within 10–100 ns releasing energy in the form of unproductive heat or photons [4, 5];
- fast backward reactions or secondary reactions that lead to undesirable by-products;
- inability to use visible light: only a small fraction of solar light (less than 5% in the case of TiO_2 anatase) can be utilized by the photocatalyst [6].

In order to resolve these deficiencies and to design high-efficiency and economical photocatalytic systems, the discovery and the development of new photocatalysts has become one of the most important topics in photocatalysis in the last years.

Ni *et al.* [5], in a review on the developments in photocatalytic water splitting using TiO_2 for hydrogen production, divided the techniques used to enhance the photocatalytic efficiency in two broad groups: photocatalyst modification techniques,

Table 15.2 Some recent photocatalysts and their applications.

Catalyst	Reaction	Year	References
Arginine-TiO_2	Selective reduction	2007	[4]
Acridine yellow G (AYG)	Total oxidation	2007	[25]
Membrane-W10	Oxidation	2006	[17]
Doped-TiO_2	Selective oxidation	2006	[8]
$ZnWO_4$	Photodegradation	2007	[24]
TaON	Total oxidation	2005	[23]
$AgGaS_2$	Hydrogen production	2007	[26]
Fe-ZSM-5	Reduction	2007	[49]
Bi_2S_3/CdS	Partial reduction	2002	[15]
Ni-doped ZnS	Hydrogen production	2000	[13]
Fe(III)-OH complexes	RedOx	2007	[21]
Co_3O_4	Photodegradation	2007	[38]
POM	Functionalization	2003	[56]
Zn phthalocyanine complexes	Photodegradation	2007	[18]
Pt, Au, Pd–doped TiO_2	Hydrogen production	2007	[9]
POM	Bromination	2007	[27]
Hydrous alumina-doped TiO_2	Reduction	2003	[14]
Bi^{3+}–doped TiO2	Reduction	2007	[12]
Activated Carbon-ZnO	Degradation	2007	[22]
$ZnB_{12}O_{20}$	Degradation (ox)	2005	[29]
La-, Cu-, Pt- doped WO_3	Selective oxidation	2005	[11]
POM	Reduction	2004	[51]
POM	Degradation (red)	2007	[28]
Au/Fe_2O_3	Degradation (ox)	2007	[10]
YVO_4	Degradation	2007	[30]

which include noble-metal loading, ion doping, catalyst sensitization, and chemical additives such as addition of electron donors or addition of carbonate salts to suppress backward reactions.

Many studies on the use of modified catalysts have been carried out [7], some of the most recent photocatalysts and their applications are reported in Table 15.2.

The main purpose of the international research is to inhibit charge recombination and to extend catalyst light absorption spectra to the visible region.

It was demonstrated that it is possible, for example, loading the catalyst surface with noble metals such as Au, Pt, Pd, Ni, Ag [6, 8–10] reducing the possibility of electron–hole recombination or doping the catalyst with metal ions such as Fe^{3+}, Co^{2+}, Cu^{2+}, Al^{3+} [11–14] that could expand its photoresponse into the visible region.

In order to suppress the recombination of the photogenerated electron–hole pairs, some researchers [6, 15] have described the photocatalytic activity of composite photocatalysts consisting of two semiconductors. In these configurations, after absorption of a photon, the transfer of the electrons from the conduction band of the photoexcited component to that of the unexcited component occurs, leading to stable semiconductor particles with separated charges that do not undergo deactivation because transfer in the reverse direction is thermodynamically forbidden.

Bonchio *et al.* [16] prepared a novel heterogeneous photo-oxidation catalysts by embedding polyoxotungstates in polymeric membranes achieving stable and recyclable photocatalytic systems with different and tunable properties depending on the nature of the polymeric materials. In particular, using PDMS and fluorinated (PVDF and Hyflon) polymers hybrid photocatalysts, with high stability in aqueous phase under turnover regime and temperatures up to 50 °C, were obtained [17].

Moreover in a certain number of recent studies are reported the use of complex catalysts, such as Zn phthalocyanine [18], Cu phthalocyanine [19], Cu porphyrin [20], Fe(III)-OH [21], activated-carbon-ZnO [22], or new synthesized photocatalysts, such us TaON [23], $ZnWO_4$ [24], Acridine yellow G [25], $AgGaS_2$ [26], polyoxometallates [27, 28], $ZnB_{12}O_{20}$ [29], YVO_4 [30], which exhibit a significant activity under visible light irradiation, as well as a retard in charge recombination, allowing a greater control of the whole photocatalytic process.

15.3 Photocatalytic Parameters

Some parameters that influence the photocatalytic process are:

- *Catalyst amount:*

The reaction rate is directly proportional to the catalyst dose, therefore optimal catalyst dosing is necessary for controlling the reaction. Nevertheless, above a certain value of mass catalyst (m) the rate of reaction becomes independent of m that corresponds to the maximum amount of catalyst in which all the surface active sites are occupied by the substrate. Moreover, when the catalyst is suspended in aqueous solution, aggregation of the catalyst particles occurs for high amounts, with a decrease of the number of surface active sites and the extent of transmitted light, due to increase of opacity and light scattering [2]. Therefore, the use of low amounts of catalyst is an useful condition not only from an economical point of view but also in order to guarantee a satisfactorily reaction efficiency.

- *pH of the aqueous solutions:*

The pH plays an important role in the efficiency of the photocatalytic process. In particular, the occurrence of aggregation phenomena involving the suspended TiO_2 catalyst particles together with their precipitation has been observed at acidic values. The last aspect determines a reduction of the catalytic active sites and therefore a decrease of the catalyst activity.

Besides, different pH values lead to the formation of different reaction products [2, 31], due to modifications in the ionization state of the catalyst surface. Depending on the substrates, an increase of the pH will have a positive or negative effect on their reaction rate because the hydrophilic/hydrophobic character of the catalyst changes with the pH. When TiO_2 is used as catalyst the strongest attractive interactions occur at pH values around the point of zero charge (PZC) (values of PZC for TiO_2 are reported in the pH range 4.5–6.2). As observed by Bekkouche *et al.* [32] in the study of adsorption of phenol on TiO_2, for example, the catalyst surface is negatively charged at alkaline value and a phenomenon of repulsion can be hypothesized that could explain the low adsorption values.

Noguchi *et al.* [14] observed an increase of the rate of BrO_3^- reduction with TiO_2 suspended in an aqueous solution by lowering the pH from 7 to 5. This behavior is attributable to an enhancement of the electrical interaction between the substrate and the positively charged surfaces of the TiO_2 photocatalyst that causes an increase in the amount of adsorbed BrO_3^- on the photocatalyst surface.

- *Wavelength and light intensity:*

Only photons with λ smaller than or equal to the absorption edge of the catalyst are effective in reaction activation. In addition, it must be checked that the reactants do not absorb light in order that the catalyst works in a truly heterogeneous catalytic regime. Besides, there is a direct proportionality between the conversion level and the radiation intensity that confirms the participation of photoinduced electrical charges (holes and electrons) to the photocatalytic mechanism.

- *Presence of others species:*

The presence of others species in the reaction environment can enhance or decrease the rate of the photocatalytic process depending on the mechanism of reaction. As reported by Kavita *et al.* [2], small quantities of some ions, such as Cu^{2+} or Mn^{2+}, increase the rate of oxidation of organic compounds thanks to the inhibition of electron–hole recombination.

On the other hand, co-dissolved ions like Cl^-, Na^+, PO_4^{3-} affect the photodegradation rate because they possibly adsorb onto TiO_2 surface competing with the substrate and hamper the formation of $OH^{\bullet}$ radicals [31].

15.4 Applications of Photocatalysis

Photocatalytic processes, occurring on a semiconductor surface, can be carried out in various media: aqueous solutions, pure organic liquid phases or gas phase. The wide application field of this technology includes a great variety of reactions such as partial or total oxidation, selective reduction, degradation of organic compounds, fuel synthesis (e.g., H_2 production through water splitting), metal-corrosion prevention, disinfection, and so on [1, 7].

Some of the most recent applications of the photocatalytic processes present in literature are reported in Table 15.3.

15.4.1 Total Oxidations

The total degradation of organic pollutants represents one of the main applications of the photocatalysis. As result of a chain of oxidation reactions, which involve primarily

Table 15.3 Applications of the photocatalytic processes.

Application	Substrate	Product	References
Total oxidation	Dyes		[10, 22, 30, 37, 38, 96]
	Pharmaceuticals		[39, 40]
	Toxic organic compounds		[18, 23–25, 29, 35]
	Pesticides		[33, 97]
	Herbicides		[34, 98, 99]
	Hormones		[100]
Partial oxidation	2-propanol	Acetone, CO_2	[8]
	4-methoxybenzyl alcohol	p-anisaldehyde	[42]
	Cyclohexane	Cyclohexanol	[43]
	Hydrocarbons	Corresponding oxigenates	[44]
	Aniline	Azobenzene	[101]
	Carbonate	Methane, methanol	[102]
	Methane	Methanol	[11]
	Benzene	Phenol	[45, 46]
	NH_3	N_2	[47]
	Herbicide		[103]
Reduction	4-nitrophenol	4-aminophenol	[4]
	Metal ions	Noble metals	[48–52]
	BrO_3^-	Br^-	[14]
	Nitrite	Ammonia	[53]
	Nitrate	N_2	[12]
	CO_2	Methane and H_2	[63]
	Dyes destruction		[28]
	p-chloronitrobenzene	p-chloroaniline	[54]
	Dyes		[15]
Redox	Cr(VI), BPA	Cr(III), oxidation products	[21]
	Cr(VI), dyes	Cr(III), oxidation products	[55]
Functionalization			
Halogenation	Arenes, Cycloalkenes	Halo-derivates, epoxide	[27, 56]
Ciclization	Amino acid	cicloderivates	[59]
Thiolation	Propene/H_2S	Propan-1-thiol	[60]
Hydrogen production	Water		[5, 26, 62]
	Ethanol/H_2O		[9]

the OH$^\bullet$, all molecules present are mineralized to inorganic species: carbon to CO_2, hydrogen to H_2O, sulfurs to sulfates, halogens to halide ions and phosphorus to phosphates, respectively.

A large number of studies reports the use of photocatalytic reactions for the mineralization of pesticides [33], herbicides [34], toxic organic compounds [18, 35, 36], dyes [22, 37, 38], pharmaceuticals [39, 40].

15.4.2
Selective Oxidations

Applications of photocatalytic oxidations as a 'green' alternative synthetic route has been investigated by several authors although these reactions have always been considered as highly nonselective processes.

Nevertheless, by selecting or modifying some photocatalytic parameters, such us the semiconductor surface or the wavelength, it is possible to control and modify the types of products and their distributions.

Palmisano *et al.* [41] in a study on the selectivity of hydroxyl radical in the partial oxidation of different benzene derivatives have investigated how the substituent group affect the distribution of the hydroxylated compounds. The reported results show that the primary photocatalytic oxidation of compounds containing an electron donor group (phenol, phenylamine, etc.) leads to a selective substitution in *ortho* and *para* positions of aromatic molecules while in the presence of an electron-withdrawing group (nitrobenzene, benzoic acid, cyanobenzene, etc.) the attack of the OH radicals is nonselective, and a mixture of all the three possible isomers is obtained.

Palmisano *et al.* [42], moreover, studied the selective oxidation of 4-methoxybenzyl alcohol to p-anisaldehyde in organic-free aqueous TiO_2 suspensions, obtaining a considerable yield of 41.5% mol. The homemade photocatalysts were obtained under mild conditions and showed to be more selective than two common commercial samples, that is, TiO_2 Degussa P25 and Merck. Nevertheless, although the reported findings are very intriguing in the light of the possibility to potentially synthesize fine chemicals in green conditions, it should be highlighted that the initial alcohol concentration used in this work (circa 1.1 mM) is quite low in comparison with those used for typical organic syntheses.

Colmenares *et al.* [8] reported the use of different metal-doped TiO_2 systems for the gas-phase selective photo-oxidation of 2-propanol to acetone. They observed that doping the catalyst with Pd, Pt or Ag caused an increase in molar conversion as compared to bare TiO_2, whereas the presence of Fe and Zr had a detrimental effect.

Du *et al.* [43] performed a study on the selective photocatalytic oxidation of cyclohexane investigating the influence of the wavelength on the product yields. They observed that the major product in the absence of the catalyst at $\lambda < 275$ nm was cyclohexanol, while the presence of the catalyst gave rise to a significant formation of cyclohexanone under the same experimental conditions. When a proper light filter (Pyrex, $\lambda > 275$ nm) was used, an almost complete selectivity was obtained (>95%).

Selective photocatalytic reactions are also used for the conversion of hydrocarbons (cyclohexane, toluene, methylcyclohexane, ethylbenzene, and cumene) to their corresponding oxygenates [44], benzene to phenol [45, 46], ammonia to nitrogen over various TiO_2 catalysts [47], methane to methanol [11], and so on.

15.4.3 Reduction Reactions

In order to photoreduce chemical compounds, the conduction band of the catalyst must be more negative than the reduction potential of the substrates.

A potential and very attractive practical application of reduction by semiconductor photocatalysis technology is the removal of harmful toxic metals and the recovery of noble metals in wastewater. Metal species, such as Hg(II), Pb(II), Cd(II), Ag(I), Ni(II) and Cr(VI), are generally nondegradable and they are very toxic when present in the environment.

Chen and Ray [48] in a study on the photocatalytic reduction of environmentally relevant metal ions on TiO_2 suspensions, observed that the presence of dissolved oxygen, acting as electron acceptor, inhibits the reduction of metal ions with low reduction potentials, while the presence of organic reductants promotes photocatalytic reduction.

Other authors reported the use of photocatalytic reduction to convert metal species as Cr(VI) to Cr(III) [49], Fe(VI) to Fe(III) [50], Pd(II) to Pd(0) [51], Hg(II) to Hg(0) [52].

The reductive potential of the photocatalysts has been exploited not only for the recovery of metallic ions but also for the degradation of other potentially toxic ions and molecules as bromate ions to Br^- [14], decomposition of nitrate to form nitrogen in water [12] or for the reductive destruction of dyes [28].

Moreover, the photocatalytic reduction has been used to convert some species in others of interest, such as nitrite to ammonia [53], or p-chloronitrobenzene to p-chloroaniline [54].

Further investigations reported the use of the redox potential of the photocatalytic processes for the simultaneous degradation of organic molecules and reduction of metallic ions.

Liu *et al.* [21] investigated the simultaneous photocatalytic reduction of Cr(VI) and oxidation of bisphenol A (BPA) in an aqueous solution in the presence of Fe(III)–OH complexes as catalysts, achieving a synergy effect of the simultaneous photocatalytic oxidation and reduction of both pollutants. Papadam *et al.* [55], instead, coupled the reduction of Cr(VI) to the oxidative degradation of an azodye, while in another study it was reported the simultaneous photocatalytic reduction of Fe(VI) and oxidation of ammonia [50].

15.4.4 Functionalization

By careful selection of the semiconductors it is possible to use heterogeneous photocatalysis as an alternative approach to more conventional synthetic pathways.

In a study on the CH_2Cl_2-assisted functionalization of cycloalkenes Maldotti *et al.* [56] proposed the use of photoexcited decatungstate (POM) for the oxidation of cyclohexene and cyclooctene in highly reactive chloro-intermediates able to induce mono-oxygenation and/or chlorination of alkenes in mild temperature and pressure conditions.

The photocatalytic ability of POM to induce bromide-assisted functionalization process was also studied by Molinari *et al.* [27] in the bromination of some aromatics and alkenes. They reported the possibility to convert phenol and anisole to the corresponding monobrominated derivates and a wide range of cycloalkenes to dibromides and bromohydrins, the last ones as intermediates for the formation of epoxides.

The obtained results are of great interest in fine chemistry and organic synthesis, since epoxides are very versatile building blocks and halogenation of alkenes is still carried out using hazardous reagents and drastic conditions.

Caronna *et al.* [57, 58] report the sunlight induced reactions of some heterocyclic bases (quinoline, quinaldine, quinoxaline, etc.) with amides or ethers in the presence of polycrystalline TiO_2. It was found that the photoreactions occurred in heterogeneous system with higher yields than in homogenous systems under the same experimental conditions.

The wide potentiality of semiconductor-mediated photocatalysis was also applied for the transformation of functional groups such as selective cyclization of amino acids in aqueous suspensions [59].

In the field of thiochemistry, the photocatalytic synthesis of mercaptans represents an interesting chemical route. Schoumacker *et al.* [60] performed the synthesis of propan-1-thiol by addition of H_2S on propene in contact with illuminated TiO_2 or CdS catalysts, according to a reaction mechanism implying photogenerated $SH^{\bullet}$ radicals.

15.4.5
Hydrogen Production

Photocatalytic water splitting, a reaction in which water molecules are reduced by the electrons to form H_2 and oxidized by the holes to form O_2, using semiconductor materials is one of the most important reactions for solving energy and environmental problems. Hydrogen is considered as an ideal fuel for the future and its synthesis from clean and renewable energy sources represents the key component in sustainable energy systems. However, as observed by Ni *et al.* [5] in a review on photocatalytic water splitting for hydrogen production, presently only about 5% of the commercial hydrogen is produced primarily via water electrolysis, while other 95% is mainly derived from fossil fuels, such as natural gas, petroleum and coal.

The early work of photoelectrochemical hydrogen production using TiO_2 as catalyst, was reported by Fujishima and Honda [61]. Subsequently, the interest for the photocatalytic processes has grown significantly, although the number of the reported photocatalysts used for water splitting is still limited.

However, recent studies demonstrate that materials, such as $Pt/SrTiO_3$ codoped with Cr and Sb or Ta, $Pt/NaInS_2$, $Pt/AgInZn_7S_9$ and Cu- or Ni-doped ZnS photocatalysts, showed high activities for H_2 evolution from aqueous solutions under visible light irradiation [62].

Another approach for H_2 photocatalytic production is reported by Mizukoshi *et al.* [9] that described the syntheses of noble metal nanoparticles TiO_2 composite photocatalysts by the sonochemical reduction method for hydrogen evolution from ethanol aqueous solutions at room temperature.

Besides, in an interesting study Sing Tan *et al.* [63] reported the possibilities to use the photocatalytic reduction of carbon dioxide with water to produce hydrogen and methane.

Using the photocatalytic process to reduce CO_2 into hydrocarbons could contribute to the control of CO_2 emission from industrial processes and both produced gases could become the key components of clean 'green' energy systems in the future.

15.4.6
Combination of Heterogeneous Photocatalysis with Other Operations

The demand to develop efficient systems for pollutant abatement and wastewater treatment has brought some researchers to develop systems in which the photocatalytic process is combined with other methods with the purpose to increase its efficiency and to decrease the reaction time thanks to a synergistic effect.

As reported by Augugliaro *et al.* [64] the photocatalysis can be combined with chemical or physical operations. In the first case, when the coupling is with ozonation [65, 66], ultrasonic irradiation, photo-Fenton reaction or electrochemical treatment, which influence the photocatalytic mechanism, an increase of the efficiency of the process is obtained.

Coupling photocatalysis with a physical technologies, such as biological treatment [67, 68], membrane reactor [39] or physical adsorption, the combination does not affect the mechanisms but increases the efficiency of the whole process.

The choice of the combination depends on the characteristics of the effluent to be treated, in other words the best solution for all cases does not exist (especially when actual effluents are under investigation), but it is mandatory to take into account all the drawbacks and the advantages before taking a decision.

15.5
Advantages and Limits of the Photocatalytic Technologies

Photocatalysis is a very promising technology that offers interesting advantages:

- The reactions occur under mild experimental conditions, that is, usually under ambient temperature and pressure without the presence of additives (only oxygen from the air) and in short times;
- It can be applied to a wide range of substrates in aqueous, solid and gaseous phase;
- The process can be adapted to destroy a variety of hazardous molecules and pollutants in different wastewater streams;

- In comparison to the simple transfer of the pollutants from one medium to another, which occurs with the conventional water-treatment methods, photocatalysis leads to the real destruction of the contaminants with the formation of innocuous products;
- The system is applicable to solutions at low concentrations;
- The process is able to recover noble metals and toxic metals that are converted to their less-toxic/nontoxic metallic states;
- Combination with other physical and chemical technologies is possible;
- It offers a good alternative to the energy-intensive conventional treatments methods with the possibility to use renewable solar energy.

Despite the great potentialities of the photocatalytic processes, their application at the industrial level is limited by different factors. Besides the problems regarding the high reactivity of the system and the low selectivity of the classical catalysts on which, as previously described, many studies have been realized in recent years, one of the main problems is the recovery of the catalyst from the reaction environment.

As observed by Choi [7], with respect to the large number of studies reported on the development and synthesis of highly efficient photocatalysts, few studies have been performed for the design of efficient photoreactors for commercial exploitation.

For the development of a continuous photocatalytic reactor, applicable at the industrial level, it is important to consider some parameters such as the catalyst configuration, the specific illuminated surface area, the UV source, the mass-transfer rate and the scale-up possibilities [69].

Regarding the catalyst configuration, two operating modes of the photoreactors can be identified: catalyst suspended or catalyst immobilized on a support.

In the supported systems the catalyst can be coated on the walls of the reactor, supported on a solid substrate or deposited around the case of the light source. Many are the supported materials used in literature, such as glass beads, and tubes [69], silica-based materials [70], hollow beads, membranes [71], optical fibers, zeolites, activated carbon, organic fibers [72], and so on.

In this way, there is not a requirement for separation steps, no particle aggregation occurs especially at high concentrations and the thin films can be easily adopted to realize continuous-flow systems. However, many studies [69, 73–75] showed that the suspended system is more efficient.

The low photocatalytic efficiency of the supported systems is due to several factors that influence the catalyst activity.

Heterogeneous catalysis is a surface phenomenon, therefore the overall kinetic parameters are dependent on the real exposed catalyst surface area. In the supported systems only a part of the photocatalyst is accessible to light and to substrate. Besides, the immobilized catalyst suffers from the surface deactivation since the support could enhance the recombination of photogenerated electron–hole pairs and a limitation of oxygen diffusion in the deeper layers is observed.

On this basis more efforts in photocatalytic engineering and reactor development are required to realize an efficient photocatalytic reactor.

15.6
Membrane Photoreactors

15.6.1
Introduction

The recovery of the photocatalyst from the reaction environment represents one of the main problems of the photocatalytic process that limits its industrial application. Although this process step can be obviated by the use of immobilized catalyst, the suspended system has more attractive features [76]. Therefore, the separation of the photocatalyst from the treated solution and its recycle is one of the challenges in further development of this technology.

A very promising method to solve this problem is coupling the photocatalysis with membrane techniques, obtaining a very powerful process with great innovation in water treatment. In fact, membrane processes, thanks to the selective property of the membranes, have been shown to be competitive with the other separation technologies for what concerns material recovery, energy costs, reduction of the environmental impact and selective or total removal of the components [77].

The photocatalytic membrane reactors (PMRs) combine the advantages of classical photoreactors (catalyst in suspension) and those of membrane processes (separation at molecular level) with a synergy for both technologies.

The membrane allows not only the recovery and reuse of the catalyst but also the selective separation of the molecules present in the reaction environment. In the conventional photoreactors the molecules and their by-products are freely transported in the final stream giving a not efficient system [73]. In a PMR, if a suitable membrane is used, it is possible to enhance the residence time of the molecules to be degraded or to obtain a selective separation of the products.

Besides, the membrane photoreactors allow operation in continuous systems [39] in which the reaction of interest and the separation of the product(s) simultaneously occur, avoiding in some cases the formation of by-products.

Several authors have proposed promising solutions involving the use of membranes to maintain the catalyst in suspension or to immobilize it on the membrane inside the reactor. Nevertheless, despite the potential advantages of these hybrid photoreactors, the research on the PMRs is not sufficiently developed yet.

15.6.2
Membrane Photoreactor Configurations

Various types of photocatalytic membrane reactors in which the catalyst was used in different modes have been built with the purpose to have an easy separation of the catalyst from the reaction environment: a photocatalyst in suspension in magnetically or mechanically agitated slurries confined by means of a membrane, fixed bed, catalyst deposited or entrapped on an inert support or in a membrane, and so on.

15.6.2.1 Pressurized Membrane Photoreactors

The majority of PMRs described in literature combines photocatalysis with a pressure-driven membrane technique, such as nanofiltration (NF), ultrafiltration (UF) and microfiltration (MF), in which the catalyst is contained in the pressurized side of the membrane.

The first studies reported in the literature on PMRs were carried out in order to identify the best techniques for confining the photocatalyst, to choose suitable polymers for the membranes stable under UV-Vis illumination and to find the influence of some photocatalytic variables on the process [77–79].

Molinari *et al.* [73] reported a study on various configurations of photocatalytic membrane reactors for the degradation of 4-nitrophenol using TiO_2 as catalyst. Two configurations have been studied: irradiation of a recirculation tank, with the suspended catalyst confined by means of the membrane, and irradiation of the cell containing the membrane with three sub-cases: catalyst deposited on the membrane, catalyst suspended, and catalyst included in a membrane during its preparation.

Among the configurations described in that paper, the irradiation carried out on the recirculation batch seems very promising since it allows high irradiation efficiency and high membrane permeate flow rate to be obtained and also it is possible to select the membrane type depending on the photocatalytic process under study.

In another study, Tsuru *et al.* [80] reported the use of porous TiO_2 membranes having pores of several nanometers for a gas-phase photocatalytic reaction of methanol as a model of volatile organic component (VOC). In this system, the titanium dioxide is immobilized in the form of a porous membrane that is capable of selective permeation and also a photocatalytic oxidation that occurs both on the surface and inside the porous TiO_2 membranes. In this way, it is possible to obtain a permeate stream oxidized with OH radicals after one-pass permeation through the TiO_2 membranes.

Nevertheless, comparing the photodegradation efficiency of the membrane photoreactor made with entrapped TiO_2 to that with suspended TiO_2 it was observed [81] that although the amount of TiO_2 used was identical, the suspended system was more efficient. This is probably due to the presence of the polymer around the particles of catalyst that reduces the real amount of photoexcited TiO_2. Besides, in this study it was observed that the rate of pollutant photodegradation was strongly affected by the UV irradiation mode. In particular, some experimental runs were performed by using two types of photoreactor: a cylindrical photoreactor with external lamp and an annular one with immersed lamp. Although the power of the immersed lamp was about four times lower than that of the external lamp, the first system was three times more efficient than the latter.

However, since these systems need a pressure, fouling can occur with a decrease of the efficiency of the overall process.

15.6.2.2 Sucked (Submerged) Membrane Photoreactors

With the aim to avoid or to reduce the membrane fouling, which causes the membrane flux decline, another type of membrane module configuration of PMR, such as the submerged membrane system has been also studied. The catalyst is suspended in an open-air reaction environment, the membrane is immersed in the batch and the permeate is sucked by means a pump.

In particular, in order to control the hydrodynamic conditions near membrane surface to prevent catalyst deposition on the membrane, the research has been addressed to develop efficient strategies such as gas sparging at the bottom of the membrane [82, 83].

In recent years some studies were performed using submerged membrane modules coupled to photocatalytic systems for the removal of organic pollutants such as fulvic acid [84], bisphenol-A [85], para-chlorobenzoate [86].

In a study on a submerged membrane photocatalysis reactor (SMPR) for the degradation of fulvic acid using a nanostructured TiO_2 catalyst [84], the effects of the operational parameters such as TiO_2 amount, pH, and airflow were investigated. The reported results show that the permeate flux rate was improved and thus the membrane fouling was reduced with addition of nanostructured TiO_2 because it has a larger particle size than P25 and therefore it can be easily separated and reused by means the membrane.

Chin *et al.* [85] used a hybrid system combining a low-pressure submerged membrane module and a TiO_2 suspension to purify water containing bisphenol-A. The factors affecting the performance of a SMPR were studied and, in particular, it was observed that the aeration, allowing a mechanical agitation, reduces the fouling of the membrane and keeps the TiO_2 well suspended in the solution, acting also on the size of catalyst aggregates. However, beyond an aeration of $0.5\,L\,min^{-1}$ no enhancement of photodegradation rate was observed, probably due to the presence of bubble clouds that could attenuate UV-light transmission in the photoreactor. Besides, in this study the possibility of using an intermittent permeation method was demonstrated with the aim to maintain high flux ($100\,L\,h^{-1}\,m^{-2}$) at low aeration rate, with low membrane fouling.

The ultrafiltration or microfiltration membrane used in the described submerged membrane photoreactors showed high fluxes and a good removal efficiency of organic molecules, nevertheless, they are not able to reject compounds with low molecular weight.

In order to solve this problem, Choi *et al.* [87] studied the performance of a submerged membrane bioreactor using NF cellulose acetate membrane for domestic wastewater treatment. The reported results underlined a DOC concentration in the permeates in the range $0.5–2.0\,mg\,L^{-1}$ for the first 130 days, and subsequently increased approximately to $3.0\,mg\,L^{-1}$, indicating that the NF submerged MBR could produce very good quality permeate for a long-term operation.

15.6.2.3 **Membrane Contactor Photoreactors**

Other types of membrane separation processes that can be useful when they are coupled with a photocatalytic system are the membrane contactors.

The separation performance in these processes is determined by the distribution coefficient of a component in two phases and the membrane acts only as an interface. They can be divided in gas–liquid (G–L) and liquid–liquid (L–L) membrane contactors. In the first type, one phase is a gas or a vapor and the other phase is a liquid, whereas in the latter both phases are liquid [88].

Membrane contactors represent interesting membrane processes that allow, for example, to separate one or most reaction products from the photocatalytic environment thanks to their different distribution coefficient between the two phases. In this way it is possible to solve the problem of secondary reactions that occur during a synthetic pathway and lead to undesirable by-products.

Pervaporation – photocatalysis In the described systems the membrane usually permeates water and rejects the reactants, enhancing their residence time in the photoreactor. However, it is known that some intermediate products of the photocatalytic degradation of organic compounds can negatively affect the reaction rate, therefore, in some cases it is useful to eliminate these by-products in order to improve the thermodynamic and/or the kinetics of the reaction.

To this purpose, in a study on the photocatalytic degradation of 4-chlorophenol, Camera-Roda and Santarelli [89] proposed an integrated system in which photocatalysis is coupled with pervaporation as process intensification for water detoxification. Pervaporation represents a useful separation process in the case of the removal of VOCs and in this study it is used to remove continuously and at higher rate the organic intermediates that are formed in the first steps of the photocatalytic degradation of the weakly permeable 4-CP.

Membrane distillation – photocatalysis To solve the problem of membrane fouling observed in the pressure-driven membrane photoreactor, Mozia *et al.* [90] studied a new type of PMR in which photocatalysis was combined with a direct contact membrane distillation (DCMD). MD can be used for the preparation of ultrapure water or for the separation and concentration of organic matter, acids and salt solutions. In the MD the feed volatile components are separated by means of a porous hydrophobic membrane thanks to a vapor-pressure difference that acts as driving force and then they are condensed in cold distillate (distilled water), whereas the nonvolatile compounds were retained on the feed side.

In this study the possibility was investigated of coupling photocatalysis and membrane distillation for degradation of organic pollutants in aqueous solution using Acid Red 18 as a model dye and TiO_2 Aeroxide P25 as photocatalyst. In particular, the rejection of the MD process towards the catalyst, the dye and its photodegradation products was tested and it was found that the nonvolatile model dye was completely rejected by the membrane and it remained in the feed side. In the case of TOC concentration in permeate, an initial significant increase of concentration during the first hour was observed, probably due to the production of some volatile by-products formed at the beginning of the process. However, these compounds can be degraded into water, CO_2 and other noncarbon species during subsequent hours and, therefore, the TOC values in the permeate became lower.

Dialysis – photocatalysis Azrague *et al.* [91] described a particular type of membrane contactor photoreactor in which a dialysis membrane (used as a contactor) was combined with a photocatalytic system for the decontamination of turbid waters. Photocatalysis is not easily applicable in the presence of solid particles that screening the radiations determine a decrease of the photocatalytic degradation. On this basis, the aim of this work was to realize a system in which the membrane process allows the solid particles to be kept in their initial compartments and also the transport of the pollutant (2,4-DHBA) by diffusion from the feed-tank compartment to the other where the photocatalytic reaction takes place, until a total mineralization, thanks to the different concentration between the two compartments.

In this way, the pollutants are extracted from the turbid water and then degraded, without the need of a transmembrane pressure, avoiding the fouling of membrane, which is an expensive problem in case of pressure-driven membrane processes.

15.6.3
Parameters Influencing the Photocatalytic Membrane Reactors (PMRs) Performance

In the development of a photocatalytic membrane reactor it is important to take into account some parameters that influence the performance of the system and its applicability to the industrial level.

One of the main objectives in the use of a membrane process coupled to a photocatalytic reaction is the possibility of recovering and reusing the catalyst. Moreover, when the process is used for the degradation of organic pollutants, the membrane must be able to reject the compounds and their intermediate products, while if the photocatalysis is applied to a synthesis, often the membrane have to separate the product(s) from the environment reaction. Therefore, in a PMR the choice of a suitable membrane is essential to obtain an efficient system.

To achieve a high membrane rejection towards the substrate it is important that the pore size of the membrane is smaller than the size of the molecules to be retained. Nevertheless, other factors influence the separation properties of a membrane, such as the shape and flexibility of the substrate and its acid–base properties, as well as the concentration-polarization phenomenon and the membrane fouling.

Sometimes, thanks to the Donnan effect, by modifying the pH it is possible to retain in the reaction ambient molecules that otherwise would pass the membrane. In fact, some membranes can become electrically charged at acidic and alkaline pHs, then repulsive or attractive interactions between the substrate molecules and the membrane surface may occur if the charges are of the same or of different sign, respectively. Repulsive interactions increase rejection values, whereas attractive ones decrease them.

In a pressure-driven membrane process the molecules are generally rejected by the membrane and therefore their concentrations in the permeate are lower than those in the feed solution. However, an accumulation of excess particles can occur at the membrane surface with the creation of a boundary layer. This phenomenon, called concentration polarization, causes a different membrane performance. In particular, with low molecular weight solutes the observed rejection will be lower than the real retention or, sometimes, it could be negative.

Moreover, in the presence of a layer deposited on the membrane surface an increase of the resistance to solvent flow occurs that reduces the permeate flux.

This problem could be obviated by creating a turbulent flow in the permeation cell, which allows catalyst and drug deposition on the membrane surface to be reduced.

Another important parameter that must be take into account when a PMR is used for the water treatment is the water permeability of the membrane.

The selected membrane must be able not only to selectively confine the pollutants and the catalyst, but also to offer a high water permeate flux in order to achieve a system for application purposes.

It is worth noting that when the photocatalytic process is applied to organic synthesis, the role of the membrane becomes essential in the separation of the products from the reaction ambient. In this case it is important that the membrane is able to selectively and quickly separate the product of interest in order to avoid subsequent reactions that would cause the formation of undesirable secondary products.

15.6.4 Future Perspectives: Solar Energy

As previously described, many efforts are addressed toward the development of photocatalyst and photocatalytic systems that exploit the sun as a source of light [92, 93].

Solar energy is important for achievement of sustainable processes because it constitutes a renewable, cheap, and clean energy source.

The possibility to use sunlight makes the photocatalytic membrane reactors promising in industrial and environmental fields, although very few studies have yet been performed in this area.

In a study on the photodegradation of lincomycin Augugliaro *et al.* [40] reported the use of a hybrid system consisting of a solar photoreactor with the catalyst suspended coupled with a membrane module. The photo-oxidation experiments were performed in a batch solar photoreactor at pilot plant scale by using compound parabolic collectors (CPC), installed at the 'Plataforma Solar of Almerìa'.

By means of some preliminary tests performed without the membrane it was determined that the photo-oxidation rate of lyncomycin followed a pseudo-first-order kinetics with respect to the substrate concentration under the used experimental conditions. The high membrane rejection values measured for lincomycin and its degradation products demonstrated that the hybrid system allowed the separation of these species and also of the photocatalyst particles, although in the experiments carried out in continuous mode, an accumulation of organic molecules in the system was observed. This finding, which was dependent on solar irradiance and initial lyncomycin concentration, can be explained by considering that the number of photons entering into the system are not sufficient to mineralize the organic carbon fed into the photoreactor.

Moreover, the experimental results obtained in continuous mode showed that the presence of the membrane allowed both the substrate and intermediates to be reduced down to very low concentrations, proving that the hybrid system could be very interesting from an economic point of view.

15.7 Case Study: Partial and Total Oxidation Reactions in PMRs

In this section some experimental results obtained in our laboratories on PMRs are reported. In particular two different photocatalytic membrane reactors, used in total and partial photocatalytic oxidations are described.

15.7.1 Degradation of Pharmaceutical Compounds in a PMR

The presence in the aquatic environment of pharmaceutically active compounds as an important group of toxic organic pollutants, has focused the attention of the international scientific community.

The design of more stable drugs, in order to enhance their persistence in the organisms, leads to a greater resistance of these molecules to the common chemical and biological degradation treatments, with a consequent increase in the environment.

The development of new systems, alternative to the traditional purification methods (not efficient and often suitable only to transfer pollutants from one phase to another) represents a great research interest.

In this context, hybrid systems based on photocatalysis coupled with separation process could represent an useful solution to these problems.

The aim of our experimental studies was to show the possibility to use the PMRs for the degradation of organic pollutants, in particular drugs, in water, considering different reacting system configurations of membrane photoreactors and investigating the effects of some parameters on the efficiency of the process.

The obtained results have shown that the configuration where the recirculation tank was irradiated and the catalyst was used in suspension appeared to be the most interesting for industrial applications [73]. Moreover, it was observed that the degradation rate was higher when an immersed lamp was used compared to a system with an external lamp [81]. Therefore, actually the studies in progress are realized in the system described elsewhere [39] consisting of a Pyrex annular photoreactor with a 125-W medium-pressure Hg lamp axially positioned inside the reactor. The separation module containing the flat-sheet membrane was connected to the photoreactor in a recirculation loop.

In a first set of runs the influence of pH on the rejection of several pharmaceuticals was studied by using different membranes. Among the tested membranes, the NTR7410 membrane, a NF commercial membrane made of sulfonated polyethersulfone, resulted in a good compromise for the drugs studied both at acidic and alkaline pHs, with an average flux of ca. 45 $L\,h^{-1}\,m^{-2}$ but with low rejection values that decreased down to zero during the photodegradation (Figure 15.2).

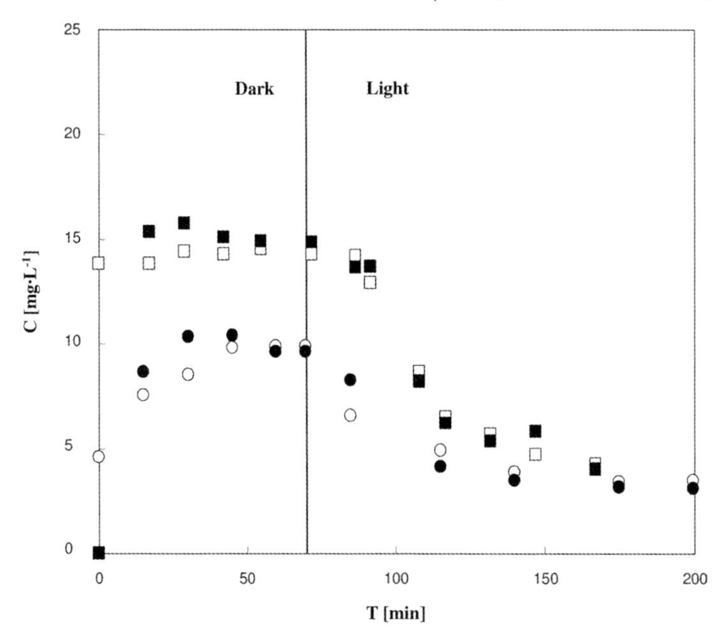

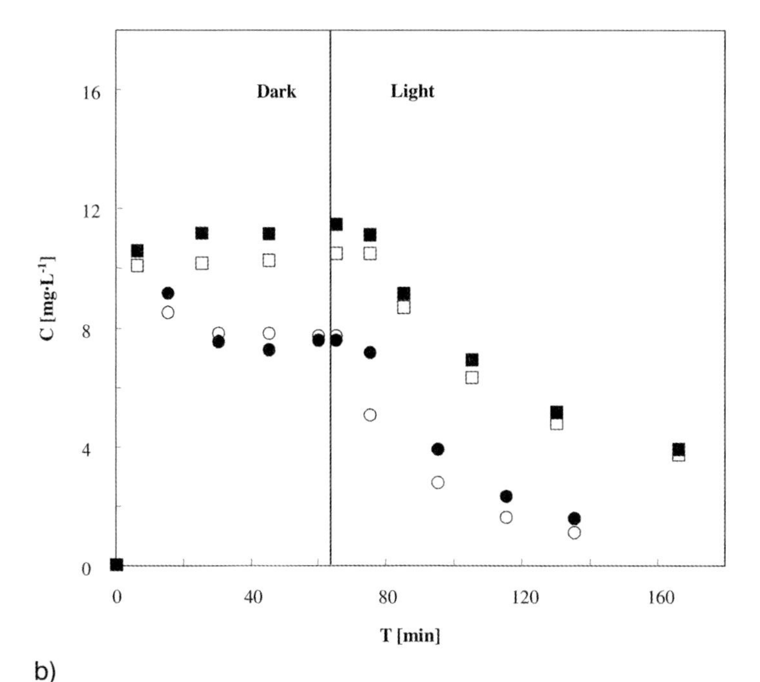

Figure 15.2 Drug concentrations versus time for runs carried out by using the hybrid system with the NTR 7410 membrane at initial pH of 3 (a) and 11 (b) ($C_{TiO2} = 1$ g/l; $C_{O2} = 22$ ppm; immersed lamp 125 W). Furosemide: (○) retentate; (●) permeate. Ranitidine: (□) retentate; (■) permeate [39].

This finding depends on different factors and, as previously described, was partially solved by enhancing the turbulent flow on the membrane surface. In this way, the deposition of the substrate and the catalyst was limited, avoiding the concentration polarization phenomenon that also affects the water flux across the membrane. This aspect is currently under study and some preliminary results obtained in a study on the Gemfibrozil degradation in the described PMR are reported in Table 15.4.

In addition, the maximum benefit when a photocatalytic membrane reactor is used for water purification consists in retaining not only the pollutants but also their intermediates.

By means of TOC analyzes carried out during photodegradation experiments in closed and continuous membrane photoreactors it was observed that some oxidation products pass through the membrane. Therefore, further investigation is in progress to analyze this aspect.

Particular attention is addressed to the permeate flux and to this purpose some preliminary experiments were realized on a different configuration of membrane photoreactor with a submerged membrane module located separately from the photoreactor. Bubbled oxygen on the membrane surface has the roles to reduce the catalyst deposition, to increase the flux through the membrane and to facilitate the photocatalytic reaction.

The submerged membrane photoreactor was more advantageous in terms of permeate flux, with values almost twice those measured with the flat-sheet membranes. Nevertheless, the UF membranes used in the submerged system were not able to reject the drug and its degradation products [94].

Consequently further studies are required to look for different types of membranes, such as for instance higher rejection NF-type or low rejection reverse-osmosis-type membranes, by taking into account the relatively low molecular weight of the drugs studied.

Table 15.4 Rejection values vs. TiO_2 concentration measured with pump flow rate of $10\,L\,h^{-1}$ and $46\,L\,h^{-1}$ [94].

$[TiO_2]$	$Q = 10\,L\,h^{-1}$ R%	$Q = 46\,L\,h^{-1}$ R%
0	75.9	88.25
0.01	56.3	/
0.02	48.2	86.7
0.04	38.9	/
0.06	31.9	80.5
0.10	−20.85	76.7
0.5	/	69.8
1	−60.0	25.9

15.7.2
Photocatalytic Production of Phenol from Benzene in a PMR

The main aim of this study is to prove the possibility to use a PMR for the synthesis and the separation of substances of industrial interest.

In particular, our research is addressed to the simultaneous one-step production and separation of phenol by selective oxidation of benzene in a membrane photoreactor using TiO_2 as catalyst and a membrane contactor for the separation process.

Phenol is an important chemical intermediate of industrial interest, used as substrate for the production of antioxidants, polymers and agrochemicals. Actually, more than of 90% of the world production is realized by the three-step cumene process that leads to the formation of acetone as by-product.

The one-step hydroxylation of benzene represents an attractive alternative pathway for the direct synthesis of phenol and many studies are performed using different processes among which the photocatalytic reaction [45, 46]. One of the main problem is the low selectivity of the process due to the higher reactivity of phenol towards the oxidation than benzene with the formation of oxidation by-products. In order to avoid these secondary products and to obtain the separation of the phenol from the oxidant reaction environment the use of a membrane system coupled with the photocatalytic process seems a useful solution.

On this basis, our research is based on the development of an experimental system that allows both high yields and good selectivity of the process, limiting the formation of undesirable by-products, and an efficient separation by the identification of a membrane with high phenol permeability and complete rejection to the catalyst.

By means of preliminary batch tests some important parameters that influence the photocatalytic oxidation of benzene to phenol were investigated. In particular, the obtained results showed an increase of phenol production depending on the pH of the aqueous TiO_2 suspensions, the catalyst concentration and the radiation intensity.

The membrane photoreactor under investigation consists of an external lamp placed on a batch reactor containing the aqueous solution with the catalyst in suspension; by means of a peristaltic pump the solution is withdrawn from the

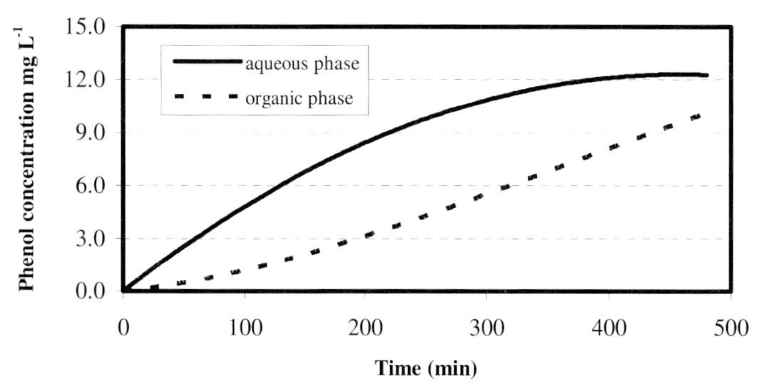

Figure 15.3 Phenol concentrations in aqueous and organic phase during an experimental run in the PMR (Molinari *et al.*, unpublished data).

photocatalytic reactor to a membrane contactor module in which a benzene solution is present as a strip phase.

The data obtained in a first set of experiments have shown a good separation of phenol from the reaction environment (Figure 15.3), although also other oxidation products passed through the membrane.

Therefore, further studies are in progress in order to obtain a better separation efficiency and a faster synthetic process.

15.8 Conclusions

Photocatalytic membrane reactors represent a very promising technology of great research and industrial interest.

The combination of heterogeneous photocatalysis with membrane processes provides many advantages in terms of output and costs thanks to their synergy.

It is well known that heterogeneous photocatalysis can be successfully used to photodegrade or to transform a wide range of molecules in liquid–solid and in gas–solid systems. Nevertheless, the knowledge of fundamentals of photocatalysis is essential to understand the mechanistic aspects and to find the parameters that influence the process under investigation. Moreover, the development of new photocatalysts and their application in the various research fields is a mandatory task.

Some drawbacks deriving from the use of a single technology can be minimized by coupling them.

Membrane processes, indeed, thanks to their selective properties, allow not only to recovery and to reuse the photocatalyst but also to enhance the residence time of the substrates to be degraded or to obtain a selective separation of the products.

Various configurations of membrane photoreactors described can be chosen to influence the performance of the photocatalytic systems and possible solutions can be found to solve some problems such as the control of the catalyst activity and the fouling, the selectivity and the rejection of the membrane.

A sustainable process can be obtained when the PMRs is used exploiting Sun as a cheap and clean source of light.

Work carried out in our laboratories using PMRs showed the possibility to apply them in processes for total or partial oxidation of organic compounds in water.

PMRs can be considered an useful 'green' system for water purification as well as for organic synthesis, although additional studies are still needed before taking advantage of their potentiality at the industrial level.

References

1 Herrmann, J.-M. (2005) *Topics in Catalysis*, **34**, 49.

2 Kavita, K., Rubina, C. and Rameshwar, L.S. (2004) *Industrial & Engineering Chemistry Research*, **43**, 7683.

3 Robertson, P.K.J. (1996) *The Journal of Cleaner Production*, **4**, 203.

4 Ahn, W.-Y., Sheeley, S.A., Rajh, T. and Cropek, D.M. (2007) *Applied Catalysis B-Environmental*, **74**, 103.
5 Ni, M., Leung, M.K.H., Leung, D.Y.C. and Sumathy, K. (2007) *Renewable and Sustainable Energy Reviews*, **11**, 401.
6 Robert, D. (2007) *Catalysis Today*, **122**, 20.
7 Choi, W. (2006) *Catalysis Surveys from Asia*, **10**, 16.
8 Colmenares, J.C., Aramendía, M.A., Marinas, A., Marinas, J.M. and Urbano, F.J. (2006) *Applied Catalysis A-General*, **306**, 120.
9 Mizukoshi, Y., Makise, Y., Shuto, T., Hu, J., Tominaga, A., Shironita, S. and Tanabe, S. (2007) *Ultrasonics Sonochemistry*, **14**, 387.
10 Wang, C.-T. (2007) *Journal of Non-Crystalline Solids*, **353**, 1126.
11 Taylor, C.E. (2005) *Topics in Catalysis*, **32**, 179.
12 Rengaraj, S. and Li, X.Z. (2007) *Chemosphere*, **66**, 930.
13 Kudo, A. and Sekizawa, M. (2000) *Chemical Communications*, 1371.
14 Noguchi, H., Nakajima, A., Watanabe, T. and Hashimoto, K. (2003) *Environmental Science & Technology*, **37**, 153.
15 Kobasa, I.M. and Tarasenko, G.P. (2002) *Theoretical Chemistry Accounts*, **38** (4), 255.
16 Bonchio, M., Carraro, M., Scorrano, G., Fontananova, E. and Drioli, E. (2003) *Advanced Synthesis and Catalysis*, **345**, 1119.
17 Bonchio, M., Carraro, M., Gardan, M., Scorrano, G., Drioli, E. and Fontananova, E. (2006) *Topics in Catalysis*, **40** (1–4), 133.
18 Marais, E., Klein, R., Antunes, E. and Nyokong, T. (2007) *Journal of Molecular Catalysis A-Chemical*, **261**, 36.
19 Mele, G., Ciccarella, G., Vasapollo, G., García-López, E., Palmisano, L. and Schiavello, M. (2002) *Applied Catalysis B-Environmental*, **38**, 309.
20 Mele, G., Del Sole, R., Vasapollo, G., García-López, E., Palmisano, L., Jun, L., Słota, R. and Dyrda, G. (2007) *Research on Chemical Intermediates*, **33**, 433.
21 Liu, Y., Deng, L., Chen, Y., Wu, F. and Deng, N. (2007) *Journal of Hazardous Materials. B*, **139**, 399.
22 Sobana, N. and Swaminathan, M. (2007) *Solar Energy Materials and Solar Cells*, **91**, 727.
23 Ito, S., Thampi, K.R., Comte, P., Liska, P. and Grätzel, M. (2005) *Chemical Communications*, 268.
24 Huang, G. and Zhu, Y. (2007) *Materials Science and Engineering B*, **139**, 201.
25 Amat, A.M., Arques, A., Galindo, F., Miranda, M.A., Santos-Juanes, L., Vercher, R.F. and Vicente, R. (2007) *Applied Catalysis B-Environmental*, **73**, 220.
26 Jang, J.S., Choi, S.H., Shin, N., Yu, C. and Lee, J.S. (2007) *Journal of Solid State Chemistry*, **180**, 1110.
27 Molinari, A., Varani, G., Polo, E., Vaccari, S. and Maldotti, A. (2007) *Journal of Molecular Catalysis A-Chemical*, **262**, 156.
28 Troupis, A., Gkika, E., Triantis, T., Hiskia, A. and Papacostantinou, E. (2007) *Journal of Photochemistry and Photobiology A-Chemistry*, **188**, 272.
29 Tang, J. and Ye, J. (2005) *Chemical Physics Letters*, **410**, 104.
30 Xu, H., Wang, H. and Yan, H. (2007) *Journal of Hazardous Materials*, **144**, 82.
31 Robert, D. and Malato, S. (2002) *The Science of the Total Environment*, **291**, 85.
32 Bekkouche, S., Bouhelassa, M., Hadj Salah, N. and Meghlaoui, F.Z. (2004) *Desalination*, **166**, 355.
33 Lhomme, L., Brosillon, S. and Wolbert, D. (2007) *Journal of Photochemistry and Photobiology A-Chemistry*, **188**, 34.
34 Singh, H.K., Saquib, M., Haque, M.M. and Muneer, M. (2007) *Journal of Hazardous Materials*, **142**, 425.
35 Waldner, G., Brüger, A., Gaikwad, N.S. and Neumann-Spallart, M. (2007) *Chemosphere*, **67**, 779.
36 Wu, C., Chang-Chien, G. and Lee, W. (2005) *Journal of Hazardous Materials. B*, **120**, 257.
37 Chen, C.-C. (2007) *Journal of Molecular Catalysis A-Chemical*, **264**, 82.

38 Lou, X., Han, J., Chu, W., Wang, X. and Cheng, Q. (2007) *Materials Science and Engineering. B*, **137**, 268.

39 Molinari, R., Pirillo, F., Loddo, V. and Palmisano, L. (2006) *Catalysis Today*, **118**, 205.

40 Augugliaro, V., García-López, E., Loddo, V., Malato-Rodríguez, S., Maldonado, I., Marcì, G., Molinari, R. and Palmisano, L. (2005) *Solar Energy*, **79**, 402.

41 Palmisano, G., Addamo, M., Augugliaro, V., Caronna, T., Di Paola, A., García López, E., Loddo, V., Marcì, G., Palmisano, L. and Schiavello, M. (2007a) *Catalysis Today*, **122**, 118.

42 Palmisano, G., Yurdakal, S., Augugliaro, V., Loddo, V. and Palmisano, L. (2007b) *Advanced Synthesis and Catalysis*, **349**, 964.

43 Du, P., Moulijin, J.A. and Mul, G. (2006) *Journal of Catalysis*, **238**, 342.

44 Gonzales, M.A., Howell, S.G. and Sikdar, S.K. (1999) *Journal of Catalysis*, **183**, 159.

45 Park, H. and Choi, W. (2005) *Catalysis Today*, **101**, 291.

46 Shimizu, K., Akahane, H., Kodama, T. and Kitayama, Y. (2004) *Applied Catalysis A-General*, **269**, 75.

47 Yamazoe, S., Okumura, T. and Tanaka, T. (2007) *Catalysis Today*, **120**, 220.

48 Chen, D. and Ray, A.K. (2001) *Chemical Engineering Science*, **56**, 1561.

49 Kanthasamy, R. and Larsen, S.C. (2007) *Microporous and Mesoporous Materials*, **100**, 340.

50 Sharma, V.K. and Chenay, B.V.N. (2005) *Journal of Applied Electrochemistry*, **35**, 775.

51 Troupis, A., Hiskia, A. and Papaconstantinou, E. (2004) *Applied Catalysis B-Environmental*, **52**, 41.

52 Wang, X., Pehkonen, S.O. and Ray, A.K. (2004) *Electrochimica Acta*, **49**, 1435.

53 Ranjit, K.T., Krishnamoorthy, R., Varadarajan, T.K. and Viswanathan, B. (1995) *Journal of Photochemistry and Photobiology A-Chemistry*, **86**, 185.

54 Zhang, T., You, L. and Zhang, Y. (2006) *Dyes and Pigments*, **68**, 95.

55 Papadam, T., Xekoukoulotakis, N.P., Poulios, I. and Mantzavinos, D. (2007) *Journal of Photochemistry and Photobiology A-Chemistry*, **186**, 308.

56 Maldotti, A., Amadelli, R., Vitali, I., Borgatti, L. and Molinari, A. (2003) *Journal of Molecular Catalysis A-Chemical*, **204**, 703.

57 Caronna, T., Gambarotti, C., Palmisano, L., Punta, C. and Recupero, F. (2003) *Chemical Communications*, 2350.

58 Caronna, T., Gambarotti, C., Palmisano, L., Punta, C. and Recupero, F. (2005) *Journal of Photochemistry and Photobiology A-Chemistry*, **171**, 241.

59 Ohtani, B., Pal, B. and Shigeru, I. (2003) *Catalysis Surveys from Asia*, **7**, 165.

60 Schoumacker, K., Geantet, C., Lacroix, M., Puzenat, E., Guillard, C. and Hermann, J.-M. (2002) *Journal of Photochemistry and Photobiology A-Chemistry*, **152**, 147.

61 Fujishima, A. and Honda, K. (1972) *Nature*, **238**, 37.

62 Kudo, A. (2003) *Catalysis Surveys from Asia*, **7** (1), 31.

63 Sing Tan, S., Zou, L. and Hu, E. (2007) *Science and Technology of Advanced Materials*, **8**, 89.

64 Augugliaro, V., Litter, M., Palmisano, L. and Soria, J. (2006) *Journal of Photochemistry and Photobiology C: Photochemistry Reviews*, **7**, 127.

65 Addamo, M., Augugliaro, V., García-López, E., Loddo, V., Marcì, G. and Palmisano, L. (2005) *Catalysis Today*, **107–108**, 612.

66 Hernández-Alonso, M.D., Coronado, J.M., Maira, A.J., Soria, J., Loddo, V. and Augugliaro, V. (2002) *Applied Catalysis B-Environmental*, **39**, 257.

67 Parra, S., Sarria, V., Malato, S., Péringer, P. and Pulgarin, C. (2000) *Applied Catalysis B-Environmental*, **27**, 153.

68 Pulgarin, C., Invernizzi, M., Parra, S., Sarria, V., Polania, R. and Péringer, P. (1999) *Catalysis Today*, **54**, 341.

69 Dijkstra, M.F.J., Buwalda, H., de Jong, A.F., Michorius, A., Wilkelman, J.G.M. and Beenackers, A.A.C.M. (2001) *Chemical Engineering Science*, **56**, 547.

70 Coronado, J.M., Soria, J., Conesa, J.C., Bellod, R., Adán, C., Yamaoka, H., Loddo, V. and Augugliaro, V. (2005) *Topics in Catalysis*, **35** (3–4), 279.

71 Rincón, A.G. and Pulgarin, C. (2003) *Applied Catalysis B-Environmental*, **44**, 263.

72 Rao, K.V.S., Subrahmanyam, M. and Boule, P. (2004) *Applied Catalysis B-Environmental*, **49**, 239.

73 Molinari, R., Palmisano, L., Drioli, E. and Schiavello, M. (2002) *Journal of Membrane Science*, **206**, 399.

74 Mascolo, G., Comparelli, R., Curri, M.L., Lovecchio, G., Lopez, A. and Agostiano, A. (2007) *Journal of Hazardous Materials*, **142**, 130.

75 Sakkas, V.A., Arabatzis, I.M., Konstantinou, I.K., Dimou, A.D., Albanis, T.A. and Falaras, P. (2004) *Applied Catalysis B-Environmental*, **49**, 195.

76 Sopajaree, K., Qasim, S.A., Basak, S. and Rajeshwar, K. (1999) *Journal of Applied Electrochemistry*, **29**, 533.

77 Molinari, R., Grande, C., Drioli, E., Palmisano, L. and Schiavello, M. (2001) *Catalysis Today*, **67**, 273.

78 Sopajaree, K., Qasim, S.A., Basak, S. and Rajeshwar, K. (1999) *Journal of Applied Electrochemistry*, **29**, 1111.

79 Molinari, R., Mungari, M., Drioli, E., Di Paola, A., Loddo, V., Palmisano, L. and Schiavello, M. (2000) *Catalysis Today*, **55**, 71.

80 Tsuru, T., Kan-no, T., Yoshioka, T. and Asaeda, M. (2003) *Catalysis Today*, **82**, 41.

81 Molinari, R., Pirillo, F., Falco, M., Loddo, V. and Palmisano, L. (2004) *Chemical Engineering and Processing*, **43**, 1103.

82 Ghosh, R. (2006) *Journal of Membrane Science*, **274**, 73.

83 Chan, C.C.V., Bérubé, P.R. and Hall, E.R. (2007) *Journal of Membrane Science*, **297**, 104.

84 Fu, J., Ji, M., Wang, Z., Jin, L. and An, D. (2006) *Journal of Hazardous Materials. B*, **131**, 238.

85 Chin, S.S., Lim, T.M., Chiang, K. and Fane, A.G. (2007) *Chemical Engineering Journal*, **130**, 53.

86 Huang, X., Meng, Y., Liang, P. and Qian, Y. (2007) *Separation and Purification Technology*, **55**, 165.

87 Choi, J.H., Fukushi, K. and Yamamoto, K. (2007) *Separation and Purification Technology*, **52**, 470.

88 Mulder, M. (1991) *Basic Principles of Membrane Technology*, 2nd edn, Kluwer Academic Publishers, Dordrecht.

89 Camera-Roda, G. and Santarelli, F. (2007) *Journal of Solar Energy Engineering-Transactions of the ASME*, **129**, 68.

90 Mozia, S., Tomaszewska, M. and Morawski, A.W. (2005) *Applied Catalysis B-Environmental*, **59**, 131.

91 Azrague, K., Aimar, P., Benoit-Marquié, F. and Maurette, M.T. (2007) *Applied Catalysis B-Environmental*, **72**, 197.

92 Fernández, P., Blanco, J., Sichel, C. and Malato, S. (2005) *Catalysis Today*, **101**, 345.

93 Sarria, V., Péringer, P., Cáceres, J., Blanco, J., Malato, S. and Pulgarin, C. (2004) *Energy*, **29**, 853.

94 Molinari, R., Caruso, A., Argurio, P. and Poerio, T. (2008) *Journal of Membrane Science*, **319**, 54.

95 Liu, X., Li, Y. and Wang, X. (2006) *Materials Letters*, **60**, 1943.

96 Kansal, S.K., Singh, M. and Sud, D. (2007) *Journal of Hazardous Materials*, **141**, 581.

97 Abu Tariq, M., Faisal, M., Muneer, M. and Bahnemann, D. (2007) *Journal of Molecular Catalysis A-Chemical*, **265**, 231.

98 Sleiman, M., Conchon, P., Ferronato, C. and Chovelon, J.M. (2007) *Applied Catalysis B-Environmental*, **71**, 279.

99 Topalov, A., Molnar-Gabor, D., Kosanic, M. and Abramovic, B. (2000) *Water Research*, **34** (5), 1473.

100 Zhang, Y., Zhou, J.L. and Ning, B. (2007) *Water Research*, **41**, 19.

101 Karunakaran, C. and Senthilvelan, S. (2006) *Electrochemistry Communications*, **8**, 95.

102 Ku, Y., Lee, W.H. and Wang, W.Y. (2004) *Journal of Molecular Catalysis A-Chemical*, **212**, 191.

103 Zertal, A., Molnar-Gabor, D., Malouki, M.A., Sehili, T. and Boule, P. (2004) *Applied Catalysis B-Environmental*, **49**, 83.

16
Wastewater Treatment by Membrane Bioreactors

TorOve Leiknes

16.1
Introduction

Water is fundamental for life, is by far the most important food item and a commodity that modern societies rely on in many aspects including potable water, agricultural water, industrial water, and recreational water. Water is essential and preservation of its safety in quantity and in quality is critical to the sustainable development of any society. Historical documents show that water has always been an important issue in all civilizations, however, sanitary engineering or environmental engineering as we know it today has a very limited history. The terms 'sanitary' engineering, or 'public health' engineering came about during the mid-nineteenth century when the close coupling between water quality, sanitation, and public health were discovered. Epidemic outbreaks of cholera, typhoid, and dysentery in European cities during that time initiated the steps, developments and implementation of what we understand today as environmental engineering. During this period investigations were done to try and understand the general ill health in urban areas. Although reports suggested that the public-health problems were mainly due to poor sanitation and unclean water supplies, it was not until the cholera epidemic in London in 1854 and the work by John Snow showing a direct connection between contaminated water and spreading of the disease that authorities were convinced [7]. Recommendations from these investigations were to secure proper urban sanitation, provide a clean water supply and a proper drainage system to remove human wastes. Following this revelation, the objectives of all developments within the field of environmental engineering were primarily to provide drinking water that looked and tasted good, to prevent the waterborne diseases, and build a system for wastewater collection. Over time the objective also included protecting the natural environment from negative impacts caused by wastewater contamination. With respect to wastewater treatment, probably the most significant development was the implementation of the 'activated sludge' process in 1913 by Arden and Lockett [1, 3, 54]. This process is the basis of biological treatment of wastewater by intensive aeration and mixing of the suspended solids formed and has found a wide application worldwide. As such, the treatment

Membrane Operations. Innovative Separations and Transformations. Edited by Enrico Drioli and Lidietta Giorno

ISBN: 978-3-527-32038-7

scheme now known as the conventional activated sludge (CAS) process has been by far the most common approach and solution for municipal wastewater treatment.

Today the world is running out of clean, safe, fresh water. By 2025 one third of humanity – almost three billion people – will face severe water scarcity. This has been described as the 'single greatest threat to health, the environment and global food security.' At the Johannesburg Summit on sustainable development (SD) 2002, two of the Millennium goals were defined as; by 2015 reduce by half the proportion of people without access to basic sanitation (2 billion people), and reduce by half the proportion of people without sustainable access to safe drinking water (1.5 billion people). Continuous extraction of water has resulted in depletion of available water sources in and around the industrial areas in many regions of the world. In addition, wastewater discharges into natural watercourses has caused surface and groundwater pollution, often leaving water unsafe for potable use and impairing industrial use without major and costly treatment. The current low-cost end-of-pipe treatment approach will become increasingly expensive as effluent discharge standards become more stringent. Meanwhile, technological advancements now make it possible to treat wastewater for a variety of end uses as direct discharge to sensitive areas, industrial or even potable-water reuse.

The development of advanced wastewater-treatment strategies is necessary to implement sustainable water management in general. Efficient but cost-effective wastewater-treatment processes are needed for two purposes; both for producing high-quality water from contaminated resources and for transforming wastewater into water able to be reused for various applications – for potable water, in agriculture and industry. Due to their unique characteristics, and mainly the possibility to adjust the retention efficiency to the level which is needed (from colloids and microorganisms to small molecules and ions), membranes will play a more and more important role in the near future. To both the problems of water management and water quality, advanced membrane technologies offer practical, cost-effective and energy-saving solutions whether for large-, medium-, or small-scale applications. Membrane-bioreactor processes represent such advanced technologies for the treatment of both municipal and industrial wastewater. Furthermore, membrane processes are suited for onsite small-scale reuse. Presently, water-piping costs represents a major part of the cost of the water and wastewater management and there is a strong tendency in this area to discuss decentralized solutions. There is also a trend towards discussing a paradigm shift in environmental engineering looking into new and alternative technologies and strategies to meet future demands. Membrane technology will have an important role in these developments.

16.2 Membranes in Wastewater Treatment

16.2.1 Background

The history of membranes applied to treatment of wastewater is relatively young, dating back to the late 1960s. Over the last 40–50 years there has been a clear evolution of what kind of membranes have been applied and what the objective of the treatment scheme has been.

Membrane technology has been applied to various types of wastewater. The largest number of installations is probably for industrial wastewater applications, however, municipal wastewater is largest in volume treated. The emphasis of wastewater treatment by membranes in this chapter will be for municipal wastewater treatment.

16.2.2 Membranes Applied to Wastewater Treatment

The evolution and possible applications of membranes in wastewater treatment is illustrated with the examples shown in Figure 16.1. Production of fresh water from brackish water or saline water sources by membrane desalination using reverse osmosis (RO) was first commercialized in the late 1950s, and in fact the starting point of membranes applied to wastewater [12, 55]. Due to the superior separation properties of the RO system and the possibility of making high-grade water, RO systems were applied as a final step in tertiary wastewater treatment [4]. In the first applications, they were preceded by several unit processes to ensure stable operation and minimum fouling. With the subsequent developments of MF/UF membranes, a more efficient pretreatment for the RO units was possible and the conventional unit processes needed to achieve tertiary treatment standards could be replaced by the combined membrane processes. Today, final treatment by RO as the last stage is considered state-of-the-art technology for reuse and recycling of wastewater.

MF/UF as a replacement for conventional sedimentation in activated-sludge processes was first reported in the late 1960s. The membrane sewage treatment (MST) process applied was based on an activated-sludge reactor coupled with a continuous withdrawal of water through rotating-drum screen followed by a UF unit. During this period other bench-scale studies were being conducted where MF/UF membranes in a side-stream configuration were used for filtration of activated sludge [10, 40, 72]. These systems were not very efficient and energy demanding, however, they succeeded in demonstrating and establishing the fact that membrane technology could be coupled with the activated-sludge process for wastewater treatment. The technology first entered the Japanese market where small systems were applied to both domestic and industrial applications. The Japanese Government subsequently joined several large companies and established programs with the aim to develop a technology with small footprint, high-quality effluent that was suitable for wastewater recycling [72]. Around this time other developments around the world were taking place and in the late 1980s to early 1990s the concept of the membrane bioreactor (MBR) was established as an alternative process to the conventional activated-sludge process [88]. As illustrated in Figure 16.1(d), wastewater treatment by membrane technology is an established alternative, particularly in sensitive areas, water-scarce regions, and in cases where wastewater reuse and recycling is required. A wide variety of products are now available with an increasing number of reference sites. Confidence in the technology is growing, and the implementation and growth of MBR technology at a significant pace is expected.

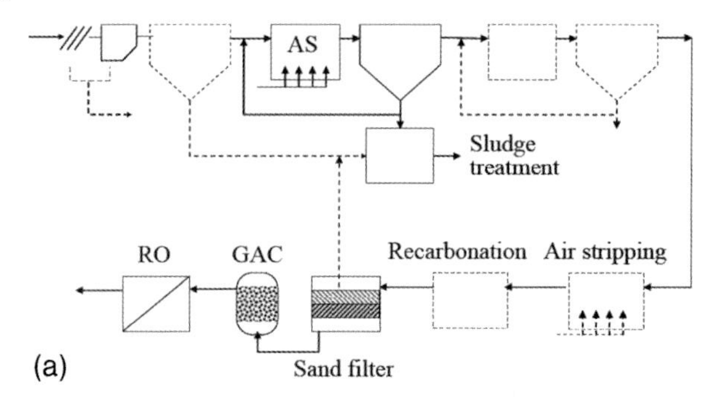

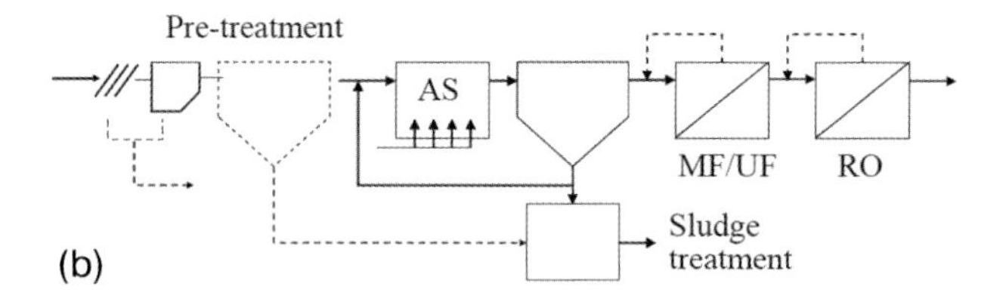

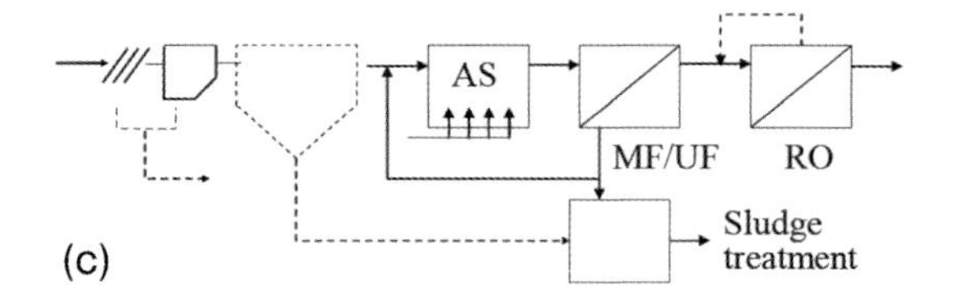

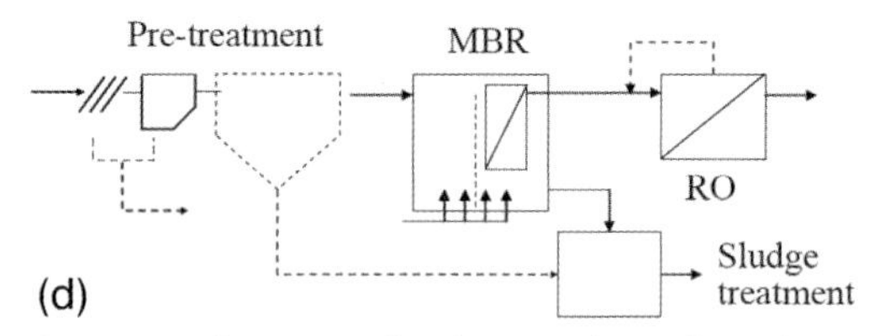

Figure 16.1 Illustration of evolution and use of membranes in wastewater treatment.

Membrane bioreactors (MBR) are commonly understood as the combination of membrane filtration and biological treatment using activated sludge (AS) where the membrane primarily serves to replace the clarifier in the wastewater-treatment system [18, 29, 79, 85]. A major advantage of the MBR system is that it can operate at a much higher solids concentration in the bioreactor than that of a conventional CAS – mixed liquor suspended solids (MLSS) concentrations typically in the range 8–12 kg/l compared to CAS that can only work at about 2–3 kg/l, because of the limitations on settling. This higher sludge concentration permits effective removal, not only of dissolved organic material but also of residual particulate solids. A comparison and assessment of MBR technology vs. the conventional activated-sludge process generally highlights the following issues:

Improved Water Quality

- Meets stringent effluent requirements;
- Filters out nearly all solids.

Allows Wastewater Reuse

- As part of a treatment scheme, provides water for potable reuse;
- Reduces wastewater discharge fees and freshwater costs;
- Provides water for nonpotable applications where fresh water is in short supply.

Lowers Capital Costs

- Clarifier is not needed;
- Biological step can be scaled down since bacteria concentration is higher.

Reduces Plant-Space Requirements

- Footprint is up to 50% smaller than conventional plant;
- Allows for expanded capacity within existing buildings.

Fewer Operational Problems

- Bulking and floating sludge problems are avoided.

Although there are clearly some benefits of the MBR process compared to CAS the technology is not void of any disadvantages. In the early years of development the process was considered to be expensive due to high membrane costs, uncertainties of membrane lifetime and anticipated membrane replacement costs. As MBR plants have been in operation for a while and experience has been gained, membrane lifetime appears to be longer than initially thought and replacement costs stipulated in the early 1990s to be 80–90% of O&M costs is now estimated to be around 10–15%. This reduction is a combination of gained confidence in the technology, better and cheaper production of membrane modules and product development in general. The main disadvantage of MBR systems is membrane fouling, an inherent phenomenon in all membrane processes. Strategies and techniques to alleviate fouling coupled with the frequency of membrane cleaning is one of the main constraints of the process. The high energy demands for aeration, both for the biological process and membrane operation, is currently recognized as another major challenge and drawback of the technology.

MBR technology is probably the membrane process that has had most success and has the best prospects for the future in wastewater treatment. Trends and developments also indicate that this technology is becoming accepted and is rapidly becoming the best available technology (BAT) for many wastewater-treatment applications. The cost of an MBR plant for secondary treatment is still higher than that for a CAS plant, but as the numbers of MBR plants increase, and as membrane costs fall, the life cycle cost differential will soon disappear, and the process advantages should lead to rapid uptake of the MBR system by the wastewater-treatment industry. This chapter will therefore focus on membrane-bioreactor technology applied to municipal wastewater.

16.3
Membrane Bioreactors (MBR)

16.3.1
Membrane-Bioreactor Configurations

16.3.1.1 Membrane Materials and Options

There is a large selection of commercial membranes that potentially can be used in MBR applications. Within the membrane industry, when classifying membranes, a distinction between polymeric membranes and inorganic membranes is made. Inorganic membranes are made either from metals or ceramics, the latter being the more common. Ceramic membranes are considered to be rather expensive and to date there is a very limited use of this material for wastewater treatment. As the costs of ceramic membranes are reduced and the membrane design is more geared towards wastewater applications one may foresee a gradual increase in the use of ceramic membrane-based systems. Polymeric membranes are by far the preferred material in wastewater treatment to date. In principle, most polymers can be used to manufacture membranes and there is a wide variety of commercially available polymeric membranes. In water and wastewater applications, however, most systems are based on a limited set of polymeric materials [12, 40, 55, 70–72]. The most common membrane materials are; polyvinyl difluoride (PVDF), polyethylsulfone (PES), polyethylene (PE) and polypropylene (PP). Depending on the manufacturing technique applied, membranes made from these polymers can be produced with various geometries and specific physical properties. In this way they can be tailored to meet the specific demands of the application for which they are intended [12, 17, 55, 60, 90].

Membranes are generally formed as flat sheets or with tubular/hollow-fiber geometry. With recent developments of manufacturing techniques, alternative products are also available on the market, for example multibore or multitube designs and self-supporting flat sheets with channels. Given these variations membranes are commonly given the following classifications:

- FS – flat sheet;
- HF – hollow fiber;
- CT – capillary tubular;
- MT – multibore or multitubular.

The membrane module and design will obviously depend on the type of membrane used. The flat-sheet membranes are commonly constructed in a plate-and-frame configuration or as spiral-wound (SW) modules. HF/CT/MT membrane types are commonly manufactured into bundles that are installed in housing units or designed to be unconfined in the fluid, that is, immersed units. The membranes are

Table 16.1 Characteristics of membranes used in MBRs (Adapted from Ref. 24.)

Characteristics	Tubular membranes	Flat-sheet membranes	Hollow-fiber membranes
Arrangement	External – recycling	External/submerged	External/submerged
Packing density	Low	Moderate	High
Energy demand	High (turbulent flow)	Low–moderate (laminar flow)	Low
Cleaning	Efficient + physical cleaning possible	Moderate	Backwashing possible
Replacement	Tubes or element	Sheets	Element

then constructed in such a way that they are self-supporting. In MBR applications the plate-and-frame FS and the HF/CT membrane modules are the preferred options (Table 16.1) [24].

A comprehensive presentation of all membrane types, modules and geometries is beyond the scope of this chapter, reference available membrane books for details [12, 17, 55, 60, 71, 77, 90]. The examples in Figure 16.2 are an illustration of a typical membrane module and installation. The most widespread FS membrane system is mounted as a spiral-wound (SW) unit. In the SW example the actual membrane module is shown together with how they are mounted inside a pressure vessel. A typical installation is shown where several pressure vessels are subsequently mounted in a stack. Pressurized HF units are typically operated as a crossflow system. In the example shown the HF modules are mounted vertically and arranged in a skid. Several variations of the theme can be found depending on the type of module and the manufacturer, where Figure 16.2 is not specific to a particular item.

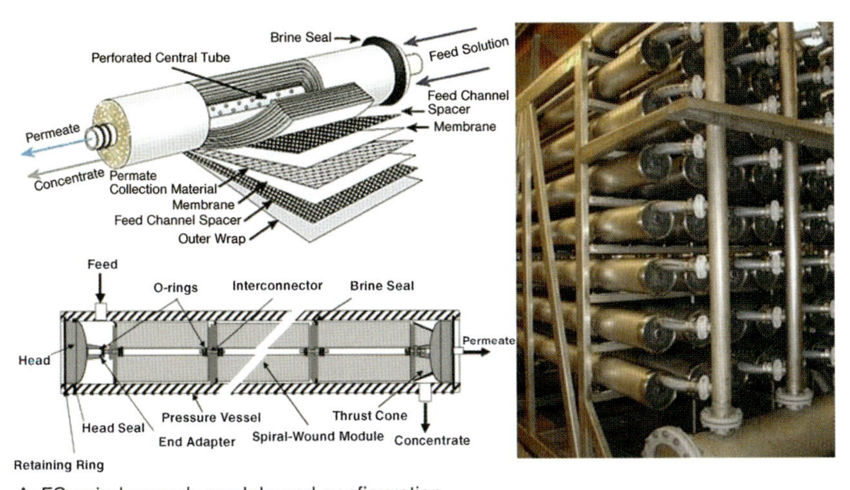

Figure 16.2 Examples of membrane types (FS and HF) and typical installation/configuration.

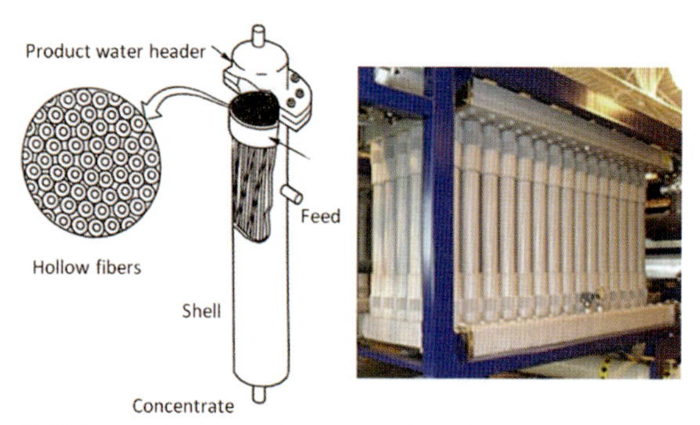

B: HF cross-flow module and configuration

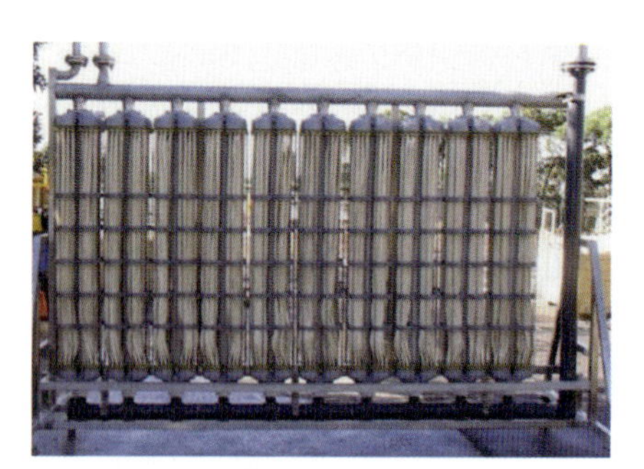

C: FS modules mounted in unit, submerged

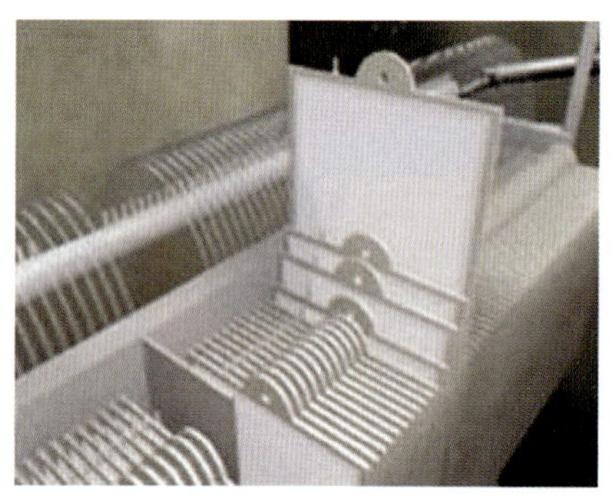

D: HF/CT membranes mounted in frame

Figure 16.2 *(Continued).*

In Table 16.2 some of the typical characteristics of the various types of membranes and configurations are given. In MBR systems SW membrane modules are not used as the channels within the spiral are prone to clogging when the feed water has high suspended-solids concentrations. Tubular membrane systems are not common either as they tend to become very expensive due to the low area to volume ratio. Commercial MBR systems today are normally based on immersed FS configurations or HF/CT configurations.

Table 16.2 Typical characteristics of different types of membrane types and configurations.

Configuration	Area/volume ratio (m^2/m^3)	Cost	Advantages	Disadvantages
Plate & frame	400–600	High	Can be dismantled for cleaning	Complicated design
				Cannot be backflushed
Spiral wound	800–1000	Low	Low energy cost, Robust and compact	Not easily cleaned Cannot be backflushed
Tubular	20–30	Very high	Easily mechanically cleaned	High capital and membrane replacement cost
			Tolerates high solids	
Capillary	600–1200	Low	(between tubular and hollow fiber)	
Hollow fiber	5000–40 000	Very low	Can be backflushed	Sensitive to pressure shocks
			Compact design Tolerates high colloid levels	

16.3.1.2 **Process Configurations**

The process configuration of a MBR plant will partly depend on the type of membrane used (FS or HF/CT) and partly on the design of the biological treatment. Figure 16.3 is an illustration of the typical configurations found in MBR treatment schemes. For immersed membrane designs the membrane modules are either inserted directly into the biological reactor or placed in a separate reactor constructed to hold the membrane modules only (schemes A and B in Figure 16.3). In a side-stream configuration the membrane modules are placed outside the biological reactor and can be operated in deadend mode or in a crossflow mode with recycling of the concentrate stream back to the biological reactor (schemes C and D in Figure 16.3). In the immersed configuration the treated water (permeate) is extracted from the membrane by vacuum (low pressure) in contrast to the side-stream option where the permeate is generally produced under pressure. The immersed systems are generally less energy intensive compared to the side-stream design where pumps are needed to maintain sufficient crossflow velocities in the membrane unit or to overcome the increasing resistance due to build up of material on the membrane with deadend operation. To offset fouling and to generate crossflow conditions along the membrane, aeration is commonly used in the immersed systems. Each configuration has its advantages and disadvantages where variations on the theme can be found from the different system manufacturers. These will be discussed in the following sections. The immersed MBR configuration is generally the preferred option, particularly for medium- to large-scale municipal wastewater treatment plants [23, 29, 39, 40, 43, 45].

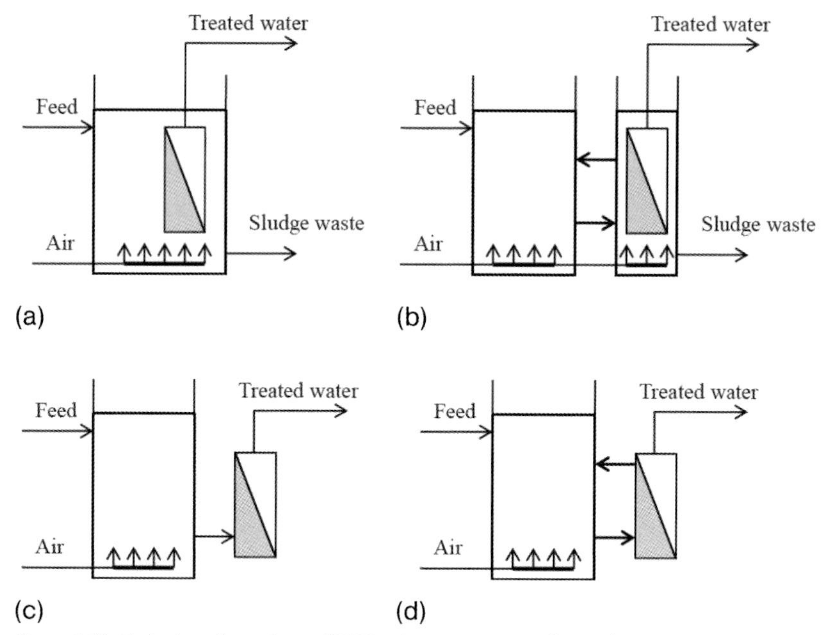

Figure 16.3 Typical configurations of MBR schemes, immersed vs. sidestream.

16.3.2
Membrane-Bioreactor Basics

The key elements, their interactions and impacts of these in the MBR process are shown in Figure 16.4. Ultimately, the feed characteristics of the wastewater to be treated will impact the design and operation of the process. The composition of the feed water as well as the required effluent quality that the treated water needs to meet will also define the treatment scheme, particularly the configuration of the biological process. In that the primary purpose of the membrane is to clarify the biologically treated water, the interaction between the biological process and the membrane process is the core of the technology. Due to the nature of the membrane filtration it is evident that components in the feed water will be retained by the membrane and as this material is captured on the membrane it will cause fouling. One of the major drawbacks of MBRs is fouling, which is common for all membrane systems, where the efficacy of the process is constrained by the accumulation of materials on the surface of or within the membrane resulting in a reduction in the membrane permeability. Several definitions of fouling can be found in the literature where both relative broad and explicit definitions are used. One definition defines fouling as a decline in time of flux during operation when all operating parameters are kept constant [12]. This has further been revised to define fouling as a long-term phenomenon where irreversible fouling that builds up over time causes a flux decline [78]. This definition has been further revised to distinguish between reversible and irreversible fouling as well as short-term and long-term fouling. As fouling is better understood other

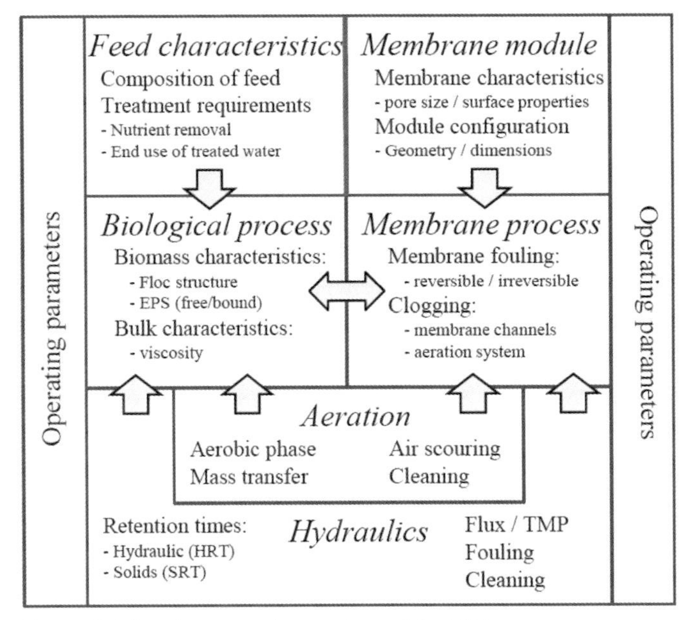

Figure 16.4 Key elements and interactions of the MBR process.

refinements of the definition are to be expected. Membrane fouling in MBR systems is caused by different substances and the mechanisms are rather complex and interrelated. Deposition of solids as a cake layer, pore plugging/clogging by colloidal particles, adsorption of soluble compounds and biofouling are some of the main forms of fouling that have been identified [21, 34, 40, 47, 56, 64]. Membrane fouling and strategies for fouling mitigation and amelioration are therefore the core of MBR operating parameters, system configurations and membrane-module designs.

Aeration is a key element in the design and operation of MBRs with multiple objectives being included in the process. With respect to the biological component of the process, aeration is necessary to fulfill the oxygen demand of the aerobic degradation of compounds by the biomass. Aeration is also used in the membrane component of the process to prevent clogging of the membrane modules from solids concentrations and as a technique and means to prevent fouling of the membranes. The design of the aeration ports, intensities, air bubble characteristics and properties will therefore need to take into consideration the primary objectives of the aeration unit. A more detailed discussion of aeration is included in the sections on membrane fouling and defining operating conditions.

The hydraulics of the MBR process will impact both the biological process as well as the membrane process. From a biological point of view the hydraulics is significant for the hydraulic retention time (HRT) that affects key operating parameters such as reactor volumes, mixed liquor suspended solids (MLSS), loading rates and so on. For the membrane process, hydraulics directly impacts the filtration process and ultimately fouling developments. The design of the membrane module and operating conditions are interlinked with flow patterns and local hydrodynamic conditions on the membrane surface, which affect different fouling mechanisms. The hydraulics in the membrane system is also important when cleaning a fouled membrane needs to be done. A more detailed discussion of hydraulics in the membrane process is included in the sections on membrane fouling and how to deal with fouling.

Overall, the MBR process needs to be designed and built with respect to two sets of operating requirements, those operating parameters needed to optimize the membrane component of the process and those necessary to achieve the desired biological conversions in the biological processes. The interactions and interdependence of the two key processes need to be understood in order to be able to develop treatment schemes that are efficient both from a treatment perspective but also from a management and operational perspective. In the following sections some of the key elements in MBR systems will be discussed in more detail.

16.3.3
Membrane Fouling[1)]

16.3.3.1 Understanding Fouling

Fouling is particularly a problem in AS-MBRs since the process deals with liquors having high concentrations of total solids as well as dissolved compounds such as extracellular polymeric substances (EPS). Fouling is defined as reversible, that is, can be removed by backwashing strategies, or as irreversible, that is, fouling that is only recoverable by chemical cleaning, where the dominating fouling mechanism subsequently determines the performance of the process. Optimizing fouling control and cleaning strategies are therefore important aspects of developing and designing MBR processes. The complex nature of fouling in MBRs makes it difficult to distinguish between which mechanisms or foulants are dominant and these may change with time during operation as well as due to variations in the feed characteristics, adjustments in operating parameters and so forth.

Figure 16.5 gives an illustration of the main fouling mechanisms identified in membrane processes. Membrane fouling is manifested in various ways and certain types of fouling (reversible) can be removed by backwashing, that is, cake formation and loose depositions, while others are permanent (irreversible), fouling that is only recoverable by chemical cleaning [47]. A lot of effort and research has been done to gain a better knowledge of the phenomenon in the last 10–15 years. A literature review shows that in recent years the average number of articles found where the registered keywords are; fouling, membrane, colloids is around 25 and 40 [2]. Suspended solids are very often identified as a main foulant [6, 21] where the significance of the submicrometer colloidal fraction in the suspended solids has been reported to correlate with membrane-fouling rates [47, 50, 69, 75, 83, 91]. In MBR processes fouling has also been attributed to extracellular polymeric substances (EPS) and

1) The basic aspects of fouling are discussed in Chapter 6, here practical aspects on fouling related to MBR are further considered.

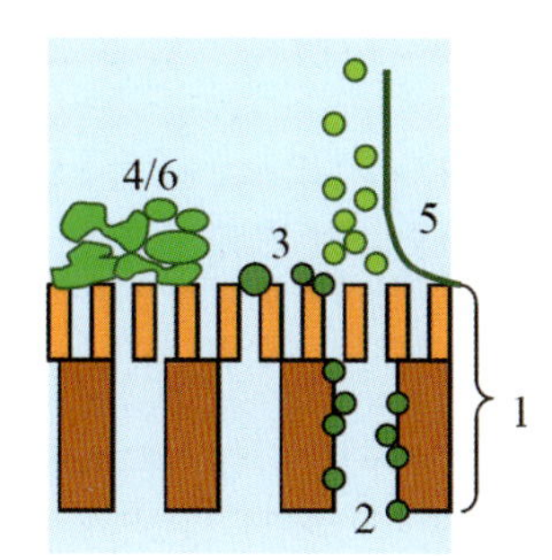

1. Membrane
2. Adsorption / scaling (reversible / irreversible)
3. Pore blocking / plugging
4. Cake deposition / formation
5. Concentration polarization
6. Bio-fouling: biofilm / EPS

Figure 16.5 Identification of potential causes for membrane fouling.

soluble microbial products (SMP) [56, 62, 69, 83]. Current research activities are very much focused on gaining a better understanding of the key fouling mechanisms, how they interact, and which factors influence their contribution to fouling as a whole.

Many studies have distinguished between the relative contributions in per cent of the main fractions that cause fouling in MBR processes. An overview of several studies shows that there is a large variation when assigning fouling to the suspended solids fraction, the colloidal fraction or the soluble fraction [40]. Although the potential for cake formation and an increase of resistance to permeation by the layer formed is proportional to the suspended solids concentration, fouling by cake-layer formation is not commonly identified as the dominating mechanism in MBRs due to operation with relatively modest fluxes. The soluble and colloidal fractions are assumed to be the dominant foulants. Several studies have reported the significance of colloidal particles as an important factor contributing to fouling development. The contribution by colloids has been estimated to be responsible for anywhere between 20% to more than 60% of total measured fouling [8, 21, 35, 37, 38, 47, 69, 83]. The colloidal material may cause pore blockage and the deposition of these small particles may form very compact layers on the membrane surface. The remaining percentage of the total measured fouling is attributed to the soluble fraction. Studies of the soluble fraction have shown that the extracellular polymeric substances (EPS) are most critical with respect to fouling potentials of this fraction [67]. EPS are biological polymers of microbial origin that are predominantly made up of polysaccharides and proteins with small contributions of nucleic acids and lipids. EPS forms protective layers around the cells and also facilitates the interactions between the cells and the environment and are therefore essential for microbial survival. The EPS is often defined as bound to the cell or soluble, which is the EPS found in the water phase due to breakup of flocs or from cell lysis [26, 27]. The soluble portion is sometimes referred to as soluble microbial products (SMP), however, this fraction not only includes the EPS substances but can also include intermediates or end products from the biological conversion as well as endogenous cell decomposition. Studies have found correlations between EPS concentrations and membrane fouling in MBR systems. The polysaccharides and proteins have been shown to be the main foulants from EPS, however, contradicting reports are found as to whether it is the polysaccharides or proteins that have the greatest effect.

16.3.3.2 Dealing with Fouling

MBR operation has aimed at trying to minimize fouling by maintaining operating parameters that give a sustainable process, that is, fouling over a long term with very low fouling rates. Within the industry, a 'sustainable flux policy' has been adapted, corresponding to a long-term flux operation where fouling rates are below an economically accepted value. Many studies have been conducted to determine what a subcritical flux operation really is. Fouling in MBR systems has recently been described as a three-stage process, each with a distinct pattern. The first stage is characterized by a short-term fouling where the systems adjust to the set operating flux and a kind of steady-state condition. This is followed by a second stage of long-term operation with relatively low fouling rates (expressed as an increase of TMP over time) until a sudden and sharp increase in TMP is observed (stage three). Several studies have reported the same observation regardless of the system studied [9, 13, 28, 64, 82]. This phenomenon has been explained by different theories, though it is not yet fully understood. However, it is apparent that defining and operating MBR systems with a sustainable flux, that is, subcritical flux, is a key aspect and strategy for dealing with fouling (Figure 16.6) [13, 28].

The concept of critical flux was introduced during the mid-1990s and defined as a flux below which fouling is absent or negligible. The basis of this concept is that for a

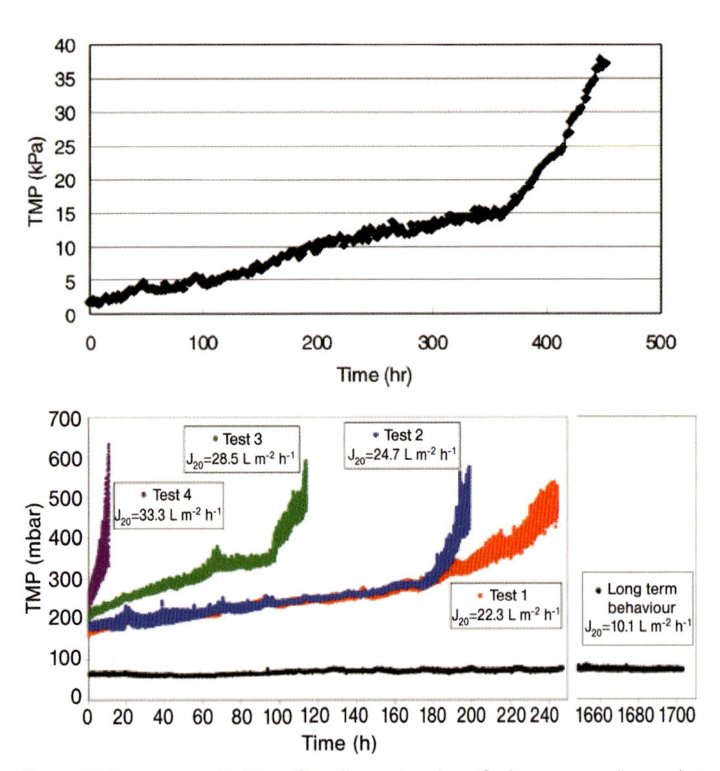

Figure 16.6 Long-term TMP profiles illustrating three fouling stages observed at subcritical flux.

given feed solution and operating condition in a crossflow mode, there is a flux at which the transport of substances to the membrane exceeds the backtransport and removal of rejected substances and fouling of the membrane begins [5, 25, 84]. The concept gives a good understanding of membrane filtration from a theoretical point of view but the concept has been prone to debate and various interpretations when complex systems such as MBR processes are considered. The basis of this concept, however, has been used to determine the optimal operating condition for complex systems as in MBRs, although the term 'critical flux' is used loosely here and should not be equated to the original concept. Coupling of the critical-flux hypothesis with the process-cost optimization has led to the so-called 'sustainable flux,' which represents the operating flux below which the fouling rate is economically acceptable for the plant operation. A stepping analysis approach has been proposed to determine at which flux one could expect a sustainable operation. Figure 16.7 shows the stepping analysis proposed to determine the critical flux (A) and an example of how this has been applied to a specific case (B) [28, 43]. It should be noted that the 'critical-flux' value obtained is very specific for each case and is dependent on the nature and properties of the feedwater, the configuration and operating conditions of the biological process, and the type of membrane modules used. The stepping analysis is, however, a tool one can use to determine the practical limitations of operation, that is, sustainable flux, for given conditions and system specifications to achieve economical and efficient operating parameters.

Aeration is one of the most important parameters in the design and operation of MBR systems. Aeration is essential for the operation and design of the aerobic stage of the biological process with specific demands and needs expressed by the biological conversion. Aeration from this perspective is discussed in more detail in the section on biological operating conditions in MBRs. As indicated previously in Figure 16.4, aeration is used in the membrane process for air scouring and cleaning of the membrane module. Aeration in submerged MBR, particularly for hollow fiber systems, also induces a lateral movement that generates a shear force on the membrane from the surrounding liquid. The overall effect is a function of the aeration intensity and how much movement is achieved [16, 23].

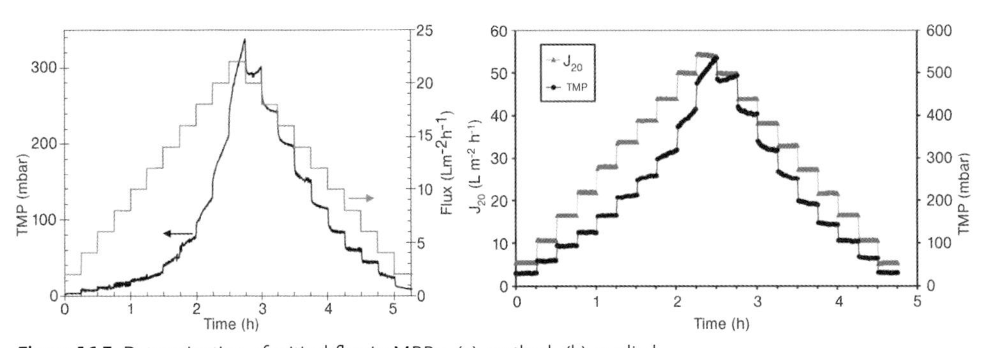

Figure 16.7 Determination of critical flux in MBRs; (a) method, (b) applied.

16.3.3.3 Cleaning Fouled Membranes

Membrane cleaning is often defined by the type of cleaning and the frequency of cleaning. The use of air scouring, periodic backwashing or back pulsing during operation are generally considered to be fouling-mitigation strategies to reduce the effect of short-term fouling. Membrane cleaning as an operational term is often associated with a more extensive action that needs to be undertaken when fouling of the membrane reaches a point where production of treated water is reduced to a critical level. Cleaning protocols are defined by applying both physical and chemical techniques and a distinction between maintenance and recovery cleaning is also made.

Maintenance cleaning is commonly used to reduce long-term fouling development and subsequently to increase the frequency between intensive cleaning necessary for recovery of permeability. Maintenance cleaning is generally performed *in situ* with the membranes kept in place (CIP). From an operation point of view, maintenance cleaning is a procedure that is done at regular intervals (3–7 days) where cleaning agents are applied at low concentrations. The frequency is determined by the feed characteristics and operating conditions, while the procedure is a manufacturer specific recommendation. For recovery cleaning the membrane is commonly removed from the process line and higher concentrations of cleaning reagents are used. Several cleaning steps are employed and can last for several hours. As this directly impacts the production capacity of the plant and operation and maintenance routines, systems are designed to try and reduce recovery cleaning frequencies between 1–3 times per year.

Physical cleaning can be performed by generating high shear forces to remove any deposits on the membrane surface, altering the hydrodynamics around the membrane, air scouring, movement (i.e., vibrations, ultrasound). Mechanical cleaning can also be induced to remove deposits depending on membrane geometry, membrane material and operating criteria. Although physical cleaning may quite effectively remove cake layers or deposited solids on the membrane surface it needs to be supplemented with chemical cleaning to fully restore the membrane permeability.

Organic and inorganic fouling caused by adsorption or deposition can only be removed through chemical cleaning. In principle, three strategies are employed; oxidation of organic compounds on the membrane (commonly targeting biofouling constituents) coupled with caustic solutions to dissolve and remove organic compounds, and acidic solutions to remove inorganic/mineral compounds. In practice, cleaning protocols are generally defined and recommended by the manufacturers and suppliers of MBR systems. With respect to recovery cleaning the choice of cleaning agent and steps is determined by whether the aim is to remove organic or inorganic foulants. A survey of cleaning protocols shows that the chemical products recommended and their concentrations are quite similar for all the systems available on the market. The backbone of the protocols include; applying hypochlorite to remove biofouling (typical concentration range 200–2000 ppm), increasing pH with sodium hydroxide (typical concentration range 150–4000 ppm) to dissolve organic compounds, and decreasing pH with acids, commonly citric acid and oxalic acid (typical concentration range up to 5000 ppm) to dissolve inorganic compounds. The choice of chemicals and cleaning steps are very much dependent on the system supplier.

There are a limited number of studies in the literature focusing on MBR cleaning protocols. In summary, the following generalized observations can be made. The membrane-cleaning protocols recommended are commonly based on the perception of what the dominant fouling mechanism or foulant has been. The overall cleaning processes generally involve many steps/stages and are often time consuming, a fact that affects operation time that may cause a loss in overall production of permeate. A combination of physical and chemical cleaning methods appears to provide the best permeability-recovery efficiency. Recoveries well over 90% and close to 100% are commonly reported. One challenge, however, is the lack of standard criteria to define membrane-cleaning parameters and measurements of flux/permeability recovery, and how these may ultimately affect membrane properties and membrane lifetime as a function of cleaning frequency. Finally, there are very limited references on waste management and handling of spent cleaning agents or the removed substances from the cleaning action [49, 51, 61, 74, 86, 87].

16.3.4 Defining Operating Conditions and Parameters in MBR Processes

16.3.4.1 Biological Operating Conditions

Biological treatment of wastewater utilizes the conversion of organic and inorganic matter into products by micro-organisms that are either easy to remove from the water (i.e., biomass growth) or converted to nonharmful substances. The biological conversions are a function of the biological community present in the process and the conditions necessary for the existence of the community, for example aerobic vs. anaerobic degradation. The biological process much depends on the treatment scheme and the target compounds to be removed by biological degradation [53]. A comprehensive assessment of the biological processes is beyond the scope of this chapter, there are many dedicated books available on the subject. There are, however, some specific details on the impact of coupling the biological process with the membrane process (Figure 16.4) that need to be highlighted with respect to implications on the operation of MBR processes.

Compared to conventional activated-sludge process, the biomass that is formed in the MBR system differs in composition and characteristics. A filtration test of sludge from CAS was found to give irreversible fouling compared to MBR sludge that gave reversible fouling [20]. One of the characteristics of the AS-MBR process is the concentration of biomass that influences the rheological properties of the sludge. During the development of MBR systems, very high sludge concentrations have been reported, however, from an energy-efficient perspective the recommended concentration has been stated to be between 10–15 g/L MLSS [1, 30]. Under these conditions the viscosity of the sludge was found to differ from CAS sludge and in general the viscosity was found to increase with increasing MLSS.

The oxygen demand is dependent on the biological process with respect to whether it is designed to meet secondary effluent standards or not. In most cases MBR systems are chosen due to their capability to produce a high-quality effluent and as such are designed to include nutrient removal. In this case the oxygen necessary to maintain a micro-organism community that degrades both organic matter and converts ammonium to nitrite/nitrate is required. The reader should refer to the appropriate literature on biological treatment for detailed explanations on how to calculate the oxygen demand for a given biological process. Aeration in MBR systems is a key issue where mass transfer of oxygen to the system can be calculated using the classical 'k_La' equation, expressed as oxygen transfer rate (OTR);

$$\mathrm{OTR} = \alpha k_L a(\beta \mathrm{DO}^* - \mathrm{DO})$$

where k_La is the overall mass-transfer coefficient and DO^* and DO are the saturated and dissolved oxygen concentrations. α and β are correction factors for the mass transfer rates and saturation concentrations commonly determined for clean water and therefore compensate for application to wastewater. Studies have shown that a decrease in OTR can be observed with increasing solids concentration where an exponential relationship between the α factor and MLSS concentration has been reported [30, 40]. The main impact has been reported to relate to bubble behavior in the liquid where higher concentrations appear to promote coalescence of bubbles and thus a reduction in the interfacial area expressed in the k_La term. Studies on bubble aeration have shown that a greater resistance to oxygen mass transfer is observed with increasing viscosity and correlations between the α-factor and viscosity have been proposed.

$$\alpha = \mu^{-x}$$

where μ is the viscosity (kg/(m s)) and x is the correlation exponential [40].

An effect of the increase in MLSS concentration in MBRs is a decrease in the α factor, where measurements at a MLSS of 12 g/L MLSS gave an α value of 0.6 compared to values of 0.8 typical of CAS sludge at concentrations of 3–5 g/L MLSS [15, 30]. The practical implication of this is that higher aeration rates and intensities are necessary in MBR systems compared to CAS systems.

In principle, the biological conversions in the MBR are performed according to what happens in the CAS process and as such the conversions achieved are not substantially different. The difference in the sludge properties do, however, affect the operational parameters related to pumping/circulation of the sludge and aeration of the sludge to maintain the necessary dissolved oxygen for the aerobic stage. Studies have also indicated that the response of the biomass to dissolved oxygen concentrations may promote or increase fouling potentials due to changes and stresses in the conversion mechanisms of the biological community. This stress has been shown to affect EPS production and composition of the EPS, one of the main foulants identified in MBR processes. Operating the biological process at optimal conditions to maximize the desired conversions while minimizing production or generation of potential foulants is therefore a key issue in the sustainable operation of MBR processes. Taking into account that aeration is an energy-demanding component and represents around 40% of the energy consumption in MBRs, aeration for the biological process in MBRs is a key operating parameter.

16.3.4.2 Membrane Filtration Operation

Membrane operation will much depend on the system configuration or mode of operation; deadend, crossflow or immersed. Although the schemes may differ, the underlying objective is to reduce fouling by the way the membrane unit is operated [11, 19, 22, 46, 52, 63, 68, 80]. Identifying and operating the system with a sustainable flux is a common theme. From a practical point of view the stepping analysis is a useful tool to identify a 'subcritical zone' of operation where it is possible to achieve long-term and reasonably stable operation of the system [32, 43, 64]. Determining the subcritical zone needs to be done experimentally for each specific condition. When the 'critical-flux' is found it is possible to identify the subcritical working area and the overall design of the MBR plant to account for daily variations in flow. An example of expressing this area of operation is illustrated in Figure 16.8. From the graph it is apparent that fouling can be kept at a minimum (expressed as an increase in TMP with increase in permeability) as long as one keeps in the subcritical zone.

Membrane aeration is a key operational parameter in that air scouring is used to keep solids from the membrane surface and to reduce fouling. Aeration is very energy demanding and the design and operation of systems in MBRs is where a lot of focus has been made in recent years to reduce this aspect of MBR operation. A key parameter for design and operation of MBR systems is therefore the specific aeration demand (SAD). A challenge within the industry is to define this parameter in such a way that different systems can be compared in a realistic manor. SAD has been normalized with respect to membrane surface area or to permeate volume produced [40]. A number of studies have demonstrated that flux increases linearly with increasing aeration rates until a threshold above which no further effect of increasing the aeration intensity can be observed [36, 44, 76]. Indeed, increasing the aeration beyond this threshold may have a negative impact on the performance of the membrane filtration

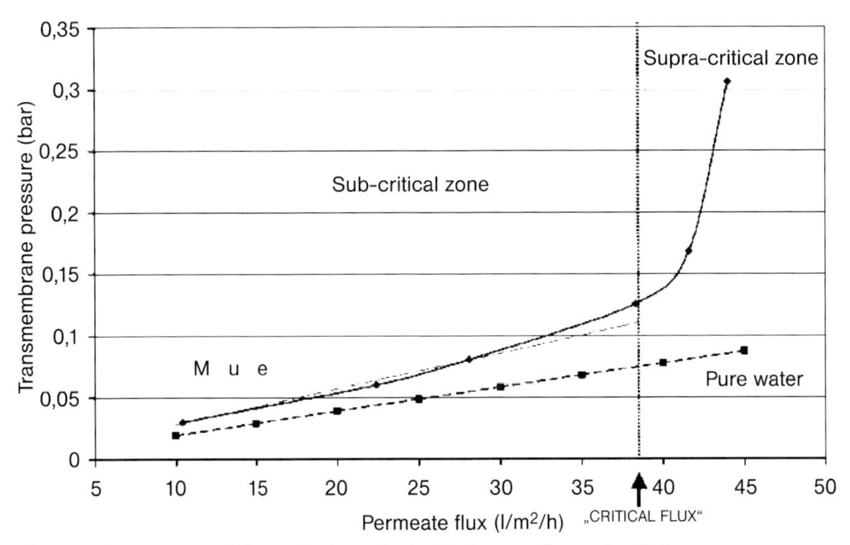

Figure 16.8 Experimental determination of subcritical zone of operation [64].

unit. Too high shear forces caused by the high aeration intensity may break up the flocs in the suspended solids, increasing the colloidal fraction that in turn increases fouling potential. Reports have also been published that show how more intensive aeration not only damages floc structures but also can release EPS-based foulants bound in the floc structure [36, 37, 40]. However, references in the literature that report correlations between aeration rates and effects on changes in colloidal-particle characteristics as a consequence of membrane aeration cannot be found. The threshold effect and the impact aeration intensity has on an increase in the colloidal fraction expressed as an increase in the differential number percentage of the submicrometer colloidal fraction is illustrated in Figure 16.9. The increase in this fraction also correlated with higher

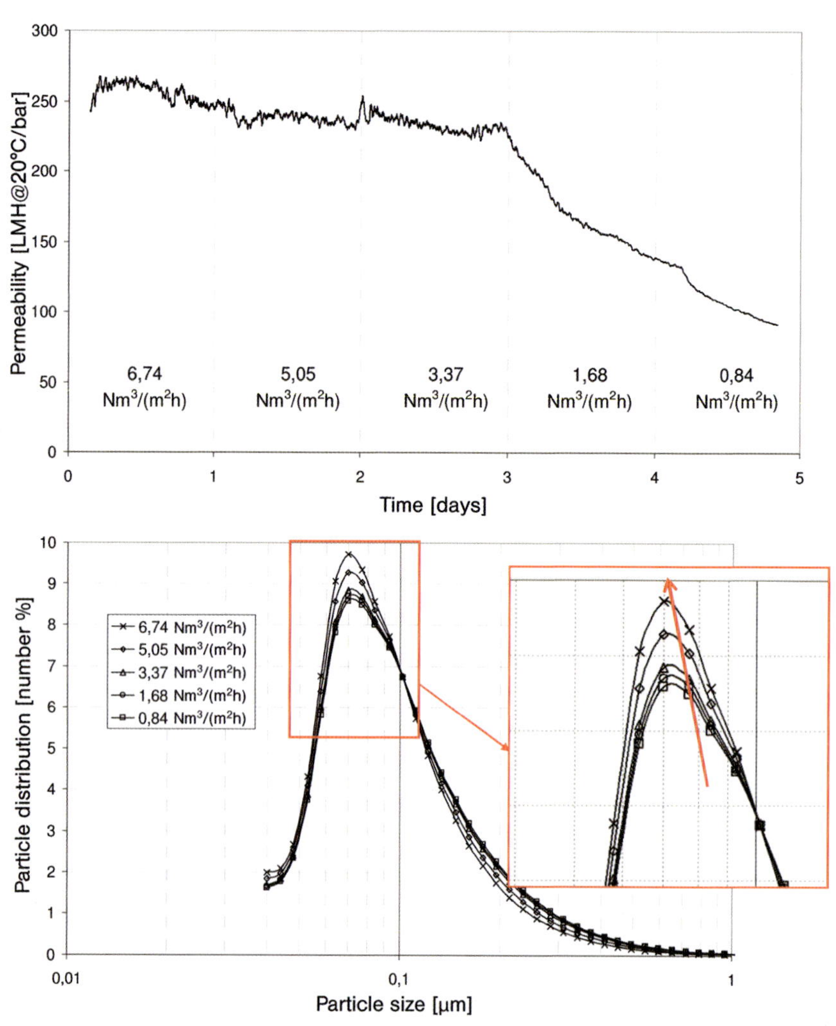

Figure 16.9 Effect of aeration intensity on membrane filtration performance [36].

fouling rates. Membrane filtration operation has also been improved by implementing alternative aeration strategies, mainly with the focus to reduce the aeration energy demand and costs. Intermittent aeration strategies and fluctuations in aeration intensities over periods are approaches that MBRs suppliers have investigated. The objective is to reduce overall energy requirements for aeration while maximizing the fouling mitigation effects of aeration. The specifics of these operating parameters are very much system dependent and each supplier has their own preferred emphasis and recommendations.

Removing or reducing the submicrometer particulate fraction is one approach to improving the membrane filtration stage. Coagulation and flocculation have been attempted as a means to improve the performance and thus gain a better understanding of dominating fouling mechanisms. Coagulation may also be used as a technique to reduce the dissolved organic matter content (particularly the EPS) and is also offered by the industry as a means to get a more stable operation of the membrane unit [14, 41, 65, 66, 81].

Other techniques such as periodic backwashing or relaxation techniques are also now standard modes of operation for most MBR suppliers. Backwashing the membranes by reversing the flow can generate enough flux to lift off deposits on the membrane and remove cake-layer fouling. FS membrane modules are not designed to be backflushed and backwashing is there predominantly applied in HF/CT systems. Backwashing in MBRs is synonymous with backwashing in filtration systems in general. Introducing backwashing makes the MBR a discontinuous process where the frequency and backwash volume ultimately affect the overall production of treated water. Backwashing adds an operational dimension to the process but reducing fouling and increasing the time between major cleaning by far outweigh the disadvantages of introducing a backwashing mode of operation. Backwashing is further enhanced when it is coupled with air scouring [31]. An alternative strategy to backwashing is relaxation, which can also be applied to the FS systems. This techniques is based on stopping the production of permeate, that is, applying vacuum, and allowing the system to rest for a short period. When air scouring is applied during relaxation the shear forces generated by the air bubbles can more efficiently remove the deposits accumulated on the membrane surface during production. The intermittent operation and frequency of relaxation is system dependent and determined by the MBR supplier. Lastly, maintenance cleaning as described in the section above on cleaning fouled membranes is also implemented as part of the normal operating parameters for commercial MBR systems.

16.3.4.3 **Optimizing MBR Operations**

When first commercialized, MBR processes were considered to be very expensive systems and only suitable for small-scale plants and for very specific applications. Capital costs have dropped drastically with the development of several commercially available systems and the treatment scheme is competitive even for large treatment plants. The operation and maintenance costs have gone through an evolution. In the infancy of the technology a major cost item was the anticipated membrane-replacement

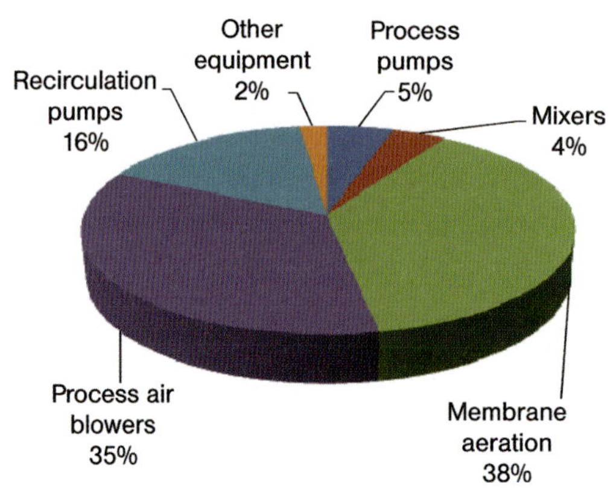

Figure 16.10 Illustration of typical energy demands in the operation of a MBR process.

costs, now this item has dropped significantly due to better and cheaper production of membrane modules as well as an increase in lifetime expectancy gained from operating experiences. Today, the energy demands are by far the largest operating costs, in particular the need for aeration both for the biological process and of the membrane process (Figure 16.10). As the aeration of the membrane unit is closely linked with fouling control and mitigation, continued studies to understand the complex nature of membrane fouling in MBR systems as well as the design and operation of membrane modules will be a major activity in the development of MBR systems in the future.

16.4 Prospects and Predictions of the MBR Process

16.4.1 Developments and Market Trends

Attempts at predicting the future needs and investments in the water and sanitation sector have been done, however, by comparing the various regions in the world and making assessments both of urban and rural needs is not easy. Some studies though have indicated the extent and challenges for the water industry in general. Investment in water quality in developing nations will help drive an estimated 5.9% annual increase (including price increases) in demand for water-treatment products through 2009, according to a new study, World Water Treatment, by research firm The Freedonia Group, Inc. The same firm has also recently made an assessment of the advancement of membrane technology in a study entitled 'Membrane Separation Technologies.' This study stipulated a yearly increase of 7.8% in the demand for membrane materials with a total value of membrane systems (including equipment such as pumps and piping) reaching USD 4.8 billion in 2004. Water and wastewater treatment has been identified as the largest end use for membrane

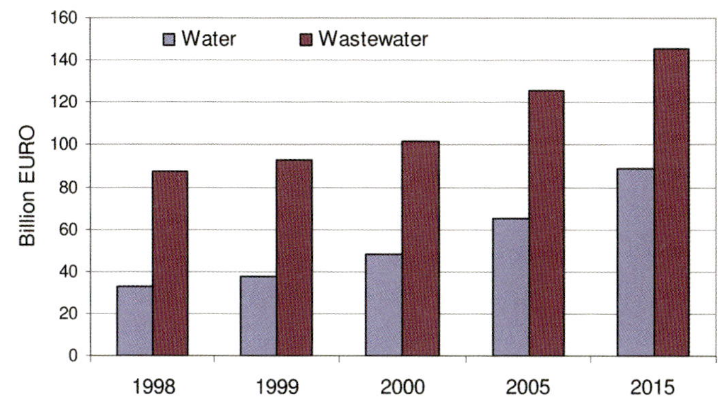

Figure 16.11 Stipulated growth in the water and wastewater industry to 2015 (Adapted from Ref. 59.)

materials, where they will continue to dominate. Implementation and expansion in consumer applications combined with replacement sales to municipal and industrial customers are suggested as the main reasons for this.

The global water market has been estimated at a total value of around 224 billion € with an anticipated annual growth of around 16–20% depending on the market segment (Figure 16.11). Drinking-water production is stipulated to have the largest growth with a doubling of the market value in the period 2000–2015. The market for wastewater treatment is the largest sector with an anticipated growth of around 43% for the same period. Membrane technology will of course play an important role in this market. Cross-flow membrane systems are expected to grow from 4.8 billion € in 2004 to 6.5 billion € in 2007 on a global basis where desalination has been reported to represent about 1/3 of this growth. The fastest growing segment, however, has been predicted to be the development of membrane-bioreactor systems for wastewater treatment with a yearly growth estimated at 15% [42, 57–59].

Looking at the water and sanitation sector in general, certain trends may be found regarding the advancement and implementation of membrane technology in environmental engineering. Microfiltration membranes (MF) are recognized as accounting for the largest portion of the market. The nature of MF separation make them by far the most widely used membrane process where they can either constitute the final separation stage or are applied as pretreatment options in, for example, reverse-osmosis (RO) systems. RO for desalination of brackish water or seawater to produce potable water is a well-established industry. The cost of desalination is expected to drop drastically (presented at Fourth World Water Forum, Mexico, 2006) and reports about the implementation of large desalination plants to produce fresh water can often be found in the news media. In areas of the world experiencing water shortages or high demands on limited fresh-water resources, wastewater recycling and reuse is becoming a necessity. Membrane technology is a central and key element in implementing sustainable solutions for wastewater recycling and reuse. Numerous examples of membrane systems success for this application can be found and there is an increasing interest in implementing the available technology.

Secondary treatment of wastewater is a large and energy-intensive process, generally based on the conventional activated-sludge process. The MBR process by itself can fulfill secondary treatment on its own – doing it much better and in less space. It is rapidly becoming accepted as the best available technology (BAT). Since commercial development of the MBR process in the late 1980s a growing reference list of installations and plant can be found as well an increasing number of manufacturers and suppliers. The list of references keeps increasing and there are currently around 3000 MBR plants in operation or under construction worldwide [48, 73, 89]. The MBR market is dominated by the earliest developers Zenon in Canada (now GE Water technologies) and Kubota in Japan, closely followed by Wehrle Werk in Germany. USFilter (now part of Siemens) and Mitsubishi Rayon have also emerged as major suppliers. There are now around 30 MBR suppliers worldwide though the markets are dominate by the larger suppliers, that is, Zenon-GE and Kubota representing 63 and 30% of the European market, respectively [48]. The growth and trends of MBR installations in the European market and the main suppliers to date are illustrated in Figure 16.12. Similarly, the situation in North America and an overview of the registered installations worldwide for some of the major suppliers is illustrated in Figure 16.13.

The first MBR installations were for relatively small treatment plants, the general impression at the time, that MBR systems were only suitable for small-scale installations. The first full-scale MBR plants were designed to treat wastewater for around 3–4000 person equivalents (i.e., the Porlock plant, UK, commissioned in 1998), however, treatment plants commissioned in the late 1990s/early 2000s showed a steady increase in size and capacity (i.e., the Nordkanal plant, Germany, commissioned in 2004 for 80 000 p.e., max. 48 000 m^3/day). The largest MBR plant announced so far is the Brightwater plant, USA, which will have a capacity of 495 000 m^3/day when it is commissioned in 2010/2011 [48, 73]. Given these trends it is clear that MBR technology will be a central and important option for advanced wastewater treatment in the future.

16.4.2
An Overview of Commercially Available Systems

The commercially available MBR systems can be classified into two distinct groups; those based on flat-sheet membranes and those using tubular or hollow-fiber membranes. In literature reviews one can find references to other membrane geometries being investigated such as multibore tubular membranes or self-supporting membrane sheets with integrated canals. There are no full-scale installations with these kinds of membranes to date. With the dynamic trend of the market, new suppliers as well as new products and novel solutions are to be expected. Given this situation, it is beyond the scope of this chapter to give a comprehensive and complete presentation of all MBR suppliers and products. In the following sections a brief overview will be given of the main systems being used, represented by the main suppliers, and the examples given are only intended to give newcomers to the technology a better understanding of the industry.

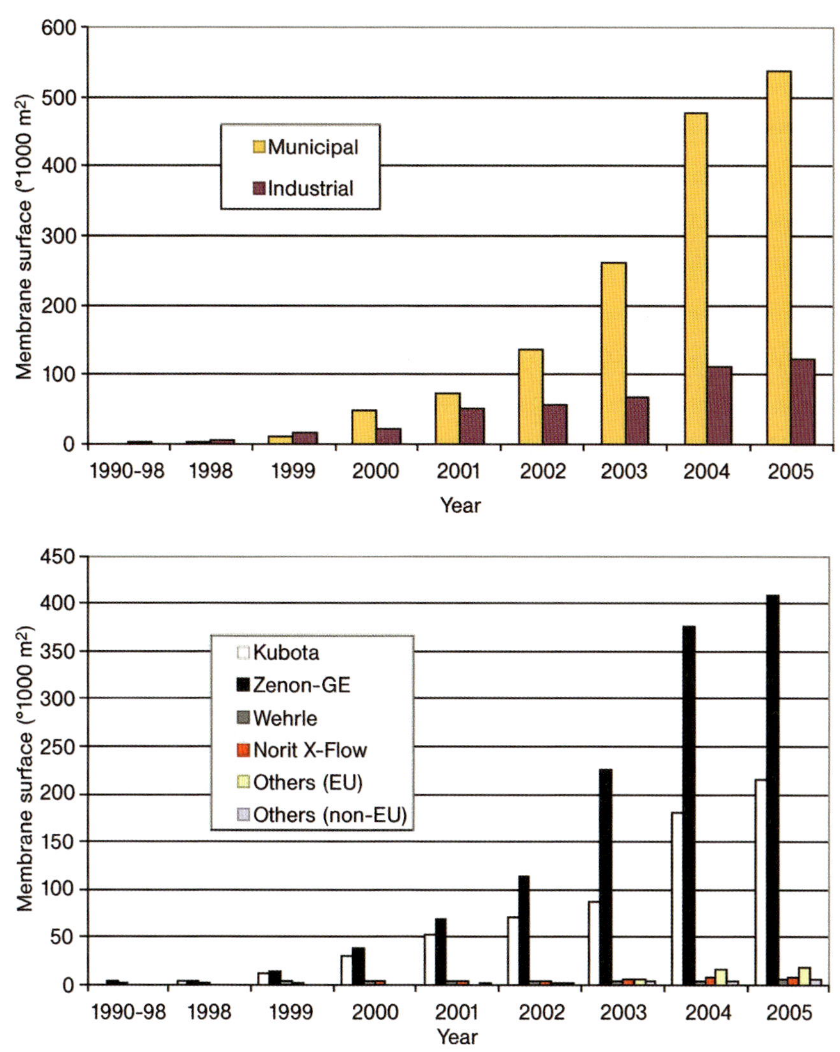

Figure 16.12 Development of the European MBR market (Adapted from Ref. 48.)

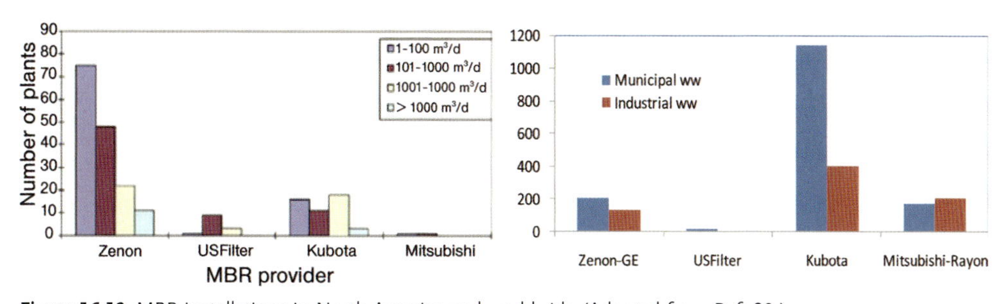

Figure 16.13 MBR installations in North America and worldwide (Adapted from Ref. 89.)

16.4.2.1 Flat-Sheet MBR Designs and Options

The flat-sheet MBR module, originally developed by Kubota, consists of two rectangular sheets of membrane that are wrapped around and welded onto a panel to make a cartridge. The support panel has a core structured for spacer material to allow water to move freely within the cartridge. These membrane cartridges are subsequently hung vertically with a defined space between each cartridge as specified by the various suppliers. Each cartridge is fitted with a nozzle and connected by tubes to a collection pipe from which the permeate is extracted by vacuum. The mode of flow is from the outside and into each cartridge. The spacing between each cartridge is optimized to allow for bubbles generated during air scouring preventing accumulation of solids between the panels causing clogging and fouling of the membrane. A view of the membrane cartridge and how they are arranged in the reactor are illustrated in Figure 16.14.

An alternative design to the membrane cartridges described above is the rotating disk unit. These membrane modules are constructed on a similar principle but are mounted on discs instead of panels. The discs are then mounted on axles enabling the stack of discs to be rotated. The permeate is extracted either through the shaft by vacuum (the Hitachi Plant design) or tubes mounted on each disc (the Huber Technology design), see Figure 16.15. One of the advantages claimed by the manufacturers' of this design is that the rotation can add to generating turbulent flow regimes that help reduce membrane fouling and potentially increases the rate of permeate production.

16.4.2.2 Tubular/Hollow-Fiber MBR Designs and Options

For the tubular or hollow-fiber MBR systems the submerged or immersed configuration is generally the preferred option. There are several suppliers offering such systems but the concept is essentially the same. Membranes are put together in modules (rectangular or circular units) that are then mounted into cassettes. The cassettes are the 'building blocks' of the system, forming the modular design typical of MBR systems. The capacity of the treatment plant is thus a function of how many modules are necessary to treat the design flow. As air scouring is an essential part of

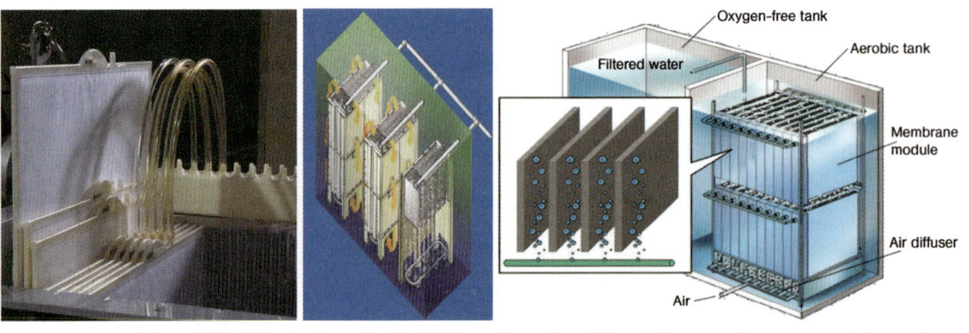

Figure 16.14 Illustration of a FS MBR configuration: FS cartridges, and arrangement in reactor.

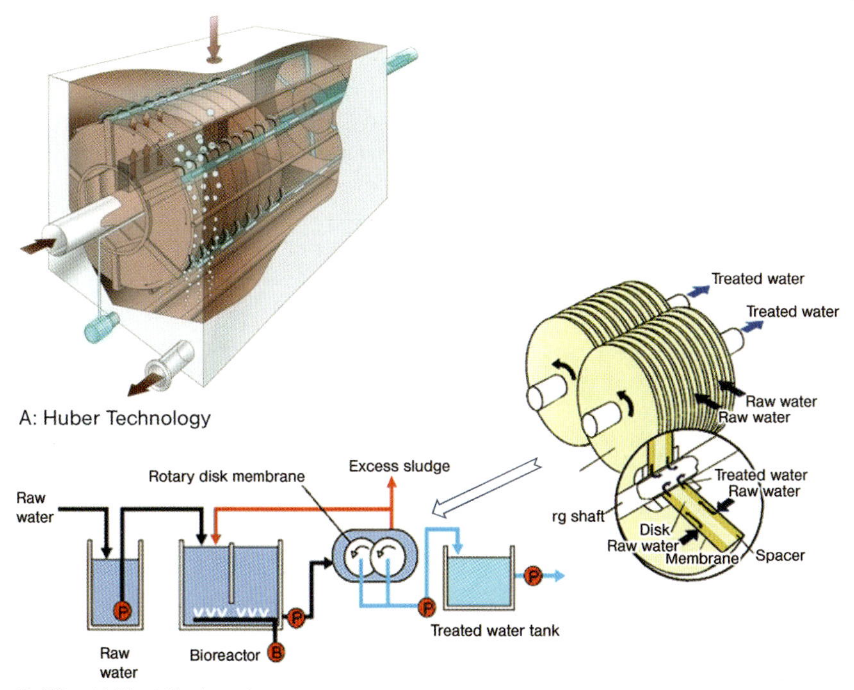

Figure 16.15 Examples of the rotating-disc membrane reactor design.

the system, the aeration devices are normally integrated into the module design. Examples of some of the available solutions are illustrated in Figure 16.16.

Examples of side-stream configurations are shown in Figure 16.17. These solutions are generally pressurized systems, though the transmembrane pressure or feed pressure is less than in conventional crossflow membrane systems. The mode of operation may be deadend or with recirculation of the concentrate. On side-stream systems a feed pump is commonly used to pressurize the membrane modules and to create the circulation of the feed water, where this applies. Some systems apply airlift principles to generate a crossflow mode of operation, typically in an inside-out configuration using tubular membranes. An added benefit of airlift systems is the effect of dual-media flow where bubbles inside the tubes help generate high shear forces that decrease fouling of the membranes. Another claimed benefit of side-stream configurations is from an operation and maintenance perspective where cleaning or replacing membrane modules is easier and more practical due to the accessibility.

Given the history and development of environmental engineering, with wastewater treatment in particular, as it is understood today, the membrane bioreactor is quite a recent invention. Commercialization and implementation of the technology has only been going on in the last couple of decades, and so, not surprisingly, MBR technology is still in a period of intense development [33]. The current MBR business

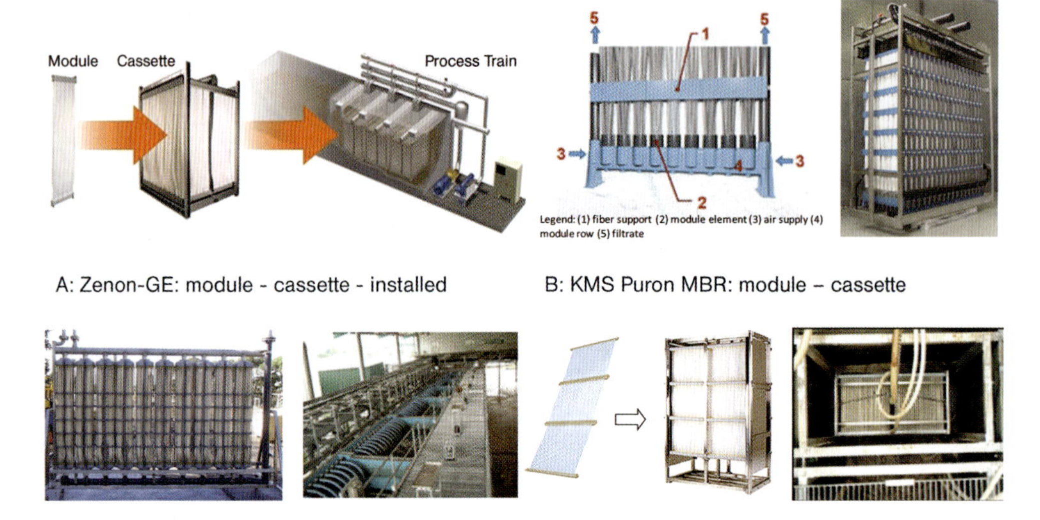

Figure 16.16 Examples of immersed tubular/hollow-fiber membrane systems.

can be found among the wastewater-equipment suppliers, either the specialist builders of wastewater treatment equipment who have acquired membrane technology (possibly by buying a membrane specialist company), or by membrane system manufacturers who have developed an expertise in wastewater treatment. Irrespective of 'point of entry' to the business, as the technology gains recognition as the best available technology (BAT) for wastewater treatment, the number of both new suppliers and new systems is expected to increase. In the foreseeable future, three key areas of system investigation can be identified independent of supplier and solution: the nature of the membrane coupled with operating energy consumption, air/gas handling and the bioreaction itself.

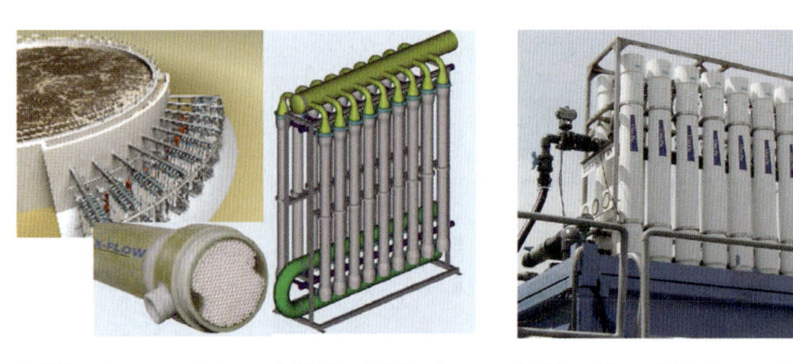

Figure 16.17 Examples of side-stream systems based on airlift or deadend mode.

Nomenclature

AS	Activated sludge
CAS	Conventional activated sludge
RO	Reverse osmosis
MBR	Membrane bioreactor
AS-MBR	Activated-sludge membrane bioreactor
FS	flat sheet
HF	Hollow fiber
MF	Microfiltration
UF	Ultrafiltration
NF	Nanofiltration
EPS	Extracellular polymeric substances
TMP	Transmembrane pressure
PVDF	Polyvinyl difluoride
PES	Polyethylsulfone
PE	Polyethylene
PP	Polypropylene
DO	Dissolved oxygen (mg/L)

References

1 Adham, S., Gagliardo, P., Boulos, L., Oppenheimer, J. and Trussel, R. (2001) Feasibility of the membrane bioreactor process for water reclamation. *Water Science and Technology*, **43** (10), 203–209.

2 Aimar, P. *Recent Progress in Understanding Particle Fouling of Filtration Membranes*, Laboratoire de Génie Chimique – CNRS – Université Paul Sabatier, http://www.membrane.unsw.edu.au/.

3 Arden, E. and Lockett, W.T. (1914) The oxidation of sewage without the aid of filter. Part I and II. *Journal of the Society of Chemical Industry*, **33** (10), 523–539, 1122–1124.

4 Alonso, E., Solis, G.J. and Riesco, P. (2001) On the feasibility of urban wastewater tertiary treatment by membranes: a comparative assessment. *Desalination*, **141**, 39–51.

5 Bacchin, P. (2004) A possible link between critical and limiting flux for colloidal systems: consideration of critical deposit formation along a membrane. *Journal of Membrane Science*, **228**, 237–241.

6 Bae, T.H. and Tak, M.T. (2005) Interpretation of fouling characteristics of ultrafiltration membranes during the filtration of membrane bioreactor mixed liquor. *Journal of Membrane Science*, **264** (1–2), 151–160.

7 Baker, M.N. (1981) *The Quest for Pure Water*, vol. **1**, 2nd edn, American Water Works Association, Denver.

8 Bouhabila, E.H., Ben Aim, R. and Buisson, H. (1998) Microfiltration of activated sludge using submerged membrane with air bubbling (application to wastewater treatment). Conference on Membranes in Drinking and Industrial Water Production, Amsterdam.

9 Brookes, A., Jefferson, B., Guglielmi, G. and Judd, S.J. (2006) Sustainable flux fouling in a membrane bioreactor: Impact of flux and MLSS. *Separation Science and Technology*, **41** (7), 1279–1291.

10 Carmen, C., Teodosiu, M., Kennedy, M.D., Van Straten, H.A. and Schippers, J.C. (1999) Evaluation of secondary refinery effluent treatment using ultrafiltration membranes. *Water Resources*, **33** (9), 2172–2180.

11 Chang, S. and Fane, A.G. (2000) Filtration of biomass with axial inter-fibre upward slug flow: performance and mechanisms. *Journal of Membrane Science*, **180**, 57–68.

12 Cheryan, M. (1998) *Ultrafiltration and Microfiltration. Handbook*, Technomic Publishing Company, Inc., USA.

13 Cho, B.D. and Fane, A.G. (2002) Fouling transients in nominally cub-critical flux operation of a membrane bioreactor. *Journal of Membrane Science*, **209** (2), 391–403.

14 Choksuchart, P., Heran, M. and Grasmick, A. (2002) Ultrafiltration enhanced by coagulation in an immersed membrane system. *Desalination*, **145**, 265–272.

15 Cornel, P., Wagner, M. and Krause, S. (2003) Investigation of oxygen transfer rates in full scale membrane bioreactors. *Water Science & Technology*, **47** (11), 313–319.

16 Cote, P., Buisson, H. and Praderie, M. (1998) Immersed membranes activated sludge process applied to the treatment of municipal wastewater. *Water Science and Technology*, **38** (4–5), 437–442.

17 Crespo, J.G. and Bödekker, K.W. (eds) (1994) *Membrane Processes in Separation and Purification*, Kluwer Academic Publishers, Dordrecht.

18 Davies, W.J., Le, M.S. and Heath, C.R. (1998) Intensified activated sludge process with submerged membrane microfiltration. *Water Science and Technology*, **38** (4–5), 421–428.

19 Defrance, L. and Jaffrin, M.Y. (1999) Comparison between filtrations at fixed transmembrane pressure and fixed permeate flux: application to a membrane bioreactor used for wastewater treatment. *Journal of Membrane Science*, **152**, 203–210.

20 Defrance, L. and Jaffrin, M.Y. (1999) Reversibility of fouling formed in activated sludge filtration. *Journal of Membrane Science*, **157**, 73–84.

21 Defrance, L., Jaffrin, M.Y., Gupta, B., Paullier, P. and Geaugey, V. (2000) Contributions of various constituents of activated sludge to membrane bioreactor fouling. *Bioresource Technology*, **73**, 105–112.

22 Duin, O., Wessels, P., van der Roest, H., Uijterlinde, C. and Schoonewilde, H. (2000) Direct nanofiltration or ultrafiltration of WTTP effluent? *Desalination*, **132**, 65–72.

23 Fane, A.G., Chang, S. and Chardon, E. (2002) Submerged hollow fibre membrane module - design options and operational considerations. *Desalination*, **146**, 231–236.

24 Fane, A. (2002) Membrane bioreactors: design and operational options. *Filtration & Separation*, **39** (5), 26–29.

25 Field, R.W., Wu, D., Howell, J.A. and Gupta, B.B. (1995) Critical flux concept for microfiltration fouling. *Journal of Membrane Science*, **100**, 259–272.

26 Flemming, H.C., Schaule, G., Griebe, T., Schmitt, J. and Tamachkiarowa, A. (1997) Biofouling - the Achilles heel of membrane processes. *Desalination*, **113**, 215–225.

27 Flemming, H.C. and Wingender, J. (2001) Relevance of microbial extracellular polymeric substances (EPSS) - part 1: Structural and ecological aspects. *Water Science and Technology*, **43**, 1–8.

28 Guglielmi, G., Chiarani, D., Judd, S.J. and Andreottola, G. (2007) Flux criticality and sustainability in a hollow fibre membrane bioreactor for municipal wastewater treatment. *Journal of Membrane Science*, **289**, 241–248.

29 Günder, B. and Krauth, K. (1998) Replacement of secondary clarification by membrane separation – results with plate and hollow fiber modules. *Water Science and Technology*, **38** (4–5), 383–393.

30 Günder, B. and Krauth, K. (2000) Excess sludge production and oxygen transfer in MBR. Proceedings ATSV conference, 8–9 February.

31 Hillis, P., Padley, M.B., Powell, N.I. and Gallagher, P.M. (1998) Effects of backwash conditions on out-to-in membrane microfiltration. *Desalination*, **118**, 197–204.

32 Howell, J.A. (1995) Subcritical flux operation of microfiltration. *Journal of Membrane Science*, **107**, 165–171.

33 Howell, J.A. (2002) Future research and developments in the membrane field. *Desalination*, **144**, 127–131.

34 Huang, L. and Morissey, M. (1998) Fouling of membranes during microfiltration of surimi wash water: Roles of pore blocking and surface cake formation. *Journal of Membrane Science*, **144**, 113–123.

35 Itonaga, T., Kimura, K. and Watanabe, Y. (2004) Influence of suspension viscosity and colloidal particles on permeability of membrane used in membrane bioreactor (MBR). *Water Science and Technology*, **50**, 301–309.

36 Ivanovic, I. and Leiknes, T. Impact of aeration rates on particle colloidal fraction in the biofilm membrane bioreactor (BF-MBR). Proceedings, IWA 4th International Membrane Technologies Conference, 15–17 May 2007, Harrogate, UK.

37 Ji, L. and Zhou, J. (2006) Influence of aeration on microbial polymers and membrane fouling in submerged membrane bioreactors. *Journal of Membrane Science*, **276**, 168–177.

38 Juang, L.C., Tseng, D.H. and Yin Lin, Y.H. (2007) Membrane processes for water reuse from the effluent of industrial park wastewater treatment plant: a study on flux and fouling of membrane. *Desalination*, **202**, 302–309.

39 Judd, S. (2002) Submerged membrane bioreactors: plate or hollow fibre? *Filtration + Separation*, **39**, 30–31.

40 Judd, S. (2006) *The MBR Book*, Elsevier, Amsterdam.

41 Kim, J.-S., Akeprathumchai, S. and Wickramasinghe, S.R. (2001) Flocculation to enhance microfiltration. *Journal of Membrane Science*, **182**, 161–172.

42 Laine, J.-M., Vial, D. and Moulat, P. (2000) Status after 10 years of operation - overview of UF technology today. *Desalination*, **13** (1), 17–25.

43 Le-Clech, P., Jefferson, B., Chang, I.-S. and Judd, S.J. (2003) Critical flux determination by the flux-step method in a submerged membrane bioreactor. *Journal of Membrane Science*, **227**, 81–93.

44 Le-Clech, P., Jefferson, B. and Judd, S.J. (2003) Impact of aeration, solids concentration and membrane characteristics on the hydraulic performance of a membrane bioreactor. *Journal of Membrane Science*, **218**, 117–129.

45 Le-Clech, P., Jefferson, B. and Judd, S.J. (2005) Comparison of submerged and side-stream tubular membrane bioreactor configurations. *Desalination*, **173**, 113–122.

46 Le-Clech, P., Fane, A.G., Leslie, G. and Childress, A. (2005) The operator's perspective. *Filtration + Separation*, **42**, 20–21.

47 Leiknes, T. and Ødegaard, H. (2007) The development of a biofilm membrane bioreactor. *Desalination*, **202**, 135–143.

48 Lesjean, B. and Huisjes, E.H. Survey of European MBR market, trends and perspectives. Proceedings, IWA 4th International Membrane Technologies Conference, 15–17 May 2007, Harrogate, UK.

49 Liao, B.Q., Catalan, L.J.J., Droppo, I.G. and Liss, S.N. (2004) Impact of chemical oxidation on sludge properties and membrane flux in membrane separation bioreactors. *Journal of Chemical Technology and Biotechnology*, **79**, 1342–1348.

50 Li, M., Xiufen, L., Du, G., Chen, J. and Shen, Z. (2005) Influence of the filtration modes on colloid adsorption on the membrane in submerged membrane bioreactor. *Colloids and Surfaces A: Physicochemical and Engineering Aspects*, **264** (1–3), 120–125.

51 Lim, A.L. and Bai, R. (2003) Membrane fouling and cleaning in microfiltration of

activated sludge wastewater. *Journal of Membrane Science*, **216**, 279–290.

52 Lodge, B., Judd, S.J. and Smith, A.J. (2004) Characterisation of dead end ultrafiltration of biotreated domestic wastewater. *Journal of Membrane Science*, **231**, 91–98.

53 Mallevialle, J., Odendaal, P.E. and Weisner, M.R. (1996) *Water Treatment Membrane Processes*, McGraw-Hill, New York.

54 Metcalf & Eddy (2003) *Wastewater Engineering: Treatment and Reuse*, 4th International edn, McGraw-Hill, New York.

55 Mulder, M. (1997) *Basic Principles of Membrane Technology*, Kluwer Academic Publishers, The Netherlands.

56 Nagaoka, H., Ueda, S. and Miya, A. (1996) Influence of bacterial extracellular polymers on the membrane separation activated sludge process. *Water Science Technology*, **34** (9), 165–172.

57 News Release, December 2000, *RO, UF, MF World Markets*, www.macilvainecompany.com, published by the McIlvaine Company.

58 News Release, July 2002, *RO, UF, MF World Markets*, www.macilvainecompany.com, published by the McIlvaine Company.

59 *News Release, 2004*, www.macilvainecompany.com, published by the McIlvaine Company.

60 Noble, R.D. and Stern, S.A. (1995) *Membrane Separations Technology. Principles and Applications*, Membrane Sciences and Technology Series, vol. **2**, Elsevier.

61 Nuengjamnong, C., Cho, J., Polprasert, C. and Ahn, K.H. (2006) Extracellular polymeric substances's influence on membrane fouling and cleaning during microfiltration process. *Water Science and Technology: Water Supply*, **6**, 141–148.

62 Ognier, S., Wisniewski, C. and Grasmick, A. (2002) Influence of macromolecule adsorption during filtration of a membrane bioreactor mixed liquor suspension. *Journal of Membrane Science*, **209**, 27–37.

63 Ognier, S., Wisniewski, C. and Grasmick, A. (2002) Membrane fouling during constant flux filtration in membrane bioreactors. *Membrane Technology*, **7**, 6–10.

64 Ognier, S., Wisniewski, C. and Grasmick, A. (2004) Membrane bioreactor fouling in sub-critical filtration conditions: a local critical flux concept. *Journal of Membrane Science*, **229**, 171–177.

65 Park, P.-K., Lee, C.-H., Choib, S.-J., Choo, K.-H., Kimd, S.-H. and Yoone, C.-H. (2002) Effect of the removal of DOMs on the performance of a coagulation-UF membrane system production. *Desalination*, **145**, 237–245.

66 Peuchot, M. and Ben Aïm, R. (1992) Improvement of crossflow microfiltration performances with flocculation. *Journal of Membrane Science*, **68** (3), 241–248.

67 Rosenberger, S. and Kraume, M. (2002) Filterability of activated sludge in membrane bioreactors. *Desalination*, **146** (1), 373–379.

68 Rosenberger, S., Witzig, R., Manz, W., Szewzyk, U. and Kraume, M. (2002) Performance of a bioreactor with submerged membranes for aerobic treatment of municipal waste water. *Water Research*, **36**, 413–420.

69 Rosenberger, S., Laabs, C., Lesjean, B., Gnirss, R., Amy, G., Jekel, M. and Schrotter, J.-C. (2006) Impact of colloidal and soluble organic material on membrane performance in membrane bioreactors for municipal wastewater treatment. *Water Research*, **40**(4), 710–720.

70 Seymour, R.B. and Carraher, C.E., Jr. (1988) *Polymer Chemistry*, Marcel Dekker Inc., New York.

71 Sirkar, K. and Winston Ho, W.S. (1992) *Membrane Handbook*, Van Nostrand Reinhold, New York.

72 Stephenson, T., Judd, S., Jefferson, B. and Brindle, K. (2000) *Membrane Bioreactors for Wastewater Treatment*, IWA Publishing, Great Britain.

73 Sutherland, K. (2007) The membrane bioreactor in sewage treatment. *Filtration + Separation*, **44**, 18–21.

74 Tao, G., Kekre, K., Wei, Z., Lee, T.C., Viswanath, B. and Seah, H. (2005)

Membrane bioreactors for water reclamation. *Water Science and Technology*, **51**, 431–440.

75 Tardieu, E., Grasmick, A., Geaugey, V. and Manem, J. (1999) Influence of hydrodynamics on fouling velocity in a recirculated MBR for wastewater treatment. *Journal of Membrane Science*, **156**, 131–140.

76 Ueda, T., Hata, K., Kikuoka, Y. and Seino, O. (1997) Effects of aeration on suction pressure in a submerged membrane bioreactor. *Water Research*, **31** (3), 489–494.

77 Allgeier, S., Alspach, B. and Vickers, J. (2003) *Membrane Filtration Guidance Manual*, United States Environmental Protection Agency (EPPA), EPA 815- D- 03-008.

78 Van den Berg, G.B. and Smolders, C.A. (1990) Flux decline in ultrafiltration processes. *Desalination*, **77-1**, 101–103.

79 van der Roest, H.F., Lawrence, D.P. and van Bentem, A.G.N. (2002) *Membrane Bioreactors for Municipal Wastewater Treatment*, STOWA Report IWA Publishing, London.

80 van Hoof, S.C.J.M., Duyvesteijn, C.P.T.M. and Vaal, P.P.R. (1998) Dead-end ultrafiltration of pretreated and untreated WWTP effluent for re-use in process water applications. *Desalination*, **118**, 249–254.

81 Wakemana, R.J. and Williams, C.J. (2002) Additional techniques to improve microfiltration. *Separation and Purification Technology*, **26**, 3–18.

82 Wen, X., Bu, Q. and Huang, X. (7–10 June 2004) Study on fouling characteristic of axial hollow fibers cross-flow microfiltration under different flux operations. in Proc of the IWA Specialty Conference -WEMT 2004, Seoul, Korea.

83 Wisniewski, C., Grasmick, A. and Cruz, A.L. (2000) Critical particle size in membrane bioreactors - case of a denitrifying bacterial suspension. *Journal of Membrane Science*, **178**, 141–150.

84 Wu, D.X., Howell, J.A. and Field, R.W. (1999) Critical flux measurement for model colloids. *Journal of Membrane Science*, **152**, 89–98.

85 Xing, C.-H., Wena, X.-H. and Tardieub, E. (2001) Microfiltration-membrane-coupled bioreactor for urban wastewater reclamation. *Desalination*, **141**, 63–73.

86 Xing, C.H., Wen, X.H., Qian, Y., Sun, D., Klose, P.S. and Zhang, X.Q. (2002) Fouling and cleaning of microfiltration membrane in municipal wastewater reclamation. *Water Science and Technology*, **47**, 263–270.

87 Xing, C.H., Wen, X.H., Qian, Y., Wu, W.Z. and Klose, P.S. (2003) Fouling and cleaning in an ultrafiltration membrane bioreactor for municipal wastewater treatment. *Separation Science and Technology*, **38**, 1773–1789.

88 Yamamoto, K., Hiasa, M., Mahmood, T. and Matsuo, T. (1989) Direct solid-liquid separation using hollow fiber membrane in an activated sludge aeration tank. *Water Science and Technology*, **21** (4/5), 43–54.

89 Yang, W., Cicek, N. and Ilg, J. (2006) State-of-the-art of membrane bioreactors: Worldwide research and commercial applications in North America. *Journal of Membrane Science*, **270**, 201–211.

90 Zeman, L.J. and Zydney, A.L. (1996) *Microfiltration and Ultrafiltration. Principles and Applications*, Marcel Dekker Inc., New York.

91 Åhl, R.M., Leiknes, T. and Ødegaard, H. (2006) Tracking particle size distributions in a moving bed biofilm membrane reactor for treatment of municipal wastewater. *Water Science and Technology*, **53** (7), 33–42.

17
Biochemical Membrane Reactors in Industrial Processes

Lidietta Giorno, Rosalinda Mazzei, and Enrico Drioli

17.1
Introduction

Biochemical membrane reactors are systems able to optimally integrate and intensify chemical transformations and transport phenomena in a single unit. The transformation is promoted by a catalyst of biological origin (commonly named biocatalyst) while the transport is governed by a membrane operation (i.e., by a driving force acting through a micro-nanostructured porous or dense membrane). Transport can be appropriately tuned so as to control the reagent supply to the catalyst and/or product removal from the reaction site.

The fundamentals of biochemical membrane reactors are reported in a previous chapter. Here, some highlights are just recalled for clarity in the subsequent discussion.

The applications presented refer to both main reactor configurations, that is, the configuration in which the membrane does not contribute to the reaction but only controls mass transport and the configuration in which the reaction also occurs at the membrane level.

The present work will mainly focus on biochemical membrane reactors operate at the production scale and give an overview of systems of potential interest studied at the laboratory level.

Despite the various fields of application (Figure 17.1), in this work industrial sectors such as pharmaceutical, food and biotechnology will be considered. Wastewater treatment and biomedical applications are discussed in other chapters.

The catalytic action of biocatalysts (enzymes, abzymes, antibodies, cells) is extremely efficient and selective compared to conventional chemical catalysts. They demonstrate higher reaction rates, milder reaction conditions and greater stereospecificity. Most of these properties come from the high molecular flexibility biocatalysts exhibit. On the other hand, this is also the origin of their major limit that holds back their application at the large scale, that is, the molecular stability, and then the catalyst lifetime.

Membrane Operations. Innovative Separations and Transformations. Edited by Enrico Drioli and Lidietta Giorno

ISBN: 978-3-527-32038-7

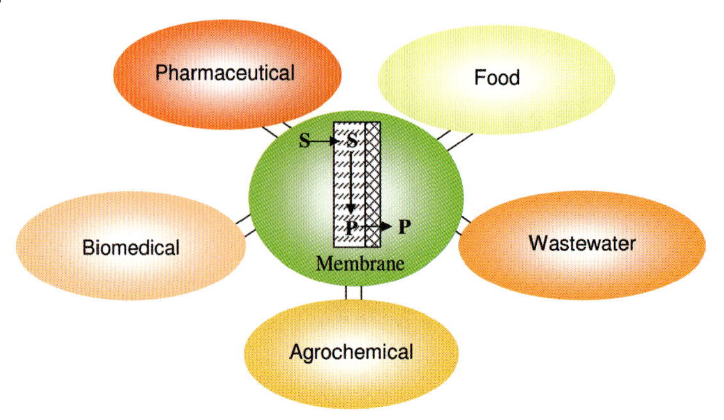

Figure 17.1 Field of biochemical membrane reactors application.

The use of biocatalysts in combination with membrane operations permits drawbacks to be overcome enabling biotransformation to be integrated into continuous production lines. These systems, being able to work at time-invariant conditions at steady state, permit a better control of reaction conditions with an increase of lifetime, productivity and economic viability of the process. In addition, the separation, purification, and concentration of the obtained product can occur in a single integrated unit operation. Thanks to the biocatalyst and membrane selectivity the mass intensity can be very high, with no by-products formation, while producing high added value coproducts.

17.2
Applications at Industrial Level

Despite their great advantages, the application of biochemical membrane reactors at the industrial scale in pharmaceutical, food, and biotechnology is still limited. Major reasons for this include the nonadequate research efforts devoted to the field, lack of predictive and holistic approach. A clear example of this situation is constituted by the commercial success submerged membrane bioreactors met in wastewater treatments. In this field, the technology was pushed by research efforts promoted to face lack of clean water and to meet regulations about wastewater discharge in the environment. More stringent regulations about ecocompatibility of industrial processes will necessarily promote technological advances also in other industrial sectors. For example, considering the mass of wastes compared to the mass of product, it appears that pharmaceutical industry used less-advanced technology than oil refineries (Figure 17.2(a)). Due to the orders of magnitude difference between the two sectors in terms of tons of productions (Figure 17.2(b)), the impact of the pharmaceutical industry is of course much lower, but it is evident that in this field there is a much higher potential for knowledge-based technologies.

Table 17.1 summarizes the most common examples of biochemical membrane reactors patented and whose robustness has been proved at the industrial production

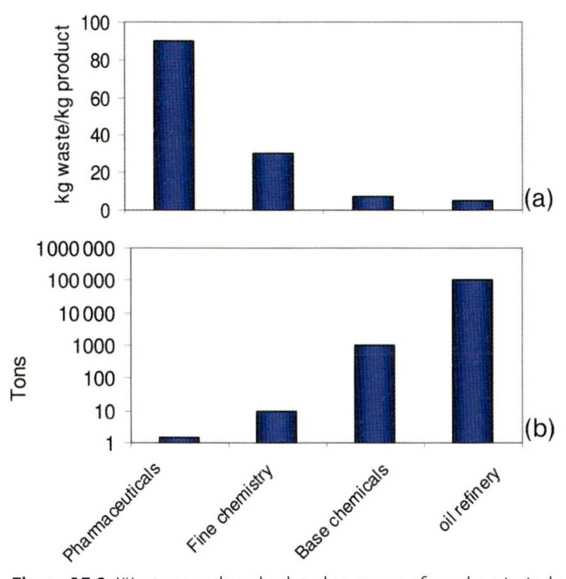

Figure 17.2 Wastes produced related to mass of product in industries [1].

scale. Table 17.1 illustrates the type of application, the biocatalyst used and the way it is used in the membrane reactor. When the enzyme is used as free, the membrane serves to separate the reaction product, whilst when it is immobilized the membrane hosts both reaction and separation.

17.2.1 Pharmaceutical Applications

The use of membrane bioreactors in the pharmaceutical field has been documented for the production of amino acids, antibiotics, anti-inflammatories, anticancer drugs, vitamins, and optically pure enantiomers [10, 29–31].

Examples at the industrial scale of membrane bioreactors in the pharmaceutical field include the production of amino acids with simultaneous regeneration of NADH, which has been commercialized by Degussa Company in Germany [10]. In Japan, the Kao Corporation investigated the so-called sandwich reactor for hydrolyzing triglycerides [32], Nitto Electric Industries immobilized cyclomaltodextrin glucanotransferasi on a hollow-fiber membrane and investigated production of cyclodestrins from starch [32].

Membrane bioreactors have been reported for the production of diltiazem chiral intermediate with a multiphase/extractive enzyme membrane reactor [15, 16]. The reaction was carried out in a two-separate phase reactor. Here, the membrane had the double role of confining the enzyme and keeping the two phases in contact while maintaining them in two different compartments. This is the case of the multiphase/extractive membrane reactor developed on a productive scale for the production of a chiral intermediate of diltiazem ((2R,3S)-methylmethoxyphenylglycidate), a drug used in the treatment of hypertension and angina [15]. The principle is illustrated in

Table 17.1 Biochemical membrane reactors in industrial processes.

Biocatalyst	Status	Application
Lactase	Immobilized	Hydrolysis of beta-D-galactosidic linkage of lactose milk (Industrial scale) [2]
Glucose isomerase	Immobilized	Conversion of D-glucose to D-fructose (Industrial scale) [3]
Penicillin acylase	Free	Production of antibiotics (Industrial scale) J. [4]
Acylase	Immobilized	Production of L-aminoacids (Industrial scale) [5]
E. Coli	Immobilized	Production of L-aspartic acid (Industrial scale) [6]
Pseudomonas dacunahe	Immobilized	Production of L-alanine (Industrial scale) [7]
Aminoacilase, and dehidrogenase	Free and immobilized	Production of L-aminoacids [8–10]
Brevibacterium ammoniagenes	Immobilized	Production of L-malic acid (Industrial scale) [11]
Pectic enzymes	Free or immobilized	Hydrolysis of pectins to improve processability (industrial scale) [12, 13]
Thermolysin	Immobilized	Production of aspartame (Industrial scale) [14]
Lipase of 360	Immobilized	Production of diltiazem chiral intermediate (industrial scale) [15, 16]
Trypsin	Free	Production of casein bioactive peptide (patented) [17, 18].
Protease	Immobilized	Hydrolysis of caroteno-proteins (patented) [19]
Acetyl transferase from Taxus	Immobilized	Production of baccatin III (patented) [20]
Lipase	Free	Production of fatty acid (patented) [21]
Cells	—	Linear or membrane-like biodevices and a bioreactor in which adhesive cells are anchored at high density (patented) [22]
Pancreatic cells	Immobilized	Artificial organs and implantable bioreactors (patented) [23, 24]
Cells	Free	Continuous cell culture (patented) [25]
Stem cells	—	Cell-expansion apparatus (patented) [26]
Viruses, virus particles, antibodies and proteins	Free	Production of a concentrated solution from biological substances (patented) [27]
Cells		Delivery of drugs or genes to individual cells (patented) [28]

Figure 17.3, the reactant was fed into the solvent, while the product was extracted in water. The lipase was immobilized by entrapment method on asymmetric PAN hollow-fiber membranes. The process was run for several years with modules for the production plant of 60 m^2 of active membrane area.

A further improvement of the multiphase reactor concept using lipase for enantioselective transformation has been recently reported, that is, an emulsion enzyme membrane reactor. Here, the organic/water interface within the pores at the enzyme level is achieved by stable oil-in-water emulsion, prepared by membrane emulsification. In this way, each pore forms a microreactor containing immobilized

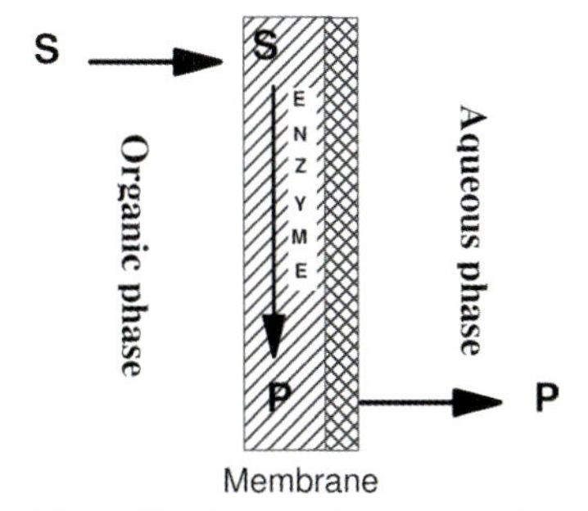

Figure 17.3 Schematic representation of multiphase membrane reactor.

enzyme (Figure 17.4) [33]. In the membrane pores, the enzyme is able to work in the same conditions as in the stirred-tank reactor, but with no shear stress due to stirring.

This configuration improved the selectivity and productivity of the biocatalytic system as well as its catalytic stability, confirming that the observed inversion relationship between activity and stability of immobilized enzyme is not a general rule.

Other biochemical membrane reactors applications include the synthesis of lovastatin with immobilized *Candida rugosa* lipase on a nylon support [34]; the synthesis of isomalto oligosaccharides and oligodextrans in a recycle membrane bioreactor by the combined use of dextransucrase and dextranase [35], the production of a derivative of kyotorphin (analgesic) in solvent media using α-chymotrypsin as catalyst and α- alumina mesoporous tubular support [36], and biodegradation of high-strength phenol solutions by *Pseudomonas putida* using microporous hollow fibers [37].

A particular application of membrane bioreactors, patented in 2005 [20], concerns the production of an antitumor substance (paclitaxel). Since a full synthesis of paclitaxel is not possible due to its low yield, a semisynthesis of 10-deacetyl-baccatin

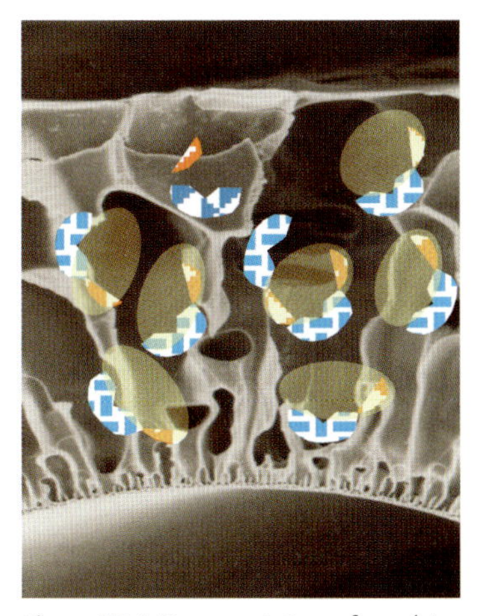

Figure 17.4 Representation of emulsion enzyme membrane.

III (10-DAB) was carried out from which baccatin III was produced in an enzyme reactor. The enzyme reactor comprised a hollow-fiber polymeric ultrafiltration membrane, with immobilized acetyl transferase from *Taxus species*. The process enabled the production of baccatin III without requiring complicated purification steps of the acetyl transferase. The purification of the baccatin III is also made distinctly easier [20].

Membrane bioreactors can be easily integrated with other systems, for example, with delivery of drugs or genes to individual cells achieved on the nanoscale using electroporation techniques. In one method developed in a recent patent, a flow-through bioreactor having an inlet and an outlet connected by a flow chamber and a nanoporous membrane positioned in the flow chamber was used [28].

Recent studies in the pharmaceutical field using MBR technology are related to optical resolution of racemic mixtures or esters synthesis. The kinetic resolution of (*R*,*S*)-naproxen methyl esters to produce (*S*)-naproxen in emulsion enzyme membrane reactors (E-EMRs) where emulsion is produced by crossflow membrane emulsification [38, 39], and of racemic ibuprofen ester [40] were developed. The esters synthesis, like for example butyl laurate, by a covalent attachment of *Candida antarctica* lipase B (CALB) onto a ceramic support previously coated by polymers was recently described [41]. An enzymatic membrane reactor based on the immobilization of lipase on a ceramic support was used to perform interesterification between castor oil triglycerides and methyl oleate, reducing the viscosity of the substrate by injecting supercritical CO_2 [42].

The production of aromatic compounds by a membrane bioreactor is widely studied and some examples are also patented. Aromatic compounds are important substances in pharmaceutical, food, and cosmetic industries due to their natural properties and because they are strong antioxidant molecules with a strong free-radical scavenging activity.

The hydrolysis of caroteno-proteins for the production of astaxanthin using protease was developed in an enzymatic membrane bioreactor [19], in which the concentration of the protein fraction by ultrafiltration and the separation of the pigments in the permeate were simultaneously carried out.

Terpene esters belong to a large family of aromatic compounds, which are important for flavoring and are widely used in the pharmaceutical and food industries. An important terpene ester is α-pinene oxide, its biotransformation to isonovalal using resting cells of *Pseudomonas fluorescens* NCIMB 11671 was evaluated in a membrane bioreactor [43]. Production of geranyl acetate, one of the best-known aromatic compounds, was studied using lipase CAL-B immobilized on polymer membranes by sorption and chemical binding [44].

17.2.2
Food Applications

The main applications of biocatalytic membrane reactors in the food sector include: reduction of the viscosity of juices by hydrolyzing pectins, reduction of the lactose content in milk and whey by its conversion into digestible sugar, treatment of musts and wine by the conversion of polyphenolic compounds and antocyanes and the removal of peroxides from diary products.

The interactions between pectins and sugars (rhamnose, arabinose, and galatose) are principally responsible for the high turbidity and viscosity of fruit juice. Pectinases immobilized in membranes are used to reduce the viscosity of fruit juice [12, 13].

One of the first cases of the application of membrane bioreactors in food processes was the production of milk with low lactose content. β-galactosidase was entrapped into cellulose acetate fibers to carry out the hydrolysis of milk and whey lactose [2] recently the system was improved by the use of microfiltration and by UV irradiation of the enzyme solution to avoid growth of micro-organisms [45].

The use of membrane reactors as continuous systems for the hydrolysis of lactose (present in whole milk or cheese whey) is an effective technique running at a large scale. Intolerance to milk is not only caused by lactose, but also by high molecular weight proteins. In fact, some children and old people have difficulty in hydrolyzing proteins with molecular weight higher than 5 kDa. In other words, they cannot digest such proteins, which induce stomach ache and can also lead to allergy. The hydrolysis of high molecular weight proteins into polypeptides lower than 5 kDa in biocatalytic membrane reactors is a new approach to produce low allergenic fresh milk with improved properties compared to the reconstituted powder milk currently used. The biocatalytic membrane reactor can be designed so that the hydrolyzed fragments equal to or lower than 5 kDa can be removed through a membrane of appropriate cutoff, while retaining the nonhydrolyzed proteins. In order to achieve high efficiency, the hydrolytic step should be part of an integrated system where up- and downstream of milk is properly considered. Biocatalytic membrane reactors can also be used to valorize coproducts of cheese-making processes. In fact, it is possible to increase the cost effectiveness of cheese-making processes and reduce waste simply by recovering and reusing compounds present in waste streams. The whey proteins (such as α-lactalbumin), which have excellent functional properties, can be recovered by ultrafiltration and hydrolyzed to produce many useful pharmaceutical intermediates. In addition, permeates from the ultrafiltered milk and whey contain lactose, which can be recovered and used in the production of glucose and galactose syrup.

Other important applications in the food industry running at a large scale are the production of L-aspartic acid with *Escherichia coli* entrapped in polyacrilamides [6], the immobilization of thermolysin for the production of aspartame [14], The production of L-alanine by Tanabe Seiyaku [7], the production of fructose concen-centrated syrup [3], the production of L-malic acid by the use of *Brevibacterium ammoniagenens* immobilized in polyacrilamide by entrapment immobilization methods [11] and L-aminoacids production by immobilized aminoacylase [5].

Biocatalytic membrane reactors are also used for the treatment of musts and wines by the conversion of polyphenolic compounds and anthocyanes. Laccase is used to oxidize polyphenols in solution and anthocianase is used immobilized on synthetic and natural polymers to hydrolyze anthocyanes.

During the maturation process, a secondary fermentation occurs that converts malic acid into lactic acid. Control of this reaction will enable the production of a product with good organoleptic properties. In comparison with batch fermentation, membrane bioreactors achieve remarkably elevated cell concentrations and productivity. In this type of bioreactor, a membrane has been installed to prevent washout of yeast when the fermented stream is withdrawn from the reactor. This apparatus makes it possible to keep the cell concentration in the reactor high while reducing product inhibition by replacing product-containing broth with fresh medium.

Recent patented works were also reported related to the design of membrane bioreactors.

A membrane biological reactor of a flat x-plate system comprising a thermostat, and a free system comprising a hydrophilic membrane with cutoff value of 30 100 kDa at 15–65 °C using lipase enzyme was developed and patented for the production of fatty acids by enzymatic hydrolysis of vegetable or animal oils or fats [21]. Membrane bioreactors were also used to improve existing systems for different applications, like cells nutrition and growth, and as delivery systems for genes and drugs [22–25, 28], to produce concentrated solution consisting of biological substances such as, for example, viruses, virus particles, antibodies, and proteins [27].

The design of food products that confer a health benefit is a relatively new trend, and recognizes the growing acceptance of the role of diet in disease prevention, treatment, and well-being. This change in attitude for product design and development has forced organizations and industries involved in formulating foods for health benefit into new areas of knowledge.

Recent works in the literature were devoted to improve the production of food similar to an existing one that is less dangerous in a compromised diet for the production of functional food or the production of nutraceuticals.

Palatinose (isomaltulose, 6-O-a-D-glucopyranosyl-D-fructose) a natural substance with a sweetening power of about 45% less than that of sucrose, and xilitol, where insulin is not involved in its methabolism, can be used as sugar substitutes in diabetic subjects. Recently a complete conversion of concentrated sucrose solutions into palatinose immobilizing in a hollow-fiber membrane reactor *Serratia plymuthica* cells was produced [46]. *Candida tropicalis*, an osmophilic strain isolated from honeycomb, was used to produce xilitol recycled in a submerged membrane bioreactor with suction pressure and air sparging, obtaining the highest productivity of xilitol $12.0\,g\,l^{-1}\,h^{-1}$ [47].

Octenylscuccinate derivatives of starch are attracting growing attention of food technologists as potential additives as emulsifying agents. The results obtained using the technology of membrane bioreactors indicated that the hydrolysis of sodium octenylsuccinate starch derivatives leads to products that reveal the surface activity, irrespective of the type as well as the amount of enzyme used in the hydrolysis process. The application of the UF membrane bioreactor to enzymatic hydrolysis could be the way of intensification of the production process [48].

A very interesting field in membrane bioreactors is the production of cyclodextrins or oligosaccharides. In general, they have applications in food pharmaceutical, cosmetic agricultural, and plastics industries as emulsifiers, antioxidant, and stabilizing agents. In the food industry cyclodextrins are employed for the preparation of cholesterol-free products. The use of enzymatic membrane reactors to produce cyclodextrins has been reported starting from different sources including soluble potato and corn starch. A recent study [49] reported also their production using enzyme membrane reactors starting from tapioca starch. The production of oligosaccharides to be used as functional food was also obtained by the immobilization of dextranase on polymeric matrix [50].

The production of substances that preserve the food from contamination or from oxidation is another important field of membrane bioreactor. For example, the production of high amounts of propionic acid, commonly used as antifungal substance, was carried out by a continuous stirred-tank reactor associated with ultrafiltration cell recycle and a nanofiltration membrane [51] or the production of gluconic acid by the use of glucose oxidase in a bioreactor using PES membranes [52]. Lactic acid is widely used as an acidulant, flavor additive, and preservative in the food, pharmaceutical, leather, and textile industries. As an intermediate product in mammalian metabolism, L(+) lactic acid is more important in the food industry than the D(−) isomer. The performance of an improved fermentation system, that is, a membrane cell-recycle bioreactors MCRB was studied [53, 54], the maximum productivity of 31.5 g/L h was recorded, 10 times greater than the counterpart of the batch-fed fermentation [54].

17.2.3 Immobilization of Biocatalysts on Membranes

The choice of reactor configuration depends on the properties of the reaction system. For example, bioconversions for which the homogeneous catalyst distribution is particularly important are optimally performed in a reactor with the biocatalyst compartmentalized by the membrane in the reaction vessel. The membrane is used to retain large components, such as the enzyme and the substrate while allowing small molecules (e.g., the reaction product) to pass through. For more labile molecules, immobilization may increase the thermal, pH and storage stability of biocatalysts.

Biocatalysts can be entrapped within the membrane, gelified on the membrane surface or bounded to the membrane surface or inner polymeric matrix [55].

The entrapment method of immobilization is based on the localization of an enzyme within a polymer membrane matrix. It is done in such a way as to retain biocatalyst, while allowing penetration of substrate.

Asymmetric hollow fibers provide an interesting support for enzyme immobilization, in this case the membrane structure allows the retention of the enzyme into the sponge layer of the fibers by crossflow filtration. The amount of biocatalyst loaded, its distribution and activity through the support and its lifetime are very important parameters to properly orientate the development of such systems. The specific effect that the support has upon the enzyme, however, greatly depend upon both the support and the enzyme involved in the immobilization as well as the method of immobilization used.

The amount of the immobilized biocatalyst is an important parameter and strongly affects the reactor performance, enzymes in fact, are not able to work at high concentration.

The amount of immobilized protein can be determined by the mass balance between initial and final solutions [56]. A combined qualitative method merged from the classical *in-situ* detection of enzyme activity and western blot analysis can be applied to determine the enzyme spatial distribution through the membrane thickness and along the membrane module and its activity after the immobilization [57–59].

The gelification of the biocatalyst on the membrane is based on one of the main drawbacks of membrane processes: fouling. Disadvantages of this systems are the reduction of the catalytic efficiency, due to mass transport limitations and the possibility of preferential pathways in the enzyme gel layer [60].

The binding of the biocatalyst to the membrane can be divided into three principal groups: ionic binding, crosslinking, and covalent linking.

In the literature there are various routes to carry enzyme immobilization creating a bound on supports, the principal strategies are based on chemical grafting or molecular recognition on porous supports. The sites involved in this chemistry are generally carboxylic acid, hydroxyls, amino or quaternary ammonium groups, which are created on the surface of porous material by various means, like direct chemical surface treatment or plasma or UV activation.

The reactive sites thus created allow the attachment of the enzyme by use of coupling reagents such as tosyl chloride, dicyclohexylcarbodiimide and glutaraldehyde.

Approaches aiming at creating biocompatible environments consist in modifying the surface of polymeric membranes by attaching functional groups like sugars, polypeptides and then to adsorb the enzymes.

Another way considered as of biomimetic inspiration and that was shown to be efficient for enzyme attachment, it consists in using the very strong and specific interaction of the small protein avidin for the biotin [61, 62]. The tetrameric structure of avidin permits itself to interact with four different molecules of biotin at the same time. Various proteins and enzyme could be easily biotinylated, and this mode of enzyme grafting has already been used for electrodes production as well as for membranes made up of conducting fibers.

Although immobilization of enzymes generally enhance their stability, one major disadvantage of random immobilization of enzymes onto polymeric microfiltration-type membranes is that the activity of the immobilized enzymes is often significantly decreased because the active site may be blocked from substrate accessibility, multiple-point binding may occur, or the enzyme may be denatured [31]. Different approaches are developed in order to accommodate site-specific immobilization of enzymes with different structural characteristics, as gene fusion to incorporate a peptidic affinity tag at the N- or C- terminus of the enzyme; post-translational modification to incorporate a single biotin moiety on enzymes; and site-directed mutagenesis to introduce unique cysteins to enzymes [63].

The selection of the membrane to be used in enzymatic membrane reactors should take into account the size of the (bio)catalyst, substrates, and products as well as the chemical species of the species in solution and of the membrane itself. An important parameter to be used in this selection is the solute-rejection coefficient, which should be zero for the product to facilitate permeation, and should be one for the enzyme to insure a complete retention of the catalyst inside the reaction system. The selectivity is normally associated with a discrimination based on size exclusion, but when a steric exclusion process may be present for molecules with size close to the pore size.

17.3 Conclusion

Currently, the major industrial application of biochemical membrane reactors is in water treatment. In the biotechnology field the development of such biohybrid catalytic systems is still at an emerging stage. The main technological difficulties in using membrane bioreactors for production on an industrial level are related with rate-limiting aspects and reproducibility on the large scale, together with the life-time of the enzyme, the availability of pure catalysts at an acceptable cost, and the necessity for biocatalysts to operate at low substrate concentration and without microbial contamination.

Many studies are oriented to the investigation of operating conditions and optimization of the various properties of membrane bioreactors. However, efforts towards the development of a predictive knowledge-based approach, able to overcome the trial and error one, is necessary to significantly advance the field. The development of membranes specifically designed to answer bioreactor needs is needed as well. The converging of technologies such as genetic engineering, bioprocess design, molecular modeling and biochemical engineering will promote innovative solutions to face the need for precise, selective, clean, safe, low energy consumption and ecofriendly processes, such as biochemical membrane bioreactors.

References

1 Sheldon, R.A. (1997) *Journal of Chemical Technology and Biotechnology*, **68**, 381.

2 Pastore, M. and Morisi, F. (1976) *Methods in Enzymology*, **44**, 822–830.

3 Carasik, W. and Carrol, J.O. (1983) *Food Technology*, **37**, 85–91.

4 Bryjak, J., Bryjak, M. and Noworyta, A. (1996) *Enzyme and Microbial Technology*, **19**, 196–201.

5 Sato, T. and Tosa, T. (1993) Optical resolution of aminoacids by aminoacylase in *Industrial Application of Immobilized Biocatalyst* (eds T. Tanaka, T. Tosa and T. Kobayashi) Marcel Dekker, New York, pp. 3–14.

6 Chibata, I. *et al.* (1974) *Applied Microbiology*, **27**, 878–885.

7 Takamatsu, S. (1982) *European Journal of Applied Microbiology and Biotechnology*, **15**, 147–149.

8 Wandrey, C. and Flaschel, E. (1979) *Advances in Biochemical Engineering*, vol. **12**, Springer Verlag, Berlin, pp. 147.

9 Wandrey, C., Wichmann, R., Leuchtenberger, W. and Kula, M.R. (1981) US Patent No. 4,304,858, Degussa AG/GBF.

10 Wandrey, C. (2004) Biochemical reaction engineering for redox reactions. *The Chemical Record*, **4**, 254–265.

11 Takata, I., Tosa, T. and Kobayashi, T. (eds.) (1993) *Industrial Application of Immobilized Biocatalysts*, Marcel Dekker, New York, pp. 53–55.

12 Alkorta, I., Garbisu, C., Llama, M.J. and Serra, J.L. (1998) *Process Biochemistry*, **33**, 21–28.

13 Giorno, L. *et al.* (1998) Fruit Processing *Separation Science and Technology*, **33**, 739–756.

14 Oyama, K. (1974) *Journal of Chemical Society*, **11**, 356–360.

15 Lopez, J.L. and Matson, S. (1997) *Journal of Membrane Science*, **125**, 189–211.

16 Matson, S.L. (1989) Patent No. US4800162.

17 Qi, H.Z. (2004) Patent No. CN1546682.

18 Trusek-Holownia, A. (2008) *Biochemical Engineering Journal*, **39**, 221–229.

19 Guerrero, L.M.I. (2004) Patent No MXPA02012838.

20 Frense, D. (2005) Patent No WO2005066353.

21 Belafine Bako, K. (2006) Patent No HU0401348.

22 Satoshi, K. and Tetsuya, M. (2006) Patent No CA2511457.

23 Shalev, A. (2006) Patent No WO2006080009.

24 Shalev, A. (2008) Patent No US2008112995.

25 Zhang, Z., Boccazzi, P., Choi, H., Jensen, K. F. and Sinskey, A. (2006) Patent No. US2006199260.

26 Antwiler Glen, D. (2008) WO2008109200.

27 Czermak, P. and Nehring, D. (2006) Patent No WO2006005305.

28 Lee, L.J., Wang, S., Xie, Y., Zeng, C., Koh, C.G. and Fei, Z. (2007) Patent No WO2007053802.

29 Giorno, L. and Drioli, E. (2000) *Trends in Biotechnology*, **18**, 339–348.

30 Giorno, L., Li, N. and Drioli, E. (2003) *Biotechnology and Bioengineering*, **84**, 677.

31 Charcosset, C. (2006) *Biotechnology Advances*, **24**, 482–492.

32 Nam, C.H. and Furusaki, S. (1991) Membrane bioreactors: Present and prospects. *Advances in Biochemical Engineering/Biotechnology*, **44**, 27–64.

33 Giorno, L., D'Amore, E., Mazzei, R., Piacentini, E., Zhang, J., Drioli, E., Cassano, R. and Picci, N. (2007) *Journal of Membrane Science*, **295**, 95–101.

34 Yang, W., Cicek, N. and Ilg, J. (1997) *Journal of Membrane Science*, **270**, 201–211.

35 Goulas, A.K., Cooper, J.M., Grandison, A.S. and Rastall, R.A. (2004) *Biotechnology and Bioengineering*, **88**, 778–787.

36 Belleville, M.P., Lonzano, P., Iborra, J.L. and Rios, G.M. (2001) *Separation and Purification Technology*, **25**, 229–233.

37 Chung, T.P., Wu, P.C. and Juang, R.S., (2005) *Journal of Membrane Science*, **317**, 55–63.

38 Giorno, L., Piacentini, E., Mazzei, R. and Drioli, E. (2008) *Journal of Membrane Science*, **317**, 19–25.

39 Li, N. and Sakaki, K. (2008) *Journal of Membrane Science*, **314**, 183–192.

40 Long, W.S., Azlina Harun, K. and Subhash, I. (2005) *Chemical Engineering Science*, **60**, 4957–4970.

41 Gumı, T., Fernandez-Delgado Albacete, J., Paolucci-Jeanjean, D., Belleville, M. and Rios, G.M. (2008) *Journal of Membrane Science*, **311**, 147–152.

42 Pomier, E., Delebecque, N., Paolucci-Jeanjean, D., Pina, M., Sarrade, S. and Rios, G.M. (2007) *Journal of Supercritical Fluids*, **41**, 380–385.

43 Boontawan, A. and Stuckey, D.C. (2006) *Applied Microbiology and Biotechnology*, **69**, 643–649.

44 Trusek-Holownia, A. and Noworyta, A. (2007) *Journal of Biotechnology*, **130**, 47–56.

45 Novalin, S., Neuhaus, W. and Kulbe, K. (2005) *Journal of Biotechology*, **119**, 212–218.

46 Krastanov, A., Blazheva, D. and Stanchev, V. (2007) *Process Biochemistry*, **42**, 1655–1659.

47 Kwona, S., Park, S. and Oh, D. (2006) *Journal of Bioscience and Bioengineering*, **101**, 13–18.

48 Prochaskaa, K., Kedzioraa, P., Le Thanhb, J. and Lewandowicz, G. (2007) *Food Hydrocolloids*, **21**, 654–659.

49 Sakinaha, A.M.M., Ismaila, A.F., Md Illiasa, R. and Hassand, O. (2007) *Desalination*, **207**, 227–242.

50 Torras, C., Nabarlatz, D., Vallot, G., Montané, D. and Garcia-Valls, R. (2008) *Chemical Engineering Journal*, **144**, 259–266.

51 Boyaval, P. and Corre, C. (1995) *Le Lait*, **75**, 453–461.

52 Liu, J. and Cui, Z. (2007) *Journal of Membrane Science*, **302**, 180–187.

53 Giorno, L., Chojnacka, K., Donato, L. and Drioli, E. (2002) *Industrial & Engineering Chemistry Research*, **41**, 433. L–440.L.

54 Xu, G., Chu, J., Wang, Y., Zhuang, Y., Zhang, S. and Peng, H. (2006) *Process Biochemistry*, **41**, 2458–2463.

55 Strathmann, H., Giorno, L. and Drioli, E. (2006) *An introduction to membrane science and technology*, CNR, Rome.

56 Giorno, L. *et al.* (1995) *Journal of Chemical Technology and Biotechnology (Oxford, Oxfordshire: 1986)*, **64**, 345–352.

57 Mazzuca, S., Giorno, L., Spadafora, A., Mazzei, R. and Drioli, E. (2006) *Journal of Membrane Science*, **285**, 152–158.

58 Crespo, J.P.S.G., Trotin, M., Hough, D. and Howell, J.A. (1999) *Journal of Membrane Science*, **155**, 209.

59 Liu, Z.M., Dubremez, J., Richard, V., Yang, Q., Xu, Z.K. and Seta, P. (2005) *Journal of Membrane Science*, **267**, 2.

60 Drioli, E. and Iorio, G. (1989) Enzyme membrane reactor and membrane fermentators, *Handbook of industrial membrane technology* (ed. C.P. Mark), Noyes Publications, Park Ridge, New Jersey, USA, pp. 401–481.

61 Amounas, M., Innocent, C., Cosnier, S. and Seta, P. (2000) *Journal of Membrane Science*, **176**, 169–176.

62 Amounas, M., Magne, V., Innocent, C., Dejean, E. and Seta, P. (2002) *Enzyme and Microbial Technology*, **31**, 171–178.

63 Butterfield, D.A., Bhattacharyya, D., Daunert, S. and Bachas, L. (2001) *Journal of Membrane Science*, **181**, 29–37.

18
Biomedical Membrane Extracorporeal Devices

Michel Y. Jaffrin and Cécile Legallais

18.1
General Introduction

18.1.1
Use of Membranes in the Medical Field

Medical applications of membranes are a fast-growing field that represents the largest consumption of membrane area per year. The first and most important application in terms of cost is the treatment of end-stage renal disease (ESRD) by hemodialysis that consumes about 10^8 m^2 of membranes per year, followed by blood oxygenators used during cardiac surgery and in the case of respiratory failure with more than 5×10^6 m^2 per year. Plasma separation and fractionation used for plasma collection from donors and in the treatment of autoimmune diseases and cholesterol removal account for another 2.5×10^6 m^2 per year. New membrane applications in artificial and bioartificial organs, for instance for liver and pancreatic function support, are presently emerging.

This chapter will focus on three types of membrane extracorporeal devices, hemodialyzers, plasma filters for fractionating blood components, and artificial liver systems. These applications share the same physical principles of mass transfer by diffusion and convection across a microfiltration or ultrafiltration membrane (Figure 18.1). A considerable amount of research and development has been undertaken by membrane and modules manufacturers for producing more biocompatible and permeable membranes, while improving modules performance by optimizing their internal fluid mechanics and their geometry.

18.1.2
Historical Perspective

Hemodialysis was first used to treat wounded British pilots from 1943 to 1945 by W. Kolff at the Groningen hospital in Holland. The patient's blood was circulated

Membrane Operations. Innovative Separations and Transformations. Edited by Enrico Drioli and Lidietta Giorno

ISBN: 978-3-527-32038-7

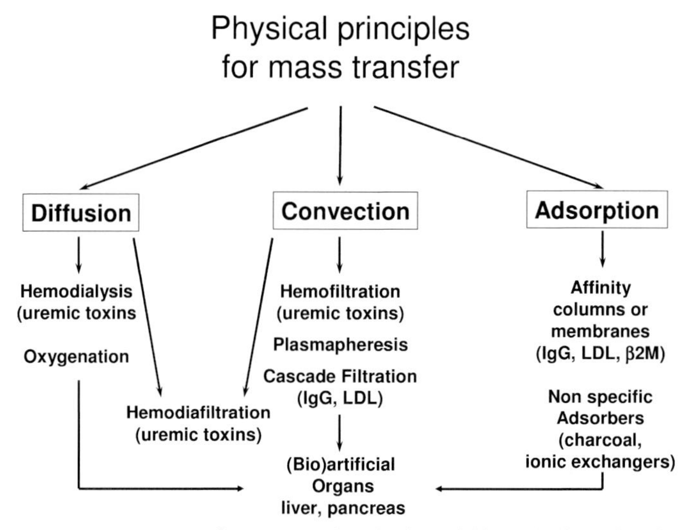

Figure 18.1 Mass transfer or removal methods available in the field of artificial organs.

inside a flat cellulose tubing normally used for wrapping sausages, wound around a wooden drum with an horizontal axis, which was rotated above a flat tank containing an ionic solution. This principle of a spiral wound membrane was retained in the first generation of coil-type disposable dialyzers produced from 1955 to 1970. The real development of hemodialysis was permitted by the arteriovenous shunt proposed by Scribner (Seattle, USA) in 1960 that permitted an easier blood access with an increased blood flow into the dialyzer. In 1966, this external shunt was replaced by the arteriovenous fistula of Ciminio (Brescia, Italy) implanted under the skin that was safer and more comfortable for the patient. Press-type disposable dialyzers replaced coil types in 1969 and the first hollow-fiber modules, more compact and lighter than the press-type ones, were introduced in 1972.

The need to transfuse blood components such as plasma, platelets, factor VIII, in addition to red blood cells (RBC) has generated the development of plasmapheresis (plasma separation from whole blood) and more generally that of apheresis (fractionation of blood components). Plasma collection from donors by centrifugation of blood bags began only in 1944. This technique was extended to therapeutic plasma purification in 1950, but RBCs were fragilized by the centrifugation and the plasma was not completely platelet-free.

Plasma separation by membrane microfiltration was proposed in 1978 by Salomon *et al.* [1] as a substitute to centrifugation and its clinical potential confirmed in 1980 by Samtleben *et al.* [2]. This technique yields a high-quality cell-free plasma that avoids for the recipient the immunological hazards of contamination by platelets and cellular fragments and is less traumatic for red cells, if precautions are taken to avoid hemolysis during filtration.

In the meanwhile, other organ replacements were investigated. In the case of liver supply, both artificial and bioartificial (using hepatic cells) approaches proposed in the 1960s and 1970s [3, 4] did not succeed due to the liver's complexity. Now, better insights into the organ physiology help in designing new and sophisticated techniques able to remove protein-bound toxins and, in the bioartificial case, to supply biotransformation, storage and synthesis functions.

18.2 Hemodialyzers

18.2.1 Introduction

Hemodialysis (HD) permits treatment of end-stage renal failure (ESRD) in patients with a residual renal capacity of less than 10% of normal one. It consists in an extracorporeal circulation of the patient's blood withdrawn from a vein through a needle, which circulates in a hemodialyzer before being returned to the patient (Figure 18.2a). The hemodialyzer extracts uremic toxins from the blood by diffusion across an ultrafiltration membrane with a cutoff of no more than 30 kDa so as to retain red blood cells and plasma proteins. These toxins are carried away by an ionic solution (the dialysate) that circulates at counter current to blood on the other side of the membrane. Treatment of ESRD by HD generally requires 3 four-hour dialysis runs per week as illustrated in Figure 18.2b and concerns now more than 1 700 000 patients worldwide. Another method of treatment, peritoneal dialysis (PD) consists in filling the peritoneal cavity with dialysate and does not necessitate an extracorporeal blood circulation as the hemodialyzer is replaced by the peritoneal membrane. Statistics of various modes of ESRD treatment are given in Table 18.1 [5]. Since hemodialyzers are used only once or reused up to 10 times, but no more than in

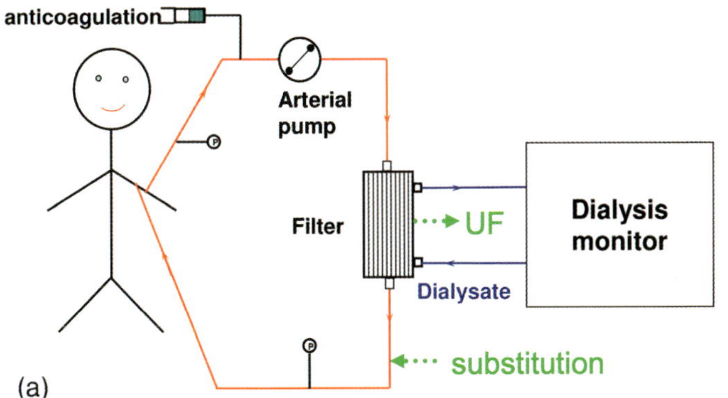

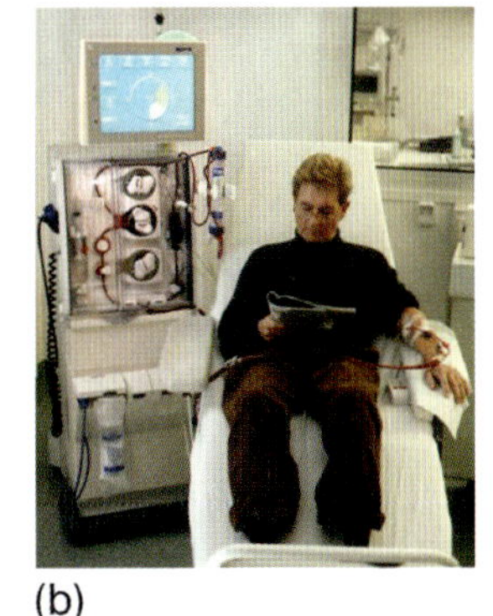

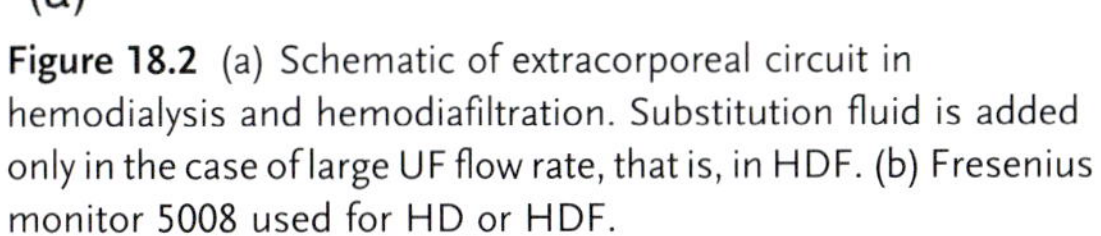
Figure 18.2 (a) Schematic of extracorporeal circuit in hemodialysis and hemodiafiltration. Substitution fluid is added only in the case of large UF flow rate, that is, in HDF. (b) Fresenius monitor 5008 used for HD or HDF.

Table 18.1 Statistics of various types of ESRD treatment.

Treatment type	France	Europe	World
Hemodialysis	26 000	260 000	1300 000
Peritoneal dialysis	3000	30 000	160 000
HDF, HF	4000	50 000	440 000
Total ESRD	33 000	340 000	1900 000

HDF: hemodiafiltration. HF: hemofiltration [5].

a few countries, hemodialysis consumes more m^2 of membrane area than all other membrane applications combined. Fortunately, due to automatized production, hemodialyzers cost from €15 to 30 a piece, which is much less per m^2 than most industrial membranes, except large spiral-wound desalination modules, when purchased in large quantities.

Hemodialyzers are generally of hollow-fiber types with an internal diameter of about 200 μm and a membrane area from 1.2 to 2 m^2. Current research is focused on augmenting the efficiency of large toxin removal by using convective transfer in addition to diffusion and on improving membrane biocompatibility. Other goals are the development of hemodiafiltration (HDF) a combination of hemodialysis with large ultrafiltration [6] by reducing its cost with online production of reinfusion fluid by microfiltration of dialysate by the generator [7].

18.2.2
Physical Principles of Hemodialysis

As blood circulates along the dialyzer membrane, uremic toxins diffuse into the dialysate that is discarded, under the action of the concentration gradient (Figure 18.3).

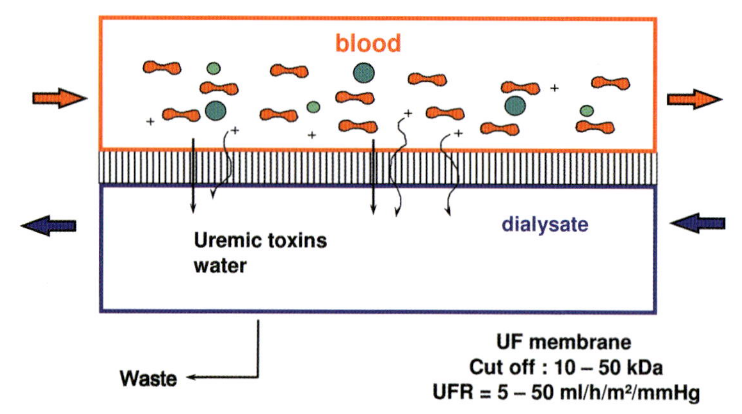

Figure 18.3 Schematic of fluid and mass transfer between blood and dialysate compartments in a hemodialyzer.

To prevent the blood from losing its ions, these ions are included in the dialysate at the same concentration as in normal plasma to stop their diffusion through the membrane. Since uremic patients produce little or no urine, their excess water is eliminated from the blood by ultrafiltration through the membrane since blood pressure is higher than dialysate pressure. This ultrafiltration also contributes to toxin removal by carrying toxins into the dialysate. This convective transfer is more efficient than diffusion for high molecular weight toxins.

18.2.3 Dialysis Requirements

The dialyzer must carry out four tasks [8]:

(a) The elimination of uremic toxins (urea, creatinin, uric acid, phosphates, etc.) mainly by diffusion in normal HD, but also by convection in HDF;
(b) The elimination of excess water, generally from 1 to 4 l, by ultrafiltration;
(c) The regulation of plasmatic ion concentrations (Na^{+}, K^{+}, Ca^{++}, Mg^{++}) by dialysate composition;
(d) The regulation of pH around 7.2 to avoid acidosis by introducing a buffer, acetate or sodium bicarbonate, in the dialysate.

The ion concentrations of normal plasma and of the plasma in the case of renal insufficiency are compared in Table 18.2 with that of dialysate. The dialysate contains fewer K^{+} and Mg^{++} ions than normal plasma, but more HCO_3^{-}. The pathologic plasma contains on average 6 times more urea and 10 times more creatinin than the normal one.

Table 18.2 Comparison of ion, urea, creatinin, glucose, and protein concentrations of normal plasma and plasma in case of renal insufficiency with those of dialysate.

Solute	Dialysate	Normal plasma	ESRD plasma
Na^{+} (mmol/L)	140	142	140
K^{+} (mmol/L)	2	4	5.5
Mg^{2+} (mmol/L)	0.5	1.5	1.5
Ca^{2+} (mmol/L)	1.75	2.5	2
Cl^{-} (mmol/L)	106.5	103	103
HCO_3^{-} (mmol/L)	35	27	21
Urea (mmol/L)	0	5	30
Creatinin (mmol/L)	0	0.1	1.0
Glucose (g/L)	0	1.0	1.0
Proteins (g/L)	0	70	70

18.2.4
Mass Transfers in a Hemodialyzer

Let us first consider, for simplicity, the case of diffusive transfer without ultrafiltration. The local mass flux through the membrane per unit area $J_s(x)$ of a specific toxin is given by

$$J_s(x) = (C_B(x) - C_D(x))/R_T \tag{18.1}$$

where $C_B(x)$ and $C_D(x)$ denote, respectively, the bulk toxin concentrations in blood and dialysate at distance x from the dialyzer inlet and R_T is the sum of diffusive resistances in the blood, membrane, and dialysate [9],

$$R_T = R_B + R_m + R_D \tag{18.2}$$

18.2.4.1 Characterization of Hemodialyzers Performance

The capacity for toxin removal of an hemodialyzer is expressed by the dialysance or the clearance which have the same unit as a flow rate. The dialysance D is defined, for each toxin as

$$D = \frac{\text{toxin mass removed per unit time}}{C_{Bi} - C_D} = \frac{Q_{Bi}C_{Bi} - Q_{Bo}C_{Do}}{C_{Bi} - C_D} \tag{18.3}$$

where Q_B denotes the blood flow rate, the subscripts i and o denote, respectively, the inlet and outlet of the dialyzer.

The clearance K is a particular case of the dialysance and is used when the inlet dialysate does not contain any toxin ($C_{Di} = 0$) which is the case of normal dialysate generators in which the dialysate circulates in open circuit. Thus, from Equation 18.3

$$K = \frac{Q_{Bi}C_{Bi} - Q_{Bo}C_{Bo}}{C_{Bi}} \tag{18.4}$$

Equations 18.3 and 18.4 are valid when an ultrafiltration flow rate Q_F is present and inlet and outlet blood flow rates are related by

$$Q_{Bi} = Q_{Bo} + Q_F \tag{18.5a}$$

Similarly, for the dialysate flow rate Q_D, we have

$$Q_{Di} = Q_{Do} - Q_F \tag{18.5b}$$

If the ultrafiltration is small, as in normal hemodialysis, $Q_F \ll Q_{Bi}$, $Q_{Bi} \approx Q_{Bo} = Q_B$ and Equation 18.4 becomes

$$K \approx Q_B(1 - C_{Bo}/C_{Bi}) \tag{18.6}$$

The clearance can be calculated in the absence of ultrafiltration (diffusive clearance K_D) by writing mass balances equations in the blood and dialysate phases and using Equation 18.6. The result is [9]

$$K_D = Q_B \frac{e^{\alpha} - 1}{e^{\alpha} - Q_B/Q_D} \tag{18.7}$$

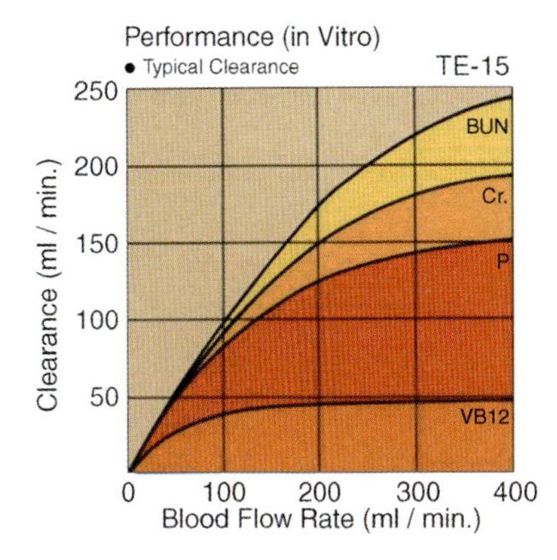

Figure 18.4 Variations of clearances of various solutes with blood flow.

with

$$\alpha = \frac{S}{R_T}\left(\frac{1}{Q_B} - \frac{1}{Q_D}\right) \tag{18.8}$$

where S is the membrane area. This diffusive clearance increases when Q_B and Q_D increase and when the molecular weight of the toxin decreases as R_T decreases. Due to the arteriovenous shunt, the blood flow rate withdrawn from the patient's vein is between 200 and 400 ml/min, while the dialysate flow rate is generally set by the generator at 500 ml/min. A typical variation of the clearance of various toxins with blood flow rate is shown in Figure 18.4. Equations 18.7 and 18.8 indicate that the diffusive clearance is independent of toxin concentration and will not vary during dialysis, if flow rates are constant.

If $C_{Di} \neq 0$, when dialysate recirculates in a closed circuit or when considering the transfer of an ion present in dialysate such as sodium, the dialysance D must be used and it is given by the same equation as Equation 18.7.

18.2.5 Hemofiltration and Hemodiafiltration

Hemofiltration (HF), proposed initially by Funck-Brentano *et al.* [10] and Henderson *et al.* [11] consists in relying only upon the ultrafiltration for toxin removal. There is no dialysate circuit and the ultrafiltration (UF) flow rate must be at least equal to 110 ml/min to produce enough urea clearance. This condition necessitates using highly permeable membranes and a large blood flow, 300 ml/min or more in the dialyzer. This high UF flow rate must be compensated by reinjection of sterile dialysate in the blood line, generally at the hemofilter outlet, in order to prevent hypovolemia and the loss of plasma ions. This reinjection flow rate is slightly below that of the UF flow rate, in order to remove excess fluid as in regular dialysis.

The clearance can be easily calculated from Equation 18.4 and the toxin mass balance on the hemodialyzer, which is

$$Q_F C_F = Q_{Bi} C_{Bi} - Q_{Bo} C_{Bo} \tag{18.9}$$

Combining Equations 18.4 and 18.9 yields

$$K_{HF} = Q_F C_F / C_{Bi} = \mathrm{Tr} Q_F \tag{18.10}$$

since C_F/C_{Bi} is the transmittance (Tr) of the membrane for the toxin considered.

This transmittance is equal to 1 for small toxins such as urea and creatinin and decreases as the molecular weight of the toxin increases, but less rapidly than the diffusive clearance of Equation 18.7. Thus, hemofiltration produces a lower clearance of small toxins than hemodialysis, but a higher one for large toxins such as myoglobin and Beta2 microglobulin. Hemofiltration has been shown to improve vascular stability and prevent hypotension. But it is a more expensive treatment than regular dialysis, due to the higher cost of highly permeable membranes and of sterile bags of reinjection fluid prepared in pharmaceutical plants. Presently, hemofiltration is mostly restricted to acute dialysis that takes place 24 h per day for one or two weeks and necessitates a smaller urea clearance than HD.

Hemodiafiltration (HDF) may be regarded as a combination of regular HD and hemofiltration as it associates a dialysate circulation and high ultrafiltration. It also necessitates the reinjection of sterile dialysate, but less than in hemofiltration, since the clearance of small toxins is also achieved by dialysis. This process, thus, combines the high urea clearance of dialysis with the high rate of large toxins removal of hemofiltration. In addition, generators now exist that can prepare simultaneously the regular dialysate from concentrate and the reinjection fluid by sterile filtration of this dialysate, therefore lowering the treatment cost.

The overall clearance of hemofiltration is more difficult to calculate than the diffusive clearance or the HF clearance, as it combines diffusive and convective transfers. An approximate equation for this clearance, obtained from an exact numerical solution has been given by Jaffrin *et al.* [12] as

$$K_{HDF} = K_D + 0.43 Q_F + 0.00083 Q_F^2 \tag{18.11a}$$

where Q_F and K_D are in ml/min, or if $Q_F < 70$ ml/min

$$K_{HDF} = K_D + 0.46 Q_F \tag{18.11b}$$

which shows that the overall clearance is generally less than the sum of diffusive and convective clearance, due to interaction between diffusive and convective transfers. Equation 18.11b can also be applied to regular hemodialysis and includes the contribution of the convective transfer to the clearance.

18.2.6 Various Types of Hemodialyzers

The most common type is the hollow-fiber cartridge, consisting in a bundle of 10 000 to 15 000 fibers of 200–220 μm i.d., placed in a polypropylene transparent housing.

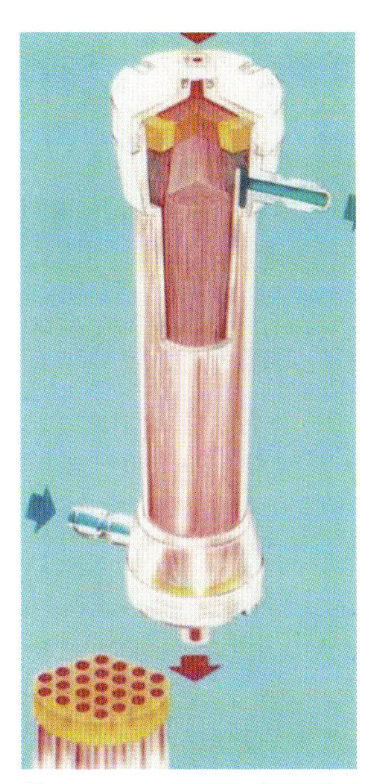

Figure 18.5 Cut out of a hollow-fiber dialyzer.

Blood circulates inside the fibers and dialysate outside (Figure 18.5). The fibers are imbedded at each end in polyurethane that ensures separation between blood and dialysate. During fabrication, the bundle ends must be carefully cut with a blade to reopen each fiber. Blood inlet and outlet are located in caps screwed on the housing, which are designed to distribute blood uniformly into all fibers, while dialysate ports are located at each end of the housing, perpendicular to the fibers.

Parallel-plate hemodialyzers using flat membranes, with several compartments in parallel, separated by plastic plates, are now only available from Hospal Co (Crystal and Hemospal models). Blood circulates between two membranes and the dialysate between the other side of membrane and the plastic plate. These parallel-plate dialyzers have a smaller blood-pressure drop than hollow-fiber ones and require less anticoagulants as flat channels are less exposed to thrombus formation than fibers, but they are heavier and bulkier and thus less popular. A recent survey of the state-of-the-art in hemodialyzers is given in [13].

18.2.6.1 Various Types of Membranes

There are two types of membranes, cellulosic and synthetic or polymeric ones. Cellulosic membranes can be in regenerated cellulose (cuprophan, Bioflux from Membrana, Germany) or modified cellulose (cellulose acetate or diacetate, from Asahi, triacetate cellulose from Baxter and Nipro, which has a high hydraulic permeability or Hemophan from Membrana). Cuprophan was originally the most common one, because of its low cost, but is no longer produced because of its lower biocompatibility and hydraulic permeability. A wide variety of polymeric membranes are now available with both high and medium hydraulic permeabilities. Only the Eval

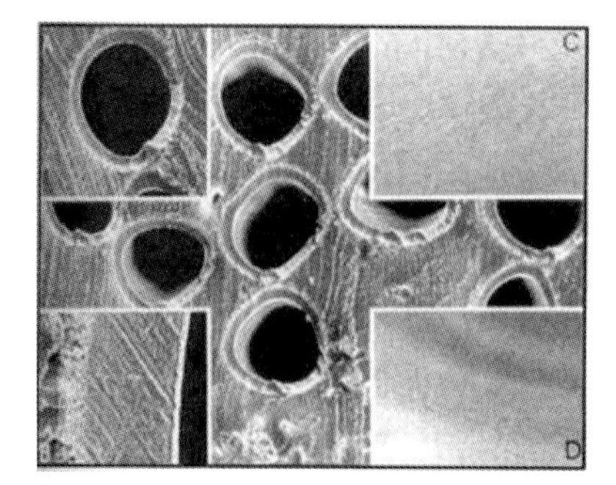

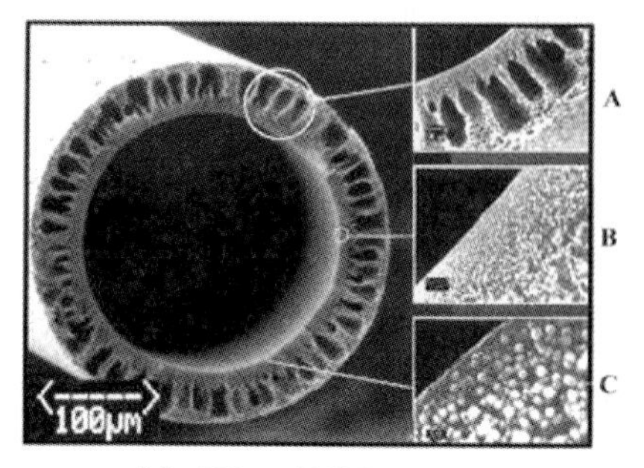

Figure 18.6 SEM pictures of hollow-fiber membranes used for HD and HDF.

(Kuraray) is naturally hydrophilic, but the other polymers can be made hydrophilic by incorporating additives during their fabrication. SEM pictures of various membranes are shown in Figure 18.6. The first high-flux membrane was the polyacrylonitrile AN69 introduced in 1973 by Hospal in France, followed by the PAN from Asahi, in Japan. These membranes, which are symmetric, were considered at this time as very biocompatible, but adsorbed proteins due to their negative charge. This adsorption was reduced by coating the AN69 with polyethylemine to reduce its negative charge in the AN69 ST. Polyamide asymmetric membranes (Gambro) were blended with PVP (polyvinylpyrrolidone) to make them hydrophilic. Polysulfone membranes introduced first by Fresenius, but also available from Asahi and Toray in Japan, had high hydraulic permeability together with excellent biocompatibility. An improved version (Helixone) with a reduced inner diameter of 185 μm to increase clearance and a mean pore size increased from 3.1 to 3.3 nm was introduced by Fresenius in 2000 as a high-flux membrane. Other synthetic membranes include the polymethylmethacrylate (PMMA, Toray), polycarbonate polyether (Gambro), and polyethersulfone (Membrana).

18.2.6.2 **Optimization of Hemodialyzer Performance**

The goal is to improve clearance without increasing membrane area or dialysate flow rate, as it would increase treatment cost. The dialysate diffusive resistance R_D can be lowered by improving dialysate flow distribution with multifilament spacer yarns between fibers or weaving these yearns around fibers. Another approach consists in giving the fibers a wavy shape by reducing the housing length below that of fibers. It is important to find the optimal packing density, which is the best compromise between a loose packing resulting in a low average dialysate velocity and a dense packing that creates preferential channels between regions of almost stagnant flow.

Decreasing the fiber inner diameter while increasing the number of fibers to keep the total area constant will both decrease the blood diffusive resistance R_B and increase diffusive clearance and also raise the blood pressure drop and the ultrafiltration in the upstream part of the membrane [14]. In fact, it is possible by creating a high UF in the first half of the filter and a back filtration in the second half to simulate HDF conditions in a hemodialyzer without the need of reinjection fluid. The fluid loss by ultrafiltration is compensated by the back filtration of the incoming fresh dialysate, which contains only a small quantity of toxins.

18.3 Plasma Separation and Purification by Membrane

18.3.1 Introduction

In contrast to hemodialysis that uses ultrafiltration membranes, plasma separation (also called plasmapheresis) requires microfiltration membranes with a pore size from 0.2 to 0.6 μm, in order to transmit all proteins and lipids, including LDL cholesterol (2000 kDa) and retain completely platelets (2 μm diameter), red blood cells (8 μm diameter) and white blood cells. Thus, membrane plasmapheresis can yield high-quality platelet-free plasma and red cells can be either continuously returned to the donor or saved in another bag for blood transfusion. But it is important, in the case of plasma collection from donors, to minimize the membrane area, in order to reduce the cost of disposable hollow-fiber filters and to avoid the risk of hemolysis (free hemoglobin release) due to RBC damage by contact at the membrane if the pressure difference across the membrane is too high.

Membrane plasmapheresis is also the first step for treatment of pathological plasma in the case of autoimmune diseases, as the patient retains his own red blood cells while his plasma is replaced by an albumin solution or fresh frozen plasma obtained from donors (plasma exchange therapy). Other more selective plasma purification techniques consist in eliminating pathologic immunoglobulins or LDL cholesterol familial hypercholesterolemia, either by a secondary filtration, chemical adsorption or immunoadsorption. A description of various applications of plasmapheresis can be found in the book edited by Smit Sibinga and Kater [15].

In France alone, about 220 000 plasmapheresis and 65 000 cytapheresis (collection of platelets, factor VIII, etc.) are performed every year, against 2 400 000 blood donations. 600 ml of plasma can be collected from the same donor every 2 weeks if needed.

18.3.2 The Baxter Autopheresis C System for Plasma Collection from Donors

This system was first introduced by Hemascience Company, Santa Ana, CA, USA. The filter consists of a cylindrical membrane of only 58 cm^2 area rotating at 3600 rpm inside a concentric cylinder of 2.9 cm inner diameter (Figure 18.7). Blood inlet and outlet are mounted tangentially, and plasma is collected through grooves molded on the inner cylinder supporting the membrane and leaves the filter through a duct in the rotation axis at the bottom.

The presence of Taylor vortices generated in the gap between the membrane and housing by the rotation, create very high shear rates at membrane, producing a large

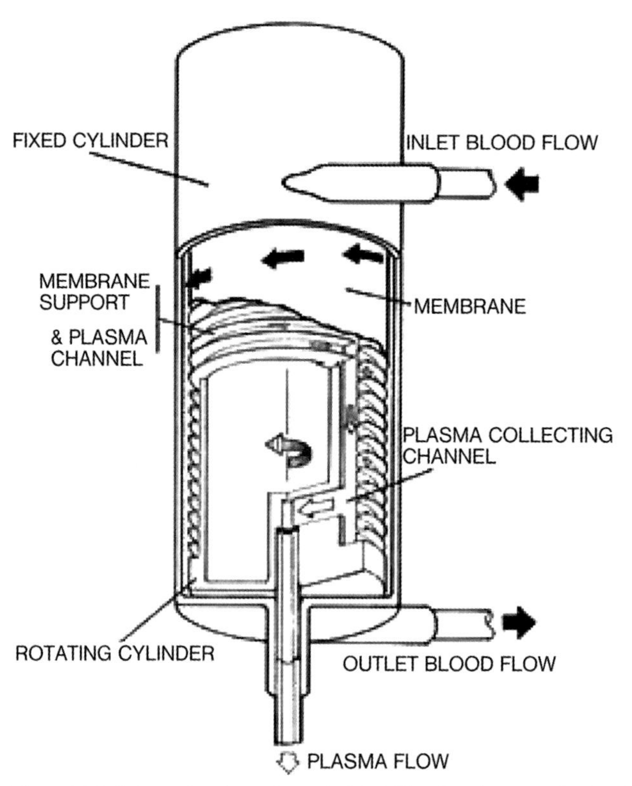

Figure 18.7 Schematic of rotating cylindrical filter for plasma collection from donors.

plasma permeate flux about 0.5 cm/min or 300 L/hm^2, which permits the use of a very small membrane area. This system was later commercialized by Baxter Company under the name Autopheresis C.

Initially, the membrane was in polycarbonate with 0.8-μm pores, but, since its peak flux occurred in a narrow transmembrane pressure range, it was replaced later by a nylon one (Pall Corp) with 0.5-μm pores, which yielded a constant permeate flux above 50 mmHg. This device was successfully applied to therapeutic plasma exchange by Kaplan *et al.* [16], who, after increasing the pressure setting, were able to collect 3 L of plasma from a patient in 90 min. An interesting modification to this device has consisted in separating first by centrifugation the blood into platelet-rich plasma using a similar rotating system without a membrane. Then the platelets were separated from the plasma using the Autopheresis C, with minimal contamination by leukocytes and red cells.

18.3.3
Therapeutic Applications of Plasma Separation

Membranes used for therapeutic plasma separation have the same characteristics as those used for plasma collection from donors, but their area is larger as the amount

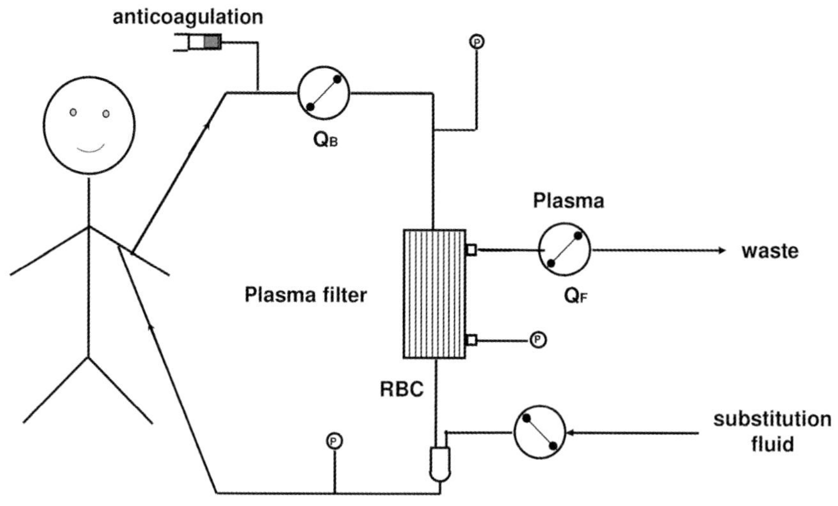

Figure 18.8 Hollow-fiber plasma filter with its extracorporeal circuit for plasma exchange.

of plasma filtered is bigger. They are generally made of synthetic hollow fibers with inner diameter of 250 to 350 μm. Their area vary from 0.2 m^2 to about 0.65 m^2 for therapeutic plasmapheresis with a unit cost from €30 to 250 depending upon their size.

18.3.3.1 Plasma Exchange

Plasma exchange can be performed by membrane filtration to avoid RBC damage, which centrifugation cannot guarantee. The removed plasma (2.5–3.5 L) is discarded and replaced by fresh frozen plasma collected from donors or albumin concentrate (Figure 18.8). Table 18.3 lists available plasmapheresis filters and their characteristics. This technique is applied in various specialties such as nephrology, haematology and neurology, such as thrombotic thrombocytopenic purpura, myasthenia gravis, chronic inflammatory demyelinating polyneuropathy, Waldenstrom macroglobulinemia, the Guillain-Barré syndrome, rheumatoid arthritis, systemic lupus erythematosus, and multiple sclerosis.

18.3.3.2 Selective Plasma Purification by Cascade Filtration

The treatment of several autoimmune diseases and of familial hypercholesterolemia generally requires the removal of a single pathogeneous molecule such as LDL, IgA or IgG for Guillain-Barre syndrome, IgM for Walderstom macroglobulinemia, and so on. Removing specific molecules avoids the need of substitution by albumin or fresh frozen plasma (FFP) to compensate protein loss as in case of total plasma exchange.

Cascade filtration, which was initially proposed by Agishi *et al.* [17], consists in filtering the collected plasma after separation on an ultrafiltration filter selected so as to retain the pathogeneous molecule, as shown in Figure 18.9. This technique is now

Table 18.3 Commercially available membrane filters for plasma separation and fractionation.

Manufacturer	Filter	Membrane	Membrane area (m^2)	Pore diam. (μm)	In fiber diam. (μm)
Plasma separators					
Baxter, USA	CPS-10	Polypropylene	0.17	0.55	320
Gambro, Sweden	PP	Polypropylene	0.38	0.5	330
Fresenius, Germany	Plasmaflux	Polypropylene	0.50	0.5	330
Dideco, Italy	Hemaplex	Polypropylene	0.20	0.5	320
Asahi, Japan	Plasmaflo 03-06-08	Polyethylene	0.3–0.6–0.8	0.2	370
Kuraray, Japan	Plasmacure	Polysulfone	0.30	0.2	300
Toray, Japan	Plasmax PS 02-05	PMMA	0.150–0.5	0.5	370
Terumo, Japan	PS-4000	Cellulose acetate	0.50	0.45	80 (height)[a]
Nippro, Japan	PEX-50	Cellulose triacetate	0.50	0.4	270
Gambro	Prisma TPE	Polypropylene	0.35	0.5	330
Plasma fractionators (secondary membranes)					
Dideco, Italy	Albusave BT902	Cellulose diacetate	1.00	0.02	350
Asahi, Japan	Cascadeflo AC1730	Cellulose diacetate	1.7	10^{6}[b]	210
	Cascadeflo AC1760	Cellulose diacetate	1.7	5×10^{6}[b]	210
Kuraray, Japan	Eval filter 2A-3A	Ethylene vinyl alcohol	1.00	0.01–0.02	200
Toray, Japan	Plasmax AS08	PMMA	0.80		370

[a]Channel height (plate-type filter).
[b]Membrane cutoff at 90% rejection.
[c]Cutoff not available.

mostly used for removal of LDL (bad cholesterol) as its molecular weight (2000 kDa) is much larger than that of HDL (good cholesterol, 400 kDa) albumin (69 kDa) and IgG (156 kDa), which are easily transmitted by the secondary membrane. This second filtration is generally carried out in quasi-dead-end mode with a retentate flow rate which is less than 15% of inlet flow rate, in order to minimize albumin losses in the discarded retentate. However, like all filtration methods, it is subject to membrane fouling, which decreases permeate flux and selectivity during the filtration and operating conditions need to be optimized in order to maximize selectivity. Thermofiltration consists in warming the plasma to 42 °C to prevent cryogel formation in the secondary filter [18]. Table 18.3 lists available filters for plasma purification or fractionation. Their membrane area is larger than that of plasma filters in order to reduce fouling and they are generally more expensive.

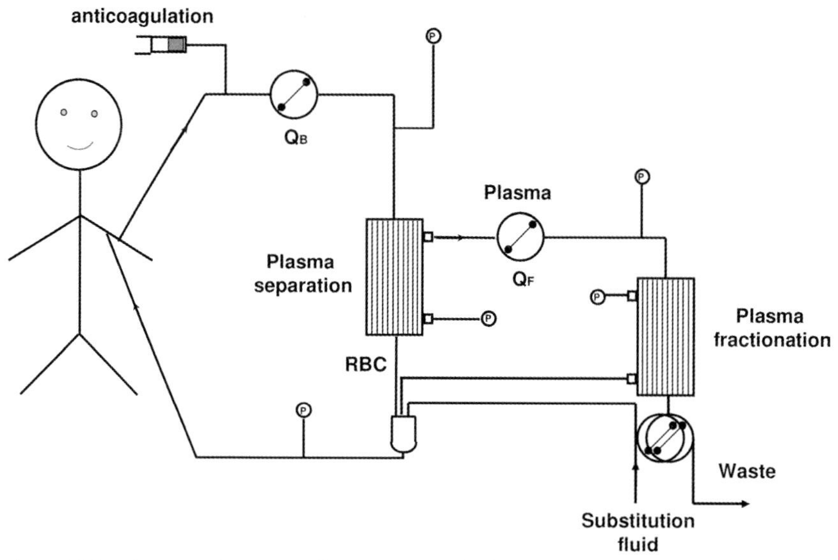

Figure 18.9 Schematic of a cascade filtration circuit for plasma purification with secondary membrane.

A review of plasma purification using secondary filtration has been presented by Siami *et al.* [19] and Table 18.4 lists the diseases treated with this technique. Diseases treated include immune-mediated disorders and familial type IIA hypercholesterolemia. These authors concluded that cryoglobulins filters were safe and effective for removing cryoproteins, did not induce complement activation and constituted one of the most promising techniques of secondary membrane application.

Table 18.4 Diseases treated with secondary membrane filtration (From Siami *et al.* [8]), with permission.)

Disease	Pathogen removed	Membrane type
Myasthenia Gravis	Anti-AChRab	2A Evaflux
Guillain Barre Syndrome	IgG antibody	2A Evaflux
IgG (IgA) Gammopathy	IgG (IgA) antibody	2A Evaflux
Idiopathic throbocytopepenic Purpura	IgG platelet antibody	2A Evaflux
Factor XI deficiency	FXI IgG inhibitor	2A Evaflux
Type I Cryoglobulinemia	Monoclonal IgG	2A Evaflux
Waldenstrom Macroglobulinemia	IgM antibody	4A Evaflux
Castleman Syndrome	IgM antibody	4A Evaflux
Familial Hypercholesterolemia	LDL-Cholesterol	4A Evaflux
Type II Cryoglobulinemia	Mono IgM and poly IgG	Cryoglobulin filter
Type III Cryoglobulinemia	Polyclonal IgM or IgG	Cryoglobulin filter

18.4
Artificial Liver

18.4.1
Introduction

The adult human liver normally weighs between 1.7–3.0 kg. It is both the second largest organ and the largest gland within the human body. The liver performs many important metabolic functions: detoxification, transformation, storage and synthesis (Figure 18.10). Loss of liver cell functions resulting in the disruption of many essential functions could lead to death. At present, transplantation is the only efficient treatment for patients suffering from acute or fulminant liver failure [20]. The shortage in specific organ donors has resulted in a high death rate among the potential patients waiting for a graft. for the past 20 years, the expanding gap between the number of patients on the waiting list and the number of liver transplants has led to the design of temporary liver support. Such an artificial organ could be employed either as a bridge to transplantation or as a means for the patient to recover native liver function [21].

As liver performs multiple and complex functions, artificial organ or bioartificial organ exploiting a synthetic cartridge to host biological components such as cells (hepatocytes in the case of a bioartificial liver) have been investigated. Among all of these potentialities, we only focus here on purely artificial systems. Membrane-based bioartificial livers (BAL) will not be described here, but could be found in other reviews [22–25] and in another chapter in the present book.

18.4.2
Physical Principles

One of the major liver functions is detoxification of substances carried by blood, which are perfused through the cellular network in the organ. These functions can

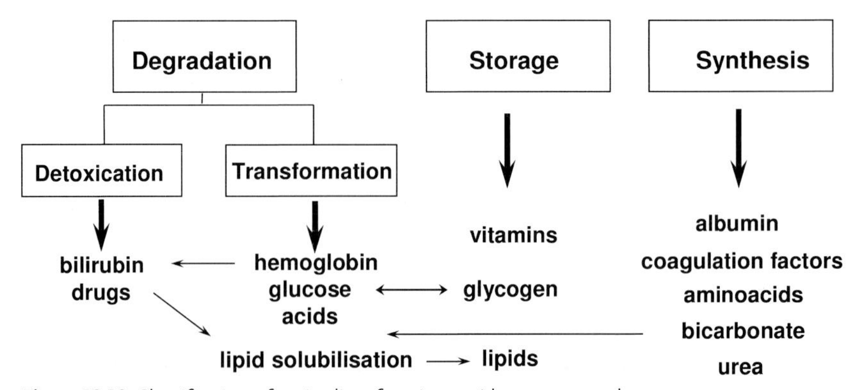

Figure 18.10 Classification of major liver functions, with some examples.

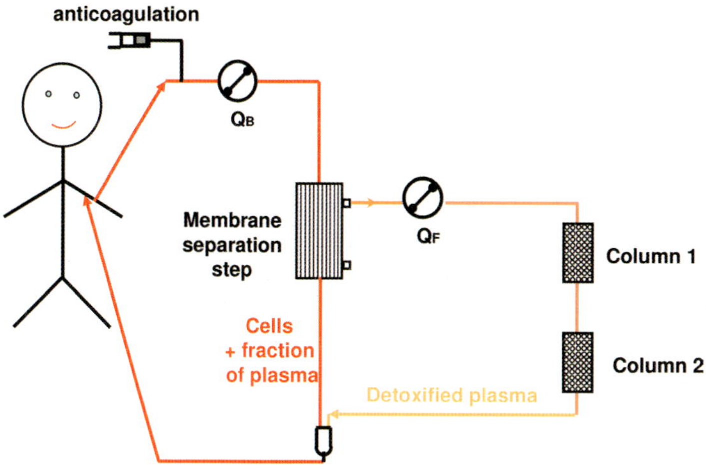

Figure 18.11 Generic representation of combined filtration and adsorption columns systems for artificial liver support.

theoretically be replaced by the three different physical principles available in extracorporeal devices (Figure 18.1), alone or in association.

Artificial liver support systems aim at the extracorporeal removal of water soluble and protein-bound toxins (albumin being the preferential binding protein) associated with hepatic failure. Albumin contains reversible binding sites for substances such as fatty acids, hormones, enzymes, dyes, trace metals and drugs [26] and therefore helps elimination by kidneys of substances that are toxic in the unbound state. It should be noticed that the range of substances to be removed is broad and not completely identified. Clinical studies showed that the critical issue of the clinical syndrome in liver failure is the accumulation of toxins not cleared by the failing liver. Based on this hypothesis, the removal of lipophilic, albumin-bound substances, such as bilirubin, bile acids, metabolites of aromatic amino acids, medium-chain fatty acids, and cytokines, should be beneficial to the clinical course of a patient in liver failure.

For this purpose, the removal procedures are mainly based on membrane separation that ideally should bring free and bound toxins to a nonspecific adsorption device (ion-exchangers and/or activated charcoal). Blood should not perfuse directly such components, due to bioincompatibity aspects. Therefore, several processes have been proposed to correctly handle toxins carried by plasma [27]. They are described in the following sections. All of them need a physical barrier between the blood cells and the adsorption system. This physical sieve is always a membrane with adequate properties, through which toxins can be transferred by diffusion or convection.

Two systems are then based on classical chromatography processes, that is, perfusion of a column hosting adsorbents, and three others are based on moving adsorbent phase.

18.4.3 Convection + Adsorption Systems

The following processes can be described as selective therapeutic plasmapheresis. In a first step, blood is withdrawn from the patient and separated by crossflow filtration in a hollow-fiber membrane cartridge: water and some plasma solutes are transferred through a semipermeable membrane under a convection process. The transmembrane pressure applied from blood to filtrate compartment ensures flow and mass transfers. Then, the filtrate perfuses the adsorption columns where toxins are retained and is finally mixed with blood cells and other plasma components before returning to the patient (Figure 18.11).

In the ASAHI KASEI Medical (Tokyo, Japan) system, the plasmapheresis step is performed by a microporous membrane (Plasmaflo) made of a copolymer of ethylene and vinyl alcohol (PEVA), with a maximum pore size of 0.3 μm. The extracted plasma flows through an activated charcoal column Hemosorba and an anion-exchange column (copolymer of styrenedivinyl benzene) Plasorba that binds bilirubin and bile acids [28]. Each column contains 350 mL of adsorbent.

In the Prometheus (Fresenius Medical Care, Bad Homburg, Germany) system, blood fractionation is achieved by means of a capillary albumin filter (AlbuFlow AF01) presenting a molecular weight cutoff of 300 000 Da. The large pores of the membrane allow albumin-rich plasma to enter the secondary plasma circuit. Albumin-bound toxins are extracted from plasma after binding to the adsorber beads. Both columns host 350 mL of neutral resin styrenedivinyl benzene copolymer beads. The inner porous structure is sponge-like and easily accessible for protein-bound liver toxins. Prometh-01 adsorbs water-insoluble compounds, such as bile acids, phenols and aromatic amino acids. Prometh-02 has anion-exchanger properties because it contains positively charged sites and is able to remove negatively charged liver toxins, such as bilirubin. The cleared filtrate returns thus to the blood main stream [29]. To remove water-soluble toxins, an additional classical dialysis step is then placed downstream. The high-flux dialysis membrane is used for the diffusive transfer of toxins from the blood to the dialysate side. The whole extracorporeal circuit is adapted from a Fresenius 4008 dialysis machine [30].

18.4.4 Diffusion + Adsorption Systems

In these cases, substances carried by blood are removed in the 'dialysate' phase, which is separated from blood by a semipermeable membrane, as described in Figure 18.2. Toxins should cross this barrier by diffusion, before being treated. Toxins that bind to albumin have proven refractory to removal by conventional hemodialysis. Such toxins can, however, be removed by adding binders to the dialysate that capture them after being dialyzed through the membrane.

In the Liver Dialysis Unit (HemoCleanse Technologies, Lafayette, USA), the adsorbents (powdered activated charcoal and cation-exchangers) are located in the dialysate-moving phase [31]. Dialysate content is adjusted so as to prevent unexpected removal of substances such as calcium, sodium ... due to diffusive effects [32].

The MARS system from Teraklin (a subsidiary of Gambro, Lund, Sweden) uses a specially designed membrane named MARS Flux dialyzer in the primary circuit. The secondary circuit consists of 600 mL of 20% human albumin solution instead of physiological buffer used in classical hemodialysis. Due to the membrane properties, protein-bound toxins and water-soluble substances from the blood side diffuse into the albumin solution [33]. The MARS Flux dialyzer presents a surface area of 2.1 m^2, a membrane thickness of 100 nm and a molecular cutoff of about 50 kDa. The specific membrane surface provides pseudobinding sites for albumin when the secondary circuit is primed with albumin solution. The albumin molecules on the 'dialysate' side of the membrane are in very close proximity to the surface of the membrane in contact with patient's blood. It is assumed that albumin-bound toxins move by physicochemical interactions between the plasma, albumin molecules bound to the dialysis side of the membrane and the circulating albumin solution. This solution is then dialyzed against a standard buffered dialysis solution to remove water-soluble substances by diffusion. The removal of the albumin-bound toxins is achieved by an activated-carbon adsorber and an anion exchanger. The concentration gradient is maintained at the first dialysis step by circulation of the online regenerated albumin [34]. A specific monitor has been designed by Teraklin Company to handle the secondary circuit and is now commercially available with Gambro dialysis machines.

The single-pass albumin dialysis (SPAD) is a simple noncommercial method of albumin dialysis using standard renal-replacement therapy machines without an additional perfusion pump system. The patient's blood flows through a circuit with a high-flux hollow-fiber hemodiafilter, identical to that used in the MARS system. The other side of this membrane is perfused with a buffered albumin solution in counterdirectional flow, which is, instead of being regenerated as in the MARS concept, discarded after passing the filter [35].

18.4.5
Future of Artificial Livers

As seen above, the artificial systems are only able to supply detoxication functions of the liver. In some cases, this might not be enough to save patients. An alternative is the design of bioartificial liver. A simplistic approach consists in considering such a device as a bioreactor based on synthetic elements able to offer an adequate environment to the liver cells. This environment would in turn lead to the maintenance of efficient functions of the cells aiming at liver supply, when placed in a bioreactor located in an extracorporeal circuit. The mandatory requirements for acceptable cell viability and functions in a bioartificial liver (BAL) are tentatively listed below, according to a biotechnological point of view:

(a) Anchorage to a support or a matrix;
(b) Effective exchanges with blood or plasma;
(c) Protection from host immunological response.

18.4.6
Conclusions

Up to now, none of the presented system can claim its ability to fully replace all liver functions in an extracorporeal circuit. On the one hand, purely artificial techniques can only cover some detoxification aspects, which is already crucial in many clinical cases to save patients. On the other hand, bioartificial livers have not proven their full efficiency yet, mainly because both regulatory and logistic aspects limit, for the moment, the inclusion of significant numbers of patients to draw statistically relevant conclusions.

It seems, nevertheless, clear that the combination of membrane-based and adsorbent techniques, perhaps in addition to bioartificial systems, present a potential supply to help the patient waiting for a graft or even for tissue regeneration. In the biomedical field, the extension of techniques previously developed for other topics, such as biochromatography for instance, has always proved to be promising. This could hopefully be the case for artificial liver support.

References

1 Solomon, B.A., Castino, F., Lysaght, M.J., Colton, C.K. and Friedman, L.I. (1978) *Transactions of the American Society for Artificial Internal Organs*, **24**, 21–26.

2 Samtleben, W., Hillebrand, G., Krummer, D. *et al.* (1980) in *Plasma Exchange* (ed. H.G. Sierbeth), Schattauer, Stuttgart, pp. 175–178.

3 Nose, Y., Mikami, J. and Kasai, S. (1963) *Transactions of the American Society for Artificial Internal Organs*, **9** (358), 362.

4 Opolon, P., Rabin, J.-R., Huquet, C., Granger, A., Delorme, J.L., Boschat, M. and Sausse, A. (1976) *Transactions of the American Society for Artificial Internal Organs*, **22**, 701–710.

5 (2006) ESRD Patients in 2005: A Global Perspective. Fresenius Medical Care document.

6 Kunimoto, T., Lowrie, E.G., Kumazawa, S., O'Brien, M., Lazarus, J.M., Gottlieb, J.M., Merril, J.P. (1977) *Transactions of American Society of Artificial Internal Organs*, **23**, 234–242.

7 Ahrenholz, P., Winkler, R.E., Ramlow, W., Tiess, M. and Mueller, W. (1997) *The International Journal of Artificial Organs*, **20**, 81–90.

8 Sargent, J.A. and Gotch, F.A. (1978) Principles and biophysics of dialysis, in *Replacement of Renal Function by Dialysis* (eds W. Von Drucker, F.M. Parsons and R.F. Maher), Nijhoff, Boston, pp. 38–68.

9 Lysaght, M.J., Ford, C.A., Colton, C.K. *et al.* (1977) Mass transfer in clinical blood ultrafiltration devices, in *Technical Aspects of Renal Dialysis* (ed. T.M. Frost), Pitman Medical, London, UK.

10 Funck-Brentano, J.L., Sausse, A., Man, N.K. *et al.* (1972) *Proceedings of European Dialysis and Transplant Association Congress*, **9**, 52–58.

11 Henderson, L.W., Colton, C.K. and Ford, C.A. (1975) *The Journal of Laboratory and Clinical Medicine*, **85**, 372–390.

12 Jaffrin, M.Y., Ding, L.H. and Laurent, J.M. (1991) *Journal of Biomechanical Engineering-Transactions of the ASME*, **112**, 212–219.

13 Uhlenbusch-Korwer, I., Bonnie-Schorn, E., Grassmann, A. and Vienken, J. (2004) *Understanding Membranes and Dialyzers*, Pabst, Berlin.

14 Ronco, C., Brendolan, A., Lupi, A., Metry, G. and Levin, N.W. (2000) *Kidney International*, **58**, 809–817.

15 Smit Sibinga, C.T. and Kater, L. (eds) (1991) *Advances in Haemapheresis*, Kluwer, Amsterdam.

16 Kaplan, A.E. and Halley, S.E. (1988) *Transactions of the American Society for Artificial Internal Organs*, **34**, 274–276.

17 Gurland, H.J., Samtleben, W. and Blumenstein, M. (1983) *Life Support Systems*, **1**, 61–70.

18 Agishi, T., Kaneko, I. Hasuo, Y.et al. (1980) *Transactions of the American Society for Artificial Internal Organs*, **26**, 406–410.

19 Siami, F.S. and Siami, G.A. (2000) *ASAIO Journal (American Society for Artificial Internal Organs: 1992)*, **46**, 383–388.

20 Chapman, R.W., Forman, D., Peto, R. and Smallwood, S. (1990) *Lancet*, **335**, 32.

21 Cao, S., Esquivel, C.O. and Keeffe, E.B. (1998) *Annual Review of Medicine*, **49**, 85.

22 Arkadopoulos, N., Detry, O., Rozga, J. and Demetriou, A.A. (1998) *The International Journal of Artificial Organs*, **21**, 781.

23 Legallais, C., David, B. and Dore, E. (2001) *Journal of Membrane Science*, **181**, 81.

24 Matsushita, M. and Nosé, Y. (1986) *Artificial Organs*, **10**, 378.

25 Nyberg, S.L. and Misra, S.P. (1998) *Mayo Clinic Proceedings*, **73** (8), 765.

26 Emerson, T.E. (1989) *Critical Care Medicine*, **17**, 690.

27 Legallais, C., Vijayalakshmi, M.A. and Moriniere, P. (2002) Biochromatography and Biomedical Applications, in *Biochromatography Theory and Practice* (ed. M.A. Vijayalakshmi), Taylor and Francis, London and New York, pp. 496.

28 Iseki, J., Touyama, K., Nakagami, K., Takagi, M., Hakamada, K., Ooba, N. and Mori, N. (1995) *Hepatogastroenter*, **42**, 394.

29 Vienken, J. and Christmann, H. (2006) *Therapeutic Apheresis and Dialysis*, **10**, 125.

30 Santoro, A., Faenza, S., Mancini, E., Ferramosca, E., Grammatico, F., Zucchelli, A., Facchini, M.G. and Pinna, A.D. (2006) *Transplantation Proceedings*, **38**, 1078.

31 Ash, S.R., Knab, W.R., Blake, D.E., Carr, D.J., Steczko, J., Harker, K.D. and Levy, H. (2000) *Therapeutic Apheresis*, **4** (3), 218.

32 Bauer, E., Gendo, A., Madl, C., Garo, F., Roth, E. and Kramer, L. (2002) *The International Journal of Artificial Organs*, **25**, 923.

33 Stange, J. and Mitzner, S. (1996) *The International Journal of Artificial Organs*, **19**, 677.

34 Klammt, S., Stange, J., Mitzner, S.R., Peszynski, P., Peters, E. and Liebe, S. (2002) *Liver*, **22**, 30.

35 Sauer, I.M., Goetz, M., Steffen, I., Walter, G., Kehr, D.C., Schwartlander, R., Hwang, Y.J., Pascher, A., Gerlach, J.C. and Neuhaus, P. (2004) *Hepatology (Baltimore, MD)*, **39**, 1408.

19
Membranes in Regenerative Medicine and Tissue Engineering

Sabrina Morelli, Simona Salerno, Antonella Piscioneri, Maria Rende, Carla Campana, Enrico Drioli, and Loredana De Bartolo

19.1
Introduction

The fields of tissue engineering and regenerative medicine aim at promoting the regeneration of tissues or replacing failing or malfunctioning organs, by means of combining a scaffold/support material, adequate cells, and bioactive molecules. The use of materials in contact with biological materials (cells, tissues/organs, physiological fluids, and biomolecules) is a current illustration of the need for interdisciplinary scientific approaches that combine the most recent advances in materials science and technology, basic sciences and life sciences. Different materials have been proposed to develop membranes to support cells and promote their differentiation and proliferation towards the formation of a new tissue. Such strategies allow for producing hybrid constructs that can be implanted in patients to induce the regeneration of tissues or replace failing or malfunctioning organs.

In recent years rapid progress has been made in the field of biomedical materials that utilize both natural and synthetic polymers and that can be used in a variety of applications, including wound closure, drug-delivery systems, novel vascular grafts, or scaffolds for *in-vitro* or *in-vivo* tissue engineering. The goal of the early or first-generation biomedical materials, during the 1960s and 1970s, was to attain suitable physical properties to match the replaced tissue with a common feature of biological 'inertness.' Second-generation biomaterials were designed to produce bioactive responses that could elicit a controlled reaction in the physiological environment. Such bioactive (ceramics, hydroxyapatite) or resorbable (polyglycolide, polylactide) materials have been successfully applied to the medical needs of many fields. Third-generation biomaterials are combining these two properties and are being designed to stimulate specific cellular responses at the molecular level [1]. Several 'smart biomaterials' for tissue engineering and regeneration are activated by either cells or genes and are designed to improve the complicated biological event of tissue repair. It was demonstrated that among polymeric materials, polymeric membranes are attractive for their selectivity and biostability characteristics in the use of biohybrid

Membrane Operations. Innovative Separations and Transformations. Edited by Enrico Drioli and Lidietta Giorno

ISBN: 978-3-527-32038-7

systems for cell culture. Semipermeable membranes act as a support for the adhesion of anchorage-dependent cells and allow the specific transport of metabolites and nutrients to cells and the removal of catabolites and specific products [2–4]. Moreover, new membrane systems that have been recently realized might also potentially contribute to regenerative medicine and tissue engineering. Membranes are used as selective barriers to entrap living cells to be transplanted avoiding contact with immunocompetent species [5, 6] or in dynamic systems used as organ typical system. New strategies in the development of new systems might lead to membranes that are able to stimulate specific cell responses and maintain differentiated functions. In addition, the growing interest in synthetic and biodegradable polymeric biomaterials for tissue engineering and human cell therapies has led to novel approaches to improve cell–biomaterial interactions: the development of new biocompatible and cytocompatible materials and modification of surface chemistry including grafting of functional groups, peptides and proteins leaving the bulk properties unaltered. For example, the incorporation of a signal peptide such as RGD (Arg-Gly-Asp) into the biomaterial was an attempt to mimic the extracellular matrix, modulate cell adhesion, and induce cell migration. An intermediate density of adhesive ligand is crucial for optimal cell migration. Cell-specific recognition factors can be incorporated into the resorbable polymer surface, including the adhesive protein fibronectin or functional domains of ECM components [7]. The polymer surface can be tailored with proteins that influence interactions with endothelium [8], synaptic development [9], and neurite stimulation [10].

On the basis of these important considerations it can be pointed out that the development of new biomaterials able to activate a specific response of the cells and to maintain cell differentiation for a long time is one of the most pertinent issues in the field of tissue engineering and regenerative medicine. Nowadays, it has been demonstrated that the suitability of polymers for tissue-engineering purposes is highly dependent on the tissue that needs to be engineered. The histological, physiological, and biomechanical properties of each tissue determine the success of the regenerative process, therefore restricting the choice of materials.

This paper reports on the development of membrane systems to be used for cell culture (e.g., hepatocytes, lymphocytes, neuronal cells) in biohybrid systems such as a therapeutic device or as *in-vitro* model systems for studying the effects of various drugs and chemicals on cell metabolism.

19.2
Membranes for Human Liver Reconstruction

Liver failure is a fatal disease and liver transplantation is the only established treatment; however, donor shortages remain problematic. Liver failure is potentially reversible because of liver regeneration [11] so considerable work has been done over many years to develop effective liver-support devices and various hepatic support systems using hepatocytes have been developed. Liver-engineered constructs can be also applied to the development of drugs to treat many diseases. The impact will be increasing for the coming decade in the design of *in-vitro* physiological models to study disease pathogenesis and in the development of molecular therapeutics.

Since isolated hepatocytes may be able to undertake the full range of known *in-vivo* biotransformation and liver-specific functions [12], they could be used *in vitro* as model system for metabolic study. Hepatocytes are anchorage-dependent cells that require adhesive substrates for their functional and phenotypic maintenance [13]. So, a lot of artificial substrates such as membranes, microcarriers and biological matrix were studied for the hepatocyte cultures. During the last few years our experience has contributed to demonstrate that semipermeable synthetic membranes, owing to their structural and physicochemical characteristics, can be used for the development of biohybrid systems for cell cultures. We demonstrated that the hydrophobic/hydrophilic properties of membranes, such as surface free-energy parameters, affect cell adhesion, cell morphology and specific metabolic function of hepatocytes [3, 14]. Previous studies showed that polymeric membranes are able to support the long-term maintenance of metabolic and biotransformation functions of isolated human hepatocytes in a biohybrid system [15, 16]. We prepared membranes from a polymeric blend of modified polyetheretherketone or PEEK-WC and polyurethane (PU) by an inverse phase technique by using the direct immersion-precipitation method. The developed PEEK-WC-PU membrane combines the advantageous properties of both polymers (i.e. biocompatibility, thermal and mechanical resistance, elasticity) with those of membranes such as permeability, selectivity and well-defined geometry. This membrane is able to support cell adhesion and differentiation in a biohybrid system constituted of human hepatocytes and PEEK-WC-PU membrane for more than 1 month.

A confocal microscopy image (Figure 19.1) evidences the 3D structure of hepatocytes after days of culture on a PEEK-WC-PU membrane; cells organize in small aggregates, which would lead to better functional maintenance, maintaining a polygonal shape, so many of the features of the liver *in vivo* are reconstituted.

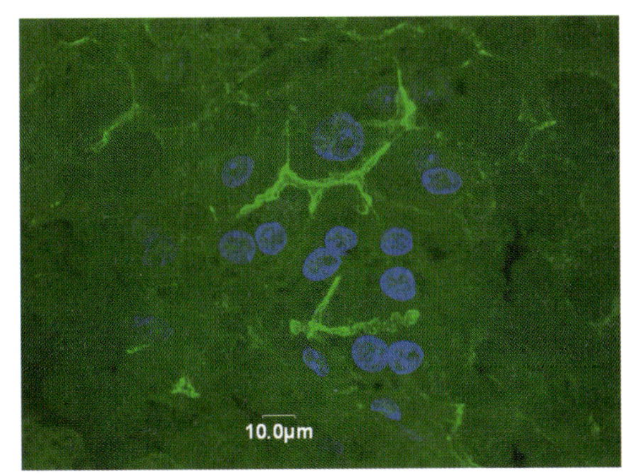

Figure 19.1 Confocal images of human hepatocytes on PEEK-WC-PU membranes by actin staining with FITC-phalloidin (green) and by nucleic acid staining with DAPI (blue). Scale bar 10 μm.

In particular, the microfilaments of actin, which is the component of the cytoskeletal system that allows movement of cells and cellular processes, are localized along the cell periphery. In addition, there are junctional complexes within the aggregates that are linked to the cortical actin. This spatial actin filament distribution is similar to that observed with hepatocytes cultured in collagen, where cells assume a cuboidal shape with extensive cell apposition. The development of a structural configuration resembling the native hepatic tissue involves the development of specific intercellular adhesions and redistribution of cell–cell and cell–surface adhesion forces, which are intimately related to the dynamic cytoskeletal organization. In human hepatocytes cultured in the membrane biohybrid system the localization of actin demonstrated the formation of cell–cell contacts that should provide better conditions for the maintenance of liver-specific functions. No spreading of cells was observed as a result of controlled interaction with the membrane surface. This study demonstrates that the PEEK-WC-PU membrane system is able to promote attachment and aggregate formation of cells outside of the body, providing a microenvironment able to elicit specific cellular responses of tissue analogues.

In the last few years the use of membrane biohybrid systems has contributed to giving important information about the effect of various drugs, such as diclofenac, rofecoxib, paracetamol the effects of which on the specific functions of human hepatocyte are not completely known [17–21].

An interesting approach to the design of membranes able to activate specific biological responses of the cells is the surface-modification technique by plasma process and biomolecule immobilization. This allows the cytocompatibility of membranes to be enhanced, leaving the bulk properties unaltered. It has been demonstrated that the immobilization of biomolecules on the membrane surface improves cell adhesion and the maintenance of differentiated functions [17, 21–24]. In particular, the RGD amino acid sequence (arginine–glycine–aspartic acid) stimulates cell adhesion on synthetic surfaces, since this oligopeptide represents the minimal adhesion domain of the majority of extracellular matrix (ECM) proteins (e.g., fibronectin, vitronectin, and collagen) [17–22].

In the case of hepatocytes the immobilization of galactose motifs on the surface enhances the specific interaction with cells owing to the specific binding between the galactose moiety and the asyaloglycoprotein receptor present on the cytoplasmatic membrane [25, 26].

We modified a polyethersulfone membrane (PES) surface with a plasma-deposited acrylic acid coating (PES-pdAA) and RGD peptide covalently immobilized through a 'spacer arm' molecule (SA), obtaining PES-pdAA-SA-RGD membranes. The same method was used to immobilize galactose in its acid form (Galactonic acid) to obtain the PES-PdAA-SA-GAL membranes.

Figure 19.2 shows the different physicochemical properties of the PES membranes modified with RGD and galactose moiety with respect to the collagen that was used as a natural substrate. Native PES membranes have a very high hydrophilic surface character, in fact, the water contact angle measured on this membrane was $30 \pm 1.4°$. Also, the modified membranes display a marked wettability even if, in this case, the

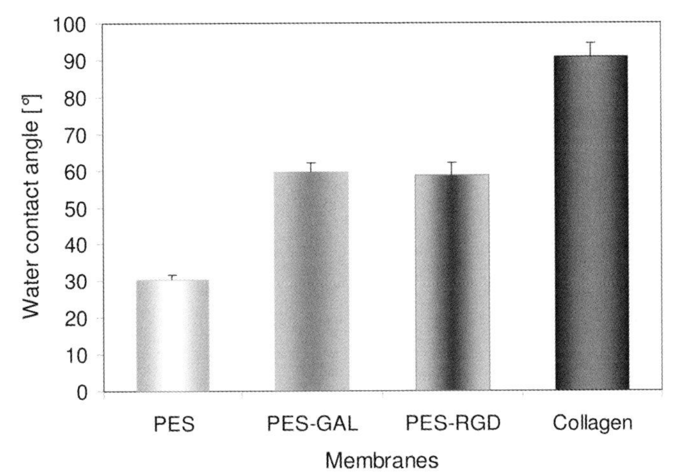

Figure 19.2 Water contact angle of unmodified, modified membrane and collagen at $t = 0$. PES-pdAA-SA-GAL and PES-pdAA-SA-RGD are abbreviated to PES-GAL and PES-RGD, respectively. The reported values are the mean of 30 measurements of different droplets on different surface regions of each sample ± standard deviation.

galactose and RGD immobilization induced a reduction in the surface hydrophilic character: the values were $60 \pm 2.3°$ and $58 \pm 3.6°$ for PES-pdAA-SA-GAL and PES-pdAA-SA-RGD membrane, respectively. Collagen film displayed a higher contact angle with respect to the membranes (Figure 19.2).

The performance of modified and unmodified membranes evaluated by analyzing the expression of liver-specific functions in terms of albumin production is shown in Figure 19.3. Hepatocytes cultured on unmodified PES membranes produced

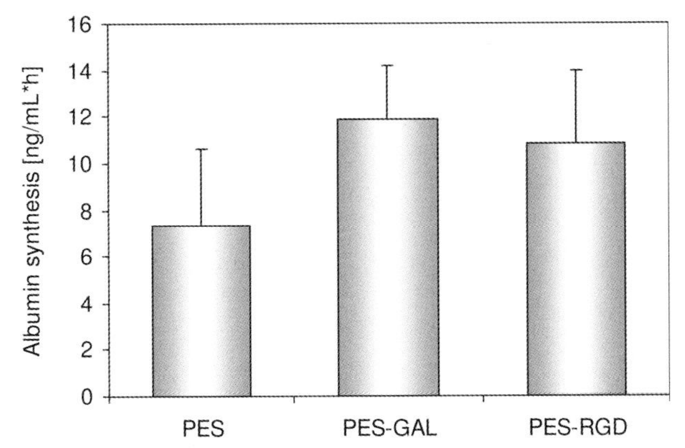

Figure 19.3 Albumin production of human hepatocytes cultured on different modified membrane surfaces. PES-pdAA-SA-GAL and PES-pdA-SA-RGD are abbreviated to PES-GAL and PES-RGD, respectively. The values are the mean of six experiments ± standard deviation.

albumin reaching values of 7.3 ng/ml h, while on modified membranes an increase of albumin production was observed. The highest levels of albumin synthesis were reached when hepatocytes were cultured on PES-pdAA-SA-GAL membranes where metabolic values of 11.9 ng/ml h were obtained. The ability of hepatocytes to synthesize albumin was also expressed at high level on PES-pdAA-SA-RGD membranes (10.8 ng/ml h). The immobilization of biomolecules, in particular with galactose moiety, improved the liver-specific functions of the hepatocytes cultured on their surface. It is interesting to note that the controlled molecular architecture of the membrane surface has a considerable impact on the liver cellular responses.

A strategy that has been developed to ameliorate the *in-vitro* retention of hepatocyte functions includes the development of bioreactors using different materials, configuration, and size [27–35]. Among these systems membrane bioreactors are particularly attractive because membranes allow the selective transport of metabolites and nutrients to cells and the removal of catabolites and specific products from cells [36, 37]. They also play the role of mechanical and chemical support for adhesion and growth of cells. In a previous study, human hepatocytes were cultured in a galactosylated polyethersufone (PES) membrane bioreactor, which permits cells to be cultured in a well-defined fluid dynamics microenvironment and metabolic rates to be easily estimated. Cells in the membrane bioreactor displayed an enhanced metabolic activity, which was maintained in the culture time at significantly higher levels with respect to the batch system [22]. In particular, the cell metabolic functions of urea synthesis and protein secretion were maintained for 21 days (Figure 19.4).

As reported in Figure 19.4 the urea synthesized by human hepatocytes reached values ranging from 28 to 60 μg/ml. Interestingly, the ability of cells to secrete proteins was also maintained for the whole period of culture demonstrating the good

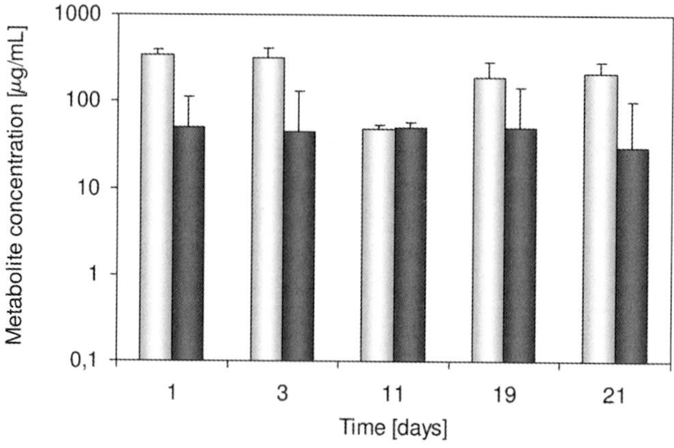

Figure 19.4 Protein secretion (shade bars) and urea synthesis (full bars) of human hepatocytes cultured in the galactosylated membrane bioreactor for 21 days. The values are the mean of six experiments ± standard deviation.

performance of the bioreactor in the long-term maintenance of differentiated functions of cells outside of the body.

This human hepatocyte galactosylated membrane bioreactor was also used as a model system to explore the modulation of the effects of a proinflammatory cytokine, Interleukin-6 (IL-6) on the liver cells at the molecular level [38] and gave evidence, for the first time, that IL-6 downregulated the gene expression and synthesis of fetuin-A by primary human hepatocytes. The human hepatocyte bioreactor behaves like the *in-vivo* liver, reproducing the same hepatic acute-phase response that occurs during the inflammation process. This hepatocyte bioreactor should find applications in drug testing, toxicological studies and in tissue engineering to help solve problems related to human diseases.

19.3 Human Lymphocyte Membrane Bioreactor

Isolated human lymphocytes are used to investigate the role of these cells in the pathogenesis of various diseases and in the autologous adoptive immunotherapy for patients with malignant disease or viral infections [39, 40]. Lymphocytes may be used also as biomarkers of target-organ susceptibility or as a marker of chemical effects and in the prediction of individual drug sensitivity alternatively to human liver biopsies [41, 42].

Generally, lymphocytes are cultured in static culture systems like T-flasks, culture bags, well plates, which imply disadvantages in nonuniform culture conditions, low cell densities and uncontrolled process parameters [43–45]. This results in the maintenance of cell viability and functions for only a short time. Differently from static culture methods, dynamic systems such as bioreactors allow the culture of cells under tissue-specific mechanical forces such as pressure, shear stress, and interstitial flow [46]. Among the various bioreactors (hollow fiber, stirred vessels, suspension bioreactor) [47–50] that have been developed to foster the retention of human lymphocytes *in vitro* one of the most used is the hollow-fiber (HF) membrane bioreactor. This bioreactor meets the main requirements for cell culture: a wide area for the exchange of oxygen/carbon dioxide and nutrient transfer, removal of catabolites and protection from shear stress. To date, the HF membrane bioreactor has mainly been used for large-scale mammalian cell culture to produce such products as monoclonal antibodies [48], for the expansion of tumor-infiltrating T lymphocytes [49]. Lamers *et al.* [47] demonstrated the usefulness of the HF bioreactor culture to produce cytokines by T lymphocytes.

We developed a PEEK-WC hollow-fiber (HF) membrane bioreactor for the maintenance of human peripheral lymphocytes as a model system for the *in-vitro* investigation of disease pathogenesis, chemical effects and individual drug sensitivity. Peripheral lymphocytes isolated from the donor's human buffy coat were cultured in the shell compartment of the PEEK-WC-HF bioreactor. Lymphocytes in the PEEK-WC-HF membrane bioreactor produced IL-2 and IL-10 throughout the culture period of 14 days (Figure 19.5).

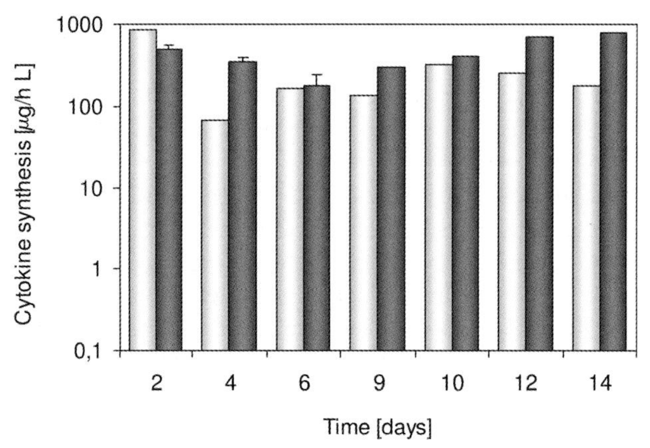

Figure 19.5 Cytokine production in human lymphocytes cultured in PEEK-WC-HF bioreactor for 14 days: (shade bars) IL-2 and (full bars) IL-10. The values are the mean of six experiments $\pm$ standard deviation.

IL-2 is the specific pattern of lymphocytes T helper 1 that increases the proliferation of T cells and activates B cells. IL-10 is a specific pattern of lymphocytes T helper 2, which is stimulatory towards certain T cells, mast cells and B cells and inhibits the synthesis of other cytokines produced by Th1. IL-10 was synthesized by cells with higher metabolic rates with respect to IL-2, suggesting a prevalent differentiation in lymphocyte immunophenotype Th2.

The obtained results demonstrated that a PEEK-WC-HF membrane bioreactor is able to support the proliferation and functions of human peripheral lymphocytes isolated from the buffy coat of healthy individuals. Therefore, the lymphocyte HF membrane bioreactor can be used as a valuable tool to maintain viable and functional lymphocytes and as an *in-vitro* model for pharmacological and adoptive immunotherapy.

19.4
Membranes for Neuronal-Tissue Reconstruction

During the last few years neuronal cell behavior on a biomaterial such as membrane has become of great interest, since it offers the advantage of developing neuronal tissue that may be used for the *in-vitro* simulation of human brain functions. This could definitely provide further insights not only into the cell but also in developing therapies in neurodegenerative disorders such as Parkinson's or Alzheimer's disease [51]. A biohybrid system using neurons could also represent a useful instrument for predictive drug testing or constitute a future model of a bioneuronal network device. For typical neuronal tissue-engineered constructs, the properties of both cell (morphology, viability functions) and material (physicochemical, morphological and transport properties) components are very important [52]. Current interest has been focused on attempts to find new biomaterials and new cell sources as well as novel designs of tissue-engineered neuronal devices to generate safer and more efficacious restored neuronal tissue. Biomaterials that have been successfully employed in the manufacture of neuronal tissue include biodegradable materials such as polyglycolic acid, poly-L-lactic acid and poly (lactide-co-glycolide) and polymeric semipermeable membranes (polyacrylonitrile-polyvinylchloride, polyethylenecovinylalcohol, cellulose acetate) in fiber and flat configurations [53–56].

In view of the widespread structural organization of all the brain regions, we focused our attention on the potential value of this biotechnological approach to a functionally key region such as the hippocampus. Indeed, the principal neurons of this brain region, that is, pyramidal cells, are actively involved in many hippocampal-dependent neurophysiological functions, such as memory and learning. This makes them a valuable tool to investigate not only their synaptic plasticity properties, but also neurodegenerative events through the distribution and quantification of microtubule-associated protein type 2. In a recent study, the reconstruction of membrane biohybrid systems, constituted of isolated cells and membranes, appears to represent a crucial step for the success of these systems [57]. Moreover, the optimization of transport, physicochemical and structural properties of the membrane as well as fluid dynamics of cellular microenvironments tend to favor cell–membrane interactions and the functional maintenance of hippocampal cells. As a consequence, the feasibility of developing a hippocampal cell membrane biohybrid system capable of regenerating a neuronal network could prove to be an important approach for studying the behavior of neuronal populations in some of the most common neurodegenerative disease such as Alzheimer's disease.

A protocol of isolation and culture of hippocampal neurons has been optimized by using a hibernating rodent, the hamster *Mesocricetus auratus* as our animal model [57]. Preliminary results demonstrated the feasibility of culturing hippocampal neurons in a membrane biohybrid system. The substrate is of great importance for the survival and differentiation of neuronal cultures. As substrate we used a gas-permeable (CO_2, O_2, and H_2O vapor) fluorocarbon foil membrane (FC), modified with a coating of poly-L-lysine in order to improve the interactions with cells.

In particular, it is interesting to note that after cells have adhered to the substrate a flattening of the cells was observed and minor processes started to emerge from several sites along the circumference of the cells (Figure 19.6(a)). With the progress of their growth process, the tiny neuronal filaments begin to acquire the definite characteristics of dendrites and axons (Figure 19.6(b)) and subsequently the synaptic contacts in this rich neuronal network. The complexity of the neuronal network increased with time: dendrites emerging from the cell body became highly branched (Figure 19.6(c)). In our membrane biohybrid system, following the evaluation of some properties such as adhesion and development of dendrites and axons, the cytoskeletal features were considered by investigating the localization of βIII-tubulin (green) in the neuronal network (Figure 19.7).

This cytoskeletal protein is present on the soma and in all neuronal processes. A complex axonal network was observed on the membrane (Figure 19.7) where the cells

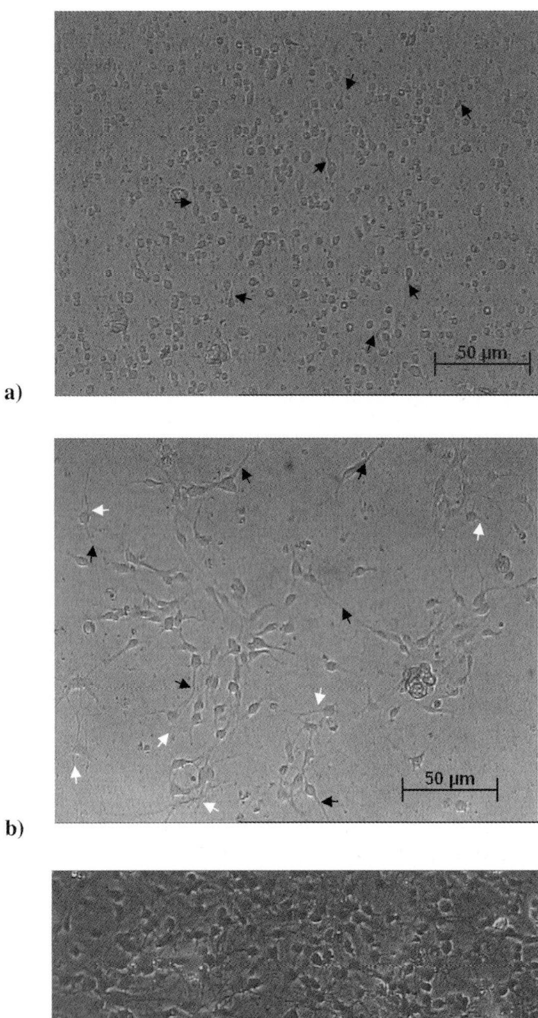

Figure 19.6 Micrographs of hippocampal neurons on FC membrane after (a) 4 hours, (b) 4 days (c) 16 days of culture. The arrows in a) indicate the emerging processes from the cell circumference; the arrows in b) indicate the (black) axon and the dendrites (white).

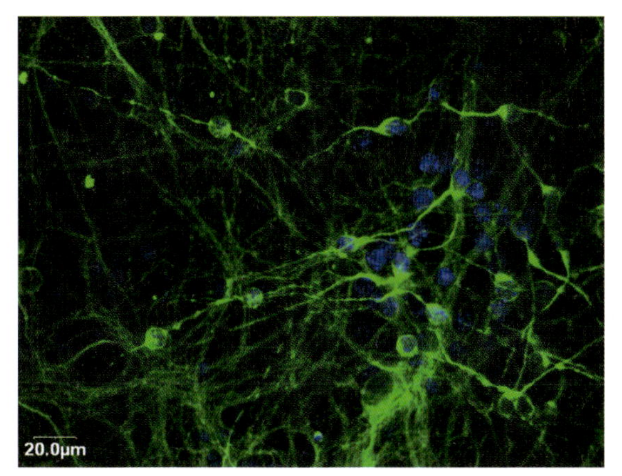

Figure 19.7 Confocal laser micrographs of hippocampal neurons after 16 days of culture on modified FC membranes. The cells were immunolabeled for βIII-tubulin (green) and cell nuclei were labeled with DAPI (blue).

showed a differentiated pyramidal somata as well as forming significantly longer neurites and more elaborated dendritic arbors. This well-defined shape of the hippocampal neurons is important for *in-vivo* studies owing to their highly remarkable synaptic plasticity capacity, which proves to be essential for the explication of some key neurophysiological functions of hippocampus, such as memory and learning [58].

These results suggest that membrane system is able to favor the differentiation of neuronal hippocampal cells. The βIII-tubulin immunoreactivity displayed the unaltered cytoskeletal characteristics of cells after their differentiation and maturational process. These results encourage the development of a membrane-engineering system of hippocampal neurons that are able to remodel and regenerate neural tissue in a well-controlled microenvironment. This experimental system might be a valuable model to investigate complex neuronal networks existing between some major neurotransmitter systems such as the histaminergic and glutammatergic system and the behavior of neuronal populations in some of the most common neurodegenerative disease such as Alzheimer's disease.

19.5
Concluding Remarks

This study reports on the potentiality of applying a membrane biohybrid system in the field of tissue engineering and regenerative medicine evidencing the crucial points in the *in-vitro* reconstruction of the physiological tissue model. A number of issues need to be addressed: the morphological and physicochemical properties of the membrane, the optimal density of immobilized cells, the interaction of cells with the membrane, the differentiation of cells as well as the maintenance of viability and metabolic functions *in vitro* membrane constructs. Interesting results were obtained using membrane biohybrid systems for the *in-vitro* liver reconstruction, neuronal tissue, and for the *in-vitro* maintenance of the patient's lymphocytes. The use of all these systems in tissue engineering can be exciting in helping to find nature's substitutes and in solving the pathogenesis of important human diseases or to select an optimal pharmaceutical treatment.

Acknowledgments

The authors acknowledge the Italian Ministry of University and Research (MIUR) for funding this research through the FIRB RBNE012B2K research project and the European Commission through the Livebiomat project STRP NMP3-CT-013653 in the FP6.

References

1 Hench, L.L. and Polak, J.M. (2002) *Science*, **295**, 1014–1017.

2 De Bartolo, L., Jarosch-Von Schweder, G., Haverich, A. and Bader, A. (2000) *Biotechnology Progress*, **16**, 102–108.

3 De Bartolo, L., Morelli, S., Bader, A. and Drioli, E. (2002) *Biomaterials*, **23**, 2485–2497.

4 Unger, R.E., Huang, Q., Peters, K., Protzer, D., Paul, D. and Kirkpatrick, C.J. (2005) *Biomaterials*, **26**, 1877–1884.

5 Gentile, F.T., Doherty, E.J., Rein, D.H., Shoichet, M.S. and Winn, S.R. (1995) *Reactive Polymers*, **25**, 207–227.

6 Li, R.H. (1998) *Advanced Drug Delivery Reviews*, **33**, 87–109.

7 Quirk, R.A., Chan, W.C., Davies, M.C., Tendler, S.J.B. and Shakesheff, K.M. (2001) *Biomaterials*, **22**, 865–872.

8 Hubbell, J.A., Massia, S.P., Desai, N.P. and Drumheller, P.D. (1991) *Biotechnology (Reading, Mass)*, **9**, 568–574.

9 Hunter, D.D., Cashman, N., Morris-Valero, R., Bulock, J.W., Adams, S.P. and Sanes, J.R. (1991) *The Journal of Neuroscience*, **11**, 3960–3971.

10 Patel, N., Padera, R., Sanders, H.W., Cannizzaro, M., Davies, C., Langer, R., Roberts, J., Tendler, J.B., Williams, M. and Shakesheff, M. (1998) *The FASEB Journal*, **12**, 1447–1454.

11 Michalopoulos, G.K. and DeFrances, M.C. (1997) *Science*, **276**, 60–66.

12 Berry, M.N. and Edwards, A.M. (2000) *The Hepatocyte Review*, Kluwer Academic Publishers, Dordrecht, pp. 365–585.

13 Dunn, J.C., Tompkins, R.G. and Yarmush, M.L. (1991) *Biotechnology Progress*, **7**, 237–244.

14 De Bartolo, L., Morelli, S., Bader, A. and Drioli, E. (2001) *Journal of Material Science: Materials in Medicine*, **12**, 959–963.

15 De Bartolo, L., Morelli, S., Rende, M., Gordano, A. and Drioli, E. (2004) *Biomaterials*, **25**, 3621–3629.

16 De Bartolo, L., Morelli, S., Gallo, M.C., Campana, C., Statti, G., Rende, M., Salerno, S. and Drioli, E. (2005) *Biomaterials*, **26**, 6625–6634.

17 De Bartolo, L., Morelli, S., Lopez, L.C., Giorno, L., Campana, C., Salerno, S., Rende, M., Favia, P., Detomaso, L., Gristina, R., d'Agostino, R. and Drioli, E. (2005) *Biomaterials*, **26**, 4432–4441.

18 De Bartolo, L., Morelli, S., Giorno, L., Campana, C., Rende, M., Salerno, S., Maida, S. and Drioli, E. (2006) *Journal of Membrane Science*, **278** (1–2), 133–143.

19 De Bartolo, L., Salerno, S., Giorno, L., Morelli, S., Barbieri, G., Curcio, E., Rende, M. and Drioli, E. (2006) *Catalysis Today*, **118**, 172–180.

20 De Bartolo, L., Morelli, S., Rende, M., Campana, C., Salerno, S., Quintiero, N. and Drioli, E. (2007) *Macromolecular Bioscience*, **7**, 671–680.

21 Memoli, B., De Bartolo, L., Favia, P., Morelli, S., Lopez, L.C., Procino, A., Barbieri, G., Curcio, E., Giorno, L., Esposito, P., Cozzolino, M., Brancaccio, D., Andreucci, V.E., d'Agostino, R. and Drioli, E. (2007) *Biomaterials*, **28**, 4836–4844.

22 De Bartolo, L., Morelli, S., Rende, M., Salerno, S., Giorno, L., Lopez, L.C., Favia, P., d'Agostino, R. and Drioli, E. (2006) *Journal of Nanoscience and Nanotechnology*, **6**, 2344–2353.

23 De Bartolo, L., Morelli, S., Piscioneri, A., Lopez, L.C., Favia, P., d'Agostino, R. and Drioli, E. (2007) *Biomolecular Engineering*, **24**, 23–26.

24 Morelli, S., Salerno, S., Rende, M., Lopez, L.C., Favia, P., Procino, A., Memoli, B., Andreucci, V.E., d'Agostino, R., Drioli, E. and De Bartolo, L., (2007) *Journal of Membrane Science*. **302**, 27–35.

25 Ying, L., Yin, C., Zhuo, R.X., Leong, K.W., Mao, H.Q., Kang, E.T. and Neoh, K.G. (2003) *Biomacromolecules*, **4**, 157–164.

26 Lopina, S.T., Wu, G., Merrill, E.W. and Griffith, C.L. (1996) *Biomaterials*, **17**, 559–567.

27 Rocha, F.G. and Whang, E.E. (2004) *Journal of Surgical Research*, **120**, 320–325.

28 Sodian, R., Hoerstrup, S.P., Sperling, J.S., Daebritz, S.H., Martin, D.P., Shoen, F.J., Vacanti, J.P. and Mayer, J.E. (2000) *The Annals of Thoracic Surgery*, **70**, 140–144.

29 Shin'oka, T., Imai, Y. and Ikada, Y. (2001) *The New England Journal of Medicine*, **3444**, 532–540.

30 Fansa, H., Dodic, T., Wolf, G., Schneider, W. and Keilhoff, G. (2003) *Microsurgery*, **23**, 72–78.

31 Parenteau, N.L., Nolte, C.M., Bilbo, P., Rosenberg, M., Wilkins, L.M., Johnson, E.W., Watson, S., Mason, V.S. and Bell, E.J., (1991) *Journal of Cellular Biochemistry*, **45**, 245–253.

32 Vacanti, C.A. and Vacanti, J.P. (1994) *Otolaryngologic Clinics of North America*, **27**, 263–269.

33 Tsai, R.J., Li, L.M. and Che, J.K. (2000) *The New England Journal of Medicine*, **343**, 86–95.

34 Jauregui, H.O. and Gann, K.L. (2002) *Tissue Engineering*, **8**, 725–732.

35 Papas, K.K., Long, R.C., Sambanis, A. and Constantinidis, I. (1999) *Biotechnology and Bioengineering*, **66**, 219–226.

36 Dionne, K.E., Cain, B.M., Li, R.H., Bell, W.J., Doherty, E.J., Rein, D.H., Lysaght, M.J. and Gentile, F.T. (1996) *Biomaterials*, **17**, 257–266.

37 Curcio, E., De Bartolo, L., Barbieri, G., Rende, M., Giorno, L., Morelli, S. and Drioli, E. (2005) *Journal of Biotechnology*, **117**, 309–318.

38 Stenvinkel, P., Ketteler, M., Johnson, R.J., Lindholm, B., Pecoits-Filho, R., Heimburger, O., Cederholm, T., Girudt, M. and Riella, M. (2005) *Kidney International*, **67**, 1216–1233.

39 Bremers, A.J.A. and Parmiani, G. (2000) *Critical Reviews in Oncology/Hematology*, **34**, 1–25.

40 Trickett, A.E., Kwan, Y.L., Cameron, B. and Dwyer, J.M. (2002) *Journal of Immunological Methods*, **262**, 71–83.

41 Krovat, B.C., Tracy, J.H. and Omiecinski, C.J. (2000) *Toxicological Sciences*, **55**, 352–360.

42 Tabatabetei, A.R., Thies, R.L., Farrell, K. and Abbott, F.S. (1997) *Fundamental and Applied Toxicology*, **37**, 181–189.

43 Muehlbauer, P.A. and Schuler, M.J. (2003) *Mutation Research*, **537**, 117–130.

44 Garland, R.J., Kaneria, S.S., Hancock, J.P., Steward, C.G. and Rowbottom, A.W. (1999) *Journal of Immunological Methods*, **227**, 53–63.

45 Woodside, K.J., Hu, M., Gugliuzza, K.K., Hunter, G.C. and Daller, J.A., (2005) *Transplantation Proceedings*, **37**, 1949–1952.

46 Martin, Y. and Vermette, P. (2005) *Biomaterials*, **26**, 7481–7503.

47 Lamers, C.H.J., Gartama, J.W., Luider-Vrieling, B., Bolhuis, R.L.H. and Bast, E.J.E.G. (1999) *Journal of Immunotherapy*, **22**, 4, 299–307.

48 Knazek, R.A., Wu, Y.W., Aebersold, P.M. and Rosenberg, S.A. (1990) *Journal of Immunological Methods*, **127**, 29–37.

49 Cadwell, J.J.S. (2004) *American Biotechnology Laboratory*, 1–5.

50 Bohnenkamp, H., Hilbert, U. and Noll, T. (2002) *Cytotechnology*, **38**, 135–145.

51 Caffesse, R.G., Najleti, C.E., Morrisson, E.C. and Sanchez, R. (1994) *Journal of Periodontology*, **65**, 583–590.

52 Zhang, N., Yan, H. and Wen, X. (2005) *Brain Research Reviews*, **49**, 48–56.

53 Young, T.H., Lin, C.W., Cheng, L.P. and Hsieh, C.C. (2001) *Biomaterials*, **22**, 1771–1779.

54 Gautier, S.E., Oudega, M., Fragoso, M., Chapon, P., Plant, G.W., Bunge, M.B. and Parel, J.M. (1998) *Journal of Biomedical Materials Research*, **42**, 642–648.

55 Patist, C.M., Mulder, M.B., Gautier, S.E., Maquet, V., Jerome, R. and Oudega, M. (2004) *Biomaterials*, **25**, 1569–1576.

56 Woerly, S., Plant, G.W. and Harvey, A.R. (1996) *Biomaterials*, **17**, 301–313.

57 De Bartolo, L., Rende, M., Giusi, G., Morelli, S., Piscioneri, A., Canonaco, M. and Drioli, E. (2007) in *Evolutionary Molecular Strategies and Plasticity* (eds M. Canonaco and R.M. Facciolo), Research Signpost, Kerala, India, pp. 379–396.

58 Lo, D.C. (1995) *Neuron*, **15** (5), 979–981.

Part Three
Membrane Contactors

This Part focuses on fundamentals and applications of membrane contactors and membrane emulsification. Special attention is given to the industrial application of membrane contactors.

Among other new unit operations involving membranes, membrane contactors are expected to play a decisive role in the development of clean, safe and sustainable processes. The key concept is to use a solid, microporous, polymeric matrix in order to create an interface for mass transfer and/or reaction between two phases: large exchange area and independent fluid dynamics allow an easily controlled operation. In general, these membrane systems use low-cost hollow fibers, and provide a high interfacial area significantly greater than most traditional absorbers between two phases to achieve high overall rates of mass transfer. In addition, whereas the design of the conventional devices is restricted by limitations in the relative flows of the fluid streams, membrane contactors give an active area, which is independent of the liquid fluid dynamics. Membrane crystallizers, membrane emulsifiers, membrane strippers and scrubbers, membrane distillation systems, membrane extractors, etc., can be devised and integrated in the production lines together with the other existing membranes operations for advanced molecular separation and chemical transformations, overcoming existing limits of the more traditional membrane processes, such as osmotic effect in reverse osmosis.

In these systems, the interface between two phases is located at the high-throughput membrane porous matrix level. Physicochemical, structural and geometrical properties of porous meso- and microporous membranes are exploited to facilitate mass transfer between two contacting immiscible phases, e.g., gas–liquid, vapor–liquid, liquid–liquid, liquid–supercritical fluid, etc., without dispersing one phase in the other (except for membrane emulsification, where two phases are contacted and then dispersed drop by drop one into another under precise controlled conditions). Separation depends primarily on phase equilibrium. Membrane-based absorbers and strippers, extractors and back extractors, supported gas membrane-based processes and osmotic distillation are examples of such processes that have already been in some cases commercialized. Membrane distillation, membrane

Membrane Operations. Innovative Separations and Transformations. Edited by Enrico Drioli and Lidietta Giorno

ISBN: 978-3-527-32038-7

crystallization, supported liquid membranes, etc., are examples of such processes that are not yet commercialized. Indeed, they are of great interest due to their inherent capabilities.

The integration of membrane contactors with other membrane operations, including membrane reactors, may lead to redesign of production lines based on intensified integrated membrane processes.

20
Basics in Membrane Contactors

Alessandra Criscuoli

20.1
Introduction

This chapter aims at providing an overview on membrane contactors, starting from their definition and main properties, going through the mass transport, and finally describing their main applications. In particular, the role of membrane contactors in reaching the goals of the process-intensification strategy is discussed in terms of new developed metrics. The objective is to give some basic information on these new systems of mass transfer and their potentialities, as well as on the main drawbacks related to their further implementation at the industrial level and on the research efforts to be made for overcoming these limitations and to extend their fields of application.

20.2
Definition of Membrane Contactors

A membrane is usually seen as a selective barrier that is able to be permeated by some species present into a feed while rejecting the others. This concept is the basis of all 'traditional' membrane operations, such as microfiltration, ultrafiltration, nanofiltration, reverse osmosis, pervaporation, gas separation. On the contrary, membrane contactors do not allow the achievement of a separation of species thanks to the selectivity of the membrane, and they use microporous membranes only as a mean for 'keeping in contact' two phases. The interface is established at the pore mouths and the transport of species from/to a phase occurs by simple diffusion through the membrane pores. In order to work with a constant interfacial area, it is important to carefully control the operating pressures of the two phases. Usually, the phase that does not penetrate into the pores must be kept at higher pressure than the other phase (Figure 20.1a and b). When the membrane is hydrophobic, polar phases can not go into the pores, whereas, if it is hydrophilic, the nonpolar/gas phase remains blocked at the pores entrance [1, 2].

Membrane Operations. Innovative Separations and Transformations. Edited by Enrico Drioli and Lidietta Giorno

ISBN: 978-3-527-32038-7

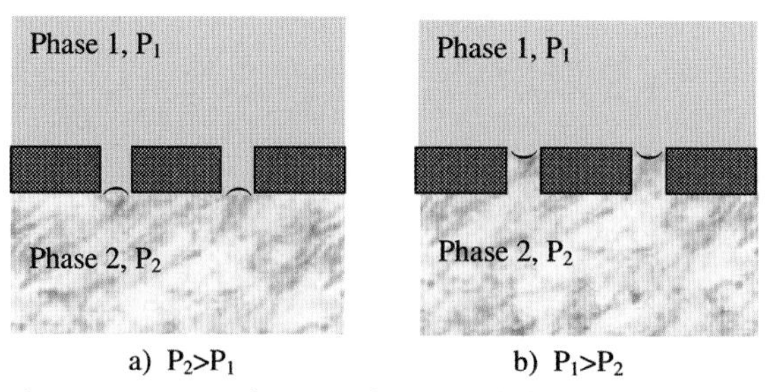

Figure 20.1 Contact between phase 1 and phase 2 through a membrane contactor. (a) phase 1 in the membrane; (b) phase 2 in the membrane.

Membrane contactors can be seen as alternative systems for carrying out gas–liquid operations (such as stripping and scrubbing) or liquid–liquid extractions. Furthermore, by applying vacuum at one membrane side, liquid–vacuum operations can be performed. Another interesting process that can be carried out is distillation (up to crystallization). In particular, depending on the way the driving force is obtained, we can distinguish between membrane and osmotic distillation. In direct contact membrane distillation (DCMD), which has been the most investigated configuration, a difference of temperature is imposed between the two streams, whereas in osmotic distillation, the driving force is achieved by using a hypertonic solution (often, a highly concentrated $CaCl_2$ solution) as strip phase. Finally, supported liquid membranes represent a particular type of membrane contactors where, usually, an organic phase fills the micropores and the transport occurs from an aqueous feed to an aqueous strip. Figure 20.2 summarizes the different types of membrane contactors described.

Membrane-contactor efficiency depends on membrane properties, membrane module, operating conditions.

Hydrophobic membranes are often used and, among the hydrophobic polymeric materials, we can mention polypropylene, PVDF, PTFE, and perfluoropolymers (e.g., hyflon). For a good performance of the system it is important that the membrane keeps its hydrophobic character, especially in long-term applications. Therefore, several studies aimed at improving the hydrophobicity values, as well as increasing the membrane stability with time, have been made [3–6]. The chemical stability of the membrane is also crucial for successful applications with organic solvent or absorbents (such as amines). Membranes with big pore sizes, high porosities and low thicknesses lead to a high transfer of the species, but big pore sizes also mean lower values of the breakthrough pressures (pressures at which the membrane is wetted by the liquid, losing its hydrophobic character), as reported in Laplace's equation (for gas–liquid operations):

$$\Delta p = (2\sigma\cos\theta)/r \tag{20.1}$$

with σ, surface tension; θ contact angle; r, pore radius.

Moreover, in direct contact membrane distillation, a minimum value of thickness is required to keep the difference of temperature across the membrane. Generally, in membrane distillation materials with low thermal conductivity are also required to reduce the heat loss through the membrane-self. Pore-size distribution also plays an

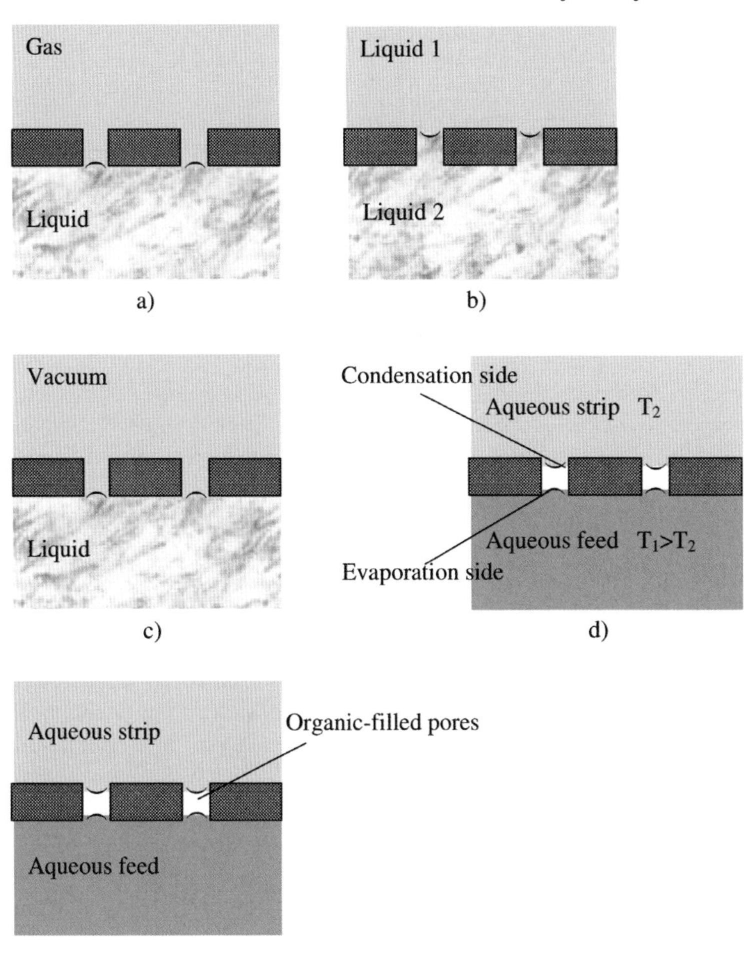

Figure 20.2 Different types of membrane contactors. (a) stripper/scrubber, driving force: difference of concentration; (b) liquid–liquid extractors, driving force: difference of concentration; (c) removal of volatiles/gases from liquids, driving force: difference of partial pressures; (d) direct contact membrane distillation, driving force: difference of partial pressures; (e) supported liquid membranes, driving force: difference of concentrations.

important role for obtaining a uniform transport along the membrane and avoiding coalescence phenomena. Table 20.1 summarizes the main effects of some membrane properties on the performance of membrane contactors.

Referring to the membrane module design, it has a big influence on the membrane-contactor efficiency, because it affects the pressure drops of the streams (and, thus, the operating pressures and flowrates), and their fluidodynamic (which means the mass and heat transport resistances of the phases). Furthermore, for hollow-fiber modules it is essential to ensure a uniform packing, in order to have

Table 20.1 Main effects of some membrane properties on the performance of membrane contactors[a].

Membrane property	Effect
Big pore size	Higher flux (+) and lower breakthrough pressure value (−)
High porosity	Higher flux (+) and higher coalescence phenomena (−)
Low thickness	Higher flux (+) and lower difference of temperature across the membrane (−), for DCMD
Low tortuosity	Higher flux (+)

[a]Symbols (+) and (−) indicates a positive and a negative effect, respectively.

an uniform flow pattern at the shell side, avoiding any phenomena of channeling, bypassing, presence of stagnant areas, which can sensibly reduce the overall performance. For these reasons there have been several studies for developing modules for membrane-contactor applications [7–13].

The right choice of operating conditions is also at the basis of a good performance of membrane contactors. Higher flowrates lead to a reduction of the mass and heat transport resistances in the phases and of the presence of stagnant zones inside the module. However, they have to be carefully defined, in order to avoid the stream pressures reaching the breakthrough values. For the same reason, the pressures of the streams to be processed must be controlled, and eventually properly varied, before sending them to the membrane-contactor unit. Streams with high viscosity or containing particles of big size should be preferably sent to the shell side of the module, and the fluid with higher affinity for the species to be transferred should fill the membrane pores, in order to reduce the membrane mass-transfer resistance.

When compared to conventional systems (such as strippers, scrubbers, distillation columns, packed towers, bubble columns, evaporators, etc.), membrane contactors present several advantages, as reported in Figure 20.3. However, some drawbacks have also to be taken into account, as shown in Figure 20.4.

20.3 Mass Transport

When a species is transferred from a phase to another phase by means of a membrane contactor, the mass-transport resistances involved are those offered by the two phases and that of the membrane (see Figure 20.5). The overall mass-transfer coefficient will, therefore, depend on the mass-transfer coefficient of the two phases and of the membrane.

If we consider a gas–liquid transfer for the species i in a hollow-fiber module with the liquid phase in the shell side and the gas phase in the lumen side of hydrophobic membranes, the interface is established at the outer diameter of the fibers and the overall mass-transfer coefficient can be calculated by [1]:

$$1/(K_l\, d_o) = 1/(k_{ils}\, d_o) + 1/(k_{im}\, H_i\, d_{lm}) + 1/(k_{igt}\, H_i\, d_i) \qquad (20.2)$$

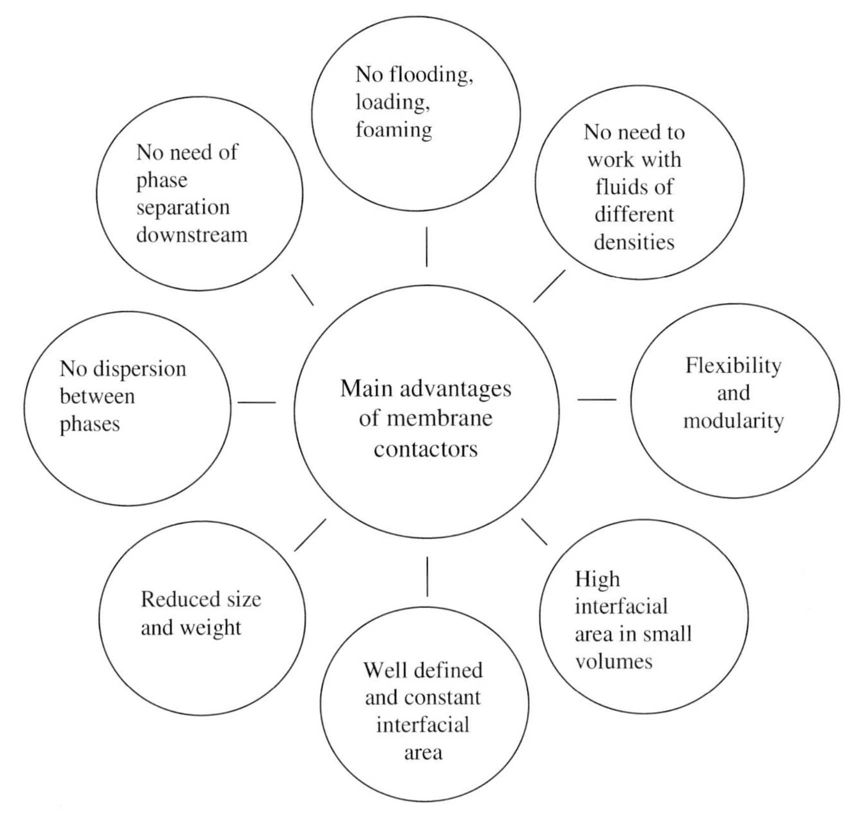

Figure 20.3 Main advantages of membrane contactors.

$$1/(K_g\, d_o) = H_i/(k_{ils}\, d_o) + 1/(k_{im}\, d_{lm}) + 1/(k_{igt}\, d_i) \qquad (20.3)$$

where K_l, liquid overall mass-transfer coefficient; K_g, gas overall mass-transfer coefficient; k_{ils}, mass-transfer coefficient for the species i in the liquid at the shell side; k_{igt}, mass-transfer coefficient for the species i in the gas at the tube side; k_{im}, mass-transfer coefficient for the species i in the membrane; d_i, inner diameter of the fiber; d_o, outer diameter of the fiber; $d_{lm,}$ logarithmic mean of the hydrophobic membrane diameters; H_i, Henry's coefficient.

The membrane mass-transfer coefficient k_{im} is usually based on Knudsen flows and can be derived by:

$$k_{im} = D_i^k \varepsilon/\tau\delta \qquad (20.4)$$

with D_i^k, Knudsen diffusion coefficient for the species i through the membrane; ε, membrane porosity; τ, membrane tortuosity; δ, membrane thickness.

The mass-transfer coefficient at the tube side is usually well obtained by the Leveque equation:

$$\mathrm{Sh} = 1.62\,(d^2\, v/(LD))^{0.33} \qquad (20.5)$$

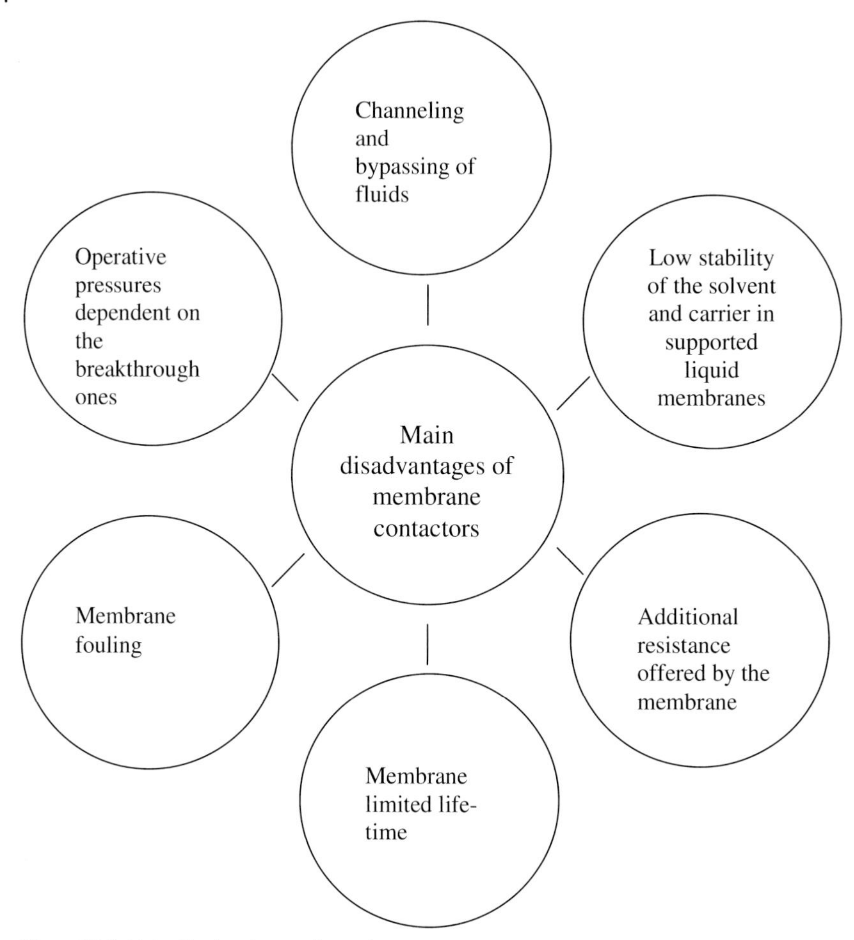

Figure 20.4 Main disadvantages of membrane contactors.

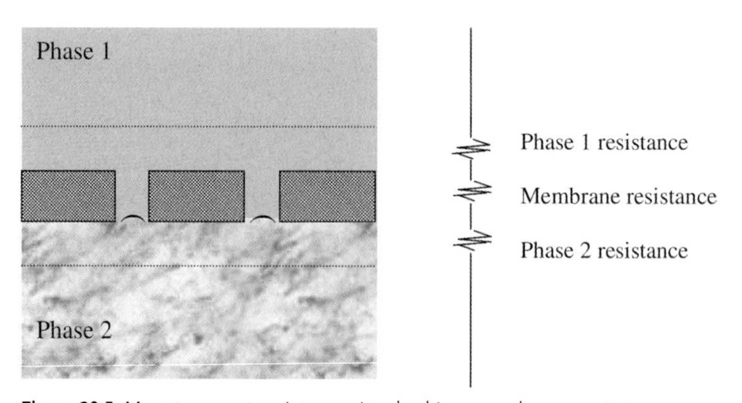

Figure 20.5 Mass-transport resistances involved in a membrane contactor.

where Sh, Sherwood number; v, fluid velocity; d, fiber diameter; L, length of the fiber; D, diffusion coefficient of the species i into the fluid.

On the contrary, no general expression is available for calculating the mass-transfer coefficient at the shell side. In the literature, in fact, different equations are proposed, depending on the type of module and on the type of flow (parallel or crossflow). Probably, this is due to the fact that the fluidodynamics of the stream sent outside the fibers is strongly affected by the phenomena of channeling or bypassing and it is not well defined as for the stream, which is sent into the fibers. Hereinafter some of the different expressions proposed are reported.

Parallel flow

$$\mathrm{Sh} = \beta\, d_e/L\,(1-\phi)\ \mathrm{Re}^{0.60}\ \mathrm{Sc}^{0.33} \quad [14] \tag{20.6}$$

$0 < \mathrm{Re} < 500$; loosely packed fibers; β is 5.8 for hydrophobic and 6.1 for hydrophilic membranes

$$\mathrm{Sh} = (0.31\phi^2 - 0.34\phi + 0.10)\mathrm{Re}^{0.90}\mathrm{Sc}^{0.33} \quad [10] \tag{20.7}$$

$\mathrm{Re} = 100$; loosely closely packed fibers

$$\mathrm{Sh} = (0.53 - 0.58\phi)\ \mathrm{Re}^{0.53}\ \mathrm{Sc}^{0.33} \quad [15] \tag{20.8}$$

$21 < \mathrm{Re} < 324$; medium-closely packed fibers with Re, Reynolds number; Sc, Schmidt number; ϕ, fiber packing; d_e, equivalent diameter.

Crossflow

$$\mathrm{Sh} = 0.15\ \mathrm{Re}^{0.80}\ \mathrm{Sc}^{0.33} \quad \mathrm{Re} > 2.5 \quad [13] \tag{20.9}$$

$$\mathrm{Sh} = 0.12\ \mathrm{Re}\ \mathrm{Sc}^{0.33} \quad \mathrm{Re} < 2.5 \quad [13] \tag{20.10}$$

$$\mathrm{Sh} = 1.38\ \mathrm{Re}^{0.34}\mathrm{Sc}^{0.33} \quad [16] \tag{20.11}$$

$1 < \mathrm{Re} < 25$; closely packed fibers

$$\mathrm{Sh} = 0.90\ \mathrm{Re}^{0.40}\mathrm{Sc}^{0.33} \quad [16] \tag{20.12}$$

$1 < \mathrm{Re} < 25$; loosely packed fibers

$$\mathrm{Sh} = \phi\mathrm{Re}^{1.2}\mathrm{Sc}^{0.33} \quad 0.03 < \mathrm{Re} < 0.3 \quad [17] \tag{20.13}$$

20.4 Applications

The variety of unit operations that can be performed by membrane contactors has led during last years to many research studies in which membrane contactors are applied to different fields.

In particular, liquid–liquid extractions, wastewater treatments, gas absorption and stripping, membrane, and osmotic distillation, are the processes more studied. For example, the VOCs removal, the extraction of aroma compounds and metal ions, the concentration of aqueous solutions, the acid-gases removal, the bubble-free oxygenation/ozonation, have been successfully carried out by using membrane contactors [1, 2].

The design of the first commercial modules has allowed the commercial application of membrane contactors for some specific operations. This is the case of the Membrana-Charlotte Company (USA) that developed the LiquiCel modules, equipped with polypropylene hollow fibers, for the water deoxygenation for the semiconductor industry. LiquiCel modules have been also applied to the bubble-free carbonation of Pepsi, in the bottling plant of West Virginia [18], and to the concentrations of fruit and vegetable juices in an osmotic distillation pilot plant at Melbourne [19]. Other commercial applications of LiquiCel are the dissolved-gases removal from water, the decarbonation and nitrogenation in breweries, and the ammonia removal from wastewater [20].

By using the rectangular transverse-flow module developed by TNO (The Netherlands), an industrial membrane gas absorption unit for ammonia recovery has been installed in The Netherlands [21]. Also, a unit for CO_2 removal has been tested [22].

Flat-membrane contactors have been specifically designed and commercialized by GVS SpA (Italy) for air dehumidification processes [23].

As previously reported, membrane contactors present interesting advantages with respect to traditional units. Moreover, they well respond to the main targets of the process intensification, such as to develop systems of production with lower equipment-size/production-capacity ratio, lower energy consumption, lower waste production, higher efficiency. In order to better identify the potentialities of membrane contactors in this logic, they have been recently compared to traditional devices for the sparkling-water production in terms of new defined indexes [24]. In particular, the comparison has been made at parity of plant capacity and quality of final product. The metrics used for the comparison between membranes and traditional units are:

(a) the productivity/size ratio (PS);
(b) the productivity/weight ratio (PW);
(c) the flexibility, as the ability to handle the changes in the operating conditions ($flexibility_1$);
(d) the flexibility, as the ability to cover different applications ($flexibility_2$);
(e) the modularity;
(f) the mass intensity ratio.

In Table 20.2 the results of the study are shown. From this, it results that when the process is carried out by membrane contactors there is a gain in terms of flexibility and modularity of the plant, and lower CO_2 consumptions and size and weight are achieved.

Another field where membrane contactors are studied in the logic of the process intensification is seawater and brackish-water desalination. The scarcity of

Table 20.2 Comparison between traditional and proposed system for sparkling-water production [24].

Traditional system (deareator + saturator)	Membrane contactor
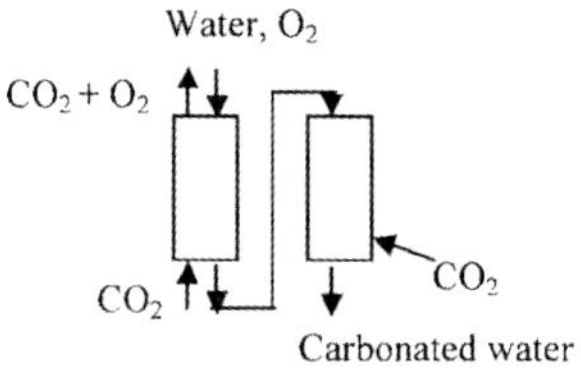 Water, O_2; $CO_2 + O_2$; CO_2; CO_2; Carbonated water	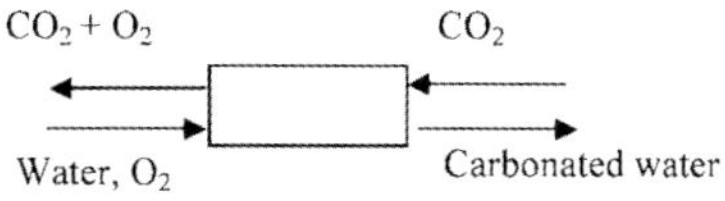 $CO_2 + O_2$; CO_2; Water, O_2; Carbonated water
PS	6.46
PW	2.33
$Flexibility_1$	1.39
$Flexibility_2$	3
Modularity	<1
Mass intensity ratio	0.85

fresh-water resources is today becoming a big issue worldwide and there are the several studies in progress for improving the existing desalination plants in terms of higher fresh-water recovery, lower use of chemicals, lower brine production, higher quality of the produced water [25–27]. In a desalination plant, membrane contactors can be introduced for different purposes:

(a) for controlling the oxygen/carbon dioxide content of the water;
(b) for disinfection/oxidation of the water;
(c) for reducing the boron content into the water, working on the reverse osmosis permeate as liquid–liquid device;
(d) for improving the fresh-water recovery factor of reverse osmosis, working as a membrane distillation unit;
(e) for producing crystals from brine, working as a membrane crystallizer.

Referring to point a, in desalination the content of oxygen and carbon dioxide in the water affects the material life of the plant (because of corrosion problems), as well as the pH and the conductivity of water. Usually, these gases are removed by stripping in a packed column and the final water pH is adjusted by means of caustic soda. By using membrane contactors, there is no need for chemicals, with a consequent reduction of the environmental impact.

Membrane contactors can be effectively used also for disinfection purposes (e.g., water ozonation) [28] or for the oxidation of species present into water, for example, arsenic. Although the content of arsenic in seawater is today within the accepted limits, it is foreseen that in the future its concentration could increase, due to the increase of pollution of rivers and groundwaters. Usually, arsenic is contained in water as As(III) and As(V) forms, in different amounts. All arsenic-removal technologies have a better performance when arsenic is present in the pentavalent form, therefore, membrane contactors can be used as an alternative system for oxidizing As(III) to As(V), by simply using air. From preliminary results obtained at the lab-scale, membrane contactors seem to be quite effective for achieving a complete oxidation [29].

The control of the boron content into water is another big issue because the existing reverse-osmosis membranes are not able to readily reject boron at natural pHs, thus several stages in series, operating at different pHs, coupled also to selective boron-exchange resins, are required for achieving the desired boron content [30]. Recently, an alternative desalination flowsheet where the permeate coming from the reverse-osmosis unit (usually containing 1 ppm of boron) is treated in a membrane-contactor device has been proposed and designed [31]. The boron removal occurs by simple diffusion across the hydrophilic microporous membranes from the feed stream to another water stream (stripping phase) that is recycled to the membrane contactor unit, after its purification in a BWRO device (see Figure 20.6). From a comparison made in terms of the new metrics, between the proposed scheme and that traditionally employed, at parity of fresh-water production and final boron content, it results that the reverse-osmosis-membrane contactor integrated system is advantageous in terms of size, energy consumption, use of chemicals, flexibility and modularity [31].

The fact that the performance of membrane distillation is not strongly affected by the concentration of the solution to be treated, allows to work with the membrane distillation unit on the reverse-osmosis concentrate, in order to further increase the

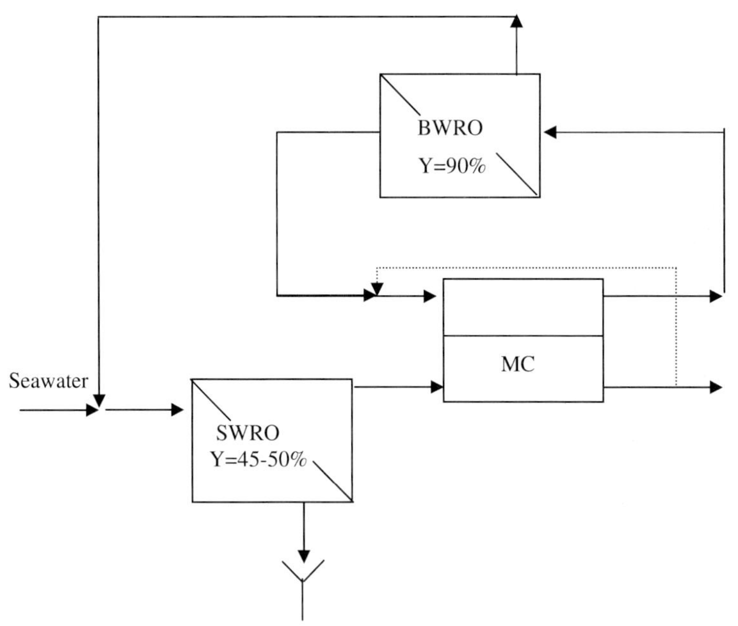

Figure 20.6 An integrated reverse-osmosis-membrane contactor system for boron removal (From Ref. [31].)

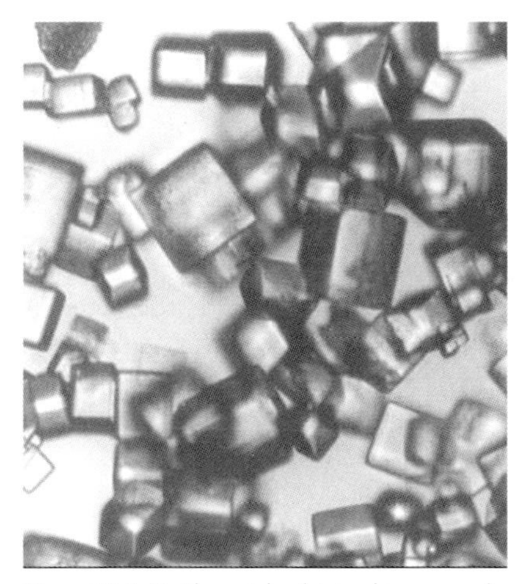

Figure 20.7 NaCl crystals obtained in a membrane crystallizer (magnification: ×10) (From Ref. [26].)

fresh-water recovery and to reduce the brine to be disposed. In particular, the membrane distillation can be used for obtaining supersaturated brines, to be further processed in membrane crystallizers for producing valuable crystals. In this way, the problem of the brine disposal is completely avoided, with a consequent reduction of the environmental impact. As example, Figures 20.7 and 20.8 show pictures of NaCl and $MgSO_4{\cdot}7H_2O$ crystals, respectively, obtained in a membrane crystallizer [26].

The membrane distillation units and membrane crystallizers will be also energy efficient if sources of energy such as solar are considered for providing the heat required for the water evaporation.

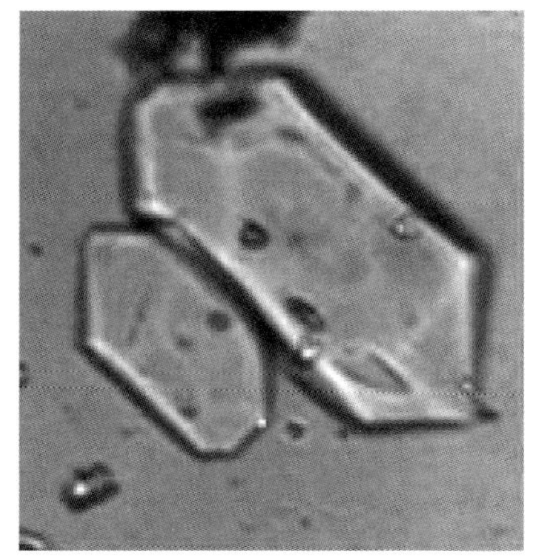

Figure 20.8 $MgSO_4{\cdot}7H_2O$ crystals obtained in a membrane crystallizer (magnification: ×20) (From Ref. [26].)

20.5 Concluding Remarks

The introduction of membrane contactors in industrial cycles might represent an interesting way to realize the rationalization of chemical productions in the logic of the process intensification. Membrane contactors are, in fact, highly efficient systems for carrying out the mass transfer between phases and achieving high removals. They also present lower size than conventional apparatus. Commercial applications are already present (e.g., the electronics industry or bubble-free carbonation lines), however, some critical points must be still overcome and several are the research efforts needed for their further implementation at industrial level, as summarized below:

(a) the membrane introduces another resistance to the mass transfer with respect to traditional systems: thin membranes have to be used;
(b) shell-side bypassing results in a loss of efficiency, especially for big contactors: design improvements are required;
(c) high hydrophobicity and methods for restoring it are required for successful applications;
(d) in order to reduce the heat loss through the membrane, materials with low thermal conductivity are required in membrane distillation operations;
(e) the operating pressures are limited by the breakthrough pressure value: membranes able to withstand high pressures will extend the range of membrane-contactor applications;
(f) more general expressions to calculate the shell-side mass-transfer coefficients are required in order to facilitate the scale-up.

References

1 Drioli, E., Criscuoli, A. and Curcio, E. (2006) *Membrane Contactors: Fundamentals, Applications and Potentialities*, Elsevier, Amsterdam, 0-444-52203-4.
2 Gabelman, A. and Hwang, S.T. (1999) *Journal of Membrane Science*, **159**, 61–106.
3 Li, K., Wang, D., Koe, C.C. and Teo, W.K. (1998) *Chemical Engineering Science*, **53** (6), 1111–1119.
4 Bhaumik, D., Majumdar, S. and Sirkar, K.K. (2000) *Journal of Membrane Science*, **167**, 107–122.
5 Majumdar, S., Bhaumik, D. and Sirkar, K.K. (2003) *Journal of Membrane Science*, **214**, 323–330.
6 Gugliuzza, A. and Drioli, E. (2007) *Journal of Membrane Science*, **300**, 51–62.
7 Ghogomu, J.N., Guigui, C., Rouch, J.C., Clifton, M.J. and Aptel, P. (2001) *Journal of Membrane Science*, **81**, 71–80.
8 Wang, K.L. and Cussler, E.L. (1993) *Journal of Membrane Science*, **85**, 265–278.
9 Seibert, A.F. and Fair, J.R. (1997) *Separation Science and Technology*, **32**, **1**, 573–583.
10 Wu, J. and Chen, V. (2000) *Journal of Membrane Science*, **172**, 59–74.
11 Lemanski, J. and Lipscomb, G.G. (2001) *Journal of Membrane Science*, **195**, 215–228.
12 Lemanski, J., Liu, B. and Lipscomb, G.G. (1999) *Journal of Membrane Science*, **153**, 33–43.
13 Wickramasinghe, S.R., Semmens, M.J. and Cussler, E.L. (1992) *Journal of Membrane Science*, **69**, 235–250.

14 Prasad, R. and Sirkar, K.K. (1998) *AICHE Journal*, **34** (2), 177–188.

15 Costello, M.J., Fane, A.G., Hogan, P.A. and Schofield, R.W. (1993) *Journal of Membrane Science*, **80**, 1–11.

16 Yang, M.-C. and Cussler, E.L. (1986) *AIChE Journal*, **32**, 1910–1915.

17 Criscuoli, A., Drioli, E. and Moretti, U. (2003) *Annals of the New York Academy of Sciences*, **984**, 1–16.

18 Peterson, P.A., Schneider, J. and Sengupta, A. (1998) Proceedings of the Workshop on 'Membrane Distillation, Osmotic Distillation and Membrane Contactors', Cetraro (CS) Italy, July 2–4.

19 Hogan, P.A., Canning, R.P., Peterson, P.A., Johnson, R.A. and Michaels, A.S. (1998) *Chemical Engineering Progress*, **94**, 49–61.

20 www.liqui-cel.com.

21 Klaassen, R., Feron, P.H.M. and Jansen, A.E. (2005) *Chemical Engineering Research & Design*, **83** (A3), 234–246.

22 Feron, P. and Jansen, A.E. (2002) *Separation and Purification Technology*, **27**, 231–242.

23 Gaeta, S.N. (2003) Proceedings of the 1st Workshop Italy–Russia, Cetraro (CS) Italy, September 17–20.

24 Criscuoli, A. and Drioli, E. (2007) *Industrial & Engineering Chemistry Research*, **46**, 2268–2271.

25 Drioli, E., Criscuoli, A. and Curcio, E. (2002) *Desalination*, **147**, 77–81.

26 Drioli, E., Curcio, E., Criscuoli, A. and Di Profio, G. (2004) *Journal of Membrane Science*, **239**, 27–38.

27 Drioli, E., Curcio, E., Di Profio, G., Macedonio, F. and Criscuoli, A. (2006) *Trans IChemE, Part A, Chem Eng Res and Des*, **84** (A3), 209–220.

28 Cissel, J., Gramer, M., Shanbhag, P.V. and Nemser, S.M. (March 5–9, 2000) Proceedings of the AIChE Spring National Meeting Atlanta, pp. 56–62.

29 Galizia, A.(May 2007) Master thesis on Treatment of water containing arsenic by means of membrane operations, University of Calabria (Italy).

30 Redondo, J., Busch, M. and De Witte, J.-P. (2003) *Desalination*, **156**, 229–238.

31 Criscuoli, A., Rossi, E., Cofone, F. and Drioli, E., (2009) accepted by *Clean Technologies and Environmental Policy*.

21
Membrane Emulsification: Principles and Applications

*Lidietta Giorno, Giorgio De Luca, Alberto Figoli,
Emma Piacentini, and Enrico Drioli*

21.1
Introduction

Emulsions and suspensions are colloidal dispersions of two or more immiscible phases in which one phase (disperse or internal phase) is dispersed as droplets or particles into another phase (continuous or dispersant phase). Therefore, various types of colloidal systems can be obtained. For example, oil/water and water/oil single emulsions can be prepared, as well as so-called multiple emulsions, which involve the preliminary emulsification of two phases (e.g., w/o or o/w), followed by secondary emulsification into a third phase leading to a three-phase mixture, such as w/o/w or o/w/o. Suspensions where a solid phase is dispersed into a liquid phase can also be obtained. In this case, solid particles can be (i) microspheres, for example, spherical particles composed of various natural and synthetic materials with diameters in the micrometer range: solid lipid microspheres, albumin microspheres, polymer microspheres; and (ii) capsules, for example, small, coated particles loaded with a solid, a liquid, a solid–liquid dispersion or solid–gas dispersion. Aerosols, where the internal phase is constituted by a solid or a liquid phase dispersed in air as a continuous phase, represent another type of colloidal system.

In emulsions and suspensions, disperse phase dimensions may vary from the molecular state to the coarse (visible) dispersion. They are commonly encountered in various productions. The average droplet/microcapsules size distribution is a key feature since they determine emulsions/suspensions properties for the intended uses and stability. For large-scale emulsion production, the most commonly employed methods are based on techniques aiming at establishing a turbulent regime in the fluid mixtures. These turbulent flows cannot be controlled or generated uniformly. The consequences are that the control of the droplet sizes is difficult and wide size distributions are commonly obtained, therefore the energy is used inefficiently in these technologies. In addition, the process scale-up is extremely difficult. The use of the ultrasonic bath yields better results with respects to the

Membrane Operations. Innovative Separations and Transformations. Edited by Enrico Drioli and Lidietta Giorno

ISBN: 978-3-527-32038-7

mentioned procedures, however, the control of the droplet dimension is still not optimal.

For these reasons, recently much attention has been put in alternative emulsification processes, such as the membrane emulsification (ME).

Membrane emulsification is an appropriate technology for production of single and multiple emulsions and suspension. It was proposed for the first time at the 1988 Autumn Conference of the Society of Chemical Engineering, Japan. Since then, the method has continued to attract attention in particular in Japan, but also in Europe [1–10].

In the early 1990s, Nakashima *et al.* [2] introduced membrane technology in emulsions preparation by a direct emulsification method, whereas, in the late 1990s, Suzuki *et al.* used premix membrane emulsification to obtain production rates higher than other membrane emulsification methods [11].

The fast progress in microengineering and semiconductor technology led at the development of microchannels, that Nakajima *et al.* applied in emulsification technology [12].

The distinguishing feature of membrane emulsification technique is that droplet size is controlled primarily by the choice of the membrane, its microchannel structure and few process parameters, which can be used to tune droplets and emulsion properties. Comparing to the conventional emulsification processes, the membrane emulsification permits a better control of droplet-size distribution to be obtained, low energy, and materials consumption, modular and easy scale-up. Nevertheless, productivity (m^3/day) is much lower, and therefore the challenge in the future is the development of new membranes and modules to keep the known advantages and maximize productivity.

Considerable progress has been achieved in understanding the technology from the experimental point of view, with the establishment of many empirical correlations. On the other hand, their theoretical interpretation by means of reliable models is not accordingly advanced. The first model devoted to membrane emulsification, based on a torque balance, was proposed in 1998 by Peng and Williams [13], that is, ten years later the first experimental work was published, and still nowadays, a theoretical study aiming at a specific description of the premix membrane emulsification process is not available.

The nonsynergistic progress of the theoretical understanding with the experimental achievements, did not refrain the technology application at the productive scale. In particular, membrane emulsification was successfully applied for preparation of emulsions and capsules having a high degree of droplet-size uniformity, obtained with low mechanical stress input [14–16]. Therefore, the application of membrane emulsification extended to various fields, such as drug delivery, biomedicine, food, cosmetics, plastics, chemistry, and some of these applications are now being developed at the commercial level. Their scale vary from large plants in the food industry, to medium-scale use in the polymer industry, and to laboratory-bench scale in biomedicine.

In this chapter, the experimental and theoretical bases as well as the applications of the technology will be discussed.

21.2 Membrane Emulsification Basic Concepts

Emulsions and suspensions are key systems for advanced formulations in various industrial sectors. Membrane emulsification is a relatively new technology in which membranes are not used as selective barriers to separate substances but as microstructures to form droplets with regular dimensions, that is, uniform or controlled droplet-size distribution (Figure 21.1). Membrane emulsifications can be generally distinguished in (Figure 21.2): (i) direct membrane emulsification (DME), in which the disperse phase is directly fed through the membrane pores to obtain the droplets, and (ii) premix membrane emulsification, in which a coarse premixed emulsion is pressed through the membrane pores to reduce and to control the droplet sizes.

In general, in the direct membrane emulsification, the disperse phase is pressed through a microporous membrane and droplets are formed at the opening of the pore on the other side of the membrane, which is in contact with the continuous phase. Here, droplets that reach a critical dimension can detach either for *spontaneous deformation* or are *sheared by the continuous phase* flowing parallel to the surface. In the former case, the driving force for the droplet formation is the surface free-energy minimization, that is, the droplet is formed by spontaneous deformation tending to form a sphere. For example, in quiescent conditions the droplets are formed by means of this mechanism. In the latter case, the shearing stress generated by the continuous phase is the driving force of the droplet detachment. For example, in the crossflow membrane emulsification (CDME) and stirred membrane emulsification droplets are formed by this mechanism.

In the premix emulsification the basic mechanism for the droplet formation is different from the direct emulsification. In fact, in this case the predominant formation mechanism is the droplet disruption within the pore.

Both direct and premix emulsification can be obtained with a continuous phase flowing along the membrane surface (i.e., crossflow, stirring) (Figure 21.2(b)). However, it is important to distinguish between the droplet-formation mechanism and the macroscopic operation procedure. In other terms, often, in the literature, the

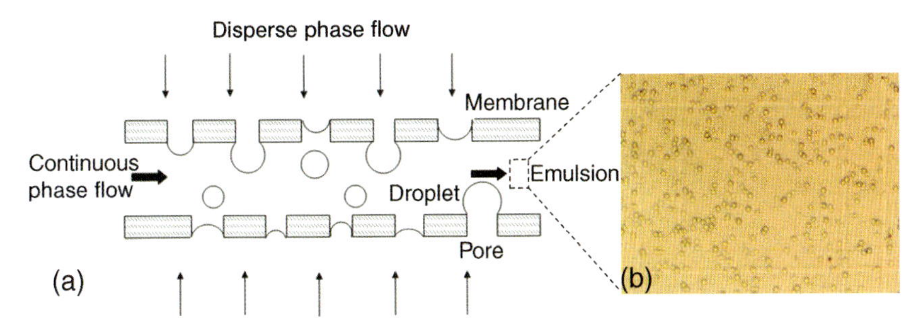

Figure 21.1 Schematic representation (a) of membrane emulsification, where the membrane works as a high-throughput device to form droplets with regular dimensions; (b) photo of an o/w emulsion

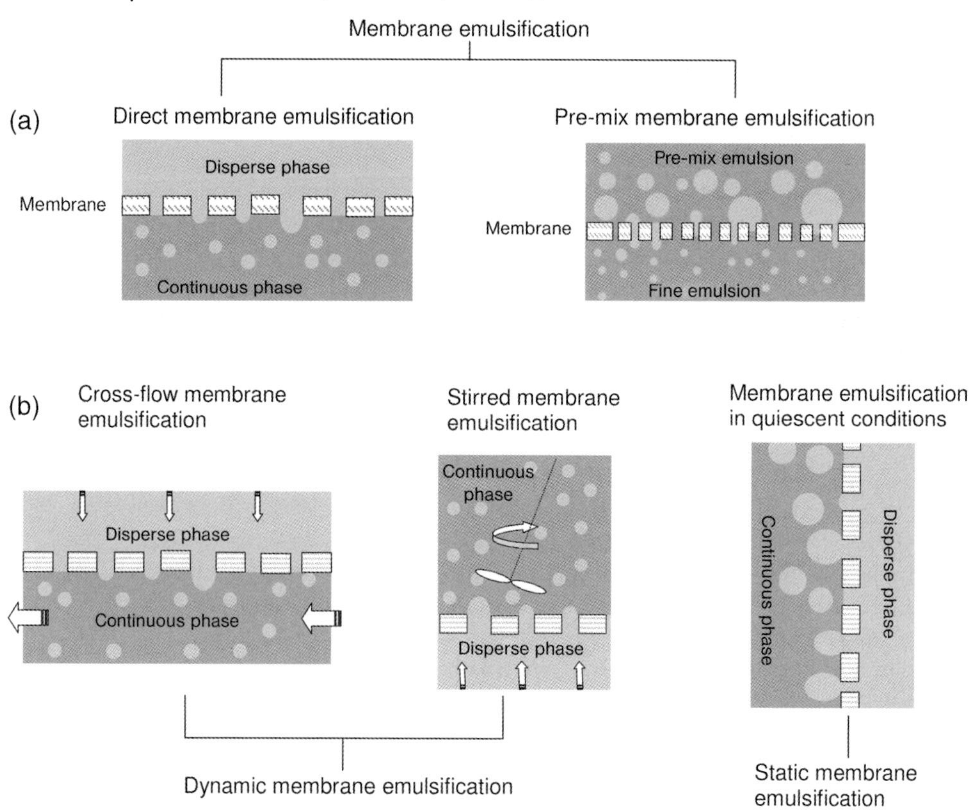

Figure 21.2 Schematic drawing of membrane emulsification: (a) mechanisms (b) operation procedures.

'crossflow' term is used to indicate that the continuous phase is flowing along the surface, but this does not guarantee that the shear stress is the driving force for the droplet detachment, as long as the appropriate conditions are not verified.

The membrane emulsification can be considered as a case of microdevice emulsification process [17, 18] in which the porous membrane is used as micro-devices. Membrane emulsification carried out in quiescent conditions is also referred to as *static membrane emulsification,* while membrane emulsification carried our in moving conditions (either the membrane, i.e., rotating module, or the phase, i.e., crossflow) is also referred to as *dynamic membrane emulsification* (Figure 21.2(b)).

A peculiar advantage of membrane emulsification is that both droplet sizes and size distributions may be carefully and easily controlled by choosing suitable membranes and focusing on some fundamental process parameters reported below. Membrane emulsification is also an efficient process, since the energy-density requirement (energy input per cubic meter of emulsion produced, in the range of 10^4–10^6 J m^{-3}) is low with respect to other conventional mechanical methods (10^6–10^8 J m^{-3}), especially for emulsions with droplet diameters smaller than 1 μm [1]. The lower energy density requirement also improves the quality and functionality of labile emulsion ingredients, such as bioactive molecules. In fact, in conventional emulsification methods, the high shear rates and the resulting increase of the process temperature have negative effects on shear- or temperature-sensitive components. The shear stresses calculated for a membrane system are much less and it is possible to process shear-sensitive ingredients.

The droplet size, its dispersion and the droplet-formation time depend on several parameters: (i) *membrane parameters*, such as pore-size distribution, pore-border morphology, number of active pores, porosity, wetting property of the membrane surface, (ii) *operating parameters*, such as crossflow velocity (i.e., wall shear stress), transmembrane pressure and disperse-phase flow, temperature, as well as the membrane module used (tubular, flat, spiral-wound); and (iii) *phase parameters*, such as dynamic interfacial tension, viscosity and density of processed phases, emulsifier types, and concentration. Such quantities combine with different magnitudes, over the ranges of operating conditions, and many of them exhibit coupling effects 4. Moreover, the production of monodisperse emulsions is essentially related to the size distribution of membrane pores and their relative spatial distribution on the membrane surface. It is worth noting that the geometry of the module in which the membrane is located is also an important parameter since it determines in conjugation to the crossflow velocity, the wall shear stress (Figure 21.3).

Droplet-size distribution and disperse-phase percentage determine the emulsion properties characterizing the final formulation for an intended use.

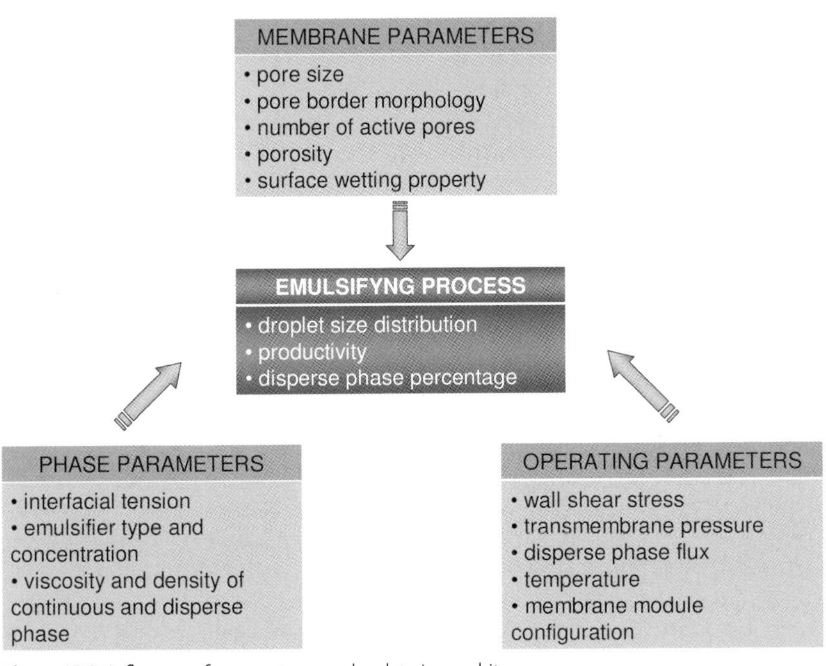

Figure 21.3 Influence of parameters on droplet size and its formation during an emulsification process.

21.3 Experimental Bases of Membrane Emulsification

In this section, an analysis of the experimental observations and empirical correlations related to membrane emulsification processes will be illustrated. The theoretical bases that support these results and predict membrane emulsification performance will be discussed in the next section.

As previously anticipated, the appropriate choice of the membrane dictates the droplet properties. Membranes employed in emulsification processes are mainly of inorganic type (ceramic, glassy, metallic), but some examples of polymeric membranes have also been applied. Tables 21.1 and 21.2 summarize some of the most common membranes used in direct and premix membrane emulsification, respectively. Most of them have been originally developed for other membrane processes, such as microfiltration, and adapted in the emulsification technology. Nowadays, the growing interest towards membrane emulsification is also promoting research efforts in the design and development of membranes specifically devoted to membrane emulsification. Shirasu porous glassy (SPG) membranes were among the first membranes specifically developed for emulsion preparation. SPG membranes are characterized by interconnected micropores, a wide spectrum of available mean pore size (0.1–20 μm) and high porosity (50–60%). Micropore metallic membranes, developed by Micropore Technologies (United Kingdom), are characterized by cylindrical pores, uniform and in a regular array with a significant distance between each pore. They are available with pore diameters in the range of 5–20 μm and exhibit very narrow pore-size distribution (Figure 21.4).

Membrane-wetting properties may be carefully considered in the membrane selection. In general, the membrane surface where the droplet is formed should not be wetted by the disperse phase. Therefore, a w/o emulsion is prepared using a hydrophobic membrane and an o/w emulsion is prepared using a hydrophilic membrane. On the other hand, w/o and o/w emulsions were successfully prepared using pretreated hydrophilic and hydrophobic membranes, respectively. The pretreatment basically consisted in absorbing the continuous phase on the membrane surface so that to render the membrane nonwetted by the disperse phase [14, 23, 25]. The presence of emulsifier in the disperse phase represents another strategy that permits the preparation of emulsions with a membrane wetted by the disperse phase.

The dispersion of droplet diameter mainly depends upon the membrane pore. In general, a linear relationship between membrane pore diameter (D_p) and droplet diameter (D_d) has been observed, especially for membranes with pore diameters larger than 0.1 micrometer. In these cases, linear coefficients varying between 2–10, depending on the operating conditions and emulsion composition, have been obtained [3, 23, 27]. Figure 21.5 summarizes the behavior of the mentioned relationships for different emulsion systems. In general, for a certain emulsion type and in comparable operating conditions, the lower the pore size the lower the droplet size.

Fluid-dynamic operating conditions, such as axial or angular velocity (i.e., shear stress that determines drag force value) and transmembrane pressure (that determines disperse-phase flux, for a given disperse-phase viscosity and membrane

Table 21.1 List of most common membranes used in direct membrane emulsification.

Membrane material/wetting property	Membrane configuration	Pore geometry/porosity (ε)	Pore diameter (μm)	Membrane producer	Application	Reference
Porous glass membrane						
SPG/hydrophilic or hydrophobic	Tubular or disk	Tortuous, interconnected cylindrical pore/50% < ε < 60%	0.1–20	SPG Technology Co., Ltd. (Japan)	o/w, w/o emulsions	[16, 19–21]
MPG/hydrophilic	Tubular or disk	Cylindrical pore/50% < ε < 60%	0.2–1.36; 10.2–16.2	ISE Chemical Industries Co. Ltd (Japan)	o/w and w/o emulsions	[21–23]
Silica glass/hydrophobic	Tubular	Cylindrical pore/ε = 61%	0.6	Lab-made	w/o emulsions	[24]
Ceramic membrane						
Mullite ceramic/hydrophilic	Disk	—	0.68	Lab-made	w/o emulsions	[25]
Alumina/hydrophilic	Tubular	ε = 35%	0.1–0.8	Westfalia Separator Membraflow (Germany)	o/w, w/o/w emulsions	[5, 26–28]
			0.5–0.2	Société des Céramiques Techniques (France)	o/w emulsions	
			0.5–0.8	Pall-Exekia (France)		
Zirconia/hydrophilic	Tubular	ε = 60%	0.1	Pall-Exekia (France)	o/w emulsions	[27, 28]
				Société des Céramiques Techniques (France)		
Polymeric membrane						
Polyamide/hydrophilic	Hollow fiber	—	10; 50 kDa (NMWCO)[a]	Forschstung Institut Berghof, (Germany)	o/w emulsions	[29]

(Continued)

Table 21.1 (*Continued*)

Membrane material/wetting property	Membrane configuration	Pore geometry/porosity (ε)	Pore diameter (μm)	Membrane producer	Application	Reference
PTFE/hydrophobic	Disk	$\varepsilon = 79\%$	0.5–50	Japan Goretex Co. (Japan)		[30]
Polycarbonate/hydrophilic	Disk	$5\% < \varepsilon < 20\%$	10	ISOPORE, Nihon Millipore Co. (Japan)		[31]
Cellulose acetate/hydrophilic	Disk	—	0.2–03	Advantec Toyo (Japan)	w/o/w emulsions	[32]
Polypropylene/hydrophobic	Hollow fiber	—	0.4	Wuppertal (Germany)	w/o emulsions	[14]

[a] Nominal molecular weight cutoff.

Table 21.2 List of most common membranes used in premix membrane emulsification.

Membrane material	Membrane configuration	Operation mode	Pore diameter (μm)	Application	Membrane producer	References
Porous glass membrane						
SPG	Tubular	Crossflow	2.7 and 4.2	o/w emulsions	SPG Technology Co., Ltd. (Japan)	[11]
		Deadend, multipass	10.7	w/o/w emulsions		[15]
Ceramic membrane						
Alumina/hydrophilic	Tubular	Deadend, multipass	3.2, 4, and 11	w/o/w emulsions	Lab-made	[33]
Polymeric membrane						
Polycarbonate	flat	Deadend, multipass	0.6, 0.8, and 3.0	o/w emulsions	Millipore corporation (United State)	[34]
Cellulose acetate	flat	Deadend	0.2, 0.45, 0.8, and 3.0	w/o/w emulsions	Advantec Toyo (Japan)	[35]
Polyamide	flat	Deadend	0.8	o/w emulsions	Whatman Intl. Ltd., Maidstone (England)	[36]

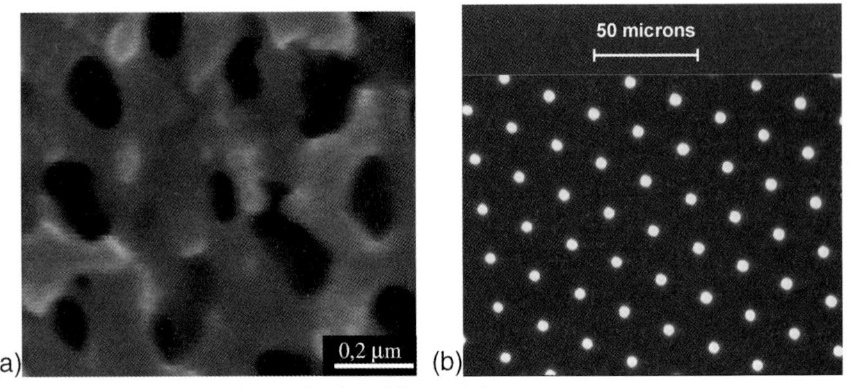

Figure 21.4 Porous membranes developed for emulsification processes. (a) Shirasu porous glassy membrane (from SPG Technology Co., LTD, Japan), (b) metallic membrane (From Micropore Technologies, United Kingdom).

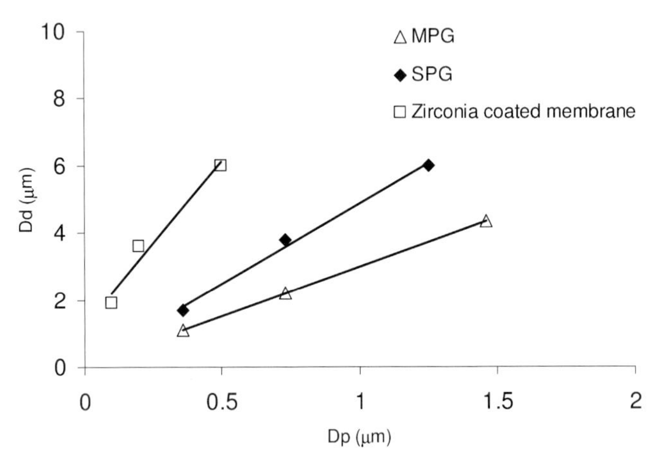

Figure 21.5 Relationship between membrane pore diameter (D_p) and droplet diameter (D_d) (Data extrapolated from Refs. [3, 23, 27]).

properties), can be properly adapted to tune emulsion properties. The commonly observed behavior of shear stress and disperse-phase flux on D_d/D_p ratio is depicted in Figures 21.6 and 21.7, respectively. The droplet size decreases with increasing shear stress at the membrane surface and decreasing of the disperse-phase flux. However, the latter influence is less predominant and depends on the droplet-formation time, which in turn is strongly affected by the interfacial dynamical tension. If the droplet-formation time is larger than the complete adsorption of the emulsifier (equilibrium interfacial tension) the lower the influence of the disperse-phase flux. Therefore, in appropriate conditions and for emulsions with droplet size above a micrometer (1–50 micrometer, so-called macroemulsions), transmembrane

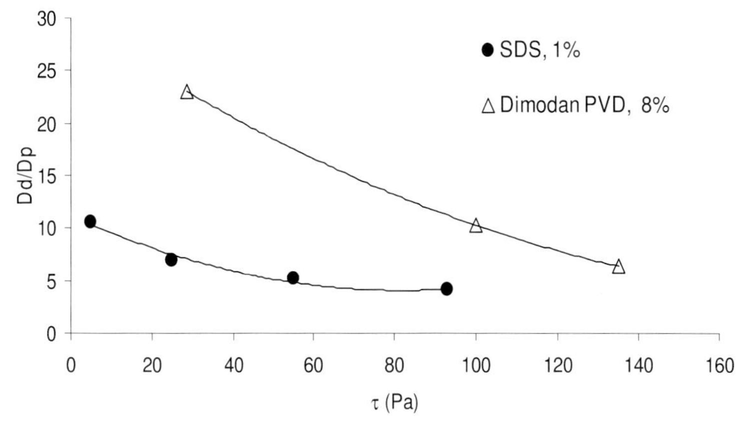

Figure 21.6 Relationship between wall shear stress (τ) and D_d/D_p (Data extrapolated from Refs. [27, 28]).

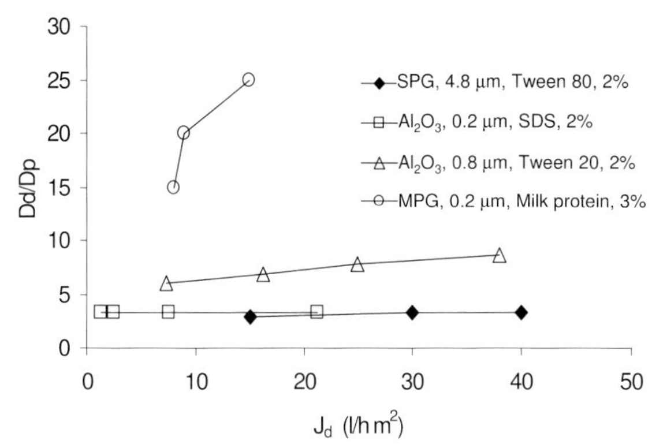

Figure 21.7 Relationship between dispersed phase flux (J_d) and D_d/D_p (Data extrapolated from Refs. [5, 16, 22, 26]).

pressure may influence the disperse-phase flux, but have little influence on changing the droplet size.

Dynamic interfacial tension, therefore the emulsifier used, and related adsorption kinetics influence the emulsification process. In general, the faster an emulsifier adsorbs to the newly formed interface, the lower the interfacial tension the smaller the droplet produced. Figure 21.8 shows a linear behavior between the D_d/D_p ratio and interfacial tension.

The axial velocity affects the droplet size by both influencing the surfactant mass transfer to the newly formed interface (that speeds up the reduction of the interfacial tension) and the drag force (that pulls droplets away from the pore mouth).

When production of submicrometer droplet size is aimed at, the continuous-phase shear stress and disperse-phase flux have to match the need for small droplet (i.e., high shear stress and low disperse-phase flux) with the need for a reliable system productivity (i.e., high disperse-phase flux).

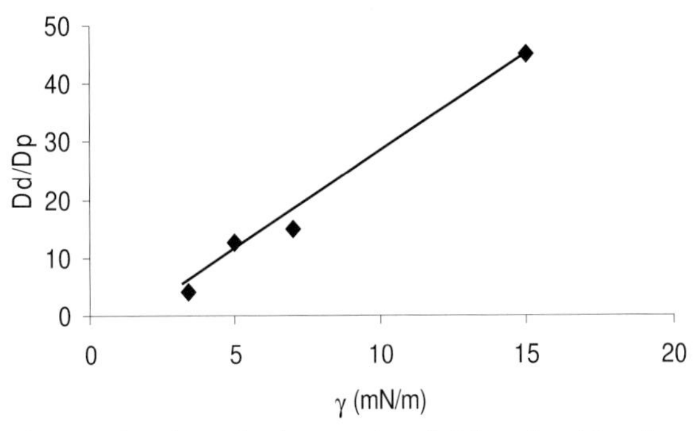

Figure 21.8 Relationship between interfacial tension (γ) and D_d/D_p (Data extrapolated from Ref. [26]).

The physical chemical properties of the phases can influence droplet formation as well as their stability in the bulk. For example, the viscosity of the continuous phase influences both the shear stress at the membrane wall and the adsorption kinetics of the emulsifier.

Concerning thermodynamically unstable emulsions, the creation of new interfaces from the disruption of the disperse phase increases the free energy of the system, which tends to return to the original two separate systems. Therefore, the use of emulsifier is necessary not only to reduce the interracial tension, but also to avoid the coalescence and the formation of macroaggregates thanks to electrostatic repulsion between adsorbed emulsifier.

21.3.1
Post-Emulsification Steps for Microcapsules Production

In this paragraph, a description of postemulsification steps needed to complete the preparation of microcapsules is reported. Microencapsulation can be described as the formation of small, coated particles loaded with a solid, a liquid, a solid–liquid dispersion, gas or solid–gas dispersion (Figure 21.9). The concept of microencapsulation originated in the 1950s and provided the means by which ink formulations used in carbonless copy paper are packaged. This application has been most successful and has led to the development of other applications like the production of microcapsules for thermal printing, optical recording, photocopy toners,

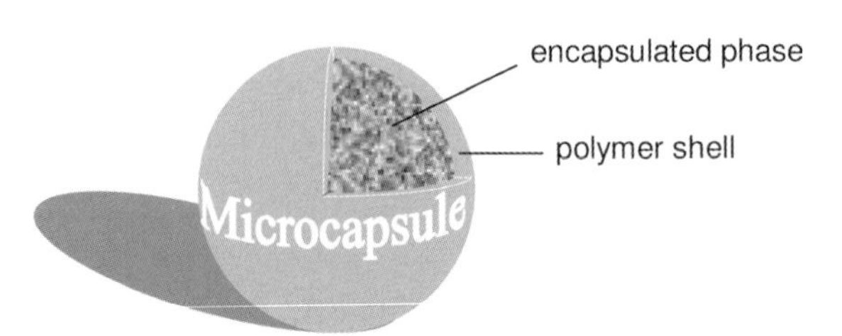

Figure 21.9 Schematic drawing of a microcapsule.

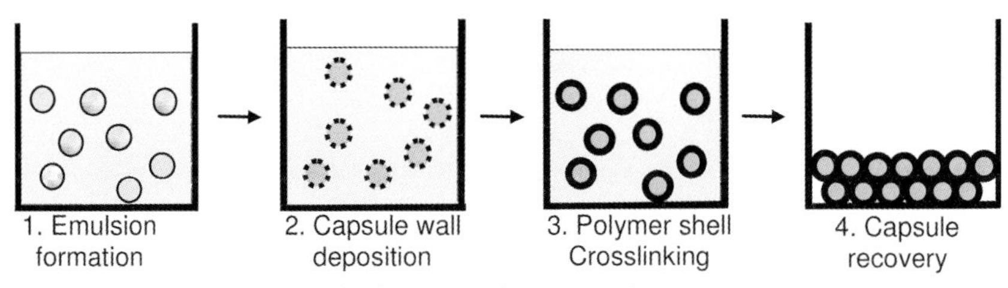

Figure 21.10 Steps involved in the formation of microcapsules.

diazo copying, herbicides, animal repellents, pesticides, oral and injectable pharmaceuticals, cosmetics, food ingredients, adhesives, curing agents, and live-cell encapsulation [38].

The size of these capsules may range from 100 nm to about 1 mm. Therefore, they can be classified as nano-, micro- and macrocapsules, depending on their size. The first commercial microcapsules were made by Green with a process called complex coacervation [37]. Since then, other methods for preparing microcapsules have been developed of which some are based exclusively on physical phenomena. Some utilize polymerization reactions to produce a capsule shell. Others combine physical and chemical phenomena. But they all have three main steps in common. The steps of the microencapulation preparation are schematically depicted in Figure 21.10. In the first step, a dispersion or emulsion has to be formed, followed by deposition of the material that forms the capsule wall (Figure 21.10, step 2). After solidification or crosslinking (step 3) of the droplets prepared, the capsules are isolated in the last step.

One of the major problems related to the capsule formation is capsule agglomeration. It involves the irreversible or largely irreversible sticking together of microcapsules that can occur during the encapsulation process and/or during the isolation step.

The microencapsulation process can be classified into two main categories (as defined by Thies [38] and reported in Table 21.3): (a) chemical process and (b) mechanical process.

Table 21.3 List of encapsulation processes (After [38]).

Chemical process	Mechanical process
Complex coacervation	Spray drying
Polymer/polymer incompatibility	Spray chilling
Interfacial polymerization in liquid media	Fluidized bed
In situ polymerization	Electrostatic deposition
In-liquid drying	Centrifugal extrusion
Thermal and ionic gelation in liquid media	Spinning disc at liquid/gas or solid gas interface
Desolvation in liquid media	Pressure extrusion or spraying into solvent extraction bath

Table 21.4 Commercial encapsulation processes and obtained capsule size (After [38]).

Process	Usual capsule size (μm)
Spray drying	5–5000
In-liquid drying or solvent evaporation	<1–1000
Polymer phase separation (coacervation)	20–1000
Rotational suspension separation	>50
Fluidized bed (Wurster)	<100

Capsules produced by a chemical process are formed entirely in a liquid-filled stirred tank or tubular reactor. Mechanical processes use a gas phase at some stage of the encapsulation process.

In Table 21.4 the typical size of capsules produced is identified by a number of processes that have been commercialized.

21.3.2 Membrane Emulsification Devices

The various membrane emulsification procedure can be practised by using appropriate membranes and devices configuration.

The crossflow membrane emulsification can be obtained either with tubular or flat-sheet membranes, which are fixed in appropriate housing modules connected to circuits controlling fluid-dynamic conditions. A schematic drawing of a crossflow plant is reported in Figure 21.11. The figure also illustrates the tubular and flat-sheet membranes and modules. SPG (Japan) and Micropore (UK) were among the first companies producing plants for crossflow membrane emulsification. Figure 21.12 shows pictures of common marketed equipments.

Emulsification devices where the membrane is immersed in a stirred vessel containing the continuous phase, so as to obtain a batch emulsification device operating in deadend emulsification mode, have also been developed (Figure 21.13). Both flat-sheet and tubular membranes are used. In this membrane emulsification device, the continuous phase kept in motion creates the shear stress at the membrane surface that detaches the forming droplets. In a different operation mode, that is, when the continuous phase is not stirred, droplet formation in quiescent conditions is obtained.

Rotating membrane emulsification is another type of batch emulsification. In this case a tubular membrane immersed in a continuous phase vessel is rotating itself and its angular velocity creates the shear stress at the membrane surface (Figure 21.14).

Both crossflow and deadend systems can be used in premix and direct membrane emulsification. In the crossflow premix system the coarse emulsion is diluted by permeation into pure continuous phase/diluted emulsion recirculating at the low-pressure side of the membrane. In the deadend system the fine emulsion is withdrawn as a product after passing through the membrane, without any recirculation and/or dilution with the continuous phase. In this process, the fine emulsion can

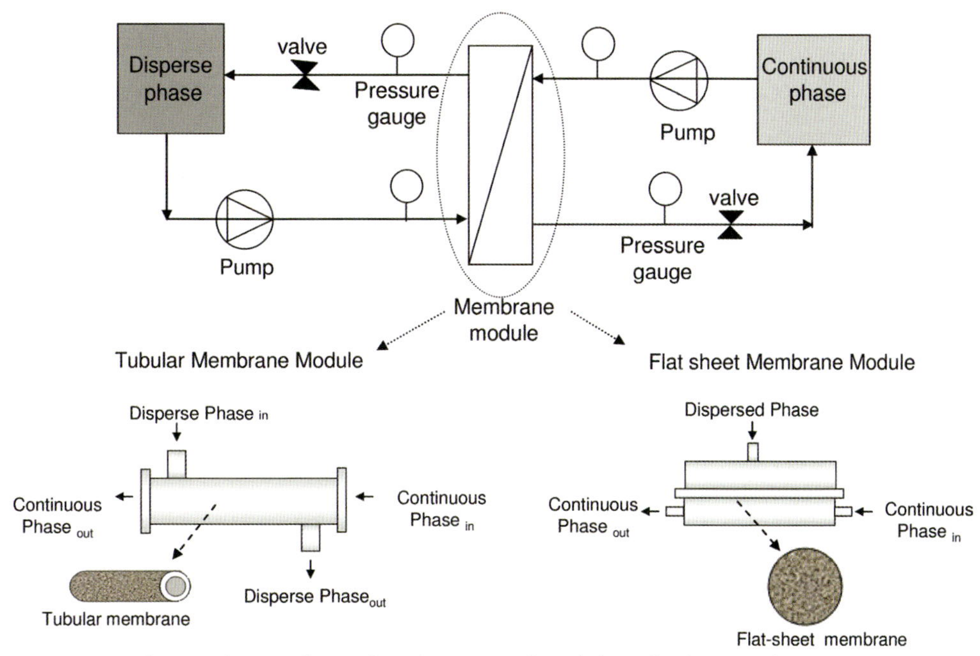

Figure 21.11 Schematic drawing of a crossflow plant, using either tubular or flat-sheet membranes.

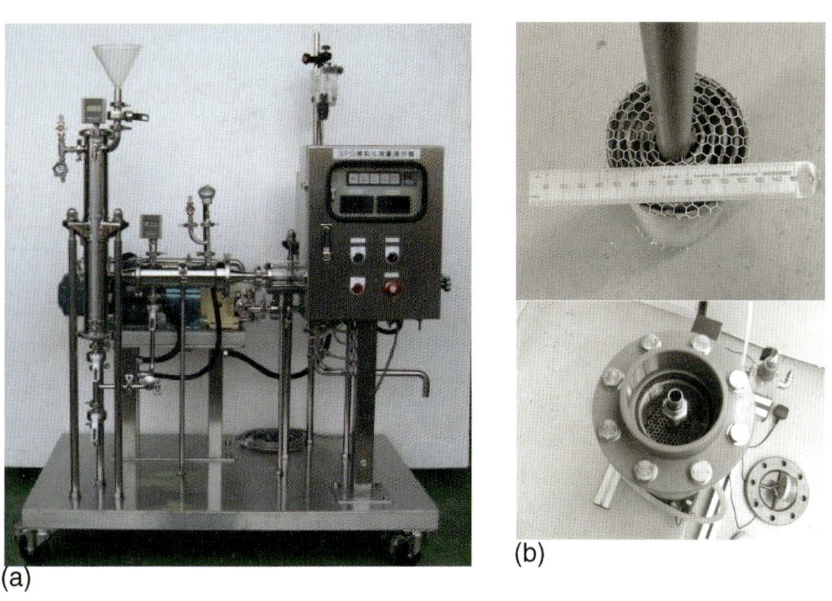

Figure 21.12 Marketed equipments for membrane emulsification. (a) Plant for crossflow membrane emulsification produced by SPG Technologies Co. Ltd (http://www.spg-techno.co.jp/); (b) spiral-wound metallic membrane module produced by Micropore Technologies (http://www.micropore.co.uk/).

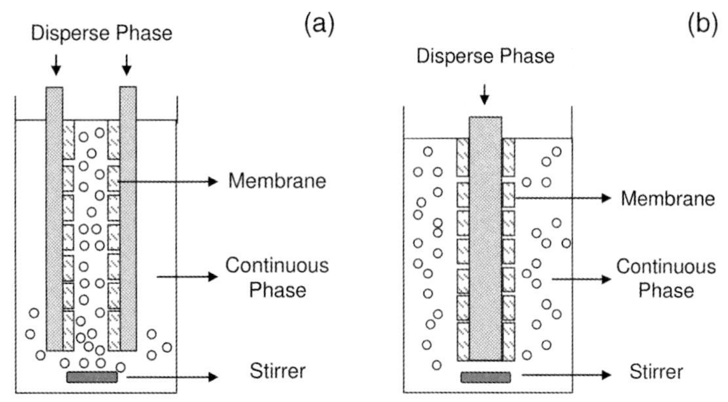

Figure 21.13 Emulsification devices where the membrane is immersed in a stirred vessel containing the continuous phase. Transmembrane pressure applied from (a) external or shell side, and (b) internal or lumen side.

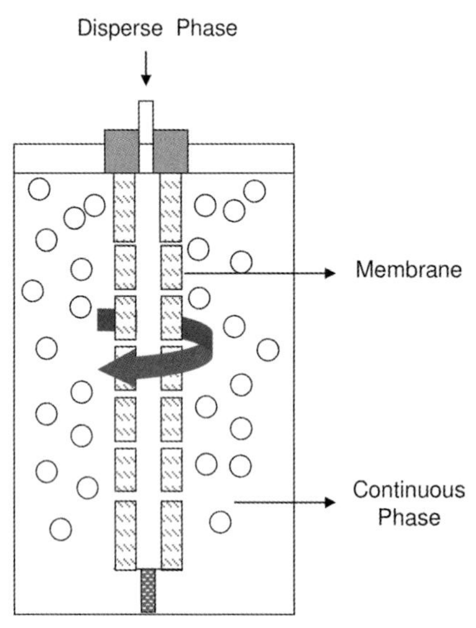

Figure 21.14 Rotating emulsification device.

be repeatedly passed through the same membrane a number of times to achieve additional droplet-size reduction and enhance size uniformity (multipass premix membrane emulsification).

Each type of device has specific advantages and disadvantages. The batch emulsification is suitable for laboratory-scale investigations. The construction of the device is simple and handling during emulsification as well as for cleaning. Crossflow membrane emulsification is used when it is important that a proper adjustment of all process parameters and larger amounts of emulsion have to be produced.

A potential disadvantage of crossflow direct membrane emulsification is the relatively low maximum disperse-phase flux through the membrane (0.01–0.1 $m^3/m^2 h$). Membrane, fluid properties, and transmembrane pressure determine the disperse-phase flux through the membrane. The opportune choice of membrane properties permits control of the flux during membrane emulsification process to be obtained. Due to the low productivity, that is, long production time, crossflow direct membrane emulsification is more suitable for the preparation of relatively diluted emulsions with disperse phase content up to 30%. Nevertheless, this process enables very narrow droplet-size distribution to be produced over a wide range of mean droplet size. Crossflow premix membrane emulsification holds several advantages over crossflow direct membrane emulsification. In fact, disperse-phase fluxes of the former emulsification process are typically above 1 $m^3/m^2 h$, which is one to two orders of magnitude higher than the latter. In addition, the mean droplet sizes that can be achieved using the same membrane and phase compositions are smaller. Also, the experimental apparatus is generally simpler and the process is easier to control and operate since the driving pressure and emulsifier properties are not so critical for the successful operation as in crossflow direct membrane emulsification. One of the disadvantages of premix membrane emulsification is a higher droplet polydispersity.

21.4 Theoretical Bases of Membrane Emulsification

From the theoretical point of view the key problem of the membrane emulsification is to explain and predict the dependence of the mean droplet diameter, D_d on the aforementioned membrane emulsification parameters. Important quantities such as droplet-formation time can thus be successively predicted by the mean droplet diameter and disperse-phase flux.

Droplet formation during direct membrane emulsification and in particular in crossflow emulsification has been described using models different in the scale and in the considered mathematical and physical phenomena, such as:

(a) balance equations involving global forces,
(b) surface free-energy minimization,
(c) microscopic modeling using computational fluid dynamics (CFD) and lattice Boltzmann approaches.

The global balance models are less accurate than the other methods, however, they are easier to handle and more instructive. The latter feature is crucial to acquire the necessary understanding of the physical causes at the basis of the droplet formation and detachment. The balance methods are versatile and permit analysis of the influence of many membrane emulsification parameters with limited computational time, useful in process optimizations. Starting from these considerations, in this section more attention will be paid to the proposed torque and force balances.

The balances approaches have to necessarily incorporate approximations and fundamental hypotheses, which reduce the prediction capability of the latter. Every hypothesis comes from a postulated droplet-formation mechanism. The formation mechanism, however, depends significantly on the mentioned *operating, membrane and phase parameters*, thus, it is very difficult to find one mechanism valid for all possible parameters values. Consequently, more accurate computation procedures, such as the microscopic modeling or methods using the minimization of the droplet surface, are necessary for the detailed description of droplet formation and accurate predictions.

21.4.1 Torque and Force Balances

This section breaks down as follows: first, the macroscopic forces acting on the droplet growing at the pore opening will be discussed, then the balance equations where these forces are involved will be dealt with. However, the accurate derivation of these forces is not reported here because it is beyond the aim of this contribution and therefore only the expressions of the forces used in balance equations will be presented.

The forces acting on a droplet attached to the pore opening can be conveniently subdivided into perpendicular and parallel direction with respect to the membrane surface. Considering the former case, the Young–Laplace F_{YL} [39] (named also static pressure force), the dynamic lift F_{DL} and buoyancy F_{BG} forces [26] are generally taken into account. They are defined as:

$$F_{YL} = \frac{\gamma}{D_d} \pi D_p^2 \tag{21.1}$$

$$F_{DL} = 0.761 \frac{\tau_{c,s}^{1.5} \rho_c^{0.5}}{\mu_c} D_d^3 \tag{21.2}$$

$$F_{BG} = \frac{1}{6} \pi g \Delta\rho D_d^3 \tag{21.3}$$

where D_p and D_d correspond to the average membrane pore and droplet diameter, respectively, γ is the liquid–liquid interfacial tension, while $\tau_{c,s}$, ρ_c and μ_c represent the shear stress, density and viscosity of the continuous phase, respectively. The quantity $\Delta\rho$ in Equation 21.3 represents the difference between the continuous- and disperse-phase densities. However, various authors [13, 26] showed that for small pores (e.g., smaller than 2 μm), the F_{DL} and F_{BG} are negligible with respect the F_{YL}. The inertial force defined by the following equation 26:

$$F_I = \int_{A_p} \rho_d v_m^2 \, dA = \rho_d A_p v_m^2 \tag{21.4}$$

caused by the disperse-phase flow, with mean velocity v_m, would be another perpendicular force to consider. Here, A_p is the pore surface. Concerning the nature of this force, recently it has been emphasized [40] that it has a predominantly viscous character rather than inertial. Starting from this observation, a more accurate expression of this hydrodynamic force has been determined. This force is explicitly dependent on the mean velocity of the disperse phase as well as of the pore diameter. The mean velocity of the disperse phase in turn depends on the effective pressure ΔP_{eff}. Neglecting the pressure drop due to the membrane pore length, ΔP_{eff} is equal to the difference between the transmembrane pressure and pressure drop necessary to overcome the capillary effect, that is the Laplace pressure [39]. To ensure monodisperse droplets and to avoid jets of the disperse-phase flux, the transmembrane pressure should never be markedly higher than the Laplace critical pressure. In these conditions, the ΔP_{eff} produces a small mean velocity and a negligible inertial force or hydrodynamic force with respect to F_{YL} and the drag force. The general expression used to consider the drag force F_{DR} [13, 26, 39] due to the continuous-phase crossflow and parallel to membrane surface is the following:

$$F_{DR} = \frac{3}{2} k_x \pi \tau_{c,s} D_d^2 \qquad (21.5)$$

where the parameter k_x is equal to 1.7 and takes into account the wall correction factor for a single sphere touching an impermeable wall [41]. In Equation 21.5, the approximation $v\mu_c \approx (1/2)\ \tau_{c,s} D_d$ is adopted, where v is the undisturbed crossflow velocity. The shear stress, evaluated at the droplet center, is assumed equal to that at the membrane surface, which is the wall shear stress. Referring to the expression 21.5, two important considerations are necessary. The first concerns the disperse-to-continuous viscosity ratio. In fact, in Equation 21.5 is only considered the viscosity of the continuous phase because this expression is Stocks's law corrected to account for the interaction with the membrane surface. Nevertheless, the disperse-to-continuous viscosity ratio can significantly affect the values of the effective drag force [42]. This consideration is connected with the droplet deformability; the solid particle approximation (Stocks's law) can be a restricted assumption in direct membrane emulsification modeling. The second observation concerning the F_{DR} is connected with the value of the wall correction factor parameter k_x. The reported value was obtained considering a solid droplet leaned on a surface, this assumption should be improved in the case of a droplet growing from a pore. Although different F_{DR} are presented in the literature [43], the wall shear stress at the membrane surface always appears explicitly in these expressions. This quantity depends on the membrane geometry and module. For simple modules (e.g., tubular or flat) consolidated expressions of the wall shear stress can be found [39]. For more complex equipments holding the membrane (e.g., rotating or vibrating systems), the evaluation of the shear stress requires more complicated analysis and calculations [43]. A particular consideration is necessary for the vibrating systems. Recently, it has been emphasized [44] that vibrations of the membrane introduces additional inertial and drag forces (secondary drag force) in a direction parallel to the membrane surface. These two forces depend on the excitation amplitude and the excitation frequency.

All presented forces are detaching forces. An increase in these forces will decrease the diameter of the droplets. On the contrary, the interfacial force caused by a uniform interfacial tension along the pore border is a holding force; increasing this force will increase the droplet size. The more simple and common expression of this force is [13, 39, 45] defined as:

$$F_\gamma = \pi D_p \gamma \sin \theta \tag{21.6}$$

where θ represents the contact angle. The capillary force 21.6 is obtained by the integration along the pore perimeter of the dF_γ, force acting on dL of the pore border. The force dF_γ has magnitude γdL and is directed towards the pore. The contact angle θ is assumed constant in the integration. This constraint could be a severe approximation. In fact, the droplet, during its formation, could counterbalance the actions due to the continuous fluid crossflow and F_{YL}, F_{DL}, F_{BG}, and F_I forces by changing its contact angle along the pore border, that is the droplet twists on the surface. The droplet inclination yields an interfacial force sufficient to keep the droplet on the membrane. Thus, the global F_γ should be rewritten in order to take into account the change of contact angle on the contact line [46]. This interfacial force can be expressed as sum of the two components $F_{\gamma i}$ and $F_{\gamma K}$. Considering the contact line Γ of generic size and shape, the interfacial force can be defined according the following two components:

$$\begin{aligned} F_{\gamma i} &= \int_\Gamma \gamma(\mathbf{M}\cdot\mathbf{m})\mathbf{m}\cdot\mathbf{i}\, d\Gamma \quad \text{parallel to the membrane surface} \\ F_{\gamma k} &= \int_\Gamma \gamma(\mathbf{M}\cdot\mathbf{k})\mathbf{k}\cdot\mathbf{k}\, d\Gamma \quad \text{perpendicular to the membrane surface} \end{aligned} \tag{21.7}$$

where **M** and **m** are the unit vectors, whose directions are indicated in Figure 21.15.

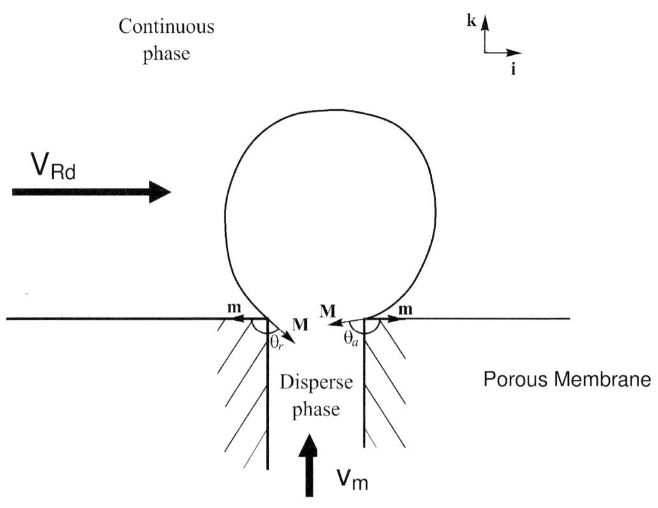

Figure 21.15 Droplet formation at the pore opening. Side view with vectors indicating the unit vectors, **M** and **m** at the pore perimeter and the advancing (θ_a) and receding (θ_r) contact angles. V_{Rd} represents the crossflow velocity of continuous phase at height equal to droplet radius, and v_m is the mean disperse-phase velocity.

Concerning the shape of Γ, it should be noted that the droplet contact line does not necessarily coincide with the pore border. In fact, depending on the affinity between the membrane and the disperse phase, the droplet can spread around the pore. Using the definition for the contact angle $\cos\theta = \mathbf{M}\cdot\mathbf{m}$, Equation 21.7 yields the components of the interfacial tension as a function of the contact angle along the contact line Γ. The integration in Equation 21.7 can be conveniently carried out by dividing the contact line into four sections: the advancing (Γ_a) and receding (Γ_r) portions, along which the contact angles assume the constant values θ_a and θ_r, respectively, and the two lines corresponding to the transition zones (TZ) in which the contact angles are not constant. Thus, Equation 21.7 can be expanded in a more easy to handle form as:

$$\begin{aligned} F_{\gamma i} &= \gamma\cos\theta_a\int_{\Gamma_a}\mathbf{m}\cdot\mathbf{i}\,d\Gamma + \gamma\cos\theta_r\int_{\Gamma_r}\mathbf{m}\cdot\mathbf{i}\,d\Gamma + 2\gamma\int_{TZ}\cos\theta(\Gamma)\mathbf{m}\cdot\mathbf{i}\,d\Gamma \\ F_{\gamma k} &= \gamma\sin\theta_a\,\Gamma_a + \gamma\sin\theta_r\,\Gamma_r + 2\gamma\int_{TZ}\sin\theta(\Gamma)\,d\Gamma \end{aligned} \tag{21.8}$$

It is worth noting that if θ_a and θ_r are equal to the equilibrium contact angle θ, then the first component becomes zero, whereas the second component reduces to the above Equation 21.6. For more information on the difference between θ_a, θ_r and equilibrium contact angle given by the Young equation the reader may refer to the original works [46, 47].

The presented forces are used in different balance equations according to mechanics assumptions. From the mechanics point of view two possible states for an immiscible droplet injected into a liquid continuous phase can be considered: (a) the droplet may maintain a spherical symmetry until it begins its detachment (rigid spherical cap configuration) and here the most appropriate mechanical model would be a torque balance, and (b) a deformed droplet at its base, where a force balance at the pore perimeter would be the most suitable model. Both the torque and force balances can be used to derive equations that will define the diameter of the droplet before its detachment. To be able to calculate the instant when the droplet has grown sufficiently to detach from the pore, Peng and Williams [13] at first suggested a torque balance around a pore edge:

$$F_{DR}h = (F_\gamma - F_{YL} - F_{BG})\frac{D_p}{2} \tag{21.9}$$

in which h is the droplet height from the membrane surface. If the droplet shape is significantly deformed towards the membrane then h can be approximated with the pore radius and Equation 21.8 reduces to:

$$F_{DR} = (F_\gamma - F_{YL} - F_{BG}) \tag{21.10}$$

In Equations 21.9 and 21.10, the interfacial force 21.6 is used by employing a contact angle equal to $\pi/2$. Generally in Equation 21.9 the droplet height is substituted by the droplet radius, $D_d/2$. These equations permit calculation of which diameter torques in clockwise and anticlockwise directions are balanced, beyond this value the droplet detaches. However, for a short period the droplet still maintains its connection with the pore through a neck. When this connection is completely broken the droplet detachment is complete 48. Kelder *et al.* and van Rijn [44, 49] used a force balance equation to predict the *final* droplet diameter, that is the droplet size at the end of its detachment. In their model the droplet is attached at the pore border by a curved neck (similar to a strap). The droplet is not leaned on the pore opening but on the membrane surface. Combining Equations 21.6 and 21.5 yields the following force-balance equation:

$$F_{DR} = F_{\gamma} \tag{21.11}$$

Although the last two balances are similar, it is worth noting that from the mechanics point of view they are different. In fact, the first derives from an approximation on the droplet height, whereas the second derives from a different droplet configuration with respect to the membrane surface. The first balances (Equations 21.9 and 21.10) estimate the droplet volume at the beginning of the detachment, whereas the second balance (Equation 21.11) estimates it at the end of the detachment (*final droplet dimension*). The F_{DR} used by Kelder *et al.* is acting on the center of the droplet, thus it is not expressed as a function of the wall shear stress. However, in both force balances, the holding force 21.6 is used as the interfacial force. De Luca and Drioli [46] at first proposed a balance force model along the droplet contact line Γ using the interfacial force 21.8 instead of Equation 21.6. Components of the interfacial force 21.8 have to counterbalance both the drag force and the forces in the direction perpendicular to membrane surface (all the detaching forces). The droplet deformability is taken into account through the evaluation of the advancing and receding contact angles along the droplet–pore contact line. The detachment is supposed to occur when the interfacial force at the droplet base is unable to counterbalance, through the droplet inclination, the actions of the detaching forces. The resultant set of force balance equations is:

$$\begin{cases} F_{\gamma i}(\theta_a, \theta_r) + F_{DR} = 0 & \text{parallel to the membrane surface} \\ F_{\gamma k}(\theta_a, \theta_r) + F_{YL} + F_{DL} + F_{BG} = 0 & \text{perpendicular to the membrane surface} \end{cases} \tag{21.12}$$

where the dependence of the interfacial tension force on the contact angles is reported for clarity. This set of equations can be solved at every droplet diameter to find the contact angles providing the equilibrium of forces. The solution of Equation 21.12 is the set of θ_a and θ_r values for any value of the droplet diameter chosen as parameter. The solution paths are in all cases closed lines lying within a minimum and maximum D_d value corresponding to the initial pore diameter ($D_0 > D_p$) and critical value denoted by D_c (*critical droplet diameter*), respectively. Since no solution exists for a droplet diameter larger than D_c, then it is concluded that this value has the meaning of droplet diameter corresponding to which the detachment of the droplet starts. In those cases where solution branches are found to be physically unacceptable, D_c is taken to be the smallest diameter corresponding to one of the two contact angles reaching the value of π. It is worth noting that if for particular θ_a and θ_r values the first equation in the balances 21.12 is not satisfied but the second equation could be satisfied, the droplet should glide along the surface without detaching from the membrane. Recently, Christov *et al.* [17] have proposed a balance-force equation similar to the balance 21.11 but for the droplet detachment in quiescent conditions. In this model the interfacial force 21.6, corrected with the Harkins-Brown factor f_d, is counterbalanced by the hydrodynamic force that substitutes the drag force F_{DR}. Using an appropriate expression for the mean velocity of the disperse phase defining the hydrodynamic force, the authors proposed a balance equation between these two forces. It is worth noting that in quiescent conditions and for a spherical disperse/water interface (i.e., a spherical sector), the Young–Laplace force 21.1 is always balanced for a contact angle equal to arcsin (D_p/D_d). Therefore, in this condition the hydrodynamic force could be the driving force for the droplet detachment.

The last consideration of this section concerns the coupling effects occurring in membrane emulsification. The adsorption of the emulsifier on the droplet interface changes the values of the interfacial tension and consequently the interfacial force [26, 50, 51]. In addition, the equilibrium contact angle θ_c changes as the emulsifier is adsorbed. As shown by [26, 50, 51] the effects of emulsifier adsorption depend on the ratio between droplet-formation time and emulsifier adsorption rate. If the droplet-formation time is large enough to permit a complete adsorption of emulsifier then the equilibrium interfacial tension can be used in the above force expressions. On the contrary, if the droplet-formation process is not large enough then the dynamic interfacial tension function has to be considered as a substitute for the scalar equilibrium interfacial tension. In this case the coupling between the droplet diameter and dynamic interfacial tension has been introduced in the correlated balance equations [50].

21.4.2
Surface-Energy Minimization

Rayner *et al.* [52] analyzed the formation mechanism of a droplet from a single pore into a quiescent continuous phase condition evaluating the dimensionless Reynolds, Bond, Weber, and the capillary numbers. The magnitude of these dimensionless numbers was calculated for the following membrane emulsification setup: 1 μm pore diameter, 5 mN/m interfacial tension, 8×10^2 kg/m^3 oil density, 5×10^{-3} Pa s viscosity and 1×10^{-3} m/s disperse-phase velocity and zero continuous-phase velocity. The above dimensionless numbers indicate that the interfacial force 21.6 absolutely dominates the emulsification process; the hydrodynamic force due to the disperse-phase flow is negligible and the drag force absent. In other terms, the holding interfacial force always balances the other involved forces until the complete detachment of the droplet due to a spontaneous deformation. Thus, the balance (Equations 21.9–21.12) is not predictive in this case because the deformation of the droplet is not included.

Starting from this consideration, Rayner *et al.* [52] analyzed the spontaneous-transformation-based (STB) droplet-formation mechanism from the point of view of the surface Gibbs free energy with the help of the Surface Evolver code. Rayner *et al.* estimated the difference of surface free energy of the droplet before (E_1) and after (E_2)

its detachment, E_2–E_1. When E_1 is larger than E_2 then the spontaneous droplet formation begins. Surface Evolver code was used to evaluate E_1 energy. In particular, for an assigned disperse–continuous–surfactant interfacial tension, Surface Evolver, for each droplet volume attached at the pore perimeter, yields the droplet surface with the minimum energy among the possible surfaces related with the given volume. This minimization must respect important constraints, that is the geometry of the pore border and the equilibrium contact angle that is set as a contact energy around the pore. Once the minimum E_1 energy is found, the E_2 energy have to be evaluated. The E_2 value is the free energy of the detached droplet having the same volume of the attached droplet plus the energy of the pore opening. The maximum stable droplet volume (MSV) is the volume of the attached droplet just before the STB droplet formation takes place, that is when $E_1 > E_2$. Surface Evolver also gives the possibility to find the maximum stable droplet volume thought the Hessian eigenvalues analysis. The occurrence of negative eigenvalues corresponds to the point at which the E_2–E_1 difference becomes negative. The MSV yields an estimation of the largest droplet that should be formed. It is well known that there is a certain volume of disperse phase remaining attached at the pore. Rayner *et al.* estimated this remaining volumes using the 'pressure pinch constraint' principle. This principle is based on the division of the droplet MSV into two parts having relative sizes that show an equal Laplace pressure across the surface of both volumes. Using this principle the authors yielded an estimation of the droplet diameters in quiescent conditions and for very low disperse-phase flow. The Rayner *et al.* approach estimates the maximum dimension achievable for the droplet. The Surface-Evolver-based simulations also showed that for pores with aspect ratio (maximum to minimum length) greater than three the necking formation should occur inside the pore. On the contrary, when the aspect ratio is smaller than three the droplet necking took place outside the membrane pore. The same authors [53] used this approach to analyze the effect of the dynamic surfactant coverage on the final droplet size coupled to the expansion rate of the continuous/disperse interface. They found that the dynamic surfactant coverage has a significant influence on the final droplet size during the analyzed membrane emulsification process.

21.4.3
Microfluid Dynamics Approaches: The Shape of the Droplets

In the microfluid dynamics approaches the continuity and Navier–Stokes equation coupled with methodologies for tracking the disperse/continuous interface are used to describe the droplet formation in quiescent and crossflow continuous conditions. Ohta *et al.* [54] used a computational fluid dynamics (CFD) approach to analyze the single-droplet-formation process at an orifice under pressure pulse conditions (pulsed sieve-plate column). Abrahamse *et al.* [55] simulated the process of the droplet break-up in crossflow membrane emulsification using an equal computational fluid dynamics procedure. They calculated the minimum distance between two membrane pores as a function of crossflow velocity and pore size. This minimum distance is important to optimize the space between two pores on the membrane surface (i.e., the membrane porosity) in order to avoid droplet coalescence on the surface. They characterized the mechanism of the droplet formation (droplet shape, pressure drop through the pore, etc.) occurring for the assigned conditions. Quite recently, Kobayashi *et al.* [56] carried out a numerical investigation on the formation of an oil droplet in water from straight-through microchannels (MC) with an elliptic cross-section and in quiescent conditions. In particular, these CFD simulations demonstrated that the neck formation considerably depends on the aspect ratio of the elliptic MC. Continuous outflow of the oil phase from the channel opening was observed for elliptic MCs below a threshold aspect ratio between 3 and 3.5. On the contrary, a droplet with neck inside the membrane pore was found for a droplet formed in the elliptic MCs exceeding the above threshold aspect ratio. This result is in agreement with the conclusion found by Rayner *et al.* [52] reported above. Cristini and Tan recently reviewed numerical simulations of droplet dynamics in complex flows [57].

The computational fluid dynamics investigations listed here are all based on the so-called volume-of-fluid method (VOF) used to follow the dynamics of the disperse/continuous phase interface. The VOF method is a technique that represents the interface between two fluids defining an *F* function. This function is chosen with a value of unity at any cell occupied by disperse phase and zero elsewhere. A unit value of *F* corresponds to a cell full of disperse phase, whereas a zero value indicates that the cell contains only continuous phase. Cells with *F* values between zero and one contain the liquid/liquid interface. In addition to the above continuity and Navier–Stokes equation solved by the finite-volume method, an equation governing the time dependence of the *F* function therefore has to be solved. A constant value of the interfacial tension is implemented in the summarized algorithm, however, the diffusion of emulsifier from continuous phase toward the droplet interface and its adsorption remains still an important issue and challenge in the computational fluid-dynamic framework.

The CFD procedures briefly presented are a valid tool for an accurate *in-silico* analysis of the droplet-formation mechanisms occurring under various membrane emulsification parameters. This knowledge can be used in the formulation and validation of the basic assumptions characterizing the aforementioned balance models. Validated computation fluid dynamic models are useful to design optimal membranes and related equipments. In other words, the CFD procedure can be used for *in-silico* experiments avoiding expensive experimental trial and error tests. Nevertheless, it is worth noting that the CFD simulations for membrane emulsification processes are time-consuming tasks. This aspect can be restrictive if many *in-silico* experiments have to be carried out. Although the CFD procedures give useful information on the droplet break up, not all phenomena are modeled on a solid physical basis, which can result in ambiguous conclusions as in the case for the modeling the contact line dynamics. Other CFD approaches that do not use the VOF procedure (e.g., level-set procedure) should be taken into account. However, this approach in the membrane emulsification is still at an early stage of development.

Lattice Boltzmann (LB) is a relatively new simulation technique and it represents an alternative numerical approach in the hydrodynamics of complex fluids. The LB method can be interpreted as an unusual finite-difference solution of the continuity and Navier–Stokes equation and it is suitable for modeling of multiphase systems. The LB is based on hypothetical fluid particles (packages of fluid) moving and colliding on a lattice according to the kinetic gas theory. One of the most important reasons why the LB algorithm works well for multiphase problems is that the interfaces appear and move automatically during the simulation. By contrast to the mentioned CDF method, it is not necessary to track the interface explicitly. In addition, the implementation of complex wetting conditions (e.g., patterned surfaces) and the dynamics of contact line turns out to be more simple and accurate with respect to the traditional CFD approaches. Moreover, the diffusion and dynamic adsorption of emulsifiers during the droplet formation is another aspect that can be correctly treated in a LB framework. In general, the LB simulations for membrane emulsification processes are less intensive (time consuming) with respect to the analogous CFD ones. Although the LB methodology has found applications in different areas of fluid dynamics, including simulations of flows in porous media and droplet formation in liquid–gas systems [58], at the moment only the work of van der Graaf *et al.* [18] is addressed to the droplet formation from a T-shaped microchannel in a liquid–liquid system. It is worth noting that the T-shaped microchannel geometry was approximated as a model of a membrane pore.

Although the premix membrane emulsification can yield larger fluxes with respect to direct membrane emulsification neither methods using surface-energy minimization nor microfluid dynamics approaches have been until now reported on the theoretical treatment of the premix membrane emulsification.

21.5 Membrane Emulsification Applications

21.5.1 Applications in the Food Industry

Emulsions play an important role in the formulation of foods, that is, o/w emulsions are used for preparation of dressings, artificial milks, cream liqueurs, and w/o emulsions are used in the production of margarines and low-fat spreads.

Food products must have appropriate texture properties. For example, it is important that mayonnaise products have thick and creamy textures, but not too high a viscosity. The rheological properties depend on their composition, such as the concentration of oil droplets or the concentration of thickening agents.

The development of membrane emulsification technologies permits production of small and uniform droplets and capsules, using mild conditions of temperature, shear stress and pressure. Furthermore, they are able to produce stable droplets with reduced stabilizers content, which will contribute to the manufacturing of improved food products with low-fat content.

In this context, the Morinaga Milk Industry (Japan) developed and commercialized a very low fat spread using membrane emulsification technology [59, 60]. The advantages in the production of low-fat spreads made the process one of the first large-scale applications of membrane emulsification. A w/o emulsion using a MPG hydrophilic membrane, previously treated with the oil phase, has been prepared by crossflow membrane emulsification. The resulting product was stable and free from aqueous phase separation, tasted smooth and melted extremely easily in the mouth.

For practical applications in the food industry, where large-volume production is conducted, it is especially important to obtain high disperse-phase flux. Abrahamse *et al.* [8] reported on the industrial-scale production of culinary cream. In this study they evaluated the required membrane area for different types of membranes: an SPG membrane, an α-Al_2O_3 membrane and a microsieve filter. The requirements for culinary cream production were: a droplet size between 1 and 3 μm and a production volume of 20 m^3/h containing 30% disperse phase. They concluded that to produce large quantities of monodisperse emulsions the most suitable was a microsieve with an area requirement of around 1 m^2.

Katoh *et al.* [3] prepared w/o emulsions composed of salt solution, polyglycerin polyricinolate (PGPR) at 2%wt and corn oil. It has been proven that the disperse-phase flux was increased 100-fold using a hydrophilic membrane pretreated by immersion in the oil phase. This made the membrane emulsification system practical for large-scale production of a w/o emulsion in food application.

Double emulsions are also very useful for food application. Sensitive food materials and flavors can be encapsulated in w/o/w emulsions. Sensory tests have indicated that there is a significant taste difference between w/o/w emulsions and o/w emulsions containing the same ingredients, and that there is a delayed release of flavor in double emulsions [61]. W/o/w or o/w/o multiple emulsions having a concentrated aqueous-soluble flavor or a concentrated oil-soluble flavor encapsulated in the internal phase can be prepared. Food products obtained with these particulates exhibit enhanced flavor perception and extended shelf-life [62].

21.5.2 Applications in the Pharmaceutical Industry

Among the applications of membrane emulsification, drug-delivery system (DDS) is one of the most attractive fields. W/o/w emulsions have been prepared to transport and deliver anticancer drug [4, 63–65]. The emulsion was directly administered into the liver using a catheter into the hepatic artery. In this way, it was possible to suppress the strong side effects of the anticancer drug and also concentrate the dosage selectively to focus on the cancer. The clinical study showed that the texture of the cancer rapidly contracted and its volume decreased to a quarter of its initial size.

Composite emulsion as carrier of hydrophilic medicine for chemotherapy was prepared by adding albumin to the internal water phase and lecithin or cholesterol to the oil phase, thus obtaining a water-in-oil emulsion. This emulsion was then pressed through Millipore membrane into an external water phase to form a w/o/w multiple emulsion. Its advantages are high size uniformity and high storage stability [66].

Nakajima *et al.* referred to membrane emulsification as a method to make functional ethanol-in-oil-in-water (e/o/w) emulsions. These e/o/w emulsions are suitable to encapsulate functional components that have a low water and oil solubility while being soluble in ethanol. An example is taxol, which is an anticancer terpenoid [67].

Vladisavljevic *et al.* reported on the production of multiple w/o/w emulsions for drug-delivery systems by extruding a coarse w/o/w emulsion five times through a SPG membrane [68].

Several studies also reported on the preparation of biodegradable polymer microcapsules to be used as drug-delivery systems due to their biodegradable nature and proven biocompatibility. The biopolymers employed are mainly poly(lactide) (PLA) [69], poly(lactic-co-glycolic acid) (PLGA) [70–74], chitosan [75, 76], and calcium alginate [77]. Such polymers have been applied for encapsulating proteins and peptides used as prophylactic and therapeutic agents in biomedical fields. So far, the delivery route is injection, which not only causes distress and inconvenience to patients, but also induces unstable curative effective and side effects. This is due to the fact that the drugs have to be given frequently, resulting in rapid increase and decrease of drug concentration in blood [75]. Therefore, a sustained delivery system for proteins and peptides is necessary not only for injection administration but also for developing an oral-administration system. The use of microspheres as a controlled release system is one of the prospective methods. In fact, it may prevent encapsulated drugs from degradation by proteolytic enzymes, prolong its half-life and improve its bioavailability *in vivo* by controlling the release rate of the drug from the microspheres.

The preparation of monodisperse hydrogel microspheres, such as poly-acrylamide-co-acrylic acid, poly(N-isopropylacrylamide-co-acrylic acid), has been performed for drug devices thanks to their biocompatibility [77, 79]. The average diameters of the microspheres were dependent on the pore sizes (from 0.33 to 1.70 μm) of SPG membranes used in the preparation procedure.

Solid lipid nanoparticles (SLN) have also been introduced as an alternative to solid particles, emulsions and liposomes in cosmetic and pharmaceutical preparations. Charcosset *et al.* reported the use of membrane emulsification for the production of SLN [80]. The lipid phase was pressed through the membrane pores into the aqueous continuous phase, at a temperature above the melting point of the lipid. The SLN are then formed by the following cooling of the preparation to room temperature. The lipids remain solid also at body temperature. The influence of process parameters on the size and the lipid-phase flux was investigated. The membranes used were supplied by Kerasep ceramic membranes with an active ZrO_2 layer on an Al_2O_3-TiO_2 support. Three different microfiltration membranes were investigated: 0.1, 0.2, and 0.45 μm mean membrane pore size. It was shown that SLN nanoparticles could be prepared with a liquid-phase flux between 0.15 and 0.35 $m^3/h\,m^2$ and mean SLN size between 70 and 215 nm.

21.5.3
Applications in the Electronics Industry

The membrane emulsification technique is also employed for the preparation of microspheres starting from monomers such as methacrylates (methylmethacrylate, cyclohexyl acrylate, etc.), polyimide prepolymer, styrene monomer [81], and so on.

The occlusion of functional materials such as the polyimide prepolymer (PIP) in uniform polymer particles, can find promising applications in sophisticated electronic devices such as adhesive spacers of liquid-crystal panel boards (after a minor screening process), adhesives or insulators for microtip circuits, and so forth. Omi *et al.* [82] showed that about 30% occlusion of polyimide prepolymer (diphenilmethane-4,4′-bis-allylnagiimide, BAN-I-M) was accomplished in the preparation of polymer particles composed of styrene, various acrylates and a crosslinking agent (ethyleneglycol dimethacrylate, EGDMA) via the emulsification technique with SPG membrane. Particles with a diameter of 6–12 micrometers were prepared. The presence of acrylates and EGDMA was essential to obtain stable lattices of styrene-based copolymers that occlude BAN-I-M. However, the presence of acrylates with longer side chains, BA and 2EHA, promoted the inclusion of BAN-I-M. In particular, the latter yielded a stable latex occluding 100% of the initial BANI-M without the crosslinking matrix and using octyl alcohol as a stabilizing agent. The lattices without a crosslinking network resulted in an excellent adhesive ability.

Guang Hui Ma *et al.* [83] prepared microcapsules with narrow size distribution, in which hexadecane (HD) was used as the oily core and poly(styrene-co-dimethylamino-ethyl metahcrylate) [P(st-DMAEMA] as the wall. The emulsion was first prepared using SPG membranes and a subsequent suspension polymerization process was performed to complete the microcapsule formation. Experimental and simulated results confirmed that high monomer conversion, high HD fraction, and addition of DMAEMA hydrophilic monomer were three main factors for the complete encapsulation of HD. The droplets were polymerized at 70 °C and the obtained microcapsules have a diameter ranging from 6 to 10 μm, six times larger than the membrane pore size of 1.4 μm.

Furthermore, such monomers can be readily emulsified by dissolving in volatile solvents such as methylene chloride and chloroform. Uniform polylactide particles, and composite polystyrene (PST) and polymethyl methacrylate (PMMA) particles were produced by solvent evaporation [84–86].

21.5.4
Other Applications

Membrane emulsification has also been applied for the preparation of oil-in-water emulsions to be used in cosmetics and/or dermatology, in particular for the treatment, protection, care, cleaning and make-up of the skin, mucous membranes and hair. The emulsion was composed by oil-phase globules having an average size less than 20 μm; it was prepared by direct membrane emulsification through a porous hydrophilic glass membrane having an average pore size ranging from 0.1 to 5 μm and preferably from 0.3 to 3 μm [87].

The technology also represents a suitable strategy for the preparation of multiphase reaction systems that use phase transfer (bio)catalysts. Giorno *et al.* [88] reported on the use of membrane emulsification to distribute lipase from *Candida rugosa* at the interface of stable oil-in-water emulsions. The enzyme itself was used as a surfactant. Shirasu Porous Glassy (SPG) membranes having a nominal pore diameter of 0.1 μm were used to prepare emulsions. Emulsions with more than 90% of organic droplets of 1.6 (±0.40) μm were obtained. The methodology allowed preservation of the catalytic performance of the biocatalyst as well as optimal enzyme distribution at the interface of stable, uniform and small oil droplets to be achieved.

Applications in the chemical field, include extrusion of an oil phase containing a photographic hydrophobic material through a microporous membrane into water [89] and emulsification of low-viscosity paraffin wax in water [90].

The polyurethane (PU) can be considered an environment-friendly material because the urethane bond resembles the amide bond, which implies possible biodegradability. It can be used in various elastomer formulations, paints, adhesives for polymers and glass, and artificial leather as well as in biomedical and cosmetic fields. Polyurethane spheres were prepared from 20/40% of PU prepolymer solution in xylene [91]. PU droplets were formed in water with the SPG membrane of different pore size (1.5–9.5 μm) and then polymerized to form the final microspheres. Finally, spherical and solid PU particles of 5 μm were obtained after the removal of the solvent. In another study, Ma *et al.* reported the formation of uniform polyurethane-vinylpolymer (PUU-VP) hybrid microspheres of about 20 μm, prepared using SPG membranes and a subsequent radical suspension polymerization process [92]. The prepolymers were solubilized in xylene and pressed through the SPG membrane into the continuous phase containing a stabilizer to form uniform droplets. The droplets were left for chain extension at room temperature for some hours with di- and triamines by suspension polymerization at 70 °C for 24 h. Solid and spherical PU-VP hybrid particles with a smooth surface and a higher destructive strength were obtained.

Ha *et al.* [93, 94] prepared monodisperse polymer microspheres from 1 to 40 μm in diameter for medical diagnostic tests, as chromatography column packing and as calibration standards. The work deals with the synthesis of large and uniform poly (butadiene-styrene) latex. The ceramic SPG membrane, with a pore diameter of 1.6 μm, was employed. The uniform particle sizes were in the diameter range of 4–6 μm.

Westover *et al.* 95 prepared lightly crosslinked nitrated poly(4-hydroxystyrene) microspheres for pH sensors. The microspheres were produced using SPG membranes followed by suspension polymerization and they showed diameters between 1 and 2 micrometers.

Figoli *et al.* [96, 97] reported the preparation of polymeric capsules combining the phase-inversion technique with the membrane process. Polyetheretherketone (PEEKWC) capsules of different size (300–800 micrometer) and morphology (asymmetric with a porous or dense layer) have been prepared. The SEM pictures of the prepared PEEKWC capsules are shown in Figure 21.16. The capsules can find application both in chemical and in food packaging fields [98].

Another field where emulsions are likely to become imperative is the production of fuel [99]. Simple and multiple emulsions represent alternative fuels for diesel engines to both increase combustion efficiency and reduce particulate emission. Considering the enormous volume of diesel that is being consumed today, a replacement of just a fraction of regular diesel by diesel emulsion could be of considerable interest to the surface chemistry community. Until now, diesel emulsions were prepared by

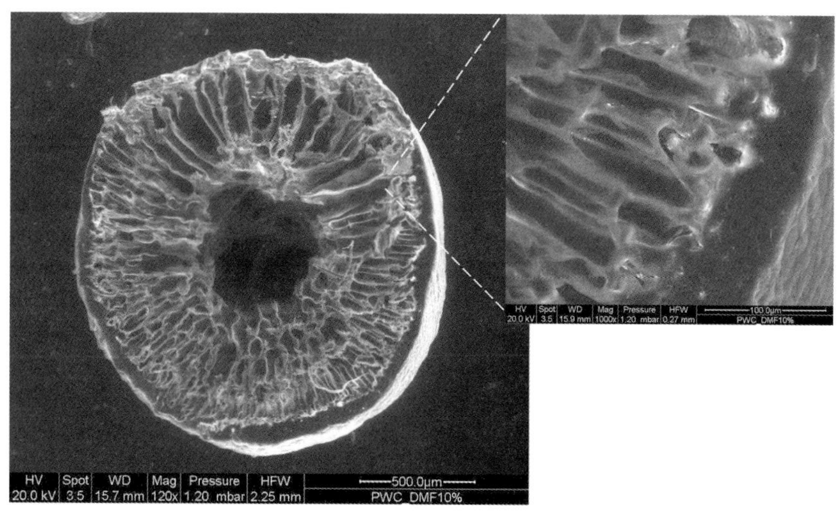

Figure 21.16 SEM pictures of the crosssection of the PEEKWC capsules prepared by the phase-inversion technique using a film with a pore size of 550 µm [97].

conventional emulsification methods but it is expected that the membrane emulsification technique will also become attractive for this application.

21.6 Conclusions

Membrane emulsification, a technology that first appeared in the early 1990s, is gaining increasing attention with many applications being explored in various fields. Nowadays, it can be considered at a developing/exploiting stage with a significant involvement of industrial and academic research effort. Many studies have been carried out, especially from the experimental point of view whereas from the theoretical point of view the knowledge is not accordingly advanced.

In this chapter, a description of membrane emulsification basic concepts, empirical correlations, theoretical studies, as well as most common applications have been discussed.

Many patents have been applied for, especially in Japan, which currently holds more than 60% of worldwide applications, in Europe and USA.

Main drivers for membrane emulsification development include high product quality – especially when labile molecules are involved, precise definition of droplet-size distribution, low energy input, equipment modularity and easy scale-up, and low equipment footprint.

Challenges in this field include the need for higher productivity, membranes and modules specifically designed for the emulsification process, modules construction standardization, and design of innovative intensified processes.

References

1 Nakashima, T., Shimizu, M. and Kukizaki, M. (1991) *Membrane Emulsification Operation Manual*, 1st edn, Industrial Research Institute of Miyazaki Prefecture, Japan.

2 Nakashima, T. and Shimizu, M. (1991) *Key Engineering Materials*, **61/62**, 513–516.

3 Katoh, R., Asano, Y., Furuya, A., Sotoyama, K. and Tomita, M. (1996) *Journal of Membrane Science*, **113** (1), 131–135.

4 Nakashima, T., Shimizu, M. and Kukizaki, M. (2000) *Advanced Drug Delivery Reviews*, **45**, 47–56.

5 Schröder, V. and Schubert, H. (1999) *Colloid and Surfaces*, **152** (1), 103–109.

6 Joscelyne, S.M. and Trägårdh, G. (2000) *Journal of Membrane Science*, **169**, 107–117.

7 Charcosset, C., Limayem, I. and Fessi, H. (2004) *Journal of Chemical Technology and Biotechnology*, **79**, 209–218.

8 Abrahamse, A.J., van der Padt, A. and Boom, R.M. (2004) *Journal of Membrane Science*, **230**, 149–159.

9 Vladisavljevi,ć G.T. and Williams, R.A. (2005) *Advances in Colloid Interface Science*, **113**, 1–20.

10 Lambrich, U. and Schubert, H. (2005) *Journal of Membrane Science*, **257**, 76–84.

11 Suzuki, K., Shuto, I. and Hagura, Y. (1996) *Food Science and Technology International Tokyo*, **2** (1), 43–47.

12 Kawakatsu, T., Kikuchi, Y. and Nakajima, M. (1997) *Journal of the American Oil Chemists Society*, **74**, 317–321.

13 Peng, S.J. and Williams, R.A. (1998) *Transactions of IChemE*, **76**, 894–901.

14 Vladisavljevi,ć G.T., Tesch, S. and Schubert, H. (2002) *Chemical Engineering and Processing*, **41**, 231–238.

15 Vladisavljevi,ć G.T., Shimizu, M. and Nakashima, T. (2004) *Journal of Membrane Science*, **244** (1–2), 97–106.

16 Vladisavljevi,ć G.T. and Schubert, H. (2003) *Journal of Membrane Science*, **223**, 15–23.

17 Christov, N.C., Danov, K.D., Danova, D.K. and Kralchevsky, P.A. (2008) *Langmuir*, **24**, 1397–1410.

18 van der Graaf, S., Nisisako, T., Schröen, C.G.P.H., van der Sman, R.G.M. and Boom, R.M. (2006) *Langmuir*, **22**, 4144–4152.

19 Cheng, C.J., Chu, L.Y. and Xie, R. (2006) *Journal of Colloids and Interface Science*, **300**, 375–382.

20 Sotoyama, K., Asano, Y., Ihara, K., Takahashi, K. and Doi, K. (1999) *Journal of Food Science*, **64** (2), 221–215.

21 Vladisavljevi,ć G.T., Kobayashi, I., Nakajima, M., Williams, R.A., Shimizu, M. and Nakashima, T. (2007) *Journal of Membrane Science*, **302**, 243–253.

22 Scherze, I., Marzilger, K. and Muschiolik, G. (1999) *Colloids and Surface B*, **12**, 213–221.

23 Mine, Y., Shimizu, M. and Nakashima, T. (1996) *Colloids and Surfaces B: Biointerfaces*, **6** (4), 261–268.

24 Fuchigami, T., Toki, M. and Nakanishi, K. (2000) *Journal of Sol-Gel Science and Technology*, **19**, 337–341.

25 Wu, J., Jing, W., Xing, W. and Xu, N. (2006) *Desalination*, **193**, 381–386.

26 Schröder, V., Behrend, O. and Schubert, H. (1998) *Journal of Colloid and Interface Science*, **202**, 334–340.

27 Joscelyne, S.M. and Trägårdh, G. (1999) *Journal of Food Engineering*, **39**, 59–64.

28 Berot, S., Giraudet, S., Riaublanc, A., Anton, M. and Popineau, Y. (2003) *Transactions of IChemE*, **81**, 1077–1082.

29 Giorno, L., Mazzei, R., Oriolo, M., De Luca, G., Davoli, M. and Drioli, E. (2005) *Journal of Colloid and Interface Science*, **287**, 612–623.

30 Yamazaki, N., Yuyama, H., Nagai, M., Ma, G.H. and Omi, S. (2002) *Journal of Dispersion Science and Technology*, **23** (1–3), 279–292.

31 Kobayashi, I., Yasuno, M., Iwamoto, S., Shono, A., Satoh, K. and Nakajima, M.

(2002) *Colloids and Surface A*, **207**, 185–196.
32 Shima, M., Kobayashi, Y., Fujii, T., Tanaka, M., Kimura, Y., Adachi, S. and Matsuno, R. (2004) *Food Hydrocolloids*, **18** (1), 61–70.
33 Correia, L.A., Pex, P.A.C., van der Padt, A. and Poortinga, A.T. (2004) Presented at 8th International Conference on Inorganic Membrane (ICIM 9), 2004, Cincinnati, USA, 18–22 July.
34 Park, S., Yamaguchi, T. and Nakao, S. (2001) *Chemical Engineering Science*, **56** (11), 3539–3548.
35 Shima, M., Kobayashi, Y., Fujii, T., Tanaka, M., Kimura, Y., Adachi, S. and Matsuno, R. (2004) *Food Hydrocolloids*, **18**, 61–70.
36 Ribeiro, H.S., Rico, L.G., Badolato, G.G. and Schubert, H. (2005) *Journal of Food Science*, **70** (2), E117–E123.
37 Kondo, A. (1979) History and classification of microencapsulation, in *Microcapsule Processing and Technology* (ed. J. Wade Van Valkenburg), Marcel Dekker, New York, NY, Chapter 4.
38 Thies, C. (1996) A survey of microencapsulation processes, in *Microencapsulation Methods and Industrial Applications* (ed. S. Benita), vol. **17**, Marcel Dekker, New York, pp. 1–21.
39 Wang, Z., Wang, S., Volker, S. and Schubert, H. (2000) *Chinese Journal of Chemical Engineering*, **8**, 108–112.
40 Danov, K.D., Danova, D.K. and Kralchevsky, P.A. (2007) *Journal of Colloid and Interface Science*, **316**(2), 844–857.
41 O'Neil, M.E. (1964) *Mathematika*, **11**, 67–74.
42 Ken, H.J. and Chen, P.Y. (2001) *Chemical Engineering Science*, **56**, 6863–6871.
43 Kosvintsev, S.R., Gasparini, G., Holdich, R.G., Cumming, I.W. and Stillwell, M.T. (2005) *Industrial & Engineering Chemistry Research*, **44**, 9323–9330.
44 Kelder, J.D.H., Janssen, J.J.M. and Boom, R.M. (2007) *Journal of Membrane Science*, **304**, 50–59.
45 Chatterjee, J. (2002) *Advances in Colloid Interface Science*, **98**, 265–283.
46 De Luca, G. and Drioli, E. (2006) *Journal of Colloid and Interface Science*, **294**, 436–448.
47 De Luca, G., Di Maio, F.P., Di Renzo, A. and Drioli, E. (2008) *Chemical Engineering and Processing*, **47**, 1150–1158.
48 Xu, J.H., Luo, G.S., Chen, G.G. and Wang, J.D. (2005) *Journal of Membrane Science*, **266**, 121–131.
49 van Rijn, C.J.M. (2000) *Nano and Micro Engineered Membrane Technology* in Membrane Science and Technology Series, vol. **10**, Amsterdam, Elsevier.
50 De Luca, G., Sindona, A., Giorno, L. and Drioli, E. (2004) *Journal of Membrane Science*, **229**, 199–209.
51 van der Graaf, S., Schröen, C.G.P.H., van der Sman, R.G.M. and Boom, R.M. (2004) *Journal of Colloid and Interface Science*, **277**, 456–463.
52 Rayner, M., Trägårdh, G., Trägårdh, C. and Dejmek, P. (2004) *Journal of Colloid and Interface Science*, **279**, 175–185.
53 Rayner, M., Trägårdh, G. and Trägårdh, C. (2005) *Colloids and Surface A: Physicochemical and Engineering, Aspects* **266**, 1–17.
54 Ohta, M., Yamamoto, M. and Suzuki, M.M. (1995) *Chemical Engineering Science*, **50**, 2923–2931.
55 Abrahamse, A.J., van der Padt, A., Boom, R.M. and de Heij, W.B.C. (2001) *AIChE Journal*, **47** (6), 1285–1291.
56 Kobayashi, I., Mukataka, S. and Nakajima, M. (2004) *Langmuir*, **20**, 9868–9875.
57 Cristini, V. and Tan, Y.-C. (2004) *Lab on a Chip*, **4** (257), 257–264.
58 Kalarakis, A.N., Burganos, V.N. and Payatakes, A.C. (2003) *Physical Review E*, **67**, 016702-1–016702-8.
59 Okonogi, S., Kato, R., Asano, Y., Yuguchi, H., Kumazawa, R., Sotoyama, K., Takahashi, K. and Fujimoto, M. (1994) US5279847.
60 Okonogi, S., Kumazawa, R., Toyama, K., Kato, M., Asano, Y., Takahashi, K. and Fujimoto, M. (1992) JP4258251.
61 van der Graaf, S., Schroën, C.G.P.H. and Boom, R.M. (2005) *Journal of Membrane Science*, **251**, 7–15.

62 Gaonkar, A.G. (1994) US 5332595.

63 Higashi, S., Shimizu, M. and Setoguchi, T. (1996) *Colloids and Surface B*, **6**, 261–268.

64 Higashi, S., Shimizu, M., Nakashima, T., Iwata, K., Uchiyaemotherapyma, F., Tateno, S., Tamura, S. and Setoguchi, T. (1995) *Cancer*, **75**, 1245–1254.

65 Higashi, S. and Setoguchi, T. (2000) *Advanced Drug Delivery Reviews*, **45**, 57–64.

66 Guanghui, M., Hui, S., Zhiguo, S. and Lianyan, W. (2005) CN1600295.

67 Nakajma, M., Nabetani, H., Ichikawa, S. and Xu, Q.Y. (2003) US6538019.

68 Vladisavljević, G.T., Shimizu, M. and Nakashima, T. (2006) *Journal of Membrane Science*, **284**, 373–383.

69 Liu, R., Ma, G.H., Wan, Y.H. and Su, Z.G. (2005) *Colloid and Surface B: Biointerfaces*, **45**, 144–153.

70 Ito, F. and Makino, K. (2004) *Colloid and Surface B: Biointerfaces*, **39**, 17–21.

71 Omi, S., Katami, K., Yamamoto, A. and Iso, M. (1994) *Journal of Applied Polymer Science*, **51**, 1–11.

72 Shiga, K., Muramatsu, N. and Kondo, T. (1996) *Journal of Pharmacy and Pharmacology*, **48**, 891–895.

73 Costa, M.S. and Cardoso, M.M. (2006) *Desalination*, **200**, 498–500.

74 Shiga, K., Muramatsu, N. and Kondo, T. (1996) *Journal of Pharmacy and Pharmacology*, **48**, 891–895.

75 Wang, L.Y., Ma, G.H. and Su, Z.G. (2005) *Journal of Controlled Release*, **106**, 62–75.

76 Wang, L.Y., Gu, Y.H., Zhou, O.Z., Ma, G.H., Wan, Y.H. and Su, Z.G. (2006) *Colloid and Surface B: Biointerfaces*, **50**, 126–135.

77 Fuchigami, T., Toki, M. and Nakanishi, K. (2000) *Journal of Sol-Gel Science and Technology*, **19**, 337–341.

78 Nagashima, S., Koide, M., Ando, S., Makino, K., Tsukamoto, T. and Ohshima, T. (1999) *Colloids and Surfaces A: Physiochemical and Engineering Aspects*, **153**, 221–227.

79 Makino, K., Agata, H. and Ohshima, H. (2000) *Journal of Colloid and Interface Science*, **230**, 128–134.

80 Charcosset, C., El-Harati, A. and Fessi, H. (2005) *Journal of Controlled Release*, **108**, 112–120.

81 Dowding, P.J., Goodwin, J.W. and Vincent, B. (2001) *Colloids and Surfaces A: Physiochemical and Engineering Aspects*, **180**, 301–309.

82 Omi, S., Matsuda, A., Imamura, K., Nagai, M. and Ma, G.H. (1999) *Colloids and Surface A: Physiochemical and Engineering Aspects*, **153**, 373–381.

83 Ma, G.H., Su, G.Z., Omi, S., Sundberg, D. and Stubbs, J. (2003) *Journal of Colloid and Interface Science*, **266**, 282–294.

84 Ma, G.H., Nagai, M. and Omi, S. (1999) *Journal of Colloid and Interface Science*, **214**, 264–282.

85 Ma, G.H., Nagai, M. and Omi, S. (1999) *Colloids Surfaces A: Physiochemical and Engineering Aspects*, **153**, 383–394.

86 Muramatsu, N. and Kondo, T. (1995) *Journal of Microencapsulation*, **12**, 129–136.

87 Roulier, V. and Quemin, E. (2000) WO0021491.

88 Giorno, L., Piacentini, E., Mazzei, R. and Drioli, E. *Journal of Membrane Science*, **317** (1–2), 19–25.

89 Kiyoshi, E. (1999) JP11242317.

90 Aryanti, N., Williams, R.A., Hou, R. and Vladisavljević, G.T. (2006) *Desalination*, **200**, 572–574.

91 Yuyama, H., Yamamoto, K., Shirafuji, K., Nagai, M., Ma, G.H. and Omi, S. (2000) *Journal of Applied Polymer Science*, **77**, 2237–2245.

92 Ma, G.H., An, C.J., Yuyama, H., Su, Z.G. and Omi, S. (2003) *Journal of Applied Polymer Science*, **89**, 163–178.

93 Ha, Y.K., Song, H.S., Lee, H.J. and Kim, J.H. (1999) *Colloids and Surface A: Physiochemical and Engineering Aspects*, **162**, 289–293.

94 Ha, Y.K., Song, H.S., Lee, H.J. and Kim, J.H. (1998) *Colloids and Surface A: Physiochemical and Engineering Aspects*, **145**, 281–284.

95 Westover, D., Seitz, W.R. and Lavine, B.K. (2003) *Microchemical Journal*, **74**, 121–129.

96 Figoli, A., De Luca, G., Lamerata, F. and Drioli, E. (2006) *Desalination*, **199**, 115–117.

97 Figoli, A., De Luca, G., Longavita, E. and Drioli, E. (2007) *Separation Science and Technology*, **42**, 2809–2827.

98 Figoli, A., De Luca, G. and Drioli, E. (2007) *Italian Journal of Food Science*, ISSN 1120-1770, **XI**, 90–96.

99 Nakajima, N., Fujiwara, M., Maeda, D. and Watanabe, K. (2006) JP2006182890.

22
Membrane Contactors in Industrial Applications

Soccorso Gaeta

Membrane contactors are devices that selectively allow mass transfer between two different streams (gas/liquid or liquid/liquid) that are in contact, without mixing, at the membrane interface. Membrane contactors can be used in several industrial areas: liquid/liquid extraction, gas absorption and stripping, biotechnology applications, pharmaceutical applications, wastewater treatment, metal-ion extraction, the electronics industry, the automobile industry, the food industry, air dehumidification, membrane distillation, membrane crystallizers [1], purification of flue gases (CO_2, SO_2, H_2S, etc.) [2], and microclimate control (relative humidity, odors, etc.). They are used as end-of-pipe technology but also integrated in the process to recover products. The driving force for the mass transfer usually is a (partial) pressure gradient, a temperature gradient, or a concentration gradient at the membrane interface. In this chapter only membrane contactors systems working with specific extractants to capture selected molecules will be described.

In Figure 22.1 the separation process is schematically described. The two phases are separated by a hydrophobic microporous membrane that acts as an interface between two streams and controls the mass transfer. The two streams (liquid or gas) flow tangentially to the membrane: one stream contains the liquid or gas to be treated and the other stream contains an extractant active towards the specific molecules that need to be separated from the first stream. The hydrophobic membrane does not allow the liquid water-based stream to pass through the membrane; while the extractant contained in the same liquid stream captures at the membrane interface a target gas or vapor molecule contained in the second (gas or liquid) stream. Thus, the membrane does not act as selective media, it just acts as an interface between the two streams. The selectivity is guaranteed by the extractant, which is characterized by a very high affinity for the target molecule. The extractant must be selected very carefully in order to accommodate the process needs. Specifically, it must be characterized by the following characteristics: high activity towards the molecules to be captured; low energy requirement for regeneration; nontoxic; noncorrosive; low volatile; low viscosity; low cost.

Membrane Operations. Innovative Separations and Transformations. Edited by Enrico Drioli and Lidietta Giorno

ISBN: 978-3-527-32038-7

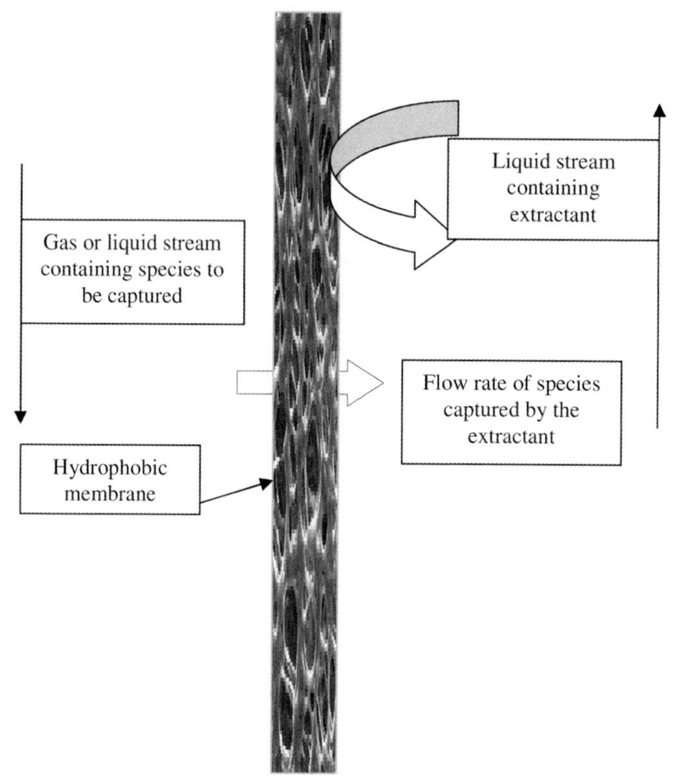

Figure 22.1 Schematic representation of a hydrophobic membrane used as interface between a gas and a liquid stream in a membrane contactor.

Membranes are polymeric microporous materials in hollow-fiber or flat-sheet configurations. The membrane properties control the contactor and the membrane contactor system performance and economy. The most important membrane properties are:

- low resistance to mass transfer: this contributes in obtaining high flow rate,
- hydrophobicity stable with time,
- high value of water breakthrough: this allows operation at higher partial pressure difference between the two sides of the membrane,
- cost low enough to have membrane contactor systems competitive with traditional equipment.

However, the engineering optimization of the equipment and of the processes is also very important, that is, all auxiliary pieces of equipment must be properly designed or selected in order to minimize parasitic costs; if not properly selected, auxiliary equipment costs can contribute strongly to the total cost.

The experimental results reported in this chapter are related to membrane contactors manufactured by GVS S.P.A by using microporous polymeric flat-sheet

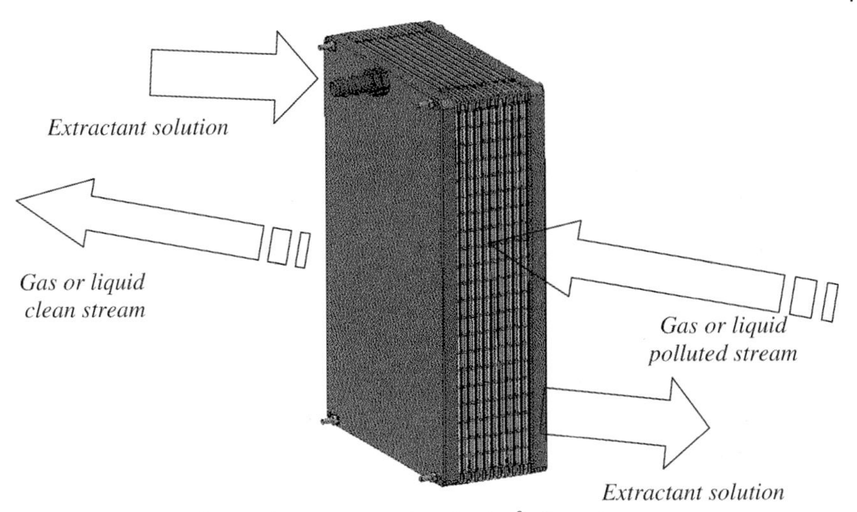

Figure 22.2 Picture of a membrane contactor housing 1 m^2 of membrane.

membranes and cassette-type membrane contactors. The flat configuration of contactors was selected to keep the pressure drops at both sides of the membrane in the contactor as low as possible and to have the possibility to use a large variety of membranes; in fact, the flat membranes commercially available are more numerous than available hollow-fiber membranes. The possibility to keep pressure drops low is very important in most applications in order to integrate membrane contactors in existing plants without changing existing auxiliary equipments: pumps, fans, heat exchangers. This contributes to minimizing assembling time and equipment cost.

In Figure 22.2 the picture of a membrane contactor manufactured by GVS and used for the demonstration tests reported in this chapter is shown. The mechanism of the capture of a specific gas molecule by the extractant is also schematically illustrated. The molecule is captured by the extractant without any contact between the stream to be treated and the extractant stream. While the gas or liquid polluted stream passes through the contactor flowing tangentially to one side of the membrane, the gas molecule to be captured passes through the pores of the membranes and is captured at the other side of membrane interface by the extractant solution. Periodically, the extractant is regenerated to release the absorbed molecules and to be reused in the system.

In Figure 22.3 a picture of various membrane contactors housing 1 and 3 m^2 of membrane is shown. The modules were developed by GVS in partnership with other institutions in the framework of the several projects partially funded by the European Union: FP6 EC integrated project 'Ultra Low CO_2 Steelmaking' — Contract No. 515960; P6 EC integrated project 'CO_2, From Capture To Storage' — Contract No. 502686; FP6 LIFE Project Novel Technology to Reduce Greenhouse Gas Emissions Contract No. LIFE 05 ENV/IT/000876. Also, part of the work was funded by the Italian government in the framework of the project PRIITT –3.1 A 09.02.2004 - Project No. 171.

Figure 22.3 Picture of two membrane contactors each housing 1 m^2 of membrane (right) and 1 contactor housing 3 m^2 of membrane (left).

These flat-type membrane contactors have been used in several demonstration tests carried out in the framework of the previously cited European projects or independently by GVS in collaboration with selected industrial end-users. The efforts have been concentrated on two sectors:

(1) air dehumidification (for short- and medium-term industrial applications),
(2) capture of CO_2 from flue gases of steel and power plants (for long-term industrial applications).

In this chapter the results related to the dehumidification of air in refrigerated trucks, the dehumidification of air in refrigerated storage rooms, and the capture of CO_2 from flue gases of steel industry are described.

A good review of industrial application is reported in [3].

In Table 22.1 the characteristics of a membrane contactor housing 1 m^2 of flat membrane and used for the demonstration tests related to air dehumidification are

Table 22.1 Characteristics of the membrane contactor used to dehumidify air and housing 1 m^2 of membrane area.

Dimension, mm	251 × 380 × 88
Membrane area, m^2	1
Specific surface, m^2/m^3	120
Pressure drop gas side, Pa	<200
Water vapor flow rate, g/h m^2	>1000
Water breakthrough - desiccant side, m H_2O	>5
Thermal resistance, °C	max 90

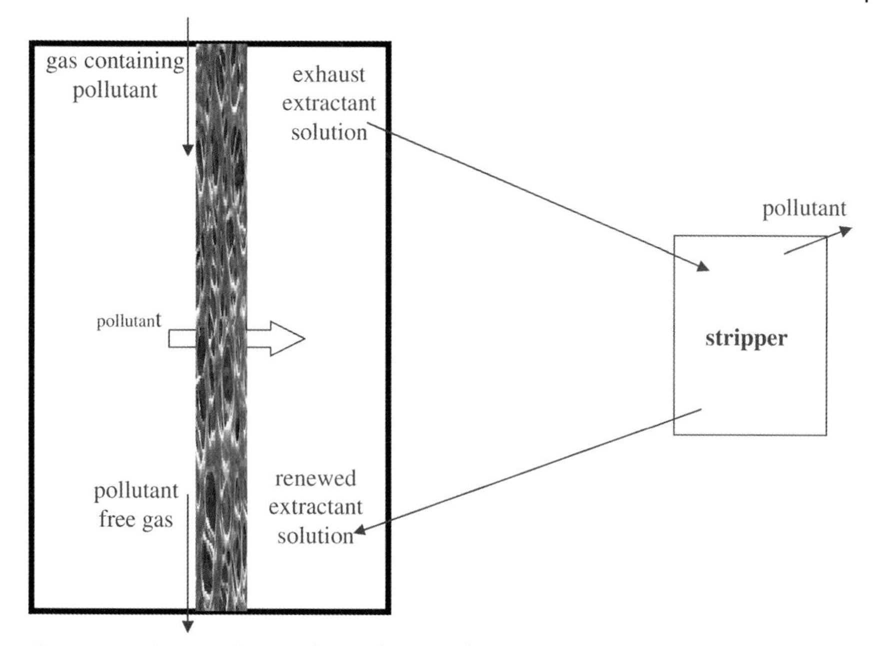

Figure 22.4 Schematic diagram showing how membrane contactors systems operate.

reported. Dimensions also include connections. The contactors are made of reinforced polypropylene (except the membrane); this suggest to keep the working temperature lower than 90 °C. Alternative materials (such as, for example, polyamides) can be used if higher thermal resistance is required. They can easily be assembled in series or parallel configurations.

In Figure 22.4 a schematic diagram showing how membrane contactors systems operate is illustrated.

In the contactor the pollutant is captured at the membrane interface; in the stripper, which could also be a membrane contactor, the exhaust extractant solution is regenerated and the pollutant released. The regenerated extractant solution is thereafter returned to the membrane contactor.

In Figure 22.5 the general layout of a system using membrane contactors is described.

An industrial system comprises a membrane contactor, a regenerator, a liquid desiccant (LD) solution, two pumps to circulate the desiccant in the membrane contactor and in the regenerator, one or two heat exchangers to preheat the diluted desiccant before the regeneration step and to cool the concentrated desiccant before entering the membrane contactor. In the membrane contactor the target molecules are absorbed. In the regenerator the desiccant is regenerated and the absorbed molecule is discharged. This is the main energy input and energy of low quality (from waste heat, natural gas, solar collectors, etc.) must be used when possible.

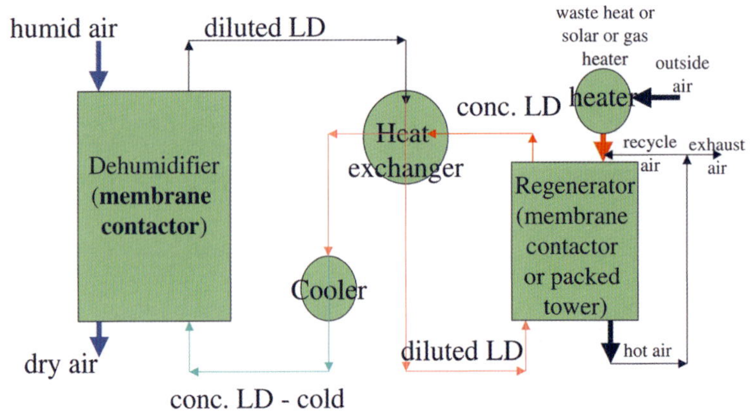

Figure 22.5 General layout of a membrane contactors system.

The advantages over traditional stripping technologies can be summarized as follows:

- there is no mixing between streams, therefore there is zero carryover,
- it is possible to independently control the flow rates of the two streams,
- the systems are very compact and the weight of the equipment is lower than currently used absorbers or strippers,
- high contact area per unit of volume of the system and precise identification of the interfacial area,
- limited pressure drop at both streams' sides,
- no channeling, foaming or by-pass phenomena occur,
- the noise level is limited,
- high recovery factors are possible,
- limited fouling of membranes,
- systems are manufactured with plastics, corrosion is limited,
- membrane contactors can be scaled up easily because they are modular system working in series and parallel: more contactors in series increase the separation efficiency, more contactors in parallel increase the capacity of the system. By using modular cassettes it is possible to design a system of any required capacity.

However, industrial applications have been limited mostly because these new membrane-based systems have not been proven extensively at the industrial level and their advantages have not been quantified in industrial terms, as is the case for traditional consolidated technologies. This fact has discouraged industries from applying these systems in large plants. Therefore, demonstration tests are needed in order to be able to fully exploit the commercial potential of the membrane contactor technology.

22.1
Air Dehumidification: Results of Demonstration Tests with Refrigerated Storage Cells and with Refrigerated Trucks

To speed up in the short and medium term the exploitation of the technology, in the last 5 years GVS has carried out several projects to demonstrate the environmental, technical, social and economic viability of processes to dehumidify air based on membrane contactors.

The objective is to quantitatively show that by using membrane contactors in refrigerated trucks and storage cell systems it is possible to save energy, decrease time and frequency of defrosting cycles, minimize ice formation on (packed) goods, improve safety of people working in the area, facilitate movement of automated or mechanical equipment, and improve the impact on environment. In fact, by using membrane contactors air at atmospheric pressure can be dehumidified without cooling, that is, independently from temperature; therefore, these systems have the potential to save energy when compared to traditional vapor-compression cycles.

In Figure 22.6 the dehumidification process based on membrane contactors is schematically represented. The humidity in the air is captured in the membrane contactor by a liquid desiccant. The liquid desiccant does not pass through the hydrophobic membrane and is stopped at the membrane interface, while the water vapor passes freely through the membrane and is captured by the liquid desiccant at the membrane interface. The driving force for the process is the difference between the vapor pressure in the air and in the desiccant aqueous solution; therefore, air temperature and RH can be controlled independently. Periodically, the desiccant solution is regenerated by stripping off the water absorbed by the desiccant by using a standard stripper or a membrane contactor. The liquid desiccant is a water solution of a highly hygroscopic salt.

This technology is important to reduce energy consumption in applications where latent loads are high.

Specifically, the demonstrations have been focused upon the dehumidification of air in refrigerated trucks at 5 °C and in intermediate cells of refrigerated storage rooms at $T = 15$ °C.

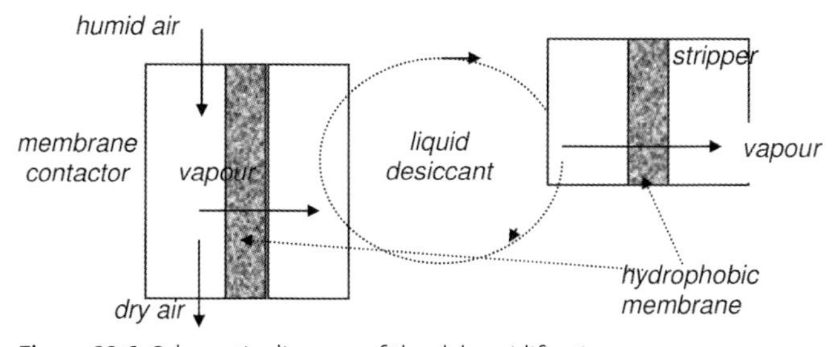

Figure 22.6 Schematic diagram of the dehumidification process.

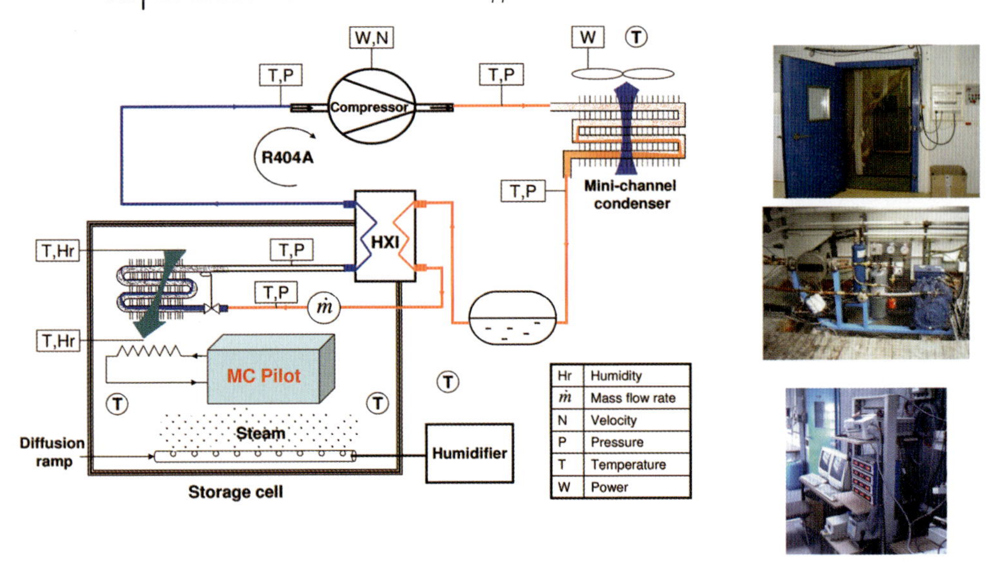

Figure 22.7 Layout of the system: membrane contactor and refrigerated cell of trucks or of storage cells.

The objective of these applications is to minimize the ice formation on evaporator surfaces of cooling equipment, on packaging, on movable parts of equipment, and on floors. This would allow less use of energy, more comfortable working places, less maintenance, and minimized loss of products.

A demonstration unit able to treat 600 m^3/h of air has been built and delivered to the demonstration site. In Figure 22.7 the layout of the system (membrane contactor and refrigerated cell of trucks or of storage cells) is described. The chamber was equipped with two refrigerating units working with R404A (reference refrigerant in industry for such applications). The unit was instrumented to measure temperature and pressure at different points of the refrigerating circuit, the refrigerant flow rate, the energy consumption of the compressor, of the evaporator and of the condenser.

On the air side calibrated thermocouples to measure temperature at different locations in the cell; two capacitive hygrometers on the evaporator; a humidity controller designed to deliver a maximum vapor flow rate of 10 kg/h have been installed. A liquid flowmeter to measure the volumetric flow rate of the aqueous desiccant solution at the membrane contactor outlet was used. All instruments are connected to a data logger.

In Figure 22.8 the layout of the membrane contactor system is described. As liquid desiccant a $CaCl_2$ solution has been used. The demonstration runs lasted more than 450 hours.

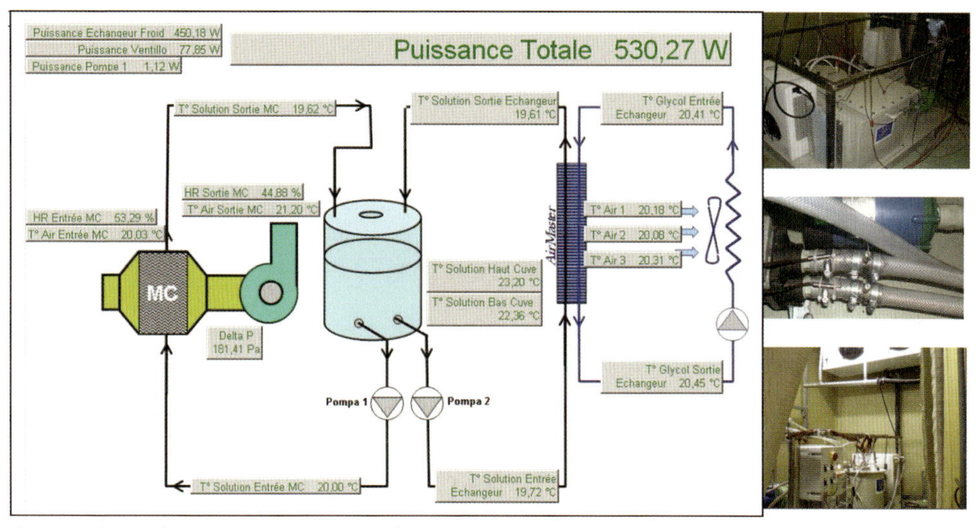

Figure 22.8 Membrane contactor and its auxiliary equipment.

22.2
Refrigerated Storage Cells

For the refrigerated storage cells application the intermediate cell (IC) application has been selected, because it has emerged that it is more convenient and effective to stop the water vapor flow rate in the intermediate cell at 15 °C before it enters the refrigerated storage cell at −25 °C. In fact, the energy required to capture the water vapor at −25 °C is remarkably higher because when the vapor enters the cold cell it must be condensed and transformed in ice at −25 °C (using energy). Therefore, it is better to capture the vapor in the IC. In Figure 22.9 the general layout of the humidity control in the IC in order to reduce the humidity to the low-temperature storage cell (SC) is reported.

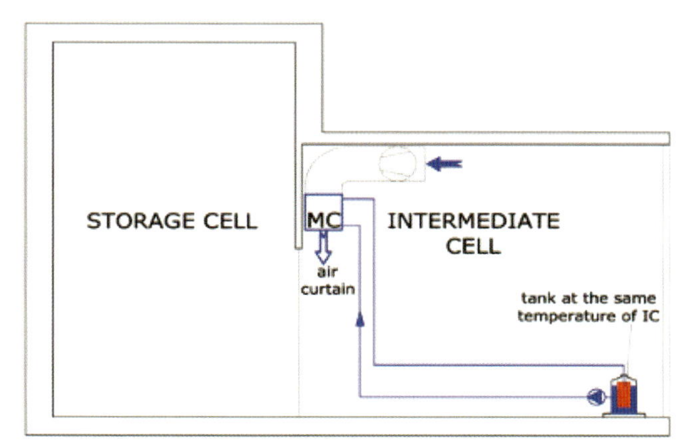

Figure 22.9 Schematic diagram of the system studied.

The air temperature of the intermediate cell is assumed to be equal to 15 °C. This intermediate cell is dehumidified by the membrane contactor system placed at the evaporator inlet. The dehumidified air is driven to an air curtain at the top of the door separating the intermediate cell from storage cell. The heat load of the system and the air circulation ratio are assumed to be constant. The temperature of $CaCl_2$ solution at the inlet of the membrane contactor is assumed to be equal to the indoor IC air temperature. The following parameters have been measured:

- outside temperature = 25 °C,
- outside R.H. = 70%,
- temperature of intermediate cell (IC) = 15 °C,
- temperature of the storage cell (SC) = −30 °C,
- ratio SC volume/IC volume = 3,
- sensible heat load of IC = 4.8 W/m^3,
- sensible heat load of SC = 1.6 W/m^3,
- air renewal outside air-IC in 12 h working time = 1 h^{-1},
- air renewal outside air-IC in no-working time = 0.05 h^{-1},
- air renewal SC-IC in 12 working time = 0.08 h^{-1},
- air renewal SC-IC in no-working time = 0.003 h^{-1},
- air curtain efficiency between SC and IC = 0.4,
- contactor membrane area = 8×10^{-3} m^2/m^3 (IC),
- air flow rate through the contactor = 1.6 $m^3/h\,m^3$(IC),
- saturated $CaCl_2$ flow (IC temperature) = 1.6 $kg/h\,m^3$(IC).
- Door IC – outdoor $H = 4$ m, $W = 3$ m,
- no-working period (12 h) ne = 0.05 [1/h],
- working period (12 h),
- 30 opening; $\Delta\tau = 360$ s.
- Door IC – SC $H = 2.5$ m, $W = 3$ m,
- no-working period (12 h) nnif = 0.005 [1/h],
- working period (12 h).

In these conditions a total energy saving of about 15% can be achieved.

22.3 Refrigerated Trucks

Applications involving transport of fresh refrigerated goods at 5 °C have been selected. In fact, these types of refrigerated transports are more demanding in terms of energy because these goods (fruit and vegetables) are generally not packed and the rate of water evaporation is quite high: this generates high energy consumption and spoilage of goods. On the contrary, generally in trucks transporting frozen products (at −25 °C) the goods are packed in plastics that prevent evaporation. Therefore, the energy consumption is quite low and the application of the membrane contactors is not very suitable. In Figure 22.10 and in Figure 22.11 the layout of the transport of fresh perishable food at +5 °C is described. The air is dehumidified in the

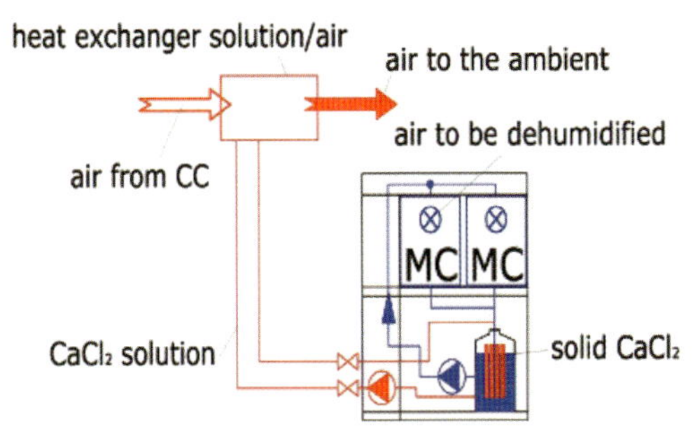

Figure 22.10 Scheme of the system considered for refrigerated trucks applications.

membrane contactors. A heat exchanger allows the control of the $CaCl_2$ solution temperature in the tank at $T = 5\,°C$.

The membrane contactor is placed at the inlet of the evaporator of the refrigerating unit and handles 10 – 20% of the total recycled air flow. This value defines the by-pass factor (BF). In the demonstrations the BF was 80–90%. After dehumidification, the air is mixed with the recycled air.

In the tests carried out the resulting energy savings are higher than 20% when captured water is more than 0.25 kg/h. In the case of optimized conditions a reduction of CO_2 equal to 20% is possible.

The surface temperature of the cooling coil has a great influence on the dehumidification process of the recycled air owing to the ice forming on the cooling coils, that is, the lower the coil temperature, the greater the frost formation, reducing the apparent efficiency of the membrane contactor system. The higher the latent load, the higher is the effectiveness of the membrane contactor.

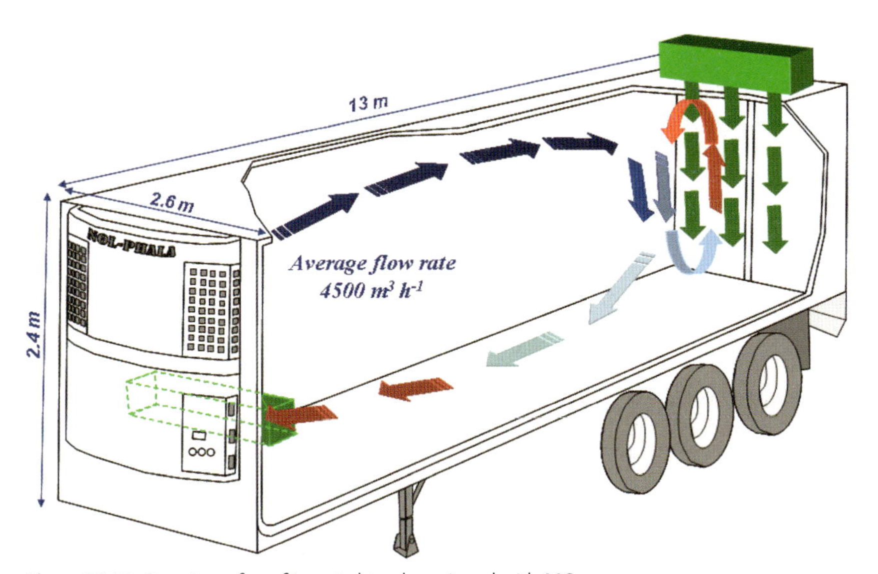

Figure 22.11 Overview of a refrigerated truck equipped with MC.

In conclusion, the demonstration runs have shown that the membrane contactor dehumidification system is effective when the total loads of the refrigerated cell do not require an evaporative temperature of the cooling coil below −10 °C. In these conditions it has been shown that energy, operating costs and CO_2 emissions can be decreased up to 20% when optimized systems are built. The optimization is related mostly to the selection of adequate pumps, fans, heat exchangers.

The system could also be further improved if membrane contactors are used in hybrid systems integrated with vapor compression inverse cycle; in this case the power saving can be as high as 50% [4].

22.4 Capture of CO_2 from Flue Gas

GVS has also dedicated important efforts to demonstrate the possibility to apply membrane contactors to capture CO_2 from flue gases of steel and power plants.

For CO_2 capture mostly PTFE membranes are used [5]; as extractants aqueous solutions of different amines are used [6]. In the past, the use of membrane contactors to capture CO_2 has been studied by TNO in the Netherlands and by Kvaerner/Gore in Norway and Germany [7]. They have worked with tubular systems. They claim that working with flue gas from a power generation plant can recover 85% of CO_2.

In the following section of the chapter results related to the recovery of CO_2 from flue gases of conventional blast furnaces of steel plants are described. The work was carried out in the framework of the FP6 EC integrated project 'Ultra Low CO_2 Steelmaking'—Contract No. 515960. GVS has used cassette-type membrane contactors as reported in the previous sections of this chapter. In Figure 22.12 a

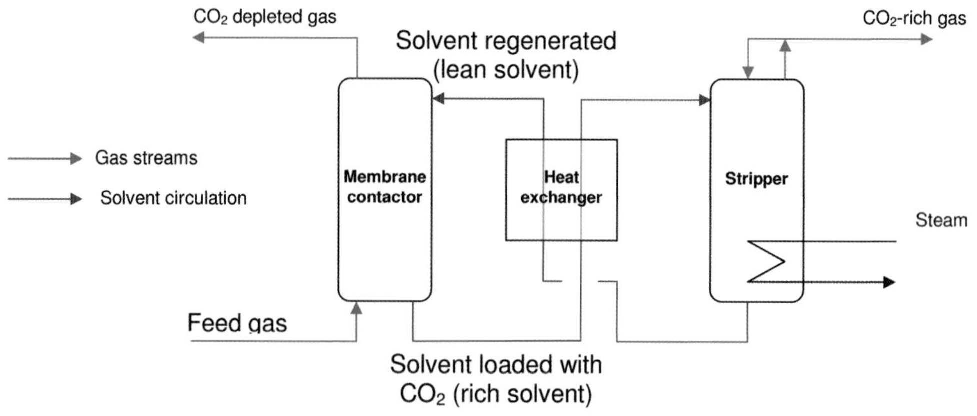

Figure 22.12 Flow sheet of a membrane contactor plant to capture CO_2 from flue gases of a steel manufacturing plant (Data developed in the framework of FP6 EC integrated project 'Ultra Low CO_2 Steelmaking'—Contract No. 515960.)

Table 22.2 Design of an industrial plant treating 300 000 m^3/h of flue gas.

Gas composition, IN	75% air + 25% CO_2
Air flow rate, IN – m^3/h	30
CO_2 flow rate, IN – m^3/h	10
CO_2 flow rate, OUT – m^3/h	4
$C0_2$ absorbed by1 contactor, $m^3/h\ m^2$	6
Density of CO_2, kg/m^3	1.8
Gas flow rate, IN - 25% CO_2, m^3/h	300 000
CO_2 flow rate, IN - m^3/h	75 000
CO_2 captured, m^3/h	66 000
CO_2 noncaptured, m^3/h	9000
Contactor membrane area, m^2	11 000
Contactor cost, €/m^2	48
Total contactor cost, €	528 000

schematic diagram of the application studied is reported. CO_2 is captured in the membrane contactor by using a 30% water solution of ethanolamine (MEA). The MEA solution is continuously regenerated in a steam-operated stripper.

Process conditions have been optimized in order to obtain the best possible efficiency and cost. It has been shown that membrane contactors can be advantageously used to capture CO_2 from flue gases containing about 25% by volume of CO_2 and to obtain in the decarbonated gas maximum 3% of CO_2 mole (i.e. 88% capture of CO_2). It has been proven that the contactors can capture up to 6 m^3/h of CO_2 per m^2 of membrane. In Table 22.2 results of a design of a potential industrial plant treating 300 000 m^3/h of flue gas are reported.

The cost of a single contactor (48 €/m^2) is based on a GVS evaluation related to the anticipated cost of contactors in the year 2013.

Table 22.3 Cost of the CO_2 captured when treating about 300 000 m^3/h of flue gas containing 25% volume of CO_2.

Cost of equipment	€1 320 000
Total steam to regenerate MEA (10 €/t steam)	200 t/h
Chemicals consumption	500 k€/y
Manpower (60 k€/man-year)	1 man/y
Variable costs/t of CO_2 captured	17 €/t
Maintenance	41 000 €/y
CO_2 recovery	88%
CO_2 purity (dry basis)	100%
Total CO_2 captured (density 1.8 kg/m^3)	66 000 m^3/h (1 040 688 t/y)
Return on investment (R/I)	12%
Amortization (15 yr linear)	7%
General overhead	1%
Total cost CO_2 captured	18.3 €/t

In Table 22.3 the cost to capture CO_2 by using a membrane contactor plant treating about 300 000 m^3/h of flue gas (25% volume of CO_2) is reported.

The cost refers only to the membrane contactor part of the schematic plant reported in Figure 22.12. The cost of remaining equipment has not been evaluated because it refers to standard pieces of equipment extensively used in industries and their cost can vary remarkably depending upon specific factors of different companies.

The resulting figure, equal to 18.3 €/ton of CO_2 recovered, is quite interesting and competitive with traditional technologies to capture CO_2, such as, for example, direct contact strippers using amines (the estimated cost to capture CO_2 in this case is about 40 €/ton) or absorption processes (VPSA) combined with a cryogenic CO_2 purification unit (the estimated cost to capture CO_2 in this case is about 30 €/ton).

References

1 Drioli, E., Curcio, E. and Di Profio, G. (2005) *Trans IchemE, Chemical Engineeging Research and Disign*, **83**, 1–11.

2 Klaassen, R., Feron, P.H.M. and Jansen, A.E. (2005) *Trans IchemE, Chemical Engineeging Research and Disign*, **83**, 234–246.

3 Gabelman, A. and Hwang, S.T. (1999) *Journal of Membrane Science*, **159**, 61–106.

4 Berghero, S., Chiari, A. and Nannei, E. (2006) Proceedings of ESDA2006, Turin Italy, July.

5 Feron, P. and Jansen, A.E. (2002) *Separation Purification Technology*, **27**, 231–242.

6 Kim, Y.S. and Yang, S.M. (2000) *Separation Purification Technology*, **21**, 101–109.

7 Svendsen, H.F., Hoff, K.A., Poplsteinova, J. and Silva, E.F. (2001) 2nd Nordic Symposium on CO_2 Capture and Storage, Goteborg, October 26.

23
Extractive Separations in Contactors with One and Two Immobilized L/L Interfaces: Applications and Perspectives

Štefan Schlosser

23.1
Introduction

Partitioning of components between two immiscible or partially miscible phases is the basis of classical solvent extraction widely used in numerous separations of industrial interest. Extraction is mostly realized in systems with dispergation of one phase into the second phase. Dispergation could be one origin of problems in many systems of interest, like entrainment of organic solvent into aqueous raffinate, formation of stable, difficult-to-separate emulsions, and so on. To solve these problems new ways of contacting of liquids have been developed. An idea to perform separations in three-phase systems with a liquid membrane is relatively new. The first papers on supported liquid membranes (SLM) appeared in 1967 [1, 2] and the first patent on emulsion liquid membrane was issued in 1968 [3]. If two miscible fluids are separated by a liquid, which is immiscible with them, but enables a mass transport between the fluids, a liquid membrane (LM) is formed. A liquid membrane enables transport of components between two fluids at different rates and in this way to perform separation. When all three phases are liquid this process is called pertraction (PT). In most processes with liquids membrane contact of phases is realized without dispergation of phases.

Separations in two-phase systems with one immobilized interface(s) are much newer. The first paper on membrane-based solvent extraction (MBSE) published Kim [4] in 1984. However, the inventions of new methods of contacting two and three liquid phases and new types of liquid membranes have led to a significant progress in the last forty years. Separations in systems with immobilized interfaces have begun to be employed in industry. New separation processes in two- and three-phase systems with one or two immobilized L/L interfaces realized with the help of microporous hydrophobic wall(s) (support) are alternatives to classical L/L extraction and are schematically shown in Figure 23.1. Membrane-based solvent extraction (MBSE) in a two-phase system with one immobilized interface feed/solvent at the mouth of microspores of hydrophobic support is depicted in Figure 23.1a and will be discussed

Membrane Operations. Innovative Separations and Transformations. Edited by Enrico Drioli and Lidietta Giorno

ISBN: 978-3-527-32038-7

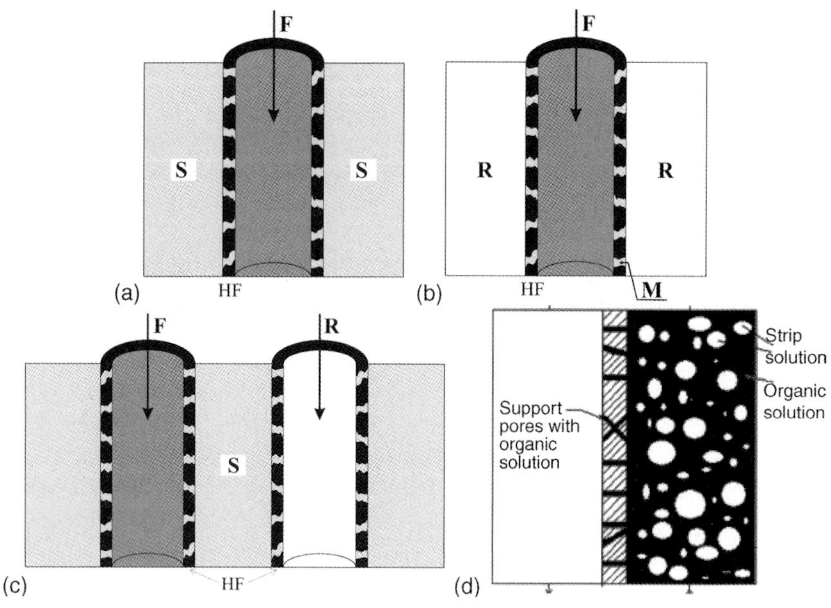

Figure 23.1 Two- and three-phase systems with one and two immobilized L/L interfaces realized with help of microporous hydrophobic wall(s) (support). (a) Two-phase system with one immobilized interface F/S, (b) Three-phase system with two immobilized interfaces F/M and M/R, organic phase in pores serves as a supported liquid membrane (SLM), (c) Three-phase system with two immobilized interfaces F/M and M/R, organic phase in pores and between fibers serves as a bulk liquid membrane (BLM), (d) Three-phase system with one immobilized interfaces F/M and emulsion of the stripping solution – bulk liquid membrane with emulsion (BLME).
F – aqueous feed, HF – hydrophobic microporous hollow fiber, M – membrane phase (organic), R – aqueous stripping solution, S – organic solvent.

in Section 23.3. Processes with two phases and one immobilized interface are de facto not membrane processes because the microporous wall only immobilize an L/L or L/G interface and has no separation function in a process.

A three-phase system with two immobilized interfaces feed/LM and LM/stripping solution is shown in Figure 23.1b, where organic phase soaked in micropores of hydrophobic hollow fiber wall serves as a supported liquid membrane (SLM). Pertraction through SLM and its applications will be discussed in Section 23.5. A three-phase system with two immobilized interfaces feed/LM and LM/stripping solution realized in two hydrophobic walls with organic phase in pores and between fibers (support walls). Organic phase serves as a bulk liquid membrane (BLM) as presented in Figure 23.1c. More information on pertraction through this type of BLM can be found in refs. [5–9]. A three-phase system can be made up of one immobilized feed/LM interface in microporous wall and a third phase, the stripping solution, emulsified in the bulk of organic phase – BLME. A pure membrane (organic) phase is in the pores of the support wall. The aqueous stripping solution cannot enter hydrophobic pores. BLM with emulsion of stripping solution is in the bulk of the organic phase as shown in Figure 23.1d and will be further designated as BLME. Pertraction through BLME will be discussed in Section 23.4.

Pertraction (PT) can be realized through a liquid membrane, but also through a nonporous polymeric membrane that was applied also industrially [10–12]. Apart from various types of SLM and BLM emulsion liquid membranes (ELM) were also widely studied just at the beginning of liquid membrane research. For example, an emulsion of stripping solution in organic phase, stabilized by surfactant, is dispersed in the aqueous feed. The continuous phase of emulsion forms ELM. Emulsion and feed are usually contacted in mixed column or mixer-settlers as in extraction. EML were applied industrially in zinc recovery from waste solution and in several pilot-plant trials [13, 14], but the complexity of the process reduced interest in this system. More information on ELM and related processes can be found in refs. [8, 13–16].

A serious problem, which has not been solved up to now, is the short lifetime of SLM. One way to overcome this is application of pertraction with bulk liquid membranes (BLM) shown in Figures 23.1c and d. This is taken into account by a doubled wall and a thicker liquid membrane layer between walls (bulk membrane) in an arrangement as in Figure 23.1c compared with SLM, which results in a higher mass-transfer resistance [17, 18]. In pertraction through BLME (Figure 23.1d) dispersion or emulsion of stripping solution is used, but the advantage of the nondispersive process is lost. In systems inclined to emulsification in stripping this may introduce a problem. The mass-transfer resistance will be lower than in the system in Figure 23.1c.

Several mechanisms to achieve transport of solute(s) through the L/L interface or through a liquid membrane can be utilized. The separation mechanism could be based on differences in the physical solubility of the solutes or their solubilization in the solvent or reverse micelles or based on the chemistry and rate of chemical or biochemical reactions occurring on L/L interface(s). The complexing or solubilization agent – extractant (carrier in the liquid membrane) forms by a reversible reaction complex(es) or aggregate(s) with the solute, which are soluble in the solvent or membrane. The chemistry of reactive extraction and stripping in MBSE and MBSS, as well as in PT, is identical to the classical solvent extraction or stripping and is presented in several books and papers, for example, [8, 19–21].

Enzymatic reactions on L/L interfaces were employed to achieve separations [22–25]. For example, the aqueous feed, a mixture of phenylacetic acid (PAA) and 6-aminopenicillanic acid and the stripping phases flow in the lumen of two bundles of hydrophobic hollow fibers with microporous walls [24]. Fibers are immersed in the immiscible organic phase forming a liquid membrane. PAA reacts at the L/L interface with alcohol added to the feed under the catalytic action of enzyme lipase *Candida rugosa* and the ester formed dissolves in the membrane and is transported through it. On the downstream interface the desertification reaction proceeds catalyzed by another lipase enzyme and PAA is deliberated to the stripping solution with a higher pH than in the feed. The second acid is not transported via this mechanism and separation of acids occurs with a separation factor of 10.

Formation of hybrid production/biotransformation – separation processes, including extractive processes, could enhance production and is of great interest, as shown in review papers [8, 21, 26–30]. The flowsheet of the extractive fermentation unit with an integrated MBSE and MBSS circuit for recovery of acid(s)/product from the fermentation broth is discussed in Section 23.3 and with integrated pertraction in Section 23.5.

The aim of this chapter is to give a short overview of recent trends on application of extractive separations in contactors with one and two immobilized L/L interfaces and their perspectives. Industrial applications and pilot-plant tests will be stressed and selected case studies highlighted.

23.2 Contactors with Immobilized L/L Interfaces

Contactors with flat-sheet and cylindrical walls are used but only hollow fiber (HF) contactors in cylindrical modules in several sizes are available commercially [31]. Flat-sheet contactors are widely used in analytical chemistry [32–34]. There are two main types of HF contactors, those with parallel flow of phases in fiber lumen and in shell or crossflow of phases. A HF contactor with crossflow of phases is shown in Figure 23.2. More details on their construction and sizes available are presented in the producer's web site [31].

HF contactors with planar elements with flowing head of fibers and crossflow of one phase in three and more phases contactor have been suggested in a patent [35] and their scheme is shown in ref. [8]. A two-phase HF contactor with planar elements was developed at TNO and tested in pilot plants [36, 37]. Reviews on two-phase HF contactors are presented in refs. [27, 38–40]. Mass-transfer characteristics of two-phase contactors are presented in ref. [30]. Three-phase HF contactors for pertraction are described in refs. [6–9, 41]. They are not produced commercially.

HF contactors have a large interfacial area per unit volume of the contactor without the requirement of dispergation of one phase, which can be advantageous in systems sensitive to emulgation [42–45]. The volume ratio of phases could be varied practically without limitations. A disadvantage of HF contactors is connected with additional mass-transfer resistance introduced by porous wall(s) immobilizing L/L interface(s). Some problems with swelling of HF and especially of potting material of HF in organic solvents may occur.

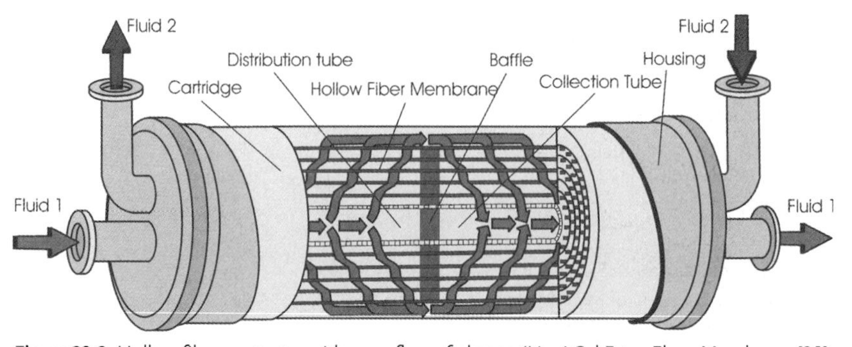

Figure 23.2 Hollow-fiber contactor with crossflow of phases (Liqui Cel Extra-Flow, Membrana [31].

Modeling of HF contactors is in most papers based on a simple diffusion resistance in series approach. In many systems with reactive extractants (carriers) it could be of importance to take into account the kinetics of extraction and stripping reactions that can influence the overall transport rate, as discussed in refs. [30, 46]. A simple short-cut method for the design and simulation of two-phase HF contactors in MBSE and MBSS with the concentration dependent overall mass-transfer and distribution coefficients taking into account also reaction kinetics in L/L interfaces has been suggested [47].

A rotary disc (RD) pertractor with BLM was suggested in refs. [48, 49]. In a RD pertractor hydrophilic discs are fixed on a rotating horizontal shaft. The lower parts of the discs are immersed in compartments, which are alternately filled with the stripping solution and the feed. The remaining parts of the discs, on which films of aqueous phases are formed due to rotation, are immersed in the membrane phase. Mass transfer occurs from the feed films into the stripping solution films through the bulk liquid membrane. Pertraction in RD contactors has been widely studied in the Boyadzhiev group for recovery of organic acids [50–53], antibiotics [54], alkaloids [55–58], biosurfactant [59] and metals [60–64].

23.3 Membrane-Based Solvent Extraction (MBSE) and Stripping (MBSS)

Membrane-based solvent extraction (MBSE) is a relatively new alternative of classical solvent extraction where mass transfer between immiscible liquids occurs from the L/L interface immobilized at the mouth of pores of a microporous wall, which is not wetted by one of the phases in contact as shown in Figure 23.1a and in more detail in Figure 23.3. Basic information on MBSE is given in refs. [18, 27, 30, 38, 39, 65]. The solvent can be regenerated by membrane-based solvent stripping (MBSS) where the solute is re-extracted into the stripping solution. Another method of regeneration could be distillation of the volatile solvent or solute, and so on, depending on the properties of the system. A schematic flow sheet of the simultaneous MBSE and

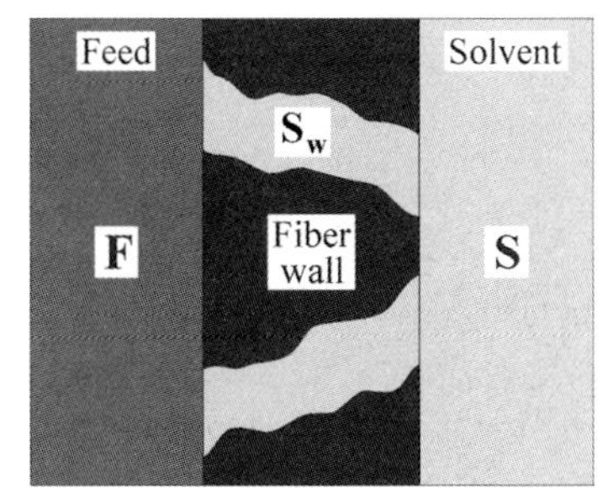

Figure 23.3 Detail view of the two-phase system in membrane-based solvent extraction (MBSE) in contactor with hydrophobic wall.

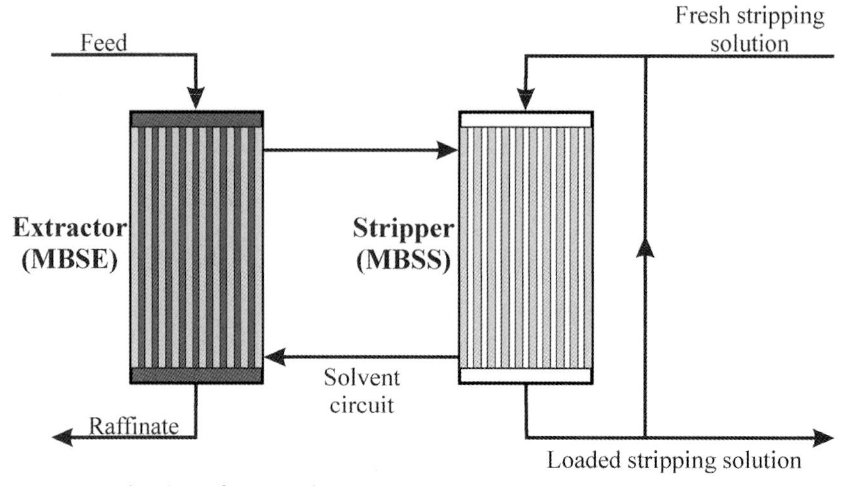

Figure 23.4 Flowsheet of MBSE with simultaneous regeneration of the solvent by MBSS in HF contactors and recirculation of the solvent to extraction. In both contactors the solvent flows in the shell.

MBSS processes with closed loop of the solvent is shown in Figure 23.4. In this way recovery of the solvent and concentration of the solute can be achieved. Preferable contactors for MBSE and MBSS are hollow-fiber contactors, which are discussed in Chapter 2. Filling the pores with gel to protect against leakage of phase through the pores has been suggested in ref. [66].

The functions of contactors in the simultaneous MBSE and MBSS with an arrangement as shown in Figure 23.4, are coupled. They react similarly as a pertractor with a SLM. The differences are only in the overall resistance, which is smaller in the pertractor where there is only one support wall. In addition, it is not necessary to pump the solvent in its circulation loop in PT, as is used in the simultaneous MBSE and MBSS process [30]. On the other hand, in pertraction through SLM its limited lifetime could be problem which is not the case in MBSE and MBSS where it is easy to keep the constant properties of the solvent phase.

Extraction into capsules with a solvent, for example, recovery of phenylethanol (a product of phenylalanine bioconversion by yeast) [67] or lactic acid from fermentation broth [68], has attracted interest recently. The polymeric core of the capsule prevents direct contact of the solvent with biomass. This process could be regarded as a batch MBSE.

An interesting variant of MBSE with a dual mechanism of separation is extraction from an ion-exchange membrane, which has been suggested by Kedem and Bromberg [69] and Isono *et al.* [70, 71]. The separation is performed by L/L partitioning and is enhanced by an electrostatic rejection in the ion-exchange membrane.

A very interesting separation of biomolecules, for example, proteins and enzymes (BSA, lysozyme), in an aqueous two-phase system (ATPS) realized in hollow fiber contactor is suggested in the paper of Riedl and Raiser [72]. Application of ATPS in

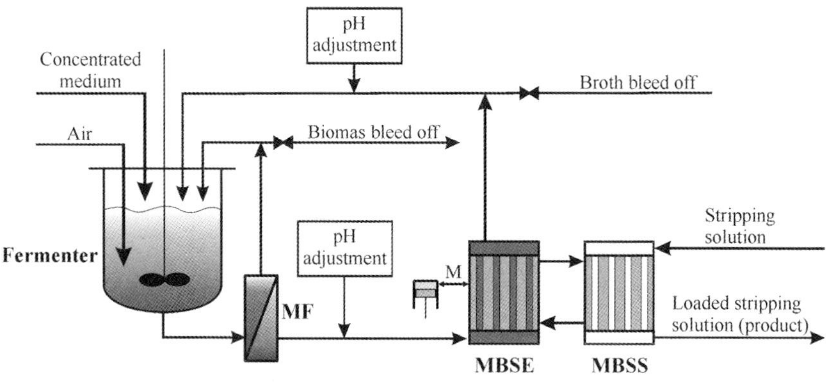

Figure 23.5 Schematic flowsheet of the fermentation unit with integrated MBSE and MBSS circuit for recovery of acids (product) from the fermentation broth [30].

MBSE opens up an attractive alternative to classical downstream processing operations as chromatography.

A schematic flowsheet of the fermentation unit with integrated MBSE and MBSS circuit for recovery of acid(s) product from the fermentation broth is presented in Figure 23.5. Marták *et al.* [73] ran a semicontinuous fermentation of lactic acid with Rhizopus arrhizus with a periodical bleed and feed operation without a decrease in LA productivity for 152 h. Such a process could be integrated with separation of lactic acid, for example, by MBSE studied in ref. [74] or by pertraction [44, 45]. Recovery of vanilline from a fermentation broth is presented in ref. [75] aiming at formation of an integrated system. A combination of MBSE of phenol from saline solution in HF contactor with bioreactor with *Pseudomonas putida* to remove phenol is studied in ref. [76].

Modeling and optimization of MBSE and MBSS of a multicomponent metallic solution in HF contactors is discussed in ref. [77]. A short-cut method for the design and simulation of two-phase HF contactors in MBSE and MBSS with the concentration-dependent overall mass-transfer and distribution coefficients taking into account also reaction kinetics was suggested by Kertész and Schlosser [47]. Comparison of performance of the MBSE and MBSS circuit with pertraction through ELM in case of phenol removal presented Reis [78] and for copper removal Gameiro [79].

An overview of selected papers on recovery/removal of organic compounds by MBSE and/or MBSS is presented in Table 23.1 and for metals in Table 23.2.

23.3.1 Case Studies

The first published information on the industrial application of a hybrid system with a HF contactor for production of the drug dilthiazem intermediate was reported by Lopez and Matson [23]. An enzymatic resolution of dilthiazem chiral intermediate is realized in an extractive enzymatic membrane reactor. The enzyme is entrapped in the macroporous sponge part of the hydrophilic hollow-fiber membrane made of a

Table 23.1 Selected papers on recovery/removal of organic compounds by MBSE and/or MBSS.

Solute	Process[a]	Solvent[a] (extractant/diluent)	Contactor type[a]	Literature
Phenylalanine	MBSE	Aliquat 336/(kerosene, isodecanol)	PF HF	[80, 81]
	MBSE, MBSS	TOMAC/(n-heptane, hexanol)	PF HF	[82]
	MBSE, MBSS	D2EHPA/kerosene	CF HF	[83–85]
	MBSE, MBSS	D2EHPA/n-alkanes	CF HF	[86]
Lactic acid	MBSE, MBSS PT	tertiary amines/(n-alkanes, isodecanol, isotridecanol)	CF HF PF HF(PT)	[74]
	MBSE	TOA/xylene	CF HF	[46]
	MBSE, MBSS	(TOA, Aliquat336)/sunflower oil	CF HF	[87]
Butyric acid	MBSE, MBSS	TOA/n-alkanes	CF HF	[47]
	MBSE, MBSS	amines/(corn oil, oleyl alcohol)	CF HF	[88]
Aspartic acid	MBSE, MBSS	D2EHPA/kerosene	CF HF	[89]
(±)-trans-methyl-methoxyphenyl-glycidate in toluene	MBSE	Aqueous buffer	PF HF	[23]
Penicillin-G	MBSE, MBSS	Amberlite LA2/(kerosene, isodecanol)	CF HF	[90, 91]
	MBSE	Amberlite LA-2/(Shellsol TK, TBP)	CF HF	[92]
Vanilline	MBSE, MBSS	butylacetate	PF HF	[75]
Phenol	MBSS	from organic feed into aqueous NaOH	CF HF	[93, 94]
	MBSE, MBSS	Cyanex 923/isoparafins	CF HF	[78]
	MBSE combined with bioreactor	kerosene	PF HF	[76]
MPCA	MBSE, MBSS	TOA/xylene	CF HF	[95, 96] [30]
DMCCA	MBSE	(TOA or Hostarex A327)/(n-alkanes, iso-decanol)	CF HF	[97, 98]
Thiols	MBSS	from kerosene into aqueous NaOH	PF HF	[99]
Aromatic compounds; trichloroethene	MBSE	Feed to reactor; n.a.	CF HF	[37, 100]
Aroma compounds (aqueous)	MBSE	sunflower oil	CF HF	[101]
		hexane	CF HF	[102–105]
Aroma compounds (in oil)	MBSE	cyclodextrins/water	PF HF	[106]
BSA, lysozym	MBSE	aqueous two-phase system	CF HF	[72]

[a]Abbreviations used are explained in the list of abbreviations.

Table 23.2 Selected papers on recovery and removal of metals by MBSE and/or MBSS.

Metal	Process[a]	Solvent[a] (extractant/diluent)	Contactor type[a]	Literature
Cu(II)	MBSE, MBSS	LIX84/n-decane	PF HF	[107]
	MBSE, MBSS	LIX64N/kerosene; Aliquat 336/kerosene	CF HF	[108]
	MBSE	D2EHPA/kerosene	CF HF	[46]
	MBSE	Cyanex 302/kerosene	CF HF	[109]
	MBSE, MBSS	LIX 54/paraffinic solvent (Shellsol T)	CF HF	[79]
Zn(II)	MBSE, MBSS	pure TBP	CF HF	[110–112]
	MBSE, MBSS	D2EHPA/n-dodecane	CF HF	[113, 114]
Cd(II)	MBSE, MBSS	Cyanex 302/kerosene	CF HF	[17, 115]
Co(II), Ni(II)	MBSE, MBSS	Aliquat 336/kerosene	CF HF	[116]
Cr/VI)	MBSE, MBSS	Aliquat 336/kerosene	CF HF	[117, 118]
	MBSE, MBSS	Aliquat 336/(kerosene, isodecanol)	CF HF	[119–121]
Cd(II), Hg(I)	MBSE	Cyanex 302/kerosene	PF HF, CF HF	[109]
Ag(I)	MBSE, MBSS	LIX79/*n*-heptane	CF HF	[122]
Au(I)	MBSE, MBSS	LIX79/n-heptane	CF HF	[123]
Th(IV)	MBSE, MBSS	di-n-hexyl octanamide/n. alkanes	PF HF	[124]
Tl(III)	MBSE	butyl acetate	PF HF	[125, 126]

[a]Abbreviations used are explained in the list of abbreviations.

polyacrylonitrile copolymer. The enzyme is loaded to the membrane during ultrafiltration of an aqueous enzyme solution flowing at the beginning in the shell of a HF contactor. A dense layer of the membrane (skin), retaining enzyme, is on the inner side of the hollow fiber. After immobilization of the enzyme in the wall, a toluene solution of reactant, racemic (±)-trans-methyl-methoxyphenylglycidate, is flowing in the shell. In the fiber lumen an aqueous buffer solution with bisulfite anion flows countercurrently. The enzymatic desertification catalyzed by lipase proceeds on the L/L interface. The deliberated (2S,3R)-methoxyphenylglycidic acid is extracted to the buffer. The required product (2R,3S)-methyl-methoxyphenylglycidate remaining in toluene is an intermediate for dilthiazem synthesis. In the commercial plant with a capacity of 75 tons of drug per year 24 contactors with a surface area of 60 m^2 each are installed.

An aromatic compound from industrial waste water from a reactor is removed by MBSE in a HF crossflow contactor. This installation, with a capacity of 15 $m^3 h^{-1}$ of water, went into operation in 1998 [37, 100]. As a solvent is used the feed into the reactor, and in this way pollutant, is recycled into the process.

The recovery of phenol from the hydrocarbon fraction with a phenol concentration of 2–4 wt.% by MBSS into an alkali solution has been applied industrially in Poland [93, 94]. The capacity of the plant with two rigs in series, each with 8 crossflow HF contactors Liqui Cel 4 × 28″ (Membrana) connected in parallel is about 650 kg h^{-1}. Both the hydrocarbon raffinate, with less than 0.02 wt.% of phenol, and the phenolate concentrate (25–30 wt. %) are recycled back to the technology producing the waste stream.

Valuable organic acids of industrial interest can be recovered from aqueous waste solutions, for example, from an enzymatic resolution process. Mass-transfer data for MBSE and MBSS in HF contactors for dimethylcyclopropanecarboxylic acid (DMCCA) [97] and 5-methyl-2-pyrazinecarboxylic acid (MPCA) [95, 96] have been estimated. Processes for recovery of DMCCA from a highly acidic waste solution with pH below 2 containing about 19 kg m^{-3} of DMCCA by MBSE and MBSS have been suggested [98]. A solvent with 0.4 kmol m^{-3} of TOA in n-alkanes (dodecane fraction) was used. A recovery of more than 90% of DMCCA and a concentrate with about 200 kg m^{-3} of DMCCA can be achieved.

Based on laboratory data on MBSE and MBSS of MPCA in the crossflow HF contactors 2.5 in. × 8 in. [95, 96] a production pilot plant was simulated and designed [30]. 0.1 $m^3 h^{-1}$ of the aqueous feed with 0.12 kmol m^{-3} MPCA, 1 kmol m^{-3} Na_2SO_4, and constant pH (~2.5) was extracted in a HF contactor by the solvent with 0.4 kmol m^{-3} TOA in xylene. The loaded solvent was regenerated in a stripping HF contactor with NaOH solution with addition of NH_4OH. The concentration factor for MPCA in the loaded stripping solution was ~10. From simulations of the pilot plant by the short-cut method suggested in ref. [47] it followed that the number of contactors (length of fibers) in MBSE and MBSS is sensitive to the increase in MPCA yield in MBSE above 90%. The velocity of the feed in fiber lumen was 1.9 cm s^{-1}. The optimum Re number for flow of the solvent in the shell and the approach to equilibrium at the raffinate end of the HF contactor is about 0.2 and 70%, respectively [30]. The number of contactors Liqui Cel 4 × 28″ (Membrana) in series, as resulted from simulations for the mentioned process data, was found to be 2 for both MBSE and MBSS, which are reasonable numbers. The technological flowsheet of a production pilot plant unit for recovery of MPCA is shown in Figure 23.6.

A large effort of the Juelich team has been devoted to development of fermentation–extraction process for production of phenylalanine (Phe) with integrated MBSE in HF contactors [83–85]. 10 v/v% of D2EHPA in kerosene was used as a reactive solvent. Starting from laboratory experiments in separatory funnels through separations of 42-l batches to a fully integrated pilot process with 300 L fermenter working in fed-batch mode of operation [83, 85]. Two HF contactors with crossflow of phases and surface areas of fibers 18.6 m^2 were used in MBSE and MBSS working in parallel, as in the scheme shown in Figure 23.5.

Based on mass-transfer data for MBSE and MBSS of Phe in a HF contactor with a surface area of 1.4 m^2 published in ref. [86] a simulation of the pilot plant for recovery of Phe was done [127]. Number of contactors needed for recovery of Phe was estimated for the unit with a feed flowrate of 100 L h^{-1}, Phe concentration of 50 mol m^{-3} in the filtered broth, yield of Phe in MBSE 70%, $Re_{shell} = 2.0$, approach to the equilibrium at the raffinate end of contactor of 60%, and concentration factor of 10 were supposed. The estimated number of contactors of Liqui-Cel type 4″ × 28″ (Membrana, with an effective length of fibers of 0.6 m and surface area of fibers 19.2 m^2) in series was 6 in MBSE and 5 in stripping. From simulations it followed that the number of contactors (length of fibers) in MBSE and MBSS is very sensitive to the

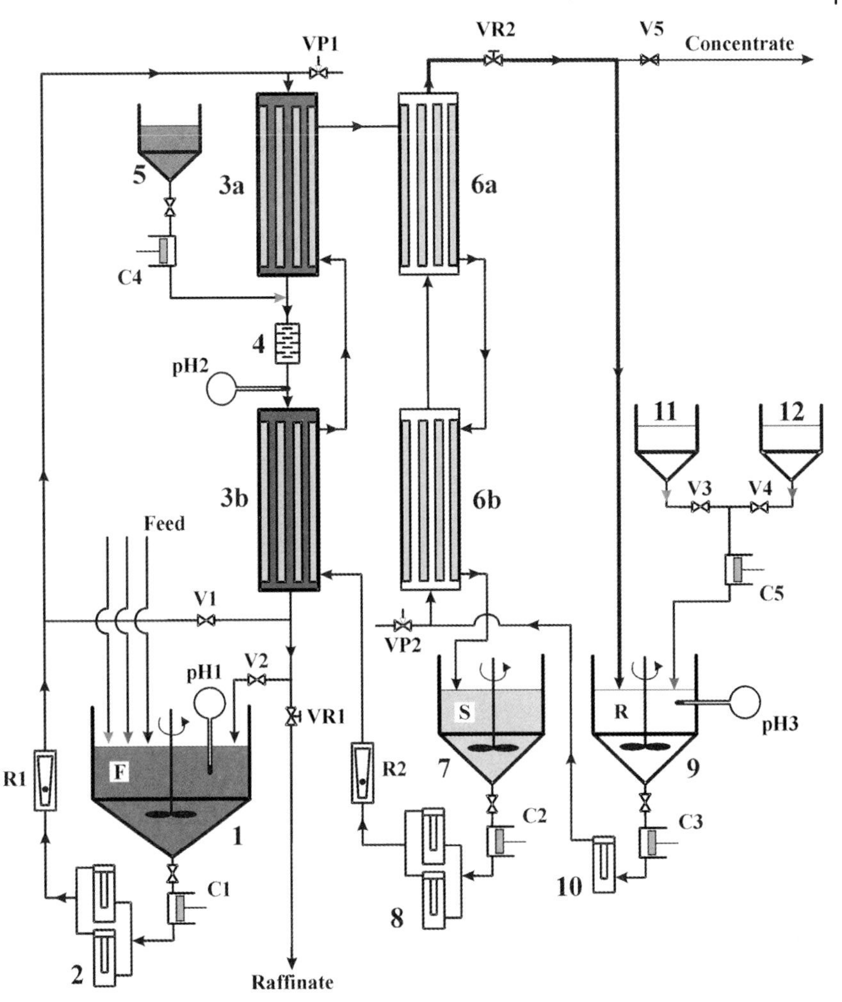

Figure 23.6 Scheme of a continuous pilot plant unit for recovery of MPCA from mother liquor (ML) by MBSE and MBSS in hollow fiber contactors [30].
1 – container of filtered feed, 2, 8, 10 – polishing (safety) filters, 3a, 3b – HF contactors for MBSE, 4 – static mixer, 5 – container of H_2SO_4 (pH adjustment), 6a, 6b – HF contactors for MBSS of the solvent, 7 – container of the regenerated solvent, 9 – container of the stripping solution, 11 – container of NaOH solution, 12 – container of NH_4OH solution, C1 to C5 – pumps, pH1, pH2, pH3 – pH sensors, R1, R2 – flowmeters (rotameters), V1 to V5, VP1, VP2 – valves, VR1, VR2 – valves for Δp adjustment.

increase of the Phe yield in MBSE above 0.70 (70%), Figure 23.7. The dependences of the total number of contactors on the concentration ratio in the solvent (actual to equilibrium concentration) at the raffinate end of the MBSE contactor (approach to the equilibrium) exhibit a minimum at about 0.60 (60%) as shown in Figure 23.8. The total number of contactors monotonously decreases with increasing Reynolds

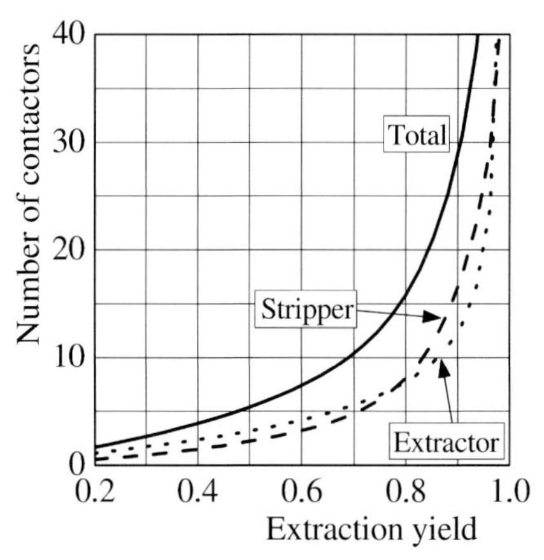

Figure 23.7 Number of contactors Liqui Cel 4″ × 28″ in MBSE and MBSS of phenylalanine, as well as the total number of contactors vs. yield of phenylalanine in MBSE [127]. The concentration ratio in the solvent at the raffinate end of the extractor was 0.60 (60%).

number. These results document the potential for application of HF contactors in recovery of organic acids from waste or fermentation solutions.

Long-term evaluation of the MBSE/MBSS process for removal of chromium from ground water is presented in refs. [120, 121]. For ground water containing 774 g m^{-3} of Cr(VI) the process showed stable performance during 700 h of operation of concentrating metal up to 20 kg m^{-3} [120]. An integrated process for simultaneous removal of chromium from groundwater by MBSE followed by ion exchange is

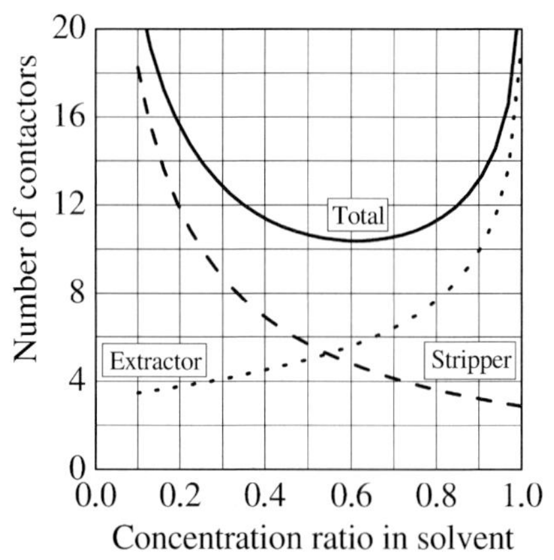

Figure 23.8 Number of contactors Liqui Cel 4″ × 28″ in MBSE and MBSS of phenylalanine, as well as the overall number of contactors vs. concentration ratio (approach to equilibrium) in the solvent at the raffinate end of the extractor [127]. Yield in MBSE was 70%.

suggested in ref. [121]. Laboratory and pilot-plant studies were performed with HF contactors with surface areas of fibers of 1.4 and 19.3 m^2, respectively. The design of the network of HF contactors and sensitivity analysis is presented in ref. [128]. Integration of the MBSE process with an ion-exchange step using Lewatit MP64 resin allowed Cr concentrations below 0.5 $g\,m^{-3}$ [121] to be achieved. Comparison of techniques for removal of chromium by MBSE, ion exchange and pertraction through BLME was carried out by Galan *et al.* [119].

23.4
Pertraction through BLME

In pertraction through BLME (Figure 23.1d) dispersion or emulsion of stripping solution in bulk of liquid membrane is used. In this way, compared to classical BLM (Figure 23.1c), one microporous wall is avoided. On the other hand, the advantage of the nondispersive process is lost. In systems inclined to emulsification in stripping this may introduce a problem. Mass-transfer resistance will be lower in BLME than in the BLM system in Figure 23.1c. The first papers on removal of copper from water by pertraction through BLM into a stripping solution emulsion were [129, 130]. A review of pertraction through BLME is given in refs. [131, 132].

PT through BLME is advantageous compared to PT through BLM or combination of MBSE with MBSS [119, 133]. Emulgation of stripping solution could be a problem in systems sensitive to formation of stable emulsions, but it is not clear how this phenomenon can influence this process. First, the nondispersive character of processes with immobilized interfaces is usually declared as an advantage. A lower loss of organic phase is connected with this. In most papers emulsion is not stabilized by a surfactant. In some papers [111, 134] a surfactant 3% v/v of Pluronic PE 3100 is used to enhance the phase separation of the aqueous stripping solution from the organic solution. By contrast, Fuad [135] added to the membrane phase emulsifier Span 80 to stabilize emulsion and increasing the concentration of surfactant improved the zinc removal rate. This observation could be related to decreased droplet size in the emulsion and their increased surface area in emulsion. This could play a role in systems with slower kinetics of the stripping reaction. Generally, the presence of surfactant decrease the mass-transfer rate through a L/L interface due to resistance of the surfactant adsorption film as shown, for example, in refs. [136, 137].

Modeling and optimization of pertraction into emulsion in HF contactors is discussed in refs. [77, 138]. The design and optimization of a network of HF contactors with minimum cost that permits the selective separation and recovery of anionic pollutants, for example, Cr(VI), using BLME process for groundwater remediation is presented in ref. [139] and for waste-water treatment in ref. [140].

Comparison of PT through BLME and MBSE with MBSS both in CF HF modules is given in ref. [138]. Galan *et al.* [119] compared techniques for removal of chromium by MBSE with ion exchange and pertraction through BLME.

An overview of selected papers on pertraction through BLME is presented in Table 23.3.

Table 23.3 Selected papers on recovery or removal of metals and organics by pertraction through BLME.

Solute	Membrane phase[a] (extractant/ diluent)	Contactor type[a]	Literature
Cu(II)	LIX 63/kerosene	PF HF	[129, 130]
	LIX 973N/(dodecane, dodecanol)	CF HF	[141–143]
	LIX622N/kerosene	CF HF	[133, 144]
	LIX 64/kerosene	PF HF	[145]
Co(II)	Cyanex 301/(Isopar, dodecanol)	CF HF	[143]
Cr(VI)	Amberlite LA-2/(dodecane, dodecanol)	CF HF	[143, 146]
	Aliquat 336/(kerosene, isodecanol)	CF HF	[119]
	Alamine 336/(Isopar L, dodecanol, surfactant)	CF HF	[111, 134]
Cr(III)	TOMAC/(n-decane, n-decanol)	CF HF	[147]
Zn(II)	Cyanex 301/(dodecane, dodecanol)	CF HF	[141–143]
	TBP (pure)	CF HF	[138]
	n.a., full-scale units	CF HF	[37, 100]
	D2EHPA/isododecane	CF HF	[135]
Cd(II)	(D2EHPA or trialkylphosphineoxide)/kerosene	flat membrane	[148]
Cu(II), Ni(II), Hg(I)	(LIX 84, D2EHPA, oleic acid)/tetradecane (without and with surfactant)	PF HF	[149]
Ni(II)	di(2-butyloctyl)monothiophosphoric acid/(dodecane, dodecanol)	CF HF	[141]
Pu(IV)	TBP/dodecane	HF	[150]
Sr(II)	(D2EHPA, 2-butyloctyl phenylphosphonic acid)/dodecane	CF HF	[143, 151]
Au(I)	LIX79/n-heptane	CF HF	[152, 153]
Phenols	kerosene with surfactant	PF HF	[154]
Penicilin G	Amberlite LA-2/Isopar	CF HF	[143]
Fumaric acid	Trialkylamine N7301/(kerosene with octanol)	PF HF	[155]

[a]Abbreviations used are explained in the list of abbreviations.

23.4.1 Case Studies

Selective removal of Zn and Fe from the passivating bath in the galvanic industry is important for preventing their accumulation in bath, and the need for bath replacement [37, 100]. Two full-scale installations of pertraction through BLME in galvanic companies proved the feasibility of this application in prolonged operation with an estimated payback time of less than 2 years [100]. Carrera *et al.* [138] reports the kinetics of zinc recovery from spent pickling solutions by pertraction through BLME in HF contactor with crossflow of phases. Pure tributyl phosphate was used as solvent and water as the stripping solution. Mass-transfer characteristics of the contactor have been estimated.

Table 23.4 Selected papers on recovery or removal of metals and organics by pertraction through SLM.

Solute	Membrane phase[a] (extractant/diluent)	Contactor type[a]	Literature
Cr(VI)	TBP/kerosene	PF HF	[181]
Cu(II)	LIX54/kerosene	CF HF	[186, 187]
	D2EHPA/kerosene; SLM renewal technique	PF HF	[182]
Cd(II)	Cyanex 302/kerosene	Flat sheet	[17, 188]
Co(II)	D2EHPA/kerosene	CF HF	[189]
	Cyanex 272/(kerosene and TBP as modifier)	Flat sheet	[190]
Ga(III), As(III)	PC88A (2-ethylhexyl phosphonic acid mono 2-ethylhexyl ester)		[191]
U(VI)	TBP/kerosene	CF HF	[192]
Zr(IV), Hf(IV)	Aliquat 336/(kerosene, 2-ethyl-1-hexanol)	PF HF	[193, 194]
Fumaric and L-malic acids	D2EHPA/dichlormethane and TOPO/ethyl acetate	Flat sheet	[195]
Lactic acid	(Aliquat 336 in carbonate form)/kerosene; Stripping solution: aqueous 1M Na_2CO_3	Flat sheet	[196]
	Cyphos IL-104/n-dodecane	Flat sheet	[44, 45]
Propranolol isomers separation	Peracylated cyclodextrin TA-b-CD/ chloroform	Flat sheet	[197]

[a]Abbreviations used are explained in the list of abbreviations.

Copper catalyst was recovered from waste water from the wet peroxide oxidation process by pertraction through BLME and copper sulfate from stripping solution can be recycled to the reactor without loss in performance [144].

Ho *et al.* [141–143, 146, 151] published a series of papers on removal/recovery of several metals from waste solutions by pertraction through BLME in HF contactors as shown in Table 23.4. For some metals, such as zinc and copper, scale-up of this system to pilot plant with a HF module with a surface area of fibers of 19 m^2 (with diameter 10.2 and length 71.1 cm) was done and mass-transfer characteristics have been estimated [142]. Separation of phases in the dispersion from the stripping was satisfactory.

23.5 Pertraction through SLM

Supported liquid membranes were one of the first types of liquid membranes published in 1967 [1, 2]. There are several types of SLM, which have an origin in three basic approaches:

a. LM is soaked in micropores of the support, which is more or less inert.

b. Liquid swells the polymer and forms a plasticized or gelled film with interpenetrating continuous LM and polymer phases. A membrane can be cast from polymer solution containing liquid carrier and volatile solvent, which after evaporation of volatiles, forms SLM. This type of SLMs are termed "plasticized membranes" [1, 156]. The most frequent terms used nowadays are "polymer inclusion membrane" (PIM) [157–160] or "activated composite membranes" [161–164]. A recent review on PIM is given in ref. [165].

c. LM is between two nonporous polymeric films [2] or films with reversed wettability, for example, organic LM between hydrophilic films [69, 166]. This type of SLM belongs more to BLM because of the thickness of the membrane.

There are several intermediate forms of these types, for example, pores of the support filled with a gel of polymer with LM [167] or SLM is covered on a surface(s) by thin film(s) [167–171].

The preferred geometry for application of SLM is a hollow-fiber contactor because of the high surface area per unit volume of contactor. Thousands of papers on separation with SLM document their potential, but there is one great problem in their larger-scale applications connected with their short lifetime. Their performance in terms of more or less sharply declining solute flux do not allowed wider utilization of SLM in practice. There are several mechanisms of flux decline, but most important is partial solubility of the carrier and other components of SLM in aqueous phases in contact with it and emulgation of organic membrane [172–177]. Even when the solubility of SLM components is only minute it destroys its performance due to the several orders of magnitude difference between the volume of SLM and liquids brought into contact with it.

There are several techniques that have been attempted to maintain stable performance of SLM, but a definite solution of this puzzle is not available yet. Covering of one or both SLM surface by a protective film was tested in refs. [167–171, 178, 179].

Modules with continuous regeneration of SLM have been proposed [180–182]. Nakano [180] suggested equipment in which the membrane phase creeps up the walls of a vertical hollow fibers. This enables a regeneration of the membrane phase in the pores. In pertraction of cobalt, the membrane broke down after about 50 h, while using a continuous regeneration the flow did not decrease even after 150 h of operation. An SLM renewal technique with dosing droplets of membrane phase into the feed flowing in fiber lumen suggest Ren *et al.* [181, 182]. This approach introduces dispergation of organic phase into the feed, which can result in membrane phase loss and at the same time contamination of the raffinate. Tailoring of hollow fibers for SLM was discussed in refs. [171, 178, 179].

Phosphonium ionic liquids can be a reactive carrier of organic acids and form effective SLM, as was found recently [183]. SLM with ionic liquid trihexyl-(tetradecyl) phosphonium bis 2,4,4-trimethylpentylphosphinate (Cyphos IL-104) had stable performance in pertraction of lactic acid for 5.3 days [44, 45], which is promising.

Pertraction through SLM is widely used in analytical chemistry for separation and preconcentration of solutes before application of selected analytical method and it is discussed in refs. [32–34, 184, 185].

An overview of selected papers on recovery/removal of metals and organic compounds by pertraction through SLM is presented in Table 23.4.

23.5.1
Case Studies

A pilot plant on recovery of copper from ammoniacal etching solutions was run with a good performance, as described in refs. [186, 187]. LIX54 in kerosene was used as the membrane phase in SLM. Large pilot-plant HF contactors Liqui cel (diam. 25.4 cm, length 71.1 cm) with surface area of 130 m^2 have also been used [186].

A simulation of the hybrid fermentation–pertraction process for production of butyric acid shows that the pH of fermentation and pertraction should be optimized independently [198]. It is advantageous to have the pH of the feed into pertraction at about 4.0 for both IL and TOA carriers. Choosing a proper carrier in the supported liquid membrane between IL and TOA should be made according to actual operation conditions, because of the different transport properties of these carriers in respect to the concentration of undisociated form of BA. While at lower BA concentrations the IL is better, at higher concentrations of above 20 kg m^{-3} and pH equal to 4.0, the membrane area needed is lower for TOA. An important factor will be the toxicity of the carrier to biomass. TOA is not very good in this respect and data for IL used are not available, but it is hoped that IL will be less toxic.

23.6
Comparison of Extractive Processes in HF Contactors and Pertraction through ELM

The advantages and disadvantages of membrane based processes and pertraction through various types of liquid membranes are summarized in Table 23.5. HF contactors are supposed in these processes with the exception of pertraction into stable emulsions (ELM) where mixed column contactors or mixer-settlers are used.

23.7
Outlook

Separation processes with one or two immobilized L/L interfaces realized in contactors with microporous or gel supports did not find wider application up to now. The present state of knowledge supports the belief that they have potential for development of successful applications. The properties of the support wall(s), despite the fact that they do not play an active role in separation itself, have great importance in the achievement of reliable lifetime of the system and should be tailored or modified for the respective extractant or carrier used in separation. Specific surface modification could be of great importance in achieving long lifetimes of separation systems. New extractants or carriers, for example, ionic liquids, could give a new impulse in this area. There is a need for advanced contactors with better stability in systems with organic phase. Deeper understanding of phenomena connected with achievement of prolonged operation of multiphase system involved in separation can help in development of successful applications. Better documented experimental

Table 23.5 Advantages and disadvantages of membrane-based processes and pertraction through various types of liquid membranes in two- and three-phase systems.

Process	Number of phases	Advantages	Disadvantages
MBSE or MBSS	2	- Nondispersive process	- Resistance of immobilizing wall. Two wall in MBSE and MBSS circuit replacing function of SLM
		- No problems with stabile performance of the solvent - One immobilizing wall - Volume ratio of phases can be varied without limitations	
PT through BLM	3	- No problems with stabile performance of BLM	- Two immobilizing walls
		- Volume ratio of phases can be varied without limitations	- Commercial contactors are not available.
PT through BLME	3	- One immobilizing wall	- Dispergation of stripping solution can introduce a problem in some systems sensitive to emulgation
		- No problems with stabile performance of LM - Larger surface are of the stripping solution droplets - Volume ratio of phases can be varied without limitations	
PT through SLM	3	One immobilizing wall for two immobilized interfaces - Volume ratio of phases can be varied without limitations - Very small volume of membrane phase	- Limited stability of SLM
PT through ELM	3	- No immobilizing wall	- Limited stability of emulsion
		- Comparatively high fluxes can be achieved	- Swelling of emulsion
		- Small volume of membrane phase	- Complexity of the process
			- Resistance of the adsorption film of surfactant

data on the long-term operation of separation systems are needed to achieve this. Knowledge accumulated in several pilot-plant experiments with larger HF modules may help to increase further the number of applications.

Development of hybrid production–separation processes may be the way to successful solutions especially in the case of higher value added compounds.

Extractive fermentations and biotransformations could be a good example of these. Also, combinations of chemical reaction(s) with separation could have a synergy effect for the production part of technology enhancing its rate and yield.

Abbreviations

(ATPS)	aqueous two-phase system
BLM	bulk liquid membrane
BLME	bulk liquid membrane with emulsion of the stripping solution
CF HF	crossflow hollow-fiber contactor
D2EHPA	di-2-ethylhexylphosphoric acid
DMCCA	dimethylcyclopropancarboxylic acid
ELM	emulsion liquid membrane
HF	hollow fiber (contactor)
Hostarex A327	mixture of trialkylamines
LM	liquid membrane
MBSE	membrane-based solvent extraction
MBSS	membrane-based solvent stripping
(ML)	mother liquor
MPCA	5-methyl-2-pyrazinecarboxylic acid
n.a.	not available
(PAA)	phenylacetic acid
PF HF	parallel flow hollow-fiber contactor
(Phe)	phenylalanine
PIM	polymer inclusion membrane
PT	pertraction
(RD)	rotary disc
SLM	supported liquid membrane
TBP	tributylphosphate
TOA	trioctylamine
TOMAC	trioctylmethylammonium chloride
TOPO	tri-*n*-octyl phosphine oxide

Acknowledgement

Support of the Slovak grant VEGA No. 1/0876/08 is acknowledged.

References

1 Bloch, R., Finkelstein, A., Kedem, O. and Vofsi, D. (1967) Metal-ion separation by dialysis through solvent membranes. *Ind Eng Chem Proc Des Develop*, **6**, 231.

2 Ward, W.J.I. and Robb, W.L. (1967) Carbon dioxide-oxygen separation: Facilitated transport of carbon dioxide across a liquid film. *Science*, **156**, 1481.

3 Li, N.N. (1968) Separating Hydrocarbons with Liquid Membranes, U.S. pat. 3,410,794.

4 Kim, B.M. (1984) Membrane-based solvent extraction for selective removal and recovery of metals. *Journal of Membrane Science*, **21**, 5.

5 Sengupta, A., Basu, R. and Sirkar, K.K. (1988) Separation of solutes from aqueous-solutions by contained liquid membranes. *AICHE Journal*, **34**, 1698.

6 Majumdar, S. and Sirkar, K.K. (1992) Hollow - fiber contained liquid membrane, in *Membrane Handbook* (eds W.S.W. Ho and K.K. Sirkar), Van Nostrand Reinhold, New York, p. 764.

7 Schlosser, Š., Rothova, I. and Frianova, H. (1993) Hollow-fiber pertractor with bulk liquid membrane. *Journal of Membrane Science*, **80**, 99.

8 Schlosser, Š. (2000) Pertraction through liquid and polymeric membranes, in *Integration of Membrane Processes into Bioconversions* (eds K. Bako, L. Gubicza and M. Mulder), Kluwer Academic, p. 73.

9 Schlosser, Š. and Sabolová, E. (2002) Three phase contactor with distributed U-shaped bundles of hollow fibers for pertraction. *Journal of Membrane Science*, **210**, 331.

10 Livingston, A.G., Arcangeli, J.P., Boam, T., Zhang, S., Marangon, M. and Freitas, L.M. (1998) Extractive membrane bioreactors for detoxification of chemical industry wastes: process development. *Journal of Membrane Science*, **151**, 29.

11 Han, S., Ferreira, F.C. and Livingston, A. (2001) Membrane aromatic recovery system (MARS) – a new membrane process for the recovery of phenols from wastewaters. *Journal of Membrane Science*, **188**, 219.

12 Ferreira, F.C., Peeva, L., Boam, A., Zhang, S. and Livingston, A. (2005) Pilot scale application of the Membrane Aromatic Recovery System (MARS) for recovery of phenol from resin production condensates. *Journal of Membrane Science*, **257**, 120.

13 Draxler, J. and Marr, R. (1986) Emulsion liquid membranes part I: Phenomenon and industrial application. *Chemical Engineering and Processing*, **20**, 319.

14 Marr, R.J. and Draxler, J. (1992) Emulsion liquid membranes: applications, in *Membrane Handbook* (eds W.S.W. Ho and K.K. Sirkar), Van Nostrand Reinhold, New York, p. 701.

15 Ho, W.S.W. and Li, N.N. (1996) Recent advances in emulsion liquid membranes, in *Chemical Separations with Liquid Membranes* (eds R.A. Bartsch and J.D. Way), ACS Symp. Series, ACS, Washington, p. 208.

16 Perera, J.M. and Stevens, G.W. (2009) Use of emulsion liquid membrane systems in chemical and biotechnological separations, in *Handbook of Membrane Separations* (eds A.K. Pabby, S.S.H. Rizvi and A.M. Sastre), CRC Press, p. 709.

17 Urtiaga, A.M., Alonso, A., Ortiz, I., Daoud, J.A., Elreefy, S.A., Deortiz, S.P. and Gallego, T. (2000) Comparison of liquid membrane processes for the removal of cadmium from wet phosphoric-acid. *Journal of Membrane Science*, **164**, 229.

18 Schlosser, Š., Sabolová, E., Kertész, R. and Kubišová, Ľ. (2001) Factors influencing transport through liquid membranes and membrane-based solvent-extraction. *Journal of Separation Science*, **24**, 509.

19 Lo, T.H., Baird, M.H.I. and Hanson, C. (1983) *Handbook of Solvent Extraction*, J. Wiley.

20 Araki, T., and Tsukube, H. (1990) *Liquid Membranes: Chemical Applications*, CRC Press, Boca Raton, Florida, p. 224.

21 Schugerl, K. (1994) *Solvent Extraction in Biotechnology*, Springer - Verlag, Berlin.

22 Rethwisch, D.G., Parida, S., Yi, G. and Dordick, J.S. (1994) Use of alcohols as cosolvents in enzyme-facilitated transport of organic-acids through a liquid

membrane. *Journal of Membrane Science*, **95**, 83.

23 López, J.L. and Matson, S.L. (1997) A multiphase/extractive enzyme membrane reactor for production of diltiazem chiral intermediate. *Journal of Membrane Science*, **125**, 189.

24 Dai, X.P., Yang, Z.F., Luo, R.G. and Sirkar, K.K. (2000) Lipase-facilitated separation of organic-acids in a hollow-fiber contained liquid membrane module. *Journal of Membrane Science*, **171**, 183.

25 Miyako, E., Maruyama, T., Kamiya, N. and Goto, M. (2003) Enzyme-facilitated enantioselective transport of (S)-ibuprofen through a supported liquid membrane based on ionic liquids. *Chemical Communications*, 2926.

26 Schugerl, K. (2000) Integrated processing of biotechnology products. *Biotechnology Advances*, **18**, 581.

27 Schlosser, Š. (2000) Membrane based processes with immobilized interface, in *Integration of Membrane Processes into Bioconversions* (eds K. Bako, L. Gubicza and M. Mulder), Kluwer Academic, p. 55.

28 Malinowski, J.J. (2001) Two-phase partitioning bioreactors in fermentation technology. *Biotechnology Advances*, **19**, 525.

29 Fernandes, P., Prazeres, D.M.F. and Cabral, J.M.S. (2003) Membrane-assisted extractive bioconversions, in *Advances in Biochemical Engineering/Biotechnology* (eds), Springer, p. 115.

30 Schlosser, S., Kertesz, R. and Martak, J. (2005) Recovery and separation of organic acids by membrane-based solvent extraction and pertraction - An overview with a case study on recovery of MPCA. *Separation and Purification Technology*, **41**, 237.

31 Membrana, Liqui-Cel® Membrane Contactors http://www.liquicel.com [approached 2008 January].

32 Jonsson, J.A. and Mathiasson, L. (2001) Membrane extraction in analytical-chemistry. *Journal of Separation Science*, **24**, 495.

33 Jonsson, J.A. (2003) Membrane extraction for sample preparation - a practical guide. *Chromatographia*, **57**, S317.

34 Belkhouche, N.E., Didi, M.A., Romero, R., Jonsson, J.A. and Villemin, D. (2006) Study of new organophosphorus derivates carriers on the selective recovery of M(II) and M(III) metals, using supported liquid membrane extraction. *Journal of Membrane Science*, **284**, 398.

35 Schlosser, Š. (1997) Method and equipment for mass and heat transfer among several liquids (in Slovak), Slovak pat. No. 278547.

36 ter Meulen, B.P. (1991) Transfer Device for the Transfer of Matter and/or Heat from One Medium Flow to Another Medium Flow, WO 91/09668.

37 Klaassen, R. and Jansen, A.E. (2001) The membrane contactor: Environmental applications and possibilities. *Environmental Progress*, **20**, 37.

38 Prasad, R. and Sirkar, K.K. (1992) Membrane-based solvent extraction, in *Membrane Handbook* (eds W.S.W. Ho and K.K. Sirkar), Van Nostrand Reinhold, New York, p. 727.

39 Reed, B.W., Semmens, M.J. and Cussler, E.L. (1995) Membrane contactors, in *Membrane Separation Technology. Principles and Applications* (eds R.D. Noble and S.A. Stern), Elsevier Science, Amsterdam, p. 467.

40 Gabelman, A. and Hwang, S.T. (1999) Hollow-fiber membrane contactors. *Journal of Membrane Science*, **159**, 61.

41 Sirkar, K.K. (1996) Hollow fiber-contained liquid membranes for separations - an overview, in *Chemical Separations with Liquid Membranes* (eds R.A. Bartsch and J.D. Way), ACS, p. 222.

42 Tong, Y.P., Hirata, M., Takanashi, H. and Hano, T. (1999) Back-extraction of lactic-acid with microporous hollow-fiber membrane. *Journal of Membrane Science*, **157**, 189.

43 Cichy, W., Schlosser, S. and Szymanowski, J. (2001) Recovery of phenol with cyanex(R)-923 in membrane extraction-stripping systems. *Solvent Extraction and Ion Exchange*, **19**, 905.

44 Marták, J. and Schlosser, Š. (2006) Pertraction of organic acids through liquid membranes containing ionic liquids. *Desalination*, **199**, 518.

45 Marták, J., Schlosser, Š. and Vlčková, S. (2007) Pertraction of lactic acid through supported liquid membranes containing phosphonium ionic liquid. Submited for publication in *Journal of Membrane Science*.

46 Juang, R.S., Chen, J.D. and Huan, H.C. (2000) Dispersion-free membrane extraction - case-studies of metal-ion and organic-acid extraction. *Journal of Membrane Science*, **165**, 59.

47 Kertész, R. and Schlosser, Š. (2005) Design and simulation of two phase hollow-fiber contactors for simultaneous membrane-based solvent extraction and stripping of organic acids and bases. *Separation and Purification Technology*, **41**, 275.

48 Bobok, D., Schlosser, Š. and Kossacský, E. (1984) Method of Separation of Solutions and Related Equipment for Realisation of this Method (in Slovak), Czechoslov. pat. 235362.

49 Schlosser, Š. and Kossaczky, E. (1986) Pertraction through liquid membranes. *Journal of Radioanalytical and Nuclear Chemistry*, **101**, 115.

50 Boyadzhiev, L. and Atanassova, I. (1991) Recovery of L-Lysine from dilute water solutions by liquid pertraction. *Biotechnology and Bioengineering*, **38**, 1059.

51 Boyadzhiev, L. and Atanassova, I. (1994) Extraction of phenylalanine from dilute-solutions by rotating film pertraction. *Process Biochemistry*, **29**, 237.

52 Boyadzhiev, L. and Kirilova, N. (2000) Extraction of tylosin from its aqueous-solutions by rotating film-pertraction. *Bioprocess Engineering*, **22**, 373.

53 Boyadzhiev, L. and Dimitrova, V. (2006) Extraction and liquid membrane preconcentration of rosmarinic acid from lemon balm (Melissa officinalis L.). *Separation Science and Technology*, **41**, 877.

54 Boyadzhiev, L., Alexandrova, S., Kirilova, N. and Saboni, A. (2003) Pertraction continue de tylosine dans un contacteur a films tournants. *Chemical Engineering Journal*, **95**, 137.

55 Boyadzhiev, L. and Yordanov, B. (2004) Pertraction of indole alkaloids from Vinca minor L. *Separation Science and Technology*, **39**, 1321.

56 Dimitrov, K., Metcheva, D. and Boyadzhiev, L. (2005) Integrated processes of extraction and liquid membrane isolation of atropine from Atropa belladonna roots. *Separation and Purification Technology*, **46**, 41.

57 Dimitrov, K., Metcheva, D., Alexandrova, S. and Boyadzhiev, L. (2006) Selective recovery of tropane alkaloids applying liquid membrane technique. *Chemical and Biochemical Engineering Quarterly*, **20**, 55.

58 Lazarova, M. and Dimitrov, K. (2009) Selective recovery of alkaloids from glaucium flavum crantz using integrated process extraction-pertraction. *Separation Science and Technology*, **44**, 227.

59 Dimitrov, K., Gancel, F., Montastruc, L. and Nikov, I. (2008) Liquid membrane extraction of bio-active amphiphilic substances: Recovery of surfactin. *Biochemical Engineering Journal*, **42**, 248.

60 Boyadzhiev, L. and Alexandrova, S. (1994) Recovery of copper from ammoniacal solutions by rotating film pertraction. *Hydrometallurgy*, **35**, 109.

61 Boyadzhiev, L. and Dimitrov, K. (1996) Silver recovery by rotating film pertraction - process modeling. *Solvent Extraction and Ion Exchange*, **14**, 105.

62 Dimitrov, K., Alexandrova, S., Saboni, A., Boyadzhiev, L. and Debray, E. (2002) Separation of metals by rotating film pertraction. *Chemical Engineering & Technology*, **25**, 823.

63 Zhivkova, S., Dimitrov, K., Kyuchoukov, G. and Boyadzhiev, L. (2004) Separation of zinc and iron by pertraction in rotating film contactor with Kelex 100 as a carrier. *Separation and Purification Technology*, **37**, 9.

64 Dimitrov, K., Rollet, V., Saboni, A. and Alexandrova, S. (2008) Recovery of nickel from sulphate media by batch pertraction in a rotating film contactor using Cyanex 302 as a carrier. *Chemical Engineering and Processing*, **47**, 1562.

65 Prasad, R. and Sirkar, K.K. (1988) Dispersion-free solvent extraction with microporous hollow-fiber modules. *AICHE Journal*, **34**, 177.

66 Ding, H. and Cussler, E.L. (1991) Fractional extraction with hollow fibers with hydrogel-filled walls. *AICHE Journal*, **37**, 855.

67 Stark, D., Kornmann, H., Muench, T., Sonnleitner, B., Marison, I.W. and von Stockar, U. (2003) Novel type of in situ extraction: use of solvent containing microcapsules for the bioconversion of 2-phenylethanol from L-Phenylalanine by Saccharomyces cerevisiae. *Biotechnology and Bioengineering*, **83**, 376.

68 Kondo, K., Otono, T. and Matsumoto, M. (2004) Preparation of microcapsules containing extractants and the application of the microcapsules to the extractive fermentation of lactic acid. *Journal of Chemical Engineering of Japan*, **37**, 1.

69 Kedem, O. and Bromberg, L. (1993) Ion-exchange membranes in extraction processes. *Journal of Membrane Science*, **78**, 255.

70 Isono, Y., Fukushima, K., Kawakatsu, T. and Nakajima, M. (1995) New selective perstraction system with charged membrane. *Journal of Membrane Science*, **105**, 293.

71 Isono, Y., Fukushima, K., Kawakatsu, T. and Nakajima, M. (1997) Integration of charged membrane into perstraction system for separation of amino-acid derivatives. *Biotechnology and Bioengineering*, **56**, 162.

72 Riedl, W. and Raiser, T. (2008) Membrane-supported extraction of biomolecules with aqueous two-phase systems. *Desalination*, **224**, 160.

73 Marták, J., Schlosser, Š., Sabolová, E., Krištofíková, L. and Rosenberg, M. (2003) Fermentation of lactic acid with Rhizopus arrhizus in a stirred tank reactor with a periodical bleed and feed operation. *Process Biochemistry*, **38**, 1573.

74 Kubišová, Ľ. and Schlosser, Š. (1996) Pertraction and extraction of lactic acid in hollow-fibre contactors. Proc. of 12th Int. Congr. Chisa 96, Praha, Czech Republic, full paper on CDROM.

75 Sciubba, L., Di Gioia, D., Fava, F. and Gostoli, C. (2008) Membrane-based solvent extraction of vanilin in hollow-fiber contactors. accepted for publication in *Desalination*.

76 Juang, R.S. and Huang, W.C. (2008) Use of membrane contactors as two-phase bioreactors for the removal of phenol in saline and acidic solutions. *Journal of Membrane Science*, **313**, 207.

77 Ortiz, I. and Irabien, A. (2009) Membrane-assisted solvent extraction for the recovery of metallic pollutants: process modeling and optimization, in *Handbook of Membrane Separations* (eds A.K. Pabby, S.S.H. Rizvi and A.M. Sastre), CRC Press, p. 1023.

78 Reis, M.T.A., de Freitas, O.M.F., Ismael, M.R.C. and Carvalho, J.M.R. (2007) Recovery of phenol from aqueous solutions using liquid membranes with Cyanex 923. *Journal of Membrane Science*, **305**, 313.

79 Gameiro, M.L.F., Ismael, M.R.C., Reis, M.T.A. and Carvalho, J.M.R. (2008) Recovery of copper from ammoniacal medium using liquid membranes, with LIX 54. *Separation and Purification Technology, Correct. Proof.*

80 Escalante, H., Alonso, A.I., Ortiz, I. and Irabien, A. (1998) Separation of L-phenylalanine by nondispersive extraction and backextraction - equilibrium and kinetic-parameters. *Separation Science and Technology*, **33**, 119.

81 Escalante, H. and Irabien, A. (2001) Separation/concentration of L-phenylalanine in hollow fibre modules. decrease of the extraction rate, in *Solvent Extraction for the 21st Century* (eds M.

Valiente and M. Hidalgo), Society of Chemical Industries, London, p. 1499.
82 Cardoso, M.M., Viegas, R.M.C. and Crespo, J.P.S.G. (1999) Extraction and reextraction of phenylalanine by cationic reversed micelles in hollow-fiber contactors. *Journal of Membrane Science*, **156**, 303.
83 Gerigk, M.R., Maass, D., Kreutzer, A., Sprenger, G., Bongaerts, J., Wubbolts, M. and Takors, R. (2002) Enhanced pilot-scale fed-batch L-phenylalanine production with recombinant escherichia coli by fully integrated reactive extraction. *Bioprocess and Biosystems Engineering*, **25**, 43.
84 Maass, D., Gerigk, M.R., Kreutzer, A., Weuster-Botz, D., Wubbolts, M. and Takors, R. (2002) Integrated L-phenylalanine separation in an E-coli fed-batch process: from laboratory to pilot scale. *Bioprocess and Biosystems Engineering*, **25**, 85.
85 Takors, R. (2004) Model-based analysis and optimization of an ISPR approach using reactive extraction for pilot-scale L-phenylalanine production. *Biotechnology Progress*, **20**, 57.
86 Kertész, R., Šimo, M. and Schlosser, Š. (2005) Membrane-based solvent extraction and stripping of phenylalanine in HF contactors. *Journal of Membrane Science*, **257**, 37.
87 Harington, T. and Hossain, M.M. (2008) Extraction of lactic acid into sunflower oil and its recovery into an aqueous solution. *Desalination*, **218**, 287.
88 Wu, Z.T. and Yang, S.T. (2003) Extractive fermentation for butyric acid production from glucose by clostridium tyrobutyricum. *Biotechnology and Bioengineering*, **82**, 93.
89 Lin, S.H., Chen, C.N. and Juang, R.S. (2006) Kinetic analysis on reactive extraction of aspartic acid from water in hollow-fiber membrane modules. *Journal of Membrane Science*, **281**, 186.
90 Rindfleisch, D., Syska, B., Lazarova, Z. and Schugerl, K. (1997) Integrated membrane extraction, enzymic conversion and electrodialysis for the synthesis of ampicilin from penicilin G. *Process Biochemistry*, **32**, 605.
91 Lazarova, Z., Syska, B. and Schugerl, K. (2002) Application of large-scale hollow-fiber membrane contactors for simultaneous extractive removal and stripping of penicillin G. *Journal of Membrane Science*, **202**, 151.
92 Hossain, M.M. and Dean, J. (2008) Extraction of penicillin G from aqueous solutions: Analysis of reaction equilibrium and mass transfer. *Separation and Purification Technology*, **62**, 437.
93 Ratajczak, W., Porebski, T., Plesnar, M., Brzozowski, R., Tomzik, S., Capala, W., Wieteska, A., Ebrowski, M., Malachowski, B. and Karabin, M. (2004) Modernization of the phenol and acetone plant at the "Orlen" Polish oil company. *Przemysl Chemiczny*, **83**, 498.
94 Ratajczak, W., Porebski, T., Plesnar, M. and Tomzik, S. (2007) Process backfitting in the Polish chemical industry. Modern unit operations. *Przemysl Chemiczny*, **86**, 262.
95 Kubišová, L., Sabolová, E., Schlosser, S., Marták, J. and Kertész, R. (2002) Membrane-based solvent extraction and stripping of a heterocyclic carboxylic acid in hollow-fiber contactors. *Desalination*, **148**, 205.
96 Kubišova, L'., Sabolová, E., Schlosser, Š., Marták, J. and Kertesz, R. (2004) Mass-transfer in membrane-based solvent extraction and stripping of 5-methyl-2-pyrazinecarboxylic acid and co-transport of sulphuric acid in HF contactors. *Desalination*, **163**, 27.
97 Schlosser, Š., Sabolová, E. and Marták, J. (2001) Pertraction and membrane-based solvent extraction of carboxylic acids in hollow-fiber contactors, in *Solvent Extraction for the 21st Century* (eds M. Valiente and M. Hidalgo), Society of Chemical Industries, London, p. 1041.
98 Schlosser, Š. (2002) Method of separation of carboxylic acids from aqueous and

nonaqueous solutions (in Slovak), Slovak pat. No. 282775.

99 Yang, X., Cao, Y.M., Wang, R. and Yuan, Q. (2007) Study on highly hydrophilic cellulose hollow-fiber membrane contactors for thiol sulfur removal. *Journal of Membrane Science*, **305**, 247.

100 Klaassen, R., Feron, P. and Jansen, A. (2008) Membrane contactors applications. *Desalination*, **224**, 81.

101 Baudot, A., Floury, J. and Smorenburg, H.E. (2001) Liquid-liquid extraction of aroma compounds with hollow-fiber contactor. *AICHE Journal*, **47**, 1780.

102 Pierre, F.X., Souchon, I. and Marin, M. (2001) Recovery of sulfur aroma compounds using membrane-based solvent-extraction. *Journal of Membrane Science*, **187**, 239.

103 Souchon, I., Athes, V., Pierre, F.X. and Marin, M. (2004) Liquid-liquid extraction and air stripping in membrane contactor: application to aroma compounds recovery. *Desalination*, **163**, 39.

104 Bocquet, S., Viladomat, E.G., Nova, C.M., Sanchez, J., Athes, V. and Souchon, I. (2006) Membrane-based solvent extraction of aroma compounds: Choice of configurations of hollow fiber modules based on experiments and simulation. *Journal of Membrane Science*, **281**, 358.

105 Younas, M., Bocquet, S.D. and Sanchez, J. (2008) Extraction of aroma compounds in a HFMC: Dynamic modelling and simulation. *Journal of Membrane Science*, **323**, 386.

106 Brose, D.J., Chidlaw, M.B., Friesen, D.T., Lachapelle, E.D. and Vaneikeren, P. (1995) Fractionation of citrus oils using a membrane-based extraction process. *Biotechnology Progress*, **11**, 214.

107 de Haan, A.B., Bartels, P.V. and de Graauw, J. (1989) Extraction of metal ions from waste water. Modelling of the mass transfer in a supported-liquid-membrane process. *Journal of Membrane Science*, **45**, 281.

108 Lin, S.H. and Juang, R.S. (2002) Kinetic modeling of simultaneous recovery of metallic cations and anions with a mixture of extractants in hollow-fiber modules. *Industrial & Engineering Chemistry Research*, **41**, 853.

109 Koopman, C. and Witkamp, G.J. (2002) Extraction of heavy metals from industrial phosphoric acid in a transverse flow hollow-fiber membrane contactor. *Separation Science and Technology*, **37**, 1273.

110 Ortiz, I., Bringas, E., San Roman, M.F. and Urtiaga, A.M. (2004) Selective separation of zinc and iron from spent pickling solutions by membrane-based solvent extraction: Process viability. *Separation Science and Technology*, **39**, 2441.

111 Ortiz, I., Bringas, E., Samaniego, H., San Roman, F. and Urtiaga, A. (2006) Membrane processes for the efficient recovery of anionic pollutants. *Desalination*, **193**, 375.

112 Samaniego, H., San Roman, M.F. and Ortiz, I. (2007) Kinetics of zinc recovery from spent pickling effluents. *Industrial & Engineering Chemistry Research*, **46**, 907.

113 Vajda, M., Kosuthova, A. and Schlosser, S. (2004) Membrane-based extraction joined with membrane-based stripping in a circulating arrangement - III. Extraction of zinc. *Chemical Papers*, **58**, 1.

114 Fouad, E.A. and Bart, H.J. (2007) Separation of zinc by a non-dispersion solvent extraction process in a hollow-fiber contactor. *Solvent Extraction and Ion Exchange*, **25**, 857.

115 Alonso, A.I., Urtiaga, A.M., Zamacona, S., Irabien, A. and Ortiz, I. (1997) Kinetic modelling of cadmium removal from phosphoric acid by non-dispersive solvent extraction. *Journal of Membrane Science*, **130**, 193.

116 Kao, H.C. and Juang, R.S. (2003) Hindered membrane diffusion in the nondispersive stripping of Co(II) from organic amine solutions with hydrochloric acid. *Industrial & Engineering Chemistry Research*, **42**, 6181.

117 Ortiz, I., Galan, B. and Irabien, A. (1996) Membrane mass-transport coefficient for the recovery of Cr(VI) in hollow-fiber extraction and back-extraction modules. *Journal of Membrane Science*, **118**, 213.

118 Alonso, A.I., Galan, B., Gonzalez, M. and Ortiz, I. (1999) Experimental and theoretical analysis of a nondispersive solvent extraction pilot plant for the removal of Cr(VI) from a galvanic process wastewaters. *Industrial & Engineering Chemistry Research*, **38**, 1666.

119 Galan, B., Castaneda, D. and Ortiz, I. (2005) Removal and recovery of Cr(VI) from polluted ground waters: A comparative study of ion-exchange technologies. *Water Research*, **39**, 4317.

120 Galan, B., Calzada, M. and Ortiz, I. (2006) Recycling of Cr(VI) by membrane solvent extraction: Long term performance with the mathematical model. *Chemical Engineering Journal*, **124**, 71.

121 Galan, B., Castaneda, D. and Ortiz, I. (2008) Integration of ion exchange and non-dispersive solvent extraction processes for the separation and concentration of Cr(VI) from ground waters. *Journal of Hazardous Materials*, **152**, 795.

122 Kumar, A., Haddad, R., Alguacil, F.J. and Sastre, A.M. (2005) Comparative performance of non-dispersive solvent extraction using a single module and the integrated membrane process with two hollow-fiber contactors. *Journal of Membrane Science*, **248**, 1.

123 Kumar, A., Haddad, R. and Sastre, A.M. (2001) Integrated membrane process for gold recovery from hydrometallurgical solutions. *AICHE Journal*, **47**, 328.

124 Patil, C.B., Mohapatra, P.K. and Manchanda, V.K. (2008) Transport of thorium from nitric acid solution by non-dispersive solvent extraction using a hollow fibre contactor. *Desalination*, **232**, 272.

125 Trtic, T.M., Vladisavljevic, G.T. and Comor, J.J. (2000) Dispersion-free solvent-extraction of thallium(III) in hollow-fiber contactors. *Separation Science and Technology*, **35**, 1587.

126 Trtic, T.M., Vladisavljevic, G.T. and Comor, J.J. (2001) Single-stage and 2-stage solvent-extraction of Tl(III) in hollow-fiber contactors under recirculation mode of operation. *Separation Science and Technology*, **36**, 295.

127 Kertész, R. and Schlosser, Š. (2004) Simulation of simultaneous membrane-based solvent extraction and stripping of phenylalanine in hollow-fiber contactors. Proc. 16th Int. Congr. CHISA 2004, Prague, full text on CD ROM.

128 Alonso, A.I. and Gruhn, G. (2002) Flexibility analysis of nondispersive solvent extraction plant. *Separation Science and Technology*, **37**, 161.

129 Schneider, K. and Rintelen, T.H. (1986) Recovery of metals with the aid of supported liquid membranes. *Chem Ing Tech*, **58**, 800.

130 Schulz, G. (1988) Separation techniques with supported liquid membranes. *Desalination*, **68**, 191.

131 Pabby, A.K. and Sastre, A.M. (2006) Hollow-fiber membrane based separation technology: performance and design perspectives, in *Solvent Extraction and Liquid Membranes: Fundamental and Application in New Materials* (eds Cortina and Aguilar), Marcel Dekker, New York.

132 Zou, J., Huang, J. and Ho, W.S.W. (2008) Facilitated transport membranes for environmental energy, and biochemical applications, in *Advanced Membrane Technology and Applications* (eds N.N. Li, A.C. Fane, W.S.W. Ho and T. Miltsuura), Wiley, p. 721.

133 Urtiaga, A., Abellan, M.J., Irabien, J.A. and Ortiz, I. (2005) Membrane contactors for the recovery of metallic compounds: Modelling of copper recovery from WPO processes. *Journal of Membrane Science*, **257**, 161.

134 Bringas, E., San Roman, M.F. and Ortiz, I. (2006) Separation and recovery of anionic pollutants by the emulsion pertraction technology. Remediation of polluted groundwaters with Cr(VI). *Industrial & Engineering Chemistry Research*, **45**, 4295.

135 Fouad, E.A. and Bart, H. (2008) Emulsion liquid membrane extraction of zinc by a hollow-fiber contactor. *Journal of Membrane Science*, **307**, 156.

136 Schlosser, Š. and Kossaczký, E. (1987) Liquid membranes stabilized by polymeric surfactants, in *Synthetic Polymeric Membranes* (ed. B. Sedlaček), W. de Gruyter, Berlin, p. 571.

137 Strzelbicki, J. and Schlosser, Š. (1989) Influence of surface-active substances on pertraction of cobalt(II) cations through bulk and emulsion liquid membranes. *Hydrometallurgy*, **23**, 67.

138 Carrera, J.A., Bringas, E., Roman, M.F.S. and Ortiz, I. (2009) Selective membrane alternative to the recovery of zinc from hot-dip galvanizing effluents. *Journal of Membrane Science*, **326**, 672.

139 Bringas, E., Karuppiah, R., San Roman, M.F., Ortiz, I. and Grossmann, I.E. (2007) Optimal groundwater remediation network design using selective membranes. *Industrial & Engineering Chemistry Research*, **46**, 5555.

140 San Roman, M.F., Bringas, E., Ortiz, I. and Grossmann, I.E. (2007) Optimal synthesis of an emulsion pertraction process for the removal of pollutant anions in industrial wastewater systems. *Comp Chem Eng*, **31**, 456.

141 Ho, W.S.W., Wang, B., Neumuller, T.E. and Roller, J. (2001) Supported liquid membranes for removal and recovery of metals from waste waters and process streams. *Environmental Progress*, **20**, 117.

142 Ho, W.S.W., Poddar, T.K. and Neumuller, T.E. (2002) Removal and recovery of copper and zinc by supported liquid membranes with strip dispersion. *Journal of the Chinese Institute of Chemical Engineers*, **33**, 67.

143 Ho, W.S.W. (2003) Removal and recovery of metals and other materials by supported liquid membranes with strip dispersion. *Annals of the New York Academy of Sciences, Advanced Membrane Technology*, 97.

144 Urtiaga, A., Abellan, M.J., Irabien, A. and Ortiz, I. (2006) Use of membrane contactors as an efficient alternative to reduce effluent ecotoxicity. *Desalination*, **191**, 79.

145 Hu, S.Y.B. and Wiencek, J.M. (1998) Emulsion-liquid-membrane extraction of copper using a hollow-fiber contactor. *AICHE Journal*, **44**, 570.

146 Ho, W.S.W. and Poddar, T.K. (2001) New membrane technology for removal and recovery of chromium from waste waters. *Environmental Progress*, **20**, 44.

147 Alguacil, F.J., Alonso, M., Lopez, F.A. and Lopez-Delgado, A. (2009) Application of pseudo-emulsion based hollow fiber strip dispersion (PEHFSD) for recovery of Cr (III) from alkaline solutions. *Separation and Purification Technology*, In Press. Corrected Proof.

148 He, D., Gu, S. and Ma, M. (2007) Simultaneous removal and recovery of cadmium (II) and CN- from simulated electroplating rinse wastewater by a strip dispersion hybrid liquid membrane (SDHLM) containing double carrier. *Journal of Membrane Science*, **305**, 36.

149 Raghuraman, B. and Wiencek, J. (1993) Extraction with emulsion liquid membranes in a hollow-fiber contactor. *AICHE Journal*, **39**, 1885.

150 Gupta, S.K., Rathore, N.S., Sonawane, J.V., Pabby, A.K., Venugopalan, A.K., Changrani, R.D., Dey, P.K. and Venkataramani, B. (2005) Application of hollow-fiber contactor in nondispersive solvent extraction of Pu(IV) by TBP. *Separation Science and Technology*, **40**, 1911.

151 Ho, W.S.W. and Wang, B. (2002) Strontium removal by new alkyl phenylphosphonic acids in supported liquid membranes with strip dispersion. *Industrial & Engineering Chemistry Research*, **41**, 381.

152 Sonawane, J.V., Pabby, A.K. and Sastre, A.M. (2007) Au(I) extraction by LIX-79/n-heptane using the pseudo-emulsion-based hollow-fiber strip dispersion (PEHFSD) technique. *Journal of Membrane Science*, **300**, 147.

153 Sonawane, J.V., Pabby, A.K. and Sastre, A.M. (2008) Pseudo-emulsion based hollow fiber strip dispersion: A novel methodology for gold recovery. *AICHE Journal*, **54**, 453.

154 Nanoti, A., Ganguly, S.K., Goswami, A.N. and Rawat, B.S. (1997) Removal of phenols from wastewater using liquid membranes in a microporous hollow fiber-membrane extractor. *Industrial & Engineering Chemistry Research*, **36**, 4369.

155 Li, S.-J., Chen, H.-L. and Zhang, L. (2009) Recovery of fumaric acid by hollow-fiber supported liquid membrane with strip dispersion using N7301 as carrier. *Separation and Purification Technology*, doi:10.1016/j.seppur.2008.12.004.

156 Paugam, M.F. and Buffle, J. (1998) Comparison of carrier-facilitated copper (II) ion transport mechanisms in a supported liquid membrane and in a plasticized cellulose triacetate membrane. *Journal of Membrane Science*, **147**, 207.

157 Schow, A.J., Peterson, R.T. and Lamb, J.D. (1996) Polymer inclusion membranes containing macrocyclic carriers for use in cation separations. *Journal of Membrane Science*, **111**, 291.

158 Gardner, J.S., Walker, J.O. and Lamb, J.D. (2004) Permeability and durability effects of cellulose polymer variation in polymer inclusion membranes. *Journal of Membrane Science*, **229**, 87.

159 Kozlowski, C.A., Kozlowska, J., Pellowski, W. and Walkowiak, W. (2006) Separation of cobalt-60, strontium-90, and cesium-137 radioisotopes by competitive transport across polymer inclusion membranes with organophosphorous acids. *Desalination*, **198**, 141.

160 Fontas, C., Tayeb, R., Dhahbi, M., Gaudichet, E., Thominette, F., Roy, P., Steenkeste, K., Fontaine-Aupart, M.P., Tingry, S., Tronel-Peyroz, E. and Seta, P. (2007) Polymer inclusion membranes: The concept of fixed sites membrane revised. *Journal of Membrane Science*, **290**, 62.

161 Oleinikova, M., Gonzalez, C., Valiente, M. and Munoz, M. (1999) Selective transport of zinc through activated composite membranes containing di(2-ethylhexyl) dithiophosphoric acid as a carrier. *Polyhedron*, **18**, 3353.

162 Gumi, T., Oleinikova, M., Palet, C., Valiente, M. and Munoz, M. (2000) Facilitated transport of lead(II) and cadmium(II) through novel activated composite membranes containing di-(2-ethyl-hexyl)phosphoric acid as carrier. *Analytica Chimica Acta*, **408**, 65.

163 Macanas, J. and Munoz, M. (2005) Mass transfer determining parameter in facilitated transport through di-(2-ethylhexyl) dithiophosphoric acid activated composite membranes. *Analytica Chimica Acta*, **534**, 101.

164 Kozlowski, C.A. (2006) Facilitated transport of metal ions through composite and polymer inclusion membranes. *Desalination*, **198**, 132.

165 Nghiem, L.D., Mornane, P., Potter, I.D., Perera, J.M., Cattrall, R.W. and Kolev, S.D. (2006) Extraction and transport of metal ions and small organic compounds using polymer inclusion membranes (PIMs). *Journal of Membrane Science*, **281**, 7.

166 Wodzki, R. and Nowaczyk, J. (2002) Propionic and acetic acid pertraction through a multimembrane hybrid system containing TOPO or TBP. *Separation and Purification Technology*, **26**, 207.

167 Kemperman, A.J.B., Damink, B., Vandenboomgaard, T. and Strathmann, H. (1997) Stabilization of supported liquid membranes by gelation with PVC. *Journal of Applied Polymer Science*, **65**, 1205.

168 Kemperman, A.J.B., Rolevink, H.H.M., Bargeman, D., Vandenboomgaard, T. and Strathmann, H. (1998) Stabilization of supported liquid membranes by interfacial polymerization top layers. *Journal of Membrane Science*, **138**, 43.

169 Wijers, M.C., Jin, M., Wessling, M. and Strathmann, H. (1998) Supported liquid membranes modification with sulfonated poly(ether ether ketone) - permeability, selectivity and stability. *Journal of Membrane Science*, **147**, 117.

170 Wang, Y.C. and Doyle, F.M. (1999) Formation of epoxy skin layers on the surface of supported liquid membranes containing polyamines. *Journal of Membrane Science*, **159**, 167.

171 He, T. (2008) Towards stabilization of supported liquid membranes: preparation and characterization of polysulfone support and sulfonated poly (ether ether ketone) coated composite hollow-fiber membranes. *Desalination*, **225**, 82.

172 Danesi, P.R., Reichley-Yinger, L. and Rickert, P.G. (1987) Lifetime of supported liquid membranes: The influence of interfacial properties, chemical composition and water transport on the long term stability of the membranes. *Journal of Membrane Science*, **31**, 117.

173 Neplenbroek, A.M., Bargeman, D. and Smolders, C.A. (1992) Supported liquid membranes – instability effects. *Journal of Membrane Science*, **67**, 121.

174 Neplenbroek, A.M., Bargeman, D. and Smolders, C.A. (1992) Mechanism of supported liquid membrane degradation – emulsion formation. *Journal of Membrane Science*, **67**, 133.

175 Kemperman, A.J.B., Bargeman, D., Vandenboomgaard, T. and Strathmann, H. (1996) Stability of supported liquid membranes - state-of-the-art. *Separation Science and Technology*, **31**, 2733.

176 Dreher, T.M. and Stevens, G.W. (1998) Instability mechanisms of supported liquid membranes. *Separation Science and Technology*, **33**, 835.

177 Zhang, B., Gozzelino, G. and Baldi, G. (2001) Membrane liquid loss of supported liquid membrane-based on N-decanol. *Colloids and Surfaces A*, **193**, 61.

178 He, T., Versteeg, L.A.M., Mulder, M.H.V. and Wessling, M. (2004) Composite hollow-fiber membranes for organic solvent-based liquid-liquid extraction. *Journal of Membrane Science*, **234**, 1.

179 Yang, Q., Chung, T.S., Xiao, Y.C. and Wang, K.Y. (2007) The development of chemically modified P84Co-polyimide membranes as supported liquid membrane matrix for Cu(II) removal with prolonged stability. *Chemical Engineering Science*, **62**, 1721.

180 Nakano, M., Takahashi, K. and Takeuchi, H. (1987) A method for continuous operation of supported liquid membranes. *Journal of Chemical Engineering of Japan*, **20**, 326.

181 Ren, Z.Q., Liu, J.T., Zhang, W.D., Du, C.S. and Ma, J.N. (2006) Treatment of wastewater containing chromiun (VI) by hollow fiber renewal liquid membrane technology. *Electroplating and Finishing*, **25**, 49.

182 Ren, Z.Q., Zhang, W.D., Liu, Y.M., Dai, Y. and Cui, C.H. (2007) New liquid membrane technology for simultaneous extraction and stripping of copper(II) from wastewater. *Chemical Engineering Science*, **62**, 6090.

183 Marták, J. and Schlosser, Š. (2007) Extraction of lactic acid by phosphonium ionic liquids. *Separation and Purification Technology*, **57**, 483.

184 Chimuka, L., Cukrowska, E. and Jonsson, J.A. (2004) Why liquid membrane extraction is an attractive alternative in sample preparation. *Pure and Applied Chemistry*, **76**, 707.

185 Bardstu, K.F., Ho, T.S., Rasmussen, K.E., Pedersen-Bjergaard, S. and Jonsson, J.A. (2007) Supported liquid membranes in hollow fiber liquid-phase microextraction (LPME) – Practical considerations in the three-phase mode. *Journal of Separation Science*, **30**, 1364.

186 Yang, Q. and Kocherginsky, N.M. (2006) Copper recovery and spent ammoniacal etchant regeneration based on hollow fiber supported liquid membrane technology: From bench-scale to pilot-scale tests. *Journal of Membrane Science*, **286**, 301.

187 Yang, Q. and Kocherginsky, N.M. (2007) Copper removal from ammoniacal wastewater through a hollow fiber supported liquid membrane system: Modeling and experimental verification. *Journal of Membrane Science*, **297**, 121.

188 Daoud, J.A., Elreefy, S.A. and Aly, H.F. (1998) Permeation of Cd(II) ions through a supported liquid membrane containing Cyanex-302 in kerosene. *Separation Science and Technology*, **33**, 537.

189 Prakorn, R., Eakkapit, S., Weerawat, P., Milan, H. and Pancharoen, U. (2006) Permeation study on the hollow-fiber supported liquid membrane for the extraction of cobalt(II). *Korean Journal of Chemical Engineering*, **23**, 117.

190 Swain, B., Jeong, J., Lee, J.C. and Lee, G.H. (2007) Extraction of Co(II) by supported liquid membrane and solvent extraction using Cyanex 272 as an extractant: A comparison study. *Journal of Membrane Science*, **288**, 139.

191 Tsai, C.Y., Chen, Y.F., Chen, W.C., Yang, F.R., Chen, J.H. and Lin, J.C. (2005) Separation of gallium and arsenic in wafer grinding extraction solution using a supported liquid membrane that contains PC88A as a carrier. *Journal of Environmental Science and Health Part A-Toxic/Hazardous Substances & Environmental Engineering*, **40**, 477.

192 Ramakul, P., Prapasawad, T., Pancharoen, U. and Pattaveekongka, W. (2007) Separation of radioactive metal ions by hollow fiber-supported liquid membrane and permeability analysis. *Journal of the Chinese Institute of Chemical Engineers*, **38**, 489.

193 Yang, X.J., Fane, A.G. and Pin, C. (2002) Separation of zirconium and hafnium using hollow fibers - Part I. Supported liquid membranes. *Chemical Engineering Journal*, **88**, 37.

194 Yang, X.J., Fane, A.G. and Soldenhoff, K. (2003) Comparison of liquid membrane processes for metal separations: Permeability, stability, and selectivity. *Industrial & Engineering Chemistry Research*, **42**, 392.

195 Bressler, E., Pines, O., Goldberg, I. and Braun, S. (2002) Conversion of fumaric acid to L-malic by sol-gel immobilized Saccharomyces cerevisiae in a supported liquid membrane bioreactor. *Biotechnology Progress*, **18**, 445.

196 Yang, Q. and Chung, T.S. (2007) Modification of the commercial carrier in supported liquid membrane system to enhance lactic acid flux and to separate L, D-lactic acid enantiomers. *Journal of Membrane Science*, **294**, 127.

197 Ferreira, Q., Coelhoso, I.M., Ramalhete, N. and Marques, H.M.C. (2006) Resolution of racemic propranolol in liquid membranes containing TA-beta-cyclodextrin. *Separation Science and Technology*, **41**, 3553.

198 Blahušiak, M., Schlosser, Š. and Marták, J. (2009) Simulation of hybrid fermentation-separation process for production of butyric acid. Submitted for publication in *Chemical Papers*.

Index

Membrane Operations. Innovative Separations and Transformations. Edited by Enrico Drioli and Lidietta Giorno

ISBN: 978-3-527-32038-7